W9-CGS-568

BIOLOGY AND PHYSIOLOGY OF THE BLOOD–BRAIN BARRIER

TRANSPORT, CELLULAR INTERACTIONS, AND BRAIN PATHOLOGIES

ADVANCES IN BEHAVIORAL BIOLOGY

Recent Volumes in This Series

Volume 35 MECHANISMS OF CEREBRAL HYPOXIA AND STROKE
Edited by George Somjen

Volume 36 NOVEL APPROACHES TO THE TREATMENT OF ALZHEIMER'S DISEASE
Edited by Edwin M. Meyer, James W. Simpkins, and Jyunji Yamamoto

Volume 37 KINDLING 4
Edited by Juhn A. Wada

Volume 38A BASIC, CLINICAL AND THERAPEUTIC ASPECTS OF ALZHEIMER'S AND PARKINSON'S DISEASES
Volume 1
Edited by Toshiharu Nagatsu, Abraham Fisher, and Mitsuo Yoshida

Volume 38B BASIC, CLINICAL AND THERAPEUTIC ASPECTS OF ALZHEIMER'S AND PARKINSON'S DISEASES
Volume 2
Edited by Toshiharu Nagatsu, Abraham Fisher, and Mitsuo Yoshida

Volume 39 THE BASAL GANGLIA III
Edited by Giorgio Bernardi, Malcolm B. Carpenter, Gaetano Di Chiara, Micaela Morelli, and Paolo Stanzione

Volume 40 TREATMENT OF DEMENTIAS: A New Generation of Progress
Edited by Edwin M. Meyer, James W. Simpkins, Jyunji Yamamoto, and Fulton T. Crews

Volume 41 THE BASAL GANGLIA IV: New Ideas and Data on Structure and Function
Edited by Gérard Percheron, John S. McKenzie, and Jean Féger

Volume 42 CALLOSAL AGENESIS: A Natural Split Brain?
Edited by Maryse Lassonde and Malcolm A. Jeeves

Volume 43 NEUROTRANSMITTERS IN THE HUMAN BRAIN
Edited by David J. Tracey, George Paxinos, and Jonathan Stone

Volume 44 ALZHEIMER'S AND PARKINSON'S DISEASES: Recent Developments
Edited by Israel Hanin, Mitsuo Yoshida, and Abraham Fisher

Volume 45 EPILEPSY AND THE CORPUS CALLOSUM 2
Edited by Alexander G. Reeves and David W. Roberts

Volume 46 BIOLOGY AND PHYSIOLOGY OF THE BLOOD–BRAIN BARRIER: Transport, Cellular Interactions, and Brain Pathologies
Edited by Pierre-Olivier Couraud and Daniel Scherman

Volume 47 THE BASAL GANGLIA V
Edited by Chihiro Ohye, Minoru Kimura, and John S. McKenzie

A Continuation Order Plan is available for this series. A continuation order will bring delivery of each new volume immediately upon publication. Volumes are billed only upon actual shipment. For further information please contact the publisher.

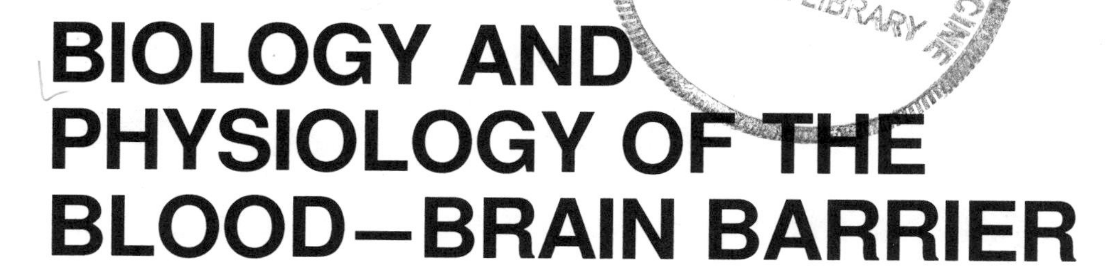

BIOLOGY AND PHYSIOLOGY OF THE BLOOD–BRAIN BARRIER

TRANSPORT, CELLULAR INTERACTIONS, AND BRAIN PATHOLOGIES

Edited by

Pierre-Olivier Couraud

Institut Cochin de Génétique Moléculaire
Paris, France

and

Daniel Scherman

CNRS–Rhône-Poulenc Rorer
Vitry sur Seine, France

PLENUM PRESS • NEW YORK AND LONDON

Library of Congress Cataloging-in-Publication Data

Biology and physiology of the blood-brain barrier : transport, cellular interactions, and brain pathologies / edited by Pierre-Olivier Couraud and Daniel Scherman.
p. cm. -- (Advances in behavioral biology ; v. 46)
"Proceedings of the Cerebral Vascular Biology Symposium, held July 10-12, 1995, in Paris, France"--T.p. verso.
Includes bibliographical references and index.
ISBN 0-306-45362-2
1. Blood-brain barrier--Congresses. 2. Blood-brain barrier disorders--Congresses. I. Couraud, Pierre-Olivier. II. Scherman, Daniel. III. Cerebral Vascular Biology Symposium (1995 : Paris, France) IV. Series.
[DNLM: 1. Blood-Brain Barrier--physiology--congresses. W3 AD215 v.46 1996 / WL 200 B615 1996]
QP375.5.B526 1996
612.8'25--dc20
DNLM/DLC
for Library of Congress 96-34815
CIP

Proceedings of the Cerebral Vascular Biology Symposium, held July 10–12, 1995, in Paris, France

ISBN 0-306-45362-2

A Division of Plenum Publishing Corporation
233 Spring Street, New York, N.Y. 10013

10 9 8 7 6 5 4 3 2 1

Printed in the United States of America

IN MEMORIAM

We dedicate this book to the memory of

Dr. Ferenc Joo, M.D., Ph.D., D.Sc.

who suddenly died on February 25, 1996

We will remember him as a cheerful scientist
and as a pioneer in the study of the blood–brain barrier

PREFACE

The endothelial cells of the cerebral vasculature constitute, together with perivascular elements (astrocytes, pericytes, basement membrane), the blood–brain barrier (BBB), which strictly limits and specifically controls the exchanges between the blood and the cerebral extracellular space.The existence of such a physical, enzymatic, and active barrier isolating the central nervous system has broad physiological, biological, pharmacological, and pathological consequences, most of which are not yet fully elucidated.

The Cerebral Vascular Biology conference (CVB '95) was organized and held at the "Carré des Sciences" in Paris on July 10–12, 1995. Like the CVB '92 conference held in Duluth, Minnesota, three years ago, the objectives were to provide a forum for presentation of the most recent progresses and to stimulate discussions in the field of the biology, physiology, and pathology of the blood–brain barrier. The Paris conference gathered more than 150 participants, including investigators in basic neuroscience, physicians, and students, who actively contributed to the scientific program by their oral or poster presentations.

This volume contains a collection of short articles that summarize most of the new data that were presented at the conference. Six thematic parts focus on physiological transports, drug delivery, multidrug resistance P-glycoprotein, signal transduction at the BBB, interactions between the immune system and the cerebral endothelial cells, and the blood–brain barrier-related pathologies in the central nervous system.

In addition, two introductory articles present new insights in the rapidly evolving topics of cerebral angiogenesis and gene transfer to the brain.

CVB '95 was a satellite conference of the BRAIN '95 meeting (Köln, Germany) and was financially supported by the Centre National de la Recherche Scientifique, the Institut National de la Santé et de la Recherche Médicale, the French Ministry of Education and Research, the Société des Neurosciences, the European Concerted Action Biomed ("Drug Delivery to the Brain: New Experimental Strategies"), the Association Naturalia et Biologia, and the following companies: Amersham, Becton Dickinson, Glaxo, Rhone Poulenc Rorer, Roussel, and Sanofi.

We would like to acknowledge the members of the CVB '95 Scientific Committee: Professors A. L. Betz, A. G. de Boer, L. E. Drewes, F. Joo, J. M. Lefauconnier, and W. Risau.

We are also indebted to Dr. M. Santarromana, Dr. D. Lechardeur, and Mrs. J. Catala; without their competent assistance, this volume could not have been published.

We are extremely grateful to all the authors and hope that this volume will prolong the creative atmosphere and enthusiasm shared by the participants to the conference.

P. O. Couraud and D. Scherman

PRÉFACE

La barrière hémato-encéphalique (BHE) est constituée par les cellules endothéliales des microvaisseaux cérébraux en étroite interaction avec des éléments périvasculaires (membrane basale, péricytes, astrocytes). Ces cellules forment une barrière physiologique qui limite les échanges entre le sang et l'espace cérébral extracellulaire. Ainsi la barrière hémato-encéphalique isole le système nerveux central du reste de l'organisme, avec d'importantes conséquences physiologiques, biologiques, et pharmacologiques qui ne sont pas encore totalement élucidées.

Le congrès "Biologie Cérébro-Vasculaire" (CVB '95) s'est déroulé au Carré des Sciences à Paris du 10 au 12 Juillet 1995. Comme le congrès CVB '92 de Duluth, Minnesota, trois ans auparavant, le congrès CVB '95 se proposait d'être un lieu d'information et d'échanges concernant les progrès les plus récents en matière de biologie, physiologie, et pathologie du système vasculaire cérébral et de la BHE, et d'ouvrir un champ de discussions actives sur ce sujet.

Le congrès CVB '95 a reuni à Paris plus de 150 participants dans le domaine des neurosciences, chercheurs, étudiants, ou cliniciens, qui ont contribué au succès du programme scientifique par leurs présentations orales ou affichées.

Ce volume regroupe une série de courts articles résumant la plupart des nouvelles données présentées au cours du congrès. Six livraisons thématiques traitent successivement du transport à travers la BHE, de la pénétration cérébrale des médicaments, de la P-glycoprotéine de resistance multidrogues, de la transduction du signal, au niveau de la BHE, des interactions entre le système immunitaire et les cellules endothéliales cérébrales, et enfin des pathologies du système nerveux central impliquant la BHE. De plus, deux articles préliminaires présentent un aperçu de deux sujets en plein développement: l'angiogénèse au niveau cérébral et le transfert de gènes dans le cerveau.

La conférence CVB '95 était un congrès satellite du congrès Brain '95. Ce congrès CVB '95 a reçu une aide financière de la part du Centre National de la Recherche Scientifique, de l'Institut National de la Santé et de la Recherche Médicale, du Ministère de l'Education Nationale, de l'Enseignement Supérieur et de la Recherche, de la Société des Neurosciences, de l'Action Concertée Européenne Biomed: "Drug Delivery to the Brain: New Experimental Strategies," et de l'Association Naturalia et Biologia. Plusieurs sociétés industrielles ont également apporté leur soutien financier à ce congrès: Amersham, Becton Dickinson, Glaxo, Rhône Poulenc Rorer, Roussel, Sanofi.

Nous tenons à remercier les membres du Comité Scientifique de la conférence CVB '95: les Professeurs A. L. Betz, A. G. De Boer, L. E. Drewes, F. Joo, J. M. Lefauconnier, W. Risau.

Nous souhaitons aussi remercier les Dr. M. Santarromana et Dr. D. Lechardeur, ainsi que Mme. J. Catala: sans leur aide, ce volume n'aurait pas vu le jour.

Nous remercions tous les auteurs et espérons que ce volume prolongera l'atmosphère créative et l'enthousiasme partagé par tous les participants de la conférence CVB '95.

P. O. Couraud et D. Scherman

CONTENTS

Introductory Overviews

Part I: Physiological Transports through the Blood-Brain Barrier

Part II: Blood-Brain Barrier and Drug Delivery to the Brain

Part IV: Signal Transduction in Brain Endothelial Cells

Part V: Interactions—Leukocytes-Brain Endothelial Cells

Part VI: Brain Pathologies and Blood-Brain Barrier

1

DEVELOPMENT OF BLOOD-BRAIN BARRIER ENDOTHELIAL CELLS

Werner Risau

Max-Planck-Institut für Physiologische und Klinische Forschung
Abteilung Molekulare Zellbiologie
W. G. Kerckhoff-Institut
D-61231 Bad Nauheim, Germany

SUMMARY

The vascular system of the central nervous system is derived from capillary endothelial cells, which have invaded the early embryonic neuroectoderm. This process is called angiogenesis and is probably regulated by brain-derived factors. Vascular endothelial cell growth factor (VEGF) is an angiogenic growth factor whose expression correlated with embryonic brain angiogenesis, i.e. expression is high in the embryonic brain when angiogenesis occurs and low in the adult brain when angiogenesis is shut off under normal physiological conditions. VEGF receptors 1 and 2 (flt-1 and flk-1) as well as another pair of receptors (tie-1 and tie-2) are receptor tyrosine kinases specifically expressed in endothelial cells. Expression of these receptors is high during brain angiogenesis but low in adult blood-brain barrier endothelium. Signal transduction by these or other receptors involved in endothelial cell growth and differentiatian may be mediated by lyn, a nonreceptor tyrosine kinase expressed in brain endothelium. Induction and maintenance of blood-brain barrier endothelial cell characteristics (complex tight junctions, low number of vesicles, specialized transport systems) are regulated by the local brain environment; e.g. astrocytes. Tight junctions between brain endothelial cells are the structural basis for the paracellular impermeability and high electrical resistance of blood-brain barrier endothelium. Association of tight junction particles with the P-face along with the intercellular adhesion forces rather than the number or branching frequency of tight junction strands correlated with BBB development and function suggesting that the cytoplasmic anchoring of the tight junctions plays an important role.

RÉSUMÉ

Le réseau vasculaire du système nerveux central dérive des cellules endothéliales capillaires, qui ont envahi le neuroectoderme embryonaire précoce. Le processus appelé

Biology and Physiology of the Blood–Brain Barrier, edited by Couraud and Scherman
Plenum Press, New York, 1996

angiogénèse est probablement régulé par des facteurs provenant du cerveau. Le facteur de croissance des cellules endothéliales vasculaires (VEGF) est un facteur angiogénique dont l'expression est corrélée à l'angiogénèse du cerveau embryonaire, c'est-à-dire que son expression est élevée dans le cerveau embryonaire quand se produit l'angiogénèse, et faible dans le cerveau adulte, où l'angiogénèse est stoppée dans les les conditions physiologiques normales. Les récepteurs 1 et 2 (flt-1 et flk-1) du VEGF, ainsi qu'une autre paire de récepteurs (tie-1 et tie-2) sont des récepteurs de tyrosine kinase spécifiquement exprimés dans les cellules endothéliales. L'expression de ces récepteurs est importante pendant l'angiogénèse du cerveau, mais est basse au niveau de l'endothélium de la BHE chez l'adulte. La transduction du signal par ces récepteurs, ou par d'autres récepteurs impliqués dans la croissance et la différenciation des cellules endothéliales, peut être médiée par lyn, une tyrosine kinase non récepteur exprimée dans l'endothélium cérébral.L'induction et le maintien des caractéristiques des cellules endothéliales de la BHE (jonctions occlusives complexes, petit nombre de vésicules, systèmes de transport spécifiques) sont régulés par l'environnement local du cerveau, par exemple les astrocytes.Les jonctions occlusives entre les cellules endothéliales cérébrales sont la structure de base pour l'imperméabilité paracellulaire et la forte résistivité électrique de l'endothélium de la BHE.

L'association des particules de jonction au niveau de la face Pabluminale faisant intervenir des forces d'adhésion intercellulaires plutôt qu'une haute fréquence de branchements ou un nombre élevé de zones de jonctions serr ées est corrélée au développement et à la fonction de la BHE, ce qui suggère que l'ancrage cytoplasmique des jonctions serrées joue un role important.

The cardiovascular system is the first organ system to develop in the embryo. During the process of gastrulation, pluripotent cells invaginate through the primitive streak. Mesodermal cells are induced during this process and migrate widely throughout the extraembryonic membranes and the embryo proper. From these mesodermal cells angioblasts differentiate which are the progenitors of endothelial cells and are defined as a cell type that has the potency of differentiating into an endothelial cell but has not yet acquired all the characteristic markers and has not yet formed a lumen. In the yolk sac splanchnopleuric mesoderm, angioblasts and hematopoietic precursor cells differentiate in close association with each other forming so called blood islands. The formation of these blood vessels from in situ differentiating endothelial cells (angioblasts) is called vasculogenesis (for review see ref. 26). The subsequent processes of blood island fusion and formation of lumina by angioblasts lead to a primordial vascular network. This vasculature is extended by sprouting of new capillaries from the preexisting network (i.e. angiogenesis) resulting in an elongated, highly branched vascular plexus.

The first molecule so far known to be expressed in a population of mesodermal cells giving rise to angioblasts is the vascular endothelial growth factor receptor-2 (VEGFR-2; also known as flk-1 in the mouse and KDR in the human). Later during embryonic development this molecule becomes restricted to endothelial cells consistent with the function of its ligand VEGF as a specific endothelial cell growth and vascular permeability factor (8, 11, 19, 39). It is also induced in the avian epiblast cells as an early response after induction by fibroblast growth factors (12). VEGF itself is expressed in the endoderm of the 7,5 day mouse and 20h avian embryo (4). Since the endoderm is adjacent to the mesoderm, a paracrine relationship between the two germ layers may exist and VEGF secreted by the endoderm may support the differentiation of VEGF-R2-expressing mesodermal cells to angioblasts.

The development of the brain vascular system begins when angioblasts invade the head region and form the perineural vascular plexus which in a 2-day chicken embryo and in a 9 day rodent embryo covers the entire surface of the neural tube. Although embryonic angioblasts are highly invasive and have the potential to migrate throughout the embryo they do not enter the neural epithelium (2). Vice versa, none of the precursors that form the perineural vascular plexus are derived from the neuroectoderm. During embryonic development the perineural vascular plexus differentiates to the leptomeningeal arteries and veins and gives rise to the intraparenchymal vessels of the central nervous system. Thus the vascular system within the nervous system does not develop by vasculogenesis. Rather, vascular sprouts originating from the perineural plexus penetrate the growing neuroectoderm commencing at 3 days of embryonic development in the chicken and 10 days in rodents. This mechanism, where new vessels form by sprouting from pre-existing vessels, is called angiogenesis (24, 28). Thus, the vasculature forming the blood-brain barrier develops entirely by angiogenesis, meaning that intraparenchymal vessels are derived from precursors outside of the neuroectodermal tissue. The precise temporal and spatial regulation of vascular development in the brain prompted us to investigate whether the embryonic brain produces factors, which are chemotactic and/or mitogenic for endothelial cells. Strong evidence has accumulated suggesting that VEGF is a key regulatory factor in brain angiogenesis. The spatial and temporal expression pattern of VEGF mRNA highly corresponds to angiogenesis during embryonic development in the mouse brain (3). By in situ hybridization we have shown that in day 17 mouse embryos VEGF mRNA is expressed in the ventricular layer of the developing neuroectoderm. In adult brain, where vascularization is complete and the rate of endothelial cell proliferation is very low (10), VEGF mRNA could not be detected any longer in the ependymal cells, which form the lining of the ventricles. Therefore, it is conceivable that during brain development VEGF is released by cells of the ventricular layer, thus promoting angiogenesis by initiating migratory and mitogenic responses in endothelial cells of the perineural vascular plexus. A concentration gradient of the diffusible VEGF isoforms declining from the ventricular layer towards the perineural vascular plexus may cause the radial ingrowth of capillaries from the vascular plexus towards the angiogenic stimulus provided by VEGF-secreting ventricular epithelial cells (3). The key role of VEGF as regulator of embryonic angiogenesis in the brain is further substantiated by the expression-pattern of two of its receptors. At day 11.5 in the mouse embryo, when the first vascular sprouts begin to radially invade the neuroectoderm from the perinerual plexus, expression of the VEGF receptors was high in the perineural vascular plexus and in structures resembling invading vascular sprouts (4, 19). In the adult brain, when angiogenesis has ceased, VEGF receptor expression was very low. Recent knockout experiments have shown that the VEGF receptors are necessary for vascular development. While endothelial cell differentiation from the mesoderm was completely abolished in VEGF-R2 deficient mice, VEGF-R1 mutation resulted in vascular dysmorphogenesis and aberrant blood vessels (14, 34). These early lethal phenotypes prevented the analysis of the role of the receptors in brain angiogenesis. Inducible gene deletion systems or other methods are necessary to provide direct evidence for the role of these receptors in brain angiogenesis in vivo.

Recently, two additional endothelial cell-specific receptor tyrosine kinases, tie-1 (tie) and tie-2 (tek) have been described that are expressed in the developing vasculature (for review see 21). Consistent with a possible role in brain angiogenesis, tie-1 and tie-2 mRNA is expressed specifically by brain endothelial cells and down-regulated in the adult organism (33). Their expression patterns therefore resemble the expression of the high affinity VEGF receptors but the ligands for the tie receptors have not been identified so far. Mice deficient for the tie receptors showed dramatic but distinct phenotypes and are both embryonic lethal. The tie-1 receptor mutants developed edema and hemorrhage indicating a role of this receptor in vascular permeability. Analysis of the tie-2 mutants revealed a defect in vascular remodeling and angiogenesis, but the early development of the vascular system, i.e. vasculogenesis was not affected. Interestingly, endothelial cells of the perineural plexus were found to be unable to invade into the neuroectoderm (31). These data raise the possibility that the two receptor systems (VEGF and tie receptors) interact to induce signals in endothelial cells that stimulate them to invade into the neuroectoderm and form new blood vessels.

In order to identify signal transduction proteins specifically expressed in the developing blood-brain barrier endothelium, we searched for proteins containing the src-homology 3 (SH3) domains. SH3-domains are structural elements typically found in signal transduction proteins. Unexpectedly, we found that all murine and bovine brain endothelial cells encoded the SH3-domain of the non-receptor tyrosine kinase lyn. Lyn has been shown to be involved in signal transduction in hematopoetic cells. We localized lyn in embryonic and early post-natal mouse brain endothelium but not in endothelium outside the brain (1). The relevance for lyn expression during brain angiogenesis remains to be determined, however its specific expression suggests that Lyn could be involved in transduction of endothelial growth, mediated possibly be the VEGF or tie receptors, or early differentiation signals critical for blood-brain barrier development.

The differentiation of the blood-brain barrier, namely the development of the selective permeability of brain capillaries is usually investigated by measuring the decreasing permeability for vascular tracers like dyes, horseradish peroxdidase or lanthanum. Blood brain barrier differentiation appears to be a gradual process which is independent of vascular proliferation in the brain, because i.e. rat brain endothelial cells *in vivo* commence barrier formation at about embryonic day 16, while many still proliferate, proliferation reaching the peak of activity between postnatal day 5 and 9 (30). Despite some controversial results obtained by using different techniques (29, 32, 37) blood-brain barrier differentiation appears to occur at different time points in different locations in the brain. Thus, not all brain capillaries become impermeable at the same time. Rather, the actual time is dependent on the anatomical location. In the mouse the barrier to proteins forms first in the spinal cord and last in the telencephalon; within the telencephalon there is an ependymal-cortical gradient of barrier differentiation, with injected protein still visible in the subependymal layer at embryonic day 16 (27). Despite the variations in the exact timing of blood-brain barrier formation, due to methodological differences, it is clear that all the capillaries that are present in the brain before embryonic day 13 in the chicken and embryonic day 15 in the rat and mouse lack a mature barrier. The transition from a leaky capillary to a capillary with features of a blood-brain barrier is characterized by several changes in endothelial cell morpholopgy, biochemistry and function that make these endothelial cells distinct from every other endothelial cell in the body. Many of these changes have been reviewed in detail previously (9).

Tight junctions between the endothelial cells of the brain microvessels have been shown by qualitative freeze-fracture and ultrathin section electron microscopic studies to be more complex than those in other endothelial cells in the body (22, 23). Tight junctions are thought to function as a seal only if they are continuous and branched (5). The complex network of tight junctions between brain endothelial cells are therefore primarily responsible for the paracellular impermeability and unlike simple tight juctions provide a high electrical resistance (about 2000 Ohm x cm^2) (7). During brain angiogenesis it has been shown in the rat that the decrease in vessel permeability to protein correlates with a conformational change of the tight junctions between endothelial cells lining the brain vessels in that the length of "unfused" junctional contacts decreased during embryonic development (36). The length of "fused" junctions increased, which can be seen by en face freeze fracture views as an increase in extensiveness and complexity of the junctions. We have performed a quantitative analysis of the structure and function of tight junctions in primary cultures of bovine brain endothelial cells using quantitative freeze-fracture electron microscoy and ion and inulin permeability (38). The complexity of tight junctions, defined as the number of branch points per unit length of tight junctional strands, decreased 5 hours after culture but thereafter remained almost constant. In contrast, the association of tight junction particles with the cytoplasmic leaflet of the endothelial membrane bilayer (P-face) decreased continuously with a major drop between 16 and 24 hours of culture. The complexity and the P-face association of tight junctions could be restored to a certain extent by co-culture of endothelial cells with astrocytes or astrocyte conditioned medium. Co-culture with fibroblasts had no effect. These data suggest that the association of tight junction particles with the P-face rather than the number of branching frequences of tight junctions correlate with blood-brain barrier function. Interestingly, tight junctions in peripheral, non-barrier endothelial cells are E-face associated *in vivo* (20, 35). These data suggest an important functional role for the P-face association and thus the cytoplasmic anchoring of the tight junction particles for brain endothelial barrier function. It is tempting to speculate that tight junction proteins such as occludin, ZO-1, ZO-2 or cingulin play a major role in this process (6, 13, 15-17).

Pathological conditions within the central nervous system like ischemia, inflammation or tumor growth lead to blood-brain barrier dysfunction, emphasizing that the permeability barrier is not simply "switched on" during embryonic development, but that a continuing regulation of its maintenance, probably provided by the tissue microenvironment, is necessary for a proper barrier function in any circumstance. In many brain tumors morphological irregularities of the perivascular ensheathment such as enlarged perivascular space, gaps in the basal lamina or deficient glial investment correlate with a breakdown of the blood-brain barrier. This again supports the concept that endothelial differentiation to form a blood-brain barrier can be modulated by the tissue microenvironment. Glioblastomas have been shown to highly express VEGF in tumor cells and VEGF receptors in tumor endothelium (25). We have demonstrated that resumption of angiogenesis is necessary for the growth of glioblastomas in the brain (18). Tumorigenesis hereby reproduces embryogenesis and apparently employs the same molecular mechanisms as angiogenesis during embryonic development. These observations strongly support the concept that angiogenesis under physiological and pathological conditions is regulated by a paracrine mechanism and identify VEGF as a tumor angiogenesis factor *in vivo*.

REFERENCES

1. Achen, M. G., M. Clauss, H. Schnürch, and W. Risau. 1995. The non-receptor tyrosine kinase Lyn is localized in the developing blood-brain barrier. *Differentiation*. 59:15-24.
2. Bär, T. 1980. The vascular system of the cerebral cortex. *Adv.Anat.Embryol.Cell Biol.* 59:1-62.
3. Breier, G., U. Albrecht, S. Sterrer, and W. Risau. 1992. Expression of vascular endothelial growth-factor during embryonic angiogenesis and endothelial-cell differentiation. *Development*. 114:521-532.
4. Breier, G., M. Clauss, and W. Risau. 1995. Coordinate expression of VEGF-receptor 1 (flt-1) and its ligand suggests a paracrine regulation of murine vascular development. *Devel. Dynamics*. in press.
5. Cereijido, M., L. González-Mariscal, and B. Contreras. 1989. Tight junction: barrier between higher organisms and environment. *NIPS*. 4:72-75.
6. Citi, S. 1993. The molecular-organization of tight junctions. *J.Cell Biol.* 121:485-489.
7. Crone, C., and S.-P. Olesen. 1982. Electrical resistance of brain microvascular endothelium. *Brain Res.* 241:49-55.
8. Eichmann, A., C. Marcelle, C. Breant, and N. M. LeDouarin. 1993. Two molecules related to the VEGF receptor are expressed in early endothelial-cells during avian embryonic development. *Mechanisms of Develop.* 42:33-48.
9. Engelhardt, B., and W. Risau. 1995. Development of the blood-brain barrier. *In* New Concepts of a blood-brain barrier. J. Greenwood, D. Begley, M. Segal, and S. Lightman, editors. Plenum, London. in press.
10. Engerman, R. L., D. Pfaffenbach, and M. D. Davis. 1967. Cell turnover of capillaries. *Lab.Invest.* 17:738-743.
11. Ferrara, N., K. Houck, L. Jakeman, and D. W. Leung. 1992. Molecular and biological properties of the vascular endothelial growth-factor family of proteins. *Endocrine Reviews*. 13:18-32.

12. Flamme, I., G. Breier, and W. Risau. 1995. Vascular endothelial growth factor (VEGF) and VEGF-receptor 2 (flk-1) are expressed during vasculogenesis and vascular differentiation in the quail embryo. *Devel. Biol.* 169:699-712.
13. Fleming, T. P., M. Hay, Q. Javed, and S. Citi. 1993. Localization of tight junction protein cingulin is temporally and spatially regulated during early mouse development. *Development* 117:1135-1144.
14. Fong, G.-H., J. Rossant, M. Gertsenstein, and M. L. Breitman. 1995. Role of the flt-1 receptor tyrosine kinase in regulating the assembly of vascular endothelium. *Nature* 376:66-70.
15. Furuse, M., T. Hirase, M. Itoh, A. Nagafuchi, S. Yonemura, and S. Tsukita. 1993. Occludin - a novel integral membrane-protein localizing at tight junctions. *J.Cell Biol.* 123:1777-1788.
16. Gumbiner, B., T. Lowenkopf, and D. Apatira. 1991. Identification of a 160-kda polypeptide that binds to the tight junction protein ZO-1. *Proc.Natl.Acad.Sc.* 88:3460-3464.
17. Itoh, M., A. Nagafuchi, S. Yonemura, T. Kitaniyasuda, and S. Tsukita. 1993. The 220-kd protein colocalizing with cadherins in nonepithelial cells is identical to ZO-1, a tight junction associated protein in epithelial-cells - cdna cloning and immunoelectron microscopy. *J.Cell Biol.* 121:491-502.
18. Millauer, B., L. K. Shawver, K. H. Plate, W. Risau, and A. Ullrich. 1994. Glioblastoma growth inhibited in vivo by a dominant-negative Flk-1 mutant. *Nature*. 367:576-9.
19. Millauer, B., S. Wizigmann-Voos, H. Schnürch, R. Martinez, N. P. H. Møller, W. Risau, and A. Ullrich. 1993. High affinity VEGF binding and developmental expression suggest Flk-1 as a major regulator of vasculogenesis and angiogenesis. *Cell* 72:835-846.
20. Mühleisen, H., H. Wolburg, and E. Betz. 1989. Freeze-fracture analysis of endothelial cell membranes in rabbit carotid arteries subjected to short-term atherogenic stimuli. *Virch.Arch.B Cell Pathol.* 56:413-417.
21. Mustonen, T., and K. Alitalo. 1995. Endothelial receptor tyrosine kinases involved in angiogenesis. *J. Cell Biol.* 129:895-898.
22. Nagy, Z., H. Peters, and I. Hüttner. 1984. Fracture faces of cell junctions in cerebral endothelium during normal and hyperosmotic conditions. *Lab.Invest.* 50:313-322.
23. Nico, B., D. Cantino, M. Bertossi, D. Ribatti, M. Sassoe, and L. Roncali. 1992. Tight endothelial junctions in the developing microvasculature - a thin-section and freeze-fracture study in the chick-embryo optic tectum. *J.Submicr.Cytol.P.* 24:85-95.
24. Noden, D. M. 1991. Development of craniofacial blood vessels. *In* The development of the vascular system. R. N. Feinberg, G. K. Sherer, and R. Auerbach, editors. Karger, Basel. 1-24.
25. Plate, K. H., G. Breier, H. A. Weich, and W. Risau. 1992. Vascular endothelial growth-factor is a potential tumor angiogenesis factor in human gliomas invivo. *Nature* 359:845-848.
26. Risau, W., and I. Flamme. 1995. Vasculogenesis. *Ann. Rev. Cell Devel. Biol.* in press.
27. Risau, W., R. Hallmann, and U. Albrecht. 1986. Differentiation-Dependent Expression of Protein in Brain Endothelium during Development of the Blood-Brain Barrier. *Dev.Biol.* 117:537-545.
28. Risau, W., and V. Lemmon. 1988. Changes in the vascular extracellular matrix during embryonic vasculogenesis and angiogenesis. *Dev.Biol.* 125:441-450.
29. Risau, W., and H. Wolburg. 1991. The importance of the blood-brain-barrier in fetuses and embryos - reply. *TINS*. 14:15.
30. Robertson, P. L., M. Du Bois, P. D. Bowman, and G. W. Goldstein. 1985. Angiogenesis in developing rat brain: an in vivo and in vitro study. *Dev.Brain Res.* 23:219-223.
31. Sato, T. N., Y. Tozawa, U. Deutsch, K. Wolburg-Buchholz, Y. Fujiwara, M. Gendron-Maguire, T. Gridley, H. Wolburg, W. Risau, and Y. Qin. 1995. Distinct roles of the receptor tyrosine kinases tie-1 and tie-2 in blood vessel formation. *Nature* 376:70-74.
32. Saunders, N. R., K. M. Dziegielewska, and K. Mollgard. 1991. The importance of the blood-brain barrier in fetuses and embryos, letter to the editor. *TINS* 14:14.
33. Schnürch, H., and W. Risau. 1993. Expression of tie-2, a member of a novel family of receptor tyrosine kinases, in the endothelial cell lineage. *Development* 119:957-968.
34. Shalaby, F., J. Rossant, T. Yamaguchi, M. Gertsenstein, X.-F. Wu, M. L. Breitman, and A. C. Schuh. 1995. Failure of blood-island formation and vasculogenesis in flk-1-deficient mice. *Nature* 376:62-66.
35. Simionescu, M., N. Ghinea, A. Fixman, M. Lasser, L. Kukes, N. Simionescu, and G. E. Palade. 1988. The cerebral microvasculature of the rat: structure and luminal surface properties during early development. *J.Submicrosc.Cytol.* 20:243-261.
36. Stewart, P. A., and E. M. Hayakawa. 1987. Interendothelial junctional changes underlie the developmental "tightening" of the blood-brain barrier. *Dev.Brain Res.* 32:271-281.
37. Wakai, S., and N. Hirokawa. 1978. Development of the blood-brain barrier to horseradish peroxidase in the chick embryo. *Cell Tissue Res.* 195:195-203.

38. Wolburg, H., J. Neuhaus, U. Kniesel, B. Krauss, E. M. Schmid, M. Ocalan, C. Farrell, and W. Risau. 1994. Modulation of tight junction structure in blood-brain-barrier endothelial-cells - effects of tissue-culture, 2nd messengers and cocultured astrocytes. *J.Cell Sci.* 107:1347-1357.
39. Yamaguchi, T. P., D. J. Dumont, R. A. Conlon, M. L. Breitman, and J. Rossant. 1993. Flk-1, an flt-related receptor tyrosine kinase is an early marker for endothelial-cell precursors. *Development.* 118:489-498.

2

PROSPECTS FOR NEUROPROTECTIVE GENE THERAPY FOR NEURODEGENERATIVE DISEASES

Marc Peschanski

INSERM U421, IM3
Faculté de Médecine
Créteil, France

Neurodegenerative diseases are still, except for a few of them like Parkinson's disease, out of reach of available therapeutics. Several breakthroughs in the past ten years have, however, opened new potential ways that are actively explored in a large number of laboratories. Two of these major advances that could be of therapeutic value are the discovery of several families of substances that are able to exert neuroprotective effects on the one hand, development of gene engineering techniques on the other hand. In our laboratory, a series of studies are ongoing to design ways, using these engineering techniques, to create biological minipumps able to secrete neuroprotective agents into the brain.

There are presently two major techniques to obtain expression of an exogenous gene of interest in the brain that are called indirect -or *ex vivo*-, or direct -*in vivo*- gene transfer. Indirect transfer relies upon the ability of cell populations to be maintained in culture and to replicate, to readily integrate exogenous genes and to survive implantation into an adult brain. As many other groups, we have started examining these possibilities using a cell line (Wojcik et al., 1993). The 1009 cell line, derived from a mouse teratocarcinoma, was chosen because these cells had been shown to differentiate *in vitro* into neurons. After implantation, they indeed stopped proliferating for most of them and presented a neuronal phenotype. They were able to express a marker gene for at least one month *in vivo*. This cell line, as well as all others studied by other groups, cannot be used in patients, however, due to the risk of tumorogenicity they present. We, therefore, turned to primary cells. The favored cells for intracerebral transplantation at this time -in gene therapy projects- are fibroblasts because they are maintained for a very long time (> 1 year) after transplantation. They are not efficient for gene expression however, for more than a few weeks because of the repression of viral promoters. We have tried to use another type, myoblasts, that has been reported to work for such a purpose. We have, indeed, not observed repression of the transgene promoter in myoblasts up to 3 months after intracerebral transplantation. In contrast, we have observed a steep decrease in the number of surviving cells. Up to now, altogether, indirect gene transfer is still looking for a reliable technique. One presently under study, in animals but also in clinical trials for amyotrophic lateral sclerosis, is the encapsulation of proliferative cells in

Biology and Physiology of the Blood–Brain Barrier, edited by Couraud and Scherman
Plenum Press, New York, 1996

polymers that allow small molecules but neither large ones nor cells to travel freely between the implanted cells and the brain.

Direct gene transfer to the brain is essentially based, at the present time, on recombinant viral vectors. Since neural cells do not proliferate, retroviral vectors are not instrumental. Other viruses are able to transfer a gene into post-mitotic cells, including the herpes virus HSV-1, the adeno-associated virus and the adenovirus. We have explored the ability of this last virus to transfer a marker gene into neural cells. Type 5 human adenoviruses, rendered replication-deficient by the deletion of the E_1A-E_1B promoters, have sustained homologous recombination with the *lacZ* gene encoding β galactosidase in *E Coli*. Injection of this vector into the brain allows long-term expression of the transgene in all cell types, including neurons, glia and ependymal cells. Injection sites are limited in volume, however, and the vector can be cytotoxic if too high titers are used (Akli et al., 1993 ; Lisovoski et al., 1994). On this basis, we have designed vectors recombinant for genes of interest, i.e. neurotrophic factors with demonstrated neuroprotective action. The first one, which is currently under study, is an adenovirus vector recombinant for the neuroactive cytokine CNTF (ciliary neurotrophic factor). Injection of this vector into the brain parenchyma provokes a specific differentiation of astrocytes, in the absence of neuronal or microglial alteration. Injected two weeks before the administration of a neurotoxin that kills striatal neurons, this CNTF adenovirus exhibits a neuroprotective effect. Because a similar protection is obtained after β gal vectors are used, but not after saline injection, we hypothesize that neuroprotection is related to the astroglial response, activated astrocytes being known to increase their detoxifying capacities.

In conclusion, various techniques are currently under study in a large number of laboratories, in order to find the optimal ways to introduce genes of interest into the central nervous system. Clinical relevance of these vectors -and for most of it, also of the genes to be transfered- is still under question. There is no doubt, however, that the enormous effort which is being made will lead, over the coming years, to major breakthroughs in a field -the neuroprotective therapeutics in neurodegenerative diseases- that was hardly existing ten years ago.

REFERENCES

Akli S, Caillaud C, Vigne E, Stratford-Perricaudet Ld, Poenaru L, Perricaudet M, Kahn A, Peschanski M. Transfer of foreign genes into the brain using adenovirus vectors. Nature Genetics,1993,3:224-228.

Lisovoski F, Cadusseau J, Akli S, Caillaud C, Vigne E, Poenaru L, Stratford-Perricaudet L, Perricaudet M, Kahn A, Peschanski M. In vivo transfer of a marker gene to study motoneuronal development. Neuroreport,1994,5:1069-1072

Wojcik B, Nothias F, Lazar M, Jouin H, Nicolas Jf, Peschanski M. Catecholaminergic neurons derived from intracerebral implantation of embryonal carcinoma cells. Proc Natl Acad Sci (USA),1993,90:1305-1309.

3

BLOOD-BRAIN BARRIER TAURINE TRANSPORT AND BRAIN VOLUME REGULATION

Richard F. Keep,[1] Walter Stummer,[1] Jianming Xiang,[1] and A. Lorris Betz[1,2]

[1] Department of Surgery (Neurosurgery)
[2] Departments of Pediatrics and Neurology
University of Michigan
Ann Arbor, Michigan 48105-0532

SUMMARY

We have investigated mechanisms that may be involved in brain taurine loss during hypo-osmotic stress using a mixture of in vivo and in vitro measurements of blood-brain and blood-CSF barrier taurine transport. Choroid plexus taurine uptake has a K_m of 230 μM, indicating that it is not saturated at normal CSF concentrations and that uptake will increase as extracellular taurine concentration increases. Choroid plexus uptake was reduced in the presence of calmodulin inhibitors suggesting that calmodulin may be involved in brain volume regulation. Choroid plexus ^{3}H-taurine efflux via a niflumic acid-sensitive pathway was stimulated directly by reductions in osmolality and also by increases in extracellular taurine. Unlike other tissues, efflux was not stimulated by hypo-osmotic stress in isolated cerebral microvessels. This may indicate that such an efflux mechanism is, if present at all, on the luminal membrane and not accessible in these experiments. Although plasma taurine was increased by hypo-osmotic stress in vivo, this was not reflected by an increase in taurine influx across the blood-brain barrier. Thus this barrier tissue also prevents taurine from being recycled back into brain.

RÉSUMÉ

Nous avons étudié les mécanismes qui pourraient expliquer la déperdition du cerveau en taurine au cours de chocs hypo-osmotiques par des expériences de mesure du transport de la taurine par la barrière entre le sang et le parenchyme cérébral (BHE) et la barrière entre le sang et le liquide céphalo-rachidien (plexus choroïdes). La captation de la taurine par les plexus choroïdes possède un Km de 230 μM, indiquant que le transport de taurine n'est pas saturé aux concentrations normales présentes dans le liquide cephalo-rachidien. De ce fait,

Biology and Physiology of the Blood–Brain Barrier, edited by Couraud and Scherman
Plenum Press, New York, 1996

une augmentation extracellulaire de la concentration en taurine s'accompagnera d'une élévation du transport de taurine au niveau des plexus choroïdes. Des inhibiteurs de la calmoduline provoquent une réduction du transport par les plexus choroïdes suggérant que la calmoduline pourrait être impliquée dans la régulation du volume cérébral. L'efflux de taurine tritiée par les plexus choroïdes, via une voie sensible à l'acide niflumique, est stimulée directement par une diminution de l'osmolarité mais aussi par une augmentation de la concentration de la taurine extracellulaire. Dans des microvaisseaux cérébraux isolés, cet efflux n'est pas stimulé par un choc hypo-osmotique.

Ces résultats pourraient indiquer que l'efflux de la taurine, s'il existe au niveau des microvaisseaux cérébraux ne se produit qu'au niveau de leur membrane luminale non accessible dans des expériences utilisant des capillaires cérébraux isolés. Bien que la concentration plasmatique de taurine augmente *in vivo* lors des chocs hypo-osmotiques, cela ne se traduit pas par une augmentation du transport de la taurine par la BHE. Ainsi, la BHE empêche aussi la taurine d'être recyclée vers le cerveau.

INTRODUCTION

Taurine, a sulphur-containing amino acid, is lost from the brain during hypo-osmotic stress as part of brain volume regulation (12; 13) . Since there is little taurine metabolism in the brain (3), it is likely that the loss of taurine involves an alteration in taurine transport at either the blood-brain or blood-CSF barriers. In the present study we used a mixture of in vivo and in vitro (isolated cerebral microvessels and choroid plexuses) preparations to examine the mechanisms of barrier taurine transport and to investigate how these mechanisms might be altered in brain volume regulation.

METHODS

Four sets of experiments were performed examining a) taurine influx into brain and CSF during osmotic stress, b) taurine uptake in isolated choroid plexus, c) taurine efflux in isolated choroid plexus and d) taurine efflux in isolated cerebral microvessels. All experiments were performed using Sprague Dawley rats anesthetized with pentobarbitol (50mg/kg; i.p.) or sacrificed under pentobarbitol anesthesia.

In Vivo: Taurine Influx during Osmotic Stress

Rats were rendered hypo-osmotic by a single injection of deionized water (10 ml/100g i.p.) or hyperosmotic by injection of 1.5 M NaCl (2 ml/100g i.p.). Isosmotic controls received a 2 ml/100g intraperitoneal injection of 0.9% saline. After 30 min, a plasma sample was taken to determine osmolality and taurine concentration. The rats then received intravenous injections of ^{3}H-taurine and ^{14}C-inulin, 5 and 3 min prior to sacrifice respectively. An influx rate constant for taurine in anterior cortex was determined by the method of Ohno et al. (7) using the ^{14}C-inulin to correct for plasma volume.

In Vitro: Choroid Plexus Taurine Uptake

Taurine uptake into isolated lateral ventricle choroid plexus was measured by a method similar to that previously described for ^{86}Rb and ^{14}C-glutamine (4; 5). In the present study uptake of ^{3}H-taurine was measured over 20 min in artificial CSF (HCO_3 buffered) at 37°C. ^{14}C-mannitol was used as an extracellular marker. In experiments examining the effect of calmodulin on taurine uptake, the calmodulin inhibitors were present for 10 min prior to and during the uptake measurement.

In Vitro: Choroid Plexus Taurine Efflux

The methodology is similar to that previously described for ^{86}Rb (5). Plexuses were incubated with ^{3}H-taurine/^{14}C-mannitol for 2 hours in artificial CSF at 37°C. The efflux rate constant for taurine was obtained by transfer to efflux buffer and serial sampling of media over 1 hour and terminal sampling of the plexuses.

In Vitro: Cerebral Microvessel Taurine Efflux

Cerebral microvessels were isolated as described by Schielke et al. (9). The microvessels were then incubated with ^{3}H-taurine/^{14}C-mannitol for 30 min. The efflux rate constant for taurine was obtained by transfer to efflux buffer and sampling of microvessels at 0 and 60 min.

Statistics

Statistical comparisons were by analysis of variance with a Newman-Keuls multiple comparisons test. All results are presented as means ± S.E.

RESULTS

Effect of Osmotic Stress on Taurine Influx into Brain

Osmotic stress had a profound effect on the cortical influx rate constant (K_1) for taurine (Fig. 1). Hypo-osmotic (258 ± 3 mOsmol/kg) and hyperosmotic (346 ± 8 mOsmol/kg) stress resulted in a 80% reduction and a 73% increase in K_1 compared to controls (288 ± 3 mOsmol/kg). These changes did not, however, result in a change in unidirectional influx (Fig. 1) as they only offset changes in plasma taurine concentration which were 400 ± 76, 100 ± 13 and 47 ± 2 μM in hypo-, iso- and hyper-osmotic rats.

In contrast, osmotic stress had no significant effect on the K_1 for taurine movement into CSF. Because of the increase in plasma taurine in hypo-osmotic stress, the lack of change in K_1 meant that the unidirectional influx into CSF increased in these rats.

In Vitro: Choroid Plexus Taurine Uptake

Taurine uptake into isolated choroid plexuses was Na-dependent. With trace taurine (8 nM) in the media, uptake was 53 ± 5 and 4 ± 1 nl/mg/min in the presence of 150 and 1 mM Na ($p < 0.001$). The K_m and V_{max} values for Na-dependent taurine uptake were 232 ± 33 μM and 6.5 ± 0.3 pmol/mg/min.

Calmodulin has been implicated in the control of taurine transport (8). In the choroid plexus, two calmodulin inhibitors, 25 μM trifluoperazine and 100 μM W-7, significantly reduced taurine uptake by 84 and 91% respectively ($p < 0.001$).

In Vitro: Choroid Plexus Taurine Efflux

The choroid plexus taurine efflux rate constant (K_o) under control conditions was 0.19 ± 0.02 h^{-1}. This efflux rate constant was stimulated in the presence of 400 μM taurine

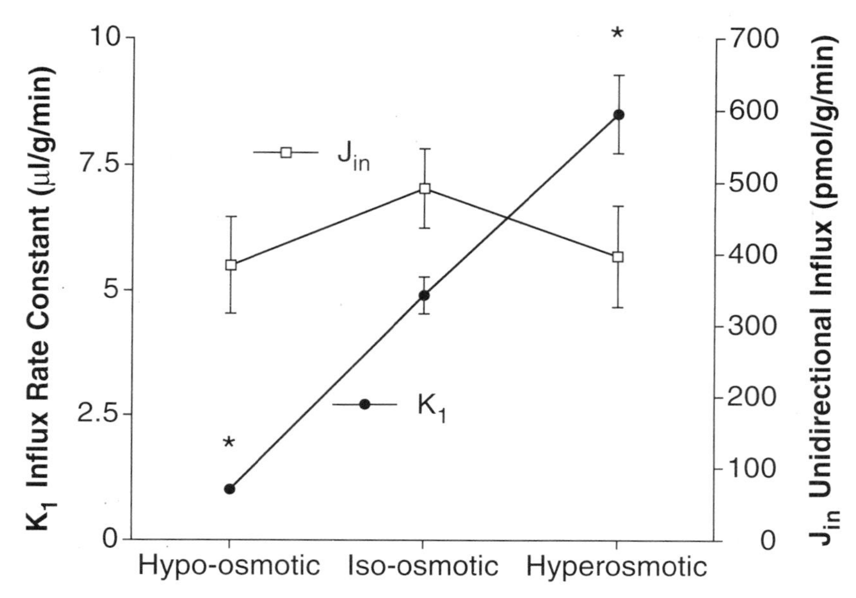

Figure 1. Changes in taurine influx rate constant (K_1) and unidirectional taurine influx (J_{in}) in anterior cortex in response to osmotic stress. Values are means ± S.E.; n=6; * indicates $p < 0.001$ vs. iso-osmotic controls.

(K_o 0.47 ± 0.05 h^{-1}; p< 0.001) and during hypo-osmotic stress (K_o 0.41 ± 0.01 h^{-1} in 245 mOsmol/kg; p<0.001). Niflumic acid (300 μM), an anion exchange inhibitor, had no significant effect on the basal efflux rate but did reduce (p<0.05) the stimulation found with both 400 μM taurine (K_o 0.30 ± 0.05 h^{-1}) and hypo-osmolality (K_o 0.29 ± 0.04 h^{-1}).

In Vitro: Cerebral Microvessel Taurine Efflux

The rate constant for taurine loss from cerebral microvessels under control conditions was 0.49 ± 0.09 h^{-1}. In contrast to the hypo-osmotically activated efflux in isolated choroid plexus, hypo-osmolality did not induce an increase in efflux in cerebral microvessels (Fig. 2). Indeed, in cerebral microvessels the efflux rate constant for taurine increased slightly with increasing osmolality (p<0.01).

DISCUSSION

Osmotic stress had a profound effect on the influx rate constant for taurine at the blood-brain barrier (BBB) in vivo. However, in terms of the rate of unidirectional influx, this effect merely offsets the changes in plasma taurine concentration in response to osmotic stress in these animals. Thus it appears that the changes in brain taurine content that occur with osmotic stress (12; 13) do not result from changes in taurine influx into the brain but rather changes in taurine clearance from the brain.

The changes in influx rate constant are important, however, in preventing the changes in plasma taurine induced by osmotic stress from being reflected in the brain. For example, if the influx rate constant was not reduced in hypo-osmotic stress, the four-fold increase in plasma taurine concentration would lead to an increased rate of taurine influx into the brain in a condition where the brain loses taurine to volume regulate. Whether, the changes in

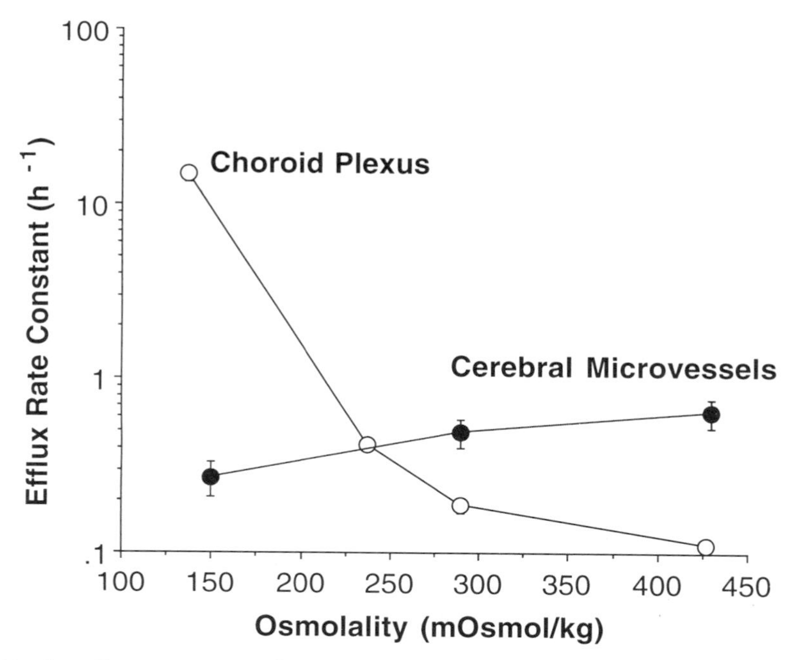

Figure 2. Taurine efflux rate constants for cerebral microvessels and choroid plexuses exposed to different osmolalities. Values are means ± S.E.

influx rate constant with osmotic stress result from saturation of taurine transport into brain or regulation of the number or activity of transporters is still uncertain.

In vitro, the choroid plexus study demonstrated a Na-dependent taurine uptake mechanism with a K_m of about 200 μM and a V_{max} of 7 pmol/mg/min, values similar to those recently measured in rabbit choroid plexus (2). Because the CSF taurine concentration is normally much lower than that in plasma (6; 10) and since there was no evidence of saturation of taurine transport from blood to CSF in vivo, it appears the transporter has an apical distribution, i.e. it moves taurine from the CSF to the plexus. The K_m for the transporter is much greater than the CSF taurine concentration (10-30 μM; (6; 10)) so that release of taurine into the CSF, as occurs in hypo-osmotic stress (6), will increase the rate of taurine transport into the choroid plexuses, the first step of clearance to the bloodstream. Tayarani et al. (11) have reported a K_m for taurine uptake into isolated microvessels of 30 μM, a value that also exceeds the brain interstitial fluid taurine concentration (3).

Taurine uptake into choroid plexus was also inhibited by two calmodulin antagonists, trifluoperazine and W-7. Christensen (1) demonstrated the presence of a stretch-activated, Ca^{2+} permeable, cation channel in amphibian choroid plexus. Hypo-osmotic stress could, therefore, lead to the influx of Ca^{2+} into the choroid plexus epithelial cell, calmodulin activation and increased taurine uptake. Thus, taurine clearance from the brain in osmotic stress may rely not only on the transport kinetics of the transporter but also its regulation.

Uptake into the choroid plexus or cerebral endothelial cells is only the first step of clearance from the brain. The second step is efflux from the barrier tissue to the blood. In isolated choroid plexus, taurine efflux was stimulated by hypo-osmotic stress. If this efflux mechanism is present on the basolateral (blood-facing) membrane it could clear taurine from plexus to blood. Our results with isolated rat choroid plexus do not allow transport localization. However, such a hypo-osmotically activated mechanism was not detectable in isolated cerebral microvessels, a finding that could be explained by a luminal (blood-facing) distribution since there may be limited access to that membrane in our preparation.

As well as being activated by hypo-osmotic stress, taurine efflux from choroid plexus was also stimulated by increases in extracellular taurine. Taurine loss from brain occurs in response to both hypo- and iso-osmotically induced cell swelling. In the latter type of condition (e.g. ischemia), activation of taurine efflux from the choroid plexus by increases in extracellular taurine may be particularly important in clearing taurine from the brain.

ACKNOWLEDGMENTS

This work was supported by the National Institute of Health grant NS-23870 and the Section of Neurosurgery, University of Michigan.

REFERENCES

1. Christensen O., 1987, Mediation of cell volume regulation by Ca^{2+} influx through stretch-activated channels, *Nature* 330:66-68.
2. Chung S. J., Ramanathan V., Giacomini K. M., and Brett C. M., 1994, Characterization of a sodium-dependent taurine transporter in rabbit choroid plexus., *Biochim. Biophys. Acta* 1193:10-16.
3. Huxtable R. J., 1989, Taurine in the central nervous system and the mammalian actions of taurine., *Prog. Neurobiol.* 32:471-533.
4. Keep R. F. and Xiang J., (in press), N-system amino acid transport at the blood-CSF barrier., *J. Neurochem.*
5. Keep R. F., Xiang J., and Betz A. L., 1994, Potassium cotransport at the rat choroid plexus., *Am. J. Physiol.* 267:C1616-C1622.
6. Lehmann A., Carlstrom C., Nagelhus E. A., and Ottersen O. P., 1991, Elevation of taurine in hippocampal extracellular fluid and cerebrospinal fluid of acutely hypoosmotic rats: contribution of influx from blood?, *J. Neurochem.* 56:690-697.
7. Ohno K., Pettigrew K. D., and Rapoport S. I., 1978, Lower limits of cerebrovascular permeability to nonelectrolytes in the conscious rat, *Am J Physiol* 235:H299-H307.
8. Ramamoorthy S., Del Monte M. A., Leibach F. H., and Ganapathy V., 1994, Molecular identity and calmodulin-mediated regulation of the taurine transporter in a human retinal pigment epithelial cell line., *Current Eye Res.* 13:523-529.
9. Schielke G. P., Moises H. C., and Betz A. L., 1990, Potassium activation of the Na,K-pump in isolated brain microvessels and synaptosomes, *Brain Res* 524:291-296.
10. Semba J. and Patsalos P. N., 1993, Milacemide effects on the temporal inter-relationship of amino acids and monoamine metabolites in rat cerebrospinal fluid., *Eur. J. Pharmacol.* 230:321-326.
11. Tayarani I., Cloez I., Lefauconnier J.-M., and Bourre J.-M., 1989, Sodium-dependent high affinity uptake of taurine by isolated rat brain capillaries., *Biochim. Biophys. Acta* 985:168-172.
12. Thurston J. H., Hauhart R. E., and Dirgo J. A., 1980, Taurine: a role in osmotic regulation of mammalian brain and possible clinical significance., *Life Sci.* 26:1561-1568.
13. Verbalis J. G. and Gullans S. R., 1991, Hyponatremia causes large sustained reductions in brain content of multiple organic osmolytes in rats., *Brain. Res.* 567:274-282.

4

VOLUME REGULATION OF THE *IN VITRO* BLOOD-BRAIN BARRIER BY OSMOREACTIVE AMINO ACIDS DURING STRESS

P. J. Gaillard, A. G. De Boer, and D. D. Breimer

Division of Pharmacology
Leiden/Amsterdam Center for Drug Research (LACDR)
Sylvius Laboratories
University of Leiden
2300 RA Leiden, The Netherlands

SUMMARY

The blood-brain barrier (BBB) is more permeable to blood-borne compounds during stress or disease. Moreover, during stress, an extra demand is put on the regulatory systems for the maintenance of cell volume. Hence, the ability of the brain capillary endothelial cells (BCEC) to regulate their volume is essential for maintaining the integrity of the BBB. In this study, the effect of stress on BCEC was investigated by means of osmolyte efflux measurements (free amino acids) into the extracellular fluid (eg., the culture medium). Extracellular concentrations of aspartic acid, glutamic acid, taurine and glycine increased several fold, while glutamine and serine showed no change in extracellular concentration after exposure to stress. This experiment showed that the volume regulatory response of stressed BCEC is, at least partially, dependent on the efflux of free amino acids. By determining the mechanism by which the volume regulation of these cells is controlled, we anticipate to have a tool by which we can modulate the permeability of the BBB in disease state (eg., vasogenic edema or CNS inflammation).

RÉSUMÉ

La barrière hémato-encéphalique (BHE) est plus perméable aux composés apportés par le sang en cas de stress ou de maladie. De plus, pendant le stress, les systèmes régulateurs sont plus sollicités, afin de maintenir le volume cellulaire.Ainsi, la faculté des cellules endothéliales capillaires cérébrales (BCEC) de réguler leur volume est indispensable à l'intégrité de la BHE. Dans cette étude, nous avons étudié l'effet du stress sur les BCEC en

Biology and Physiology of the Blood–Brain Barrier, edited by Couraud and Scherman
Plenum Press, New York, 1996

mesurant l'efflux d'osmolytes (acides aminés libres) dans le liquide extracellulaire (par exemple le milieu de culture). Les teneurs extracellulaires en acides aspartique et glutamique, en taurine et glycine augmentaient de plusieurs ordres de grandeur, tandis que les taux de glutamine et de sérine n'étaient pas modifiés dans le milieu extracellulaire après stress. Cette expérience a montré que la régulation du volume des cellules BCEC stressées dépend, au moins partiellement, de l'efflux des acides aminés libres. En déterminant par quel mécanisme s'opère cette régulation, nous pensons disposer d'un outil permettant de moduler la perméabilité de la BHE dans les états pathogènes (par exemple oedème vasogène ou inflammation du SNC).

The blood-brain barrier (BBB) is situated at the level of the cerebral microvasculature and is formed by a complex cellular system of endothelial cells, astrocytes, pericytes, perivascular macrophages and a basal lamina. The tight junctions between the endothelial cells restrict the paracellular transport of cells, polar solutes and ions from the blood to the brain [(1)]. Changes in the permeability of this barrier is implicated in many different diseases [(2)] and stress [(3)].

In stressful situations, an extra demand is put on the regulatory mechanisms for the maintenance of cell volume. Especially endothelial cells need to regulate their volume under the continuously changing intra- and extracellular environment, since volume changes induce friction at the tight junctions and consequently change paracellular transport, as is shown for epithelial cells by Noach and co-workers [(4)]. Thus, the ability of the endothelial cells of the cerebral microcapillaries to regulate their volume, is essential for maintaining the integrity of the BBB.

One of the general mechanisms by which cells can regulate their volume, besides in changing their ion concentration, is by uptake or release of osmolytes. Osmolytes are organic substances which, ideally, change their cytosolic concentration in concert with the osmolarity of the cell exterior. Free amino acids, like taurine and glutamate, meet the requirements for regulatory osmolytes almost perfectly. Furthermore, taurine has been described to be the most abundant free amino acid involved in cell volume regulation in many different cell types (for reviews see: Huxtable [(5, 6)]).

In this study, the effect of stress on BCEC was investigated by means of osmolyte efflux measurements (free amino acids) into the extracellular fluid (eg., the culture medium).

Microcapillaries from bovine brain were isolated from which endothelial cells were cultured to confluent monolayers, as described previously in our laboratory by de Vries and co-workers [(7)] (with minor modifications). The cells were plated in 24-wells plates at a density of 1.10^5cells/well and grown to confluence.

Stress was induced by exposing the monolayers to hyposmolar (Hypo) or high potassium (K^+) medium for 15 minutes. The amino acid concentration in culture medium was determined, by a RP-HPLC system (fluorometric detection with OPA derivatization). The release of amino acids from monolayers exposed to hyposmolar or high potassium medium is calculated relative to the release in isosmotic medium (Iso) after 15 minutes. The osmolarity of the culture medium was decreased from ± 300 mOsm to ± 230 mOsm by lowering the NaCl concentration. A high potassium medium was prepared by replacing NaCl with an isosmolar concentration of KCl (60 mM). After the experiment the cells were checked for their viability by means of the trypan blue exclusion method.

Exposure of endothelial monolayers to hyposmotic medium increased the extracellular concentration of several osmoreactive, as well as neuroactive, amino acids drastically. Extracellular concentrations of aspartate, glutamate, taurine and glycine increased several fold (8.2, 6.4, 4.4 and 2.2 times basal level, respectively), while glutamine and serine showed no change in extracellular concentration (figure 1). Exposure to a high potassium medium increased the extracellular concentration of the same amino acids, but to a lesser extent (2.7,

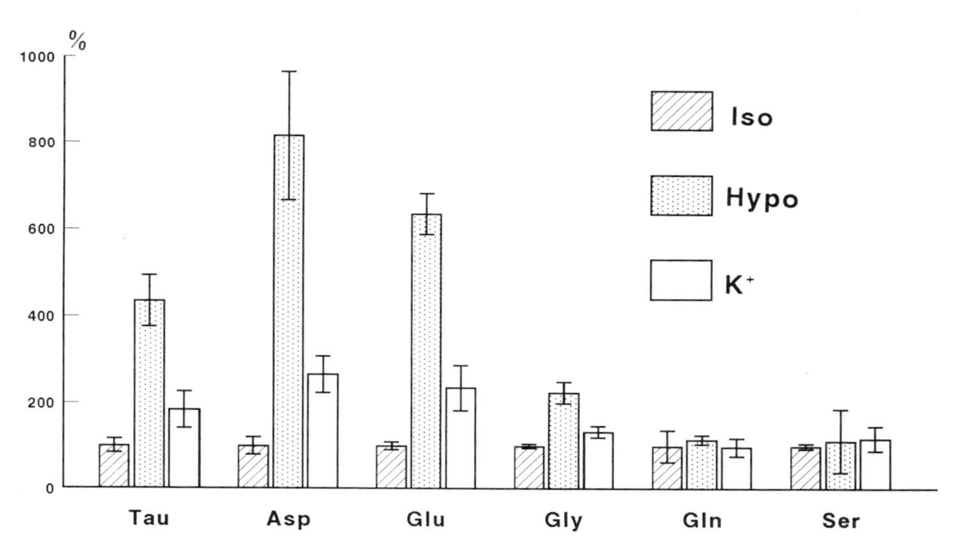

Figure 1. Relative amino acid release in the extracellular space (eg. culture medium) after exposure to different media (mean, percentage of control ± SD). For details see text.

2.6, 1.8 and 1.3 times basal level, respectively, see figure 1). The cells stayed intact, as determined by trypan blue exclusion.

This experiment showed that the volume regulatory response of stressed BCEC is, at least partially, dependent on the efflux of taurine. Furthermore, aspartate, glutamate and glycine contribute to the volume regulatory response of BCEC as well, while glutamine and serine do not. By determining the mechanism by which the volume regulation is controlled, we anticipate to have a tool by which we can modulate the permeability of the blood-brain barrier in disease state (resulting in, for instance, vasogenic edema or CNS inflammation).

REFERENCES

1. MW Bradbury. In: *The concept of a blood-brain barrier.* John Wiley & Sons, London, 1979.
2. PJ Luthert. Opening of the barrier in cerebral pathology. In: *Physiology and pharmacology of the blood-brain barrier,* Ed. MW Bradbury, Springer-Verslag Berlin Heidelberg, Ch 18: 439-457, 1992.
3. HS Sharma, J Cervós-Narvarro, PK Dey. Increased blood-brain barrier permeability following acute short-term swimming exercise in conscious normotensive young rats. *Neuroscience.* 10: 211-221, 1991.
4. ABJ Noach, M Sakai, MCM Blom-Roosemalen, HR de Jonge, AG de Boer, DD Breimer. Effect of anisotonic conditions on the transport of hydrophilic model compounds across monolayers of human colonic cell lines. *JPET.* 270: 1373-1380, 1994.
5. RJ Huxtable. Taurine in the central nervous system and the mammalian actions of taurine. *Prog. Neurobiol.* 32: 471-533, 1989.
6. RJ Huxtable. Physiological Actions of Taurine. *Physiological Reviews.* 72(1): 101-163, 1992.
7. HE de Vries, J Kuiper, AG de Boer, ThJC van Berkel, DD Breimer. Characterization of the scavenger receptor on bovine cerebral endothelial cells in vitro. *J. Neurochem.* 61: 1813-1821, 1995.

5

A REDUCTION IN THE TRANSFER OF AMINO ACIDS ACROSS THE BLOOD-BRAIN BARRIER MIGHT NOT BE THE SOLE MECHANISM BY WHICH VASOPRESSIN AFFECTS AMINO ACID LEVELS WITHIN THE BRAIN

A. Reichel,[1] D. J. Begley,[1] and A. Ermisch[2]

[1] Physiology Group, Biomedical Sciences Division
King's College London
London WC2R 2LS, United Kingdom
[2] Section of Biosciences
University of Leipzig
04103 Leipzig, Germany

SUMMARY

Arginine vasopressin (AVP) changes the kinetic parameters of the blood-brain barrier (BBB) transport of large neutral amino acids (LNAA). The effect is believed to result from the occupation of V_1 receptors at the luminal surface of brain endothelial cells which in turn induces allosteric changes in the LNAA transporter, the *L* system. These changes result in a diminished unidirectional influx of all LNAA to the brain. In addition, circulating AVP is also known to reduce amino acid levels in brain extracellular fluid (ECF) as measured in cerebrospinal fluid (CSF). In the present paper we discuss the extent to which changes in BBB transport might account for the observed changes in the CSF concentrations of LNAA. A comparative analysis of our data suggests that the reduced influx of LNAA across the BBB cannot fully explain the observed reduction in CSF levels. Thus other effects induced by the peptide such as changes in the production rate of ECF and CSF or a stimulation of amino acid uptake by brain cells might be involved.

RÉSUMÉ

La vasopressine arginine (VPA) modifie les paramètres cinétiques du transport des "grands" amino-acides neutres (GAAN) par la barrière hémato-encéphalique. L'occupation

Biology and Physiology of the Blood–Brain Barrier, edited by Couraud and Scherman
Plenum Press, New York, 1996

des récepteurs V1 présents sur la membrane luminale des cellules endothéliales cérébrales par la VPA semble induire une modification de type allostérique du transporteur des GAAN, le système L. Il en résulte une diminution de la pénétration cérébrale de tous les GAAN. De plus, après mesure dans le liquide céphalo-rachidien (LCR), la VAP circulante induit une diminution de la concentration du liquide extracellulaire cérébral (LEC) en amino-acides. Dans cette étude, nous avons étudié dans quelle mesure la VPA, en modifiant les paramètres du transport des acides aminés par le système L au niveau de la BHE, pouvait modifier la concentration du LCR en GAANs. Nos resultats suggèrent que ce phénomène ne peut à lui seul expliquer la réduction de la concentration du LCR en GAANs. Ainsi d'autres effets connus de la VPA, tels que des modifications de la quantité de LCR et de LEC secrétés ou une stimulation de la captation par le cerveau des acides aminés, pourraient être impliqués dans ce phénomène.

Alterations in the kinetic parameters of the blood-brain barrier (BBB) transport of large neutral amino acids (LNAA) induced by arginine vasopresin have been reported in a number of studies using a variety of methods. Circulating AVP induces changes in the blood-to-brain transfer of LNAA as demonstrated for leucine[1], phenylalanine[2], methionine[3], tyrosine[4] and valine[4]. Using methods such as the single pass technique[5], the integral technique[6] and a positron emission technique[3] it has been shown that physiological concentrations of AVP decrease the kinetic parameters of the blood-to-brain transfer of these LNAA. AVP induced a transport of higher affinity but lower capacity for all LNAA investigated, exhibiting a negative correlation between the substrate affinity of the respective LNAA to the transporter and the corresponding magnitude of the transport alterations induced by the peptide[4]. The effects were not dependent on hemodynamic changes and are believed to be a consequence of occupation of V_1 receptors located at the luminal surface of cerebral endothelial cells forming the BBB. The current hypothesis is that intracellular messengers released in response to V_1 receptor activation induce allosteric changes in the LNAA transporter which in turn produce alterations of the transport characterisitics of the LNAA transporter at the BBB, the *L* system[7].

Recently we have shown that moderately elevating the levels of circulating AVP, by intravenous infusion over periods of 60 minutes, reduces the concentrations of LNAA in the brain ECF as measured in the CSF[8]. These findings provide the first evidence that the vasopressin-receptor interaction at the luminal surface of the BBB is not only related to the amino acid supply of the brain endothelium but is of relevance to amino acid levels in the brain and hence to the availability of amino acids for brain metabolism and/or osmoregulation.

Given that AVP reduces the blood-to-brain transport of a number of LNAA, the question arises as to whether changes in the extraction of amino acids at the BBB are reflected in corresponding changes in LNAA concentrations in the CSF.

Intravenous infusion of the peptide over a time period of 60 min elevated the AVP blood level by a factor of about 2.5 to 5 pg/ml and reduced the CSF concentrations of LNAA by about 25 % (mean) ranging from 10 % (Leu) to 32 % (Trp) (Fig. 1).

Experimental data have been used to calculate percentage reductions in the unidirectional influx of LNAA across the BBB in response to intracarotid injection of AVP[4]. The mean reduction is about 32 % ranging from 25 % (Ile) to 50 % (Phe) (Fig. 2). Taking into consideration that only a small proportion of the total CSF volume is contributed by the brain tissue, in contrast to the much greater proportion secreted by the choroid plexus, the reductions in LNAA levels in the CSF seem larger than might be expected.

Using published data for the total CSF volume, secretion rate, and the extrachoroidal contribution of CSF in the rat[9], we have estimated to what extent the reduction of the unidirectional influx of LNAA across the BBB might be expected to account for the observed

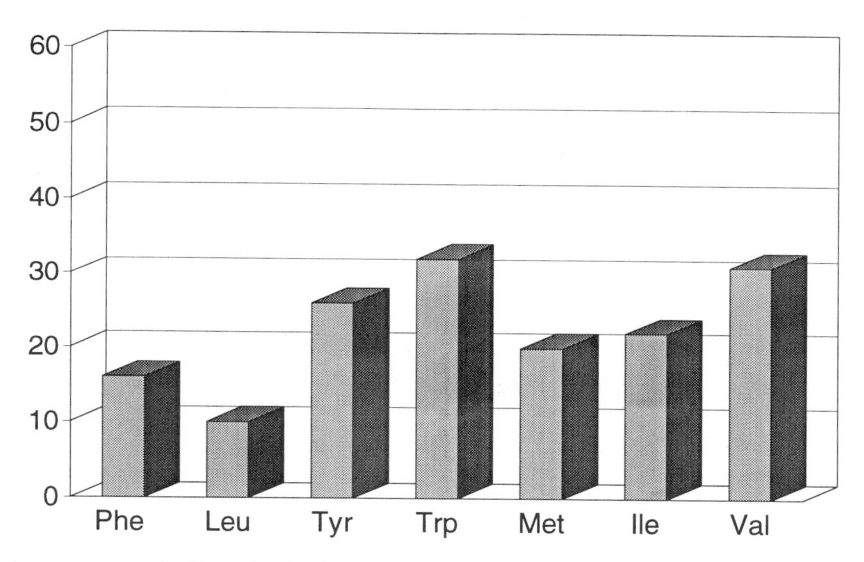

Figure 1. Percentage reduction of levels of LNAA in CSF in response to intravenous infusion of AVP over 60 min producing a final blood level of about 5 pg/ml[8].

changes in CSF levels of the respective LNAA in response to infused AVP over 60 min. All estimations are based on the assumption that the reduction in the BBB transport of LNAA as measured by the single pass technique[5] are equivalent to the effects obtained with infusion. The total CSF volume of the rat is about 200 - 250 l. The overall CSF production rate has been estimated at about 2.5 l/min with an extrachoroidal contribution of 0.25 l/ml which is about 10 % of the total[9]. Therefore during the infusion period of 60 min about 15 l might enter the ventricles contributed as interstitial fluid. Given the reduced content of LNAA by

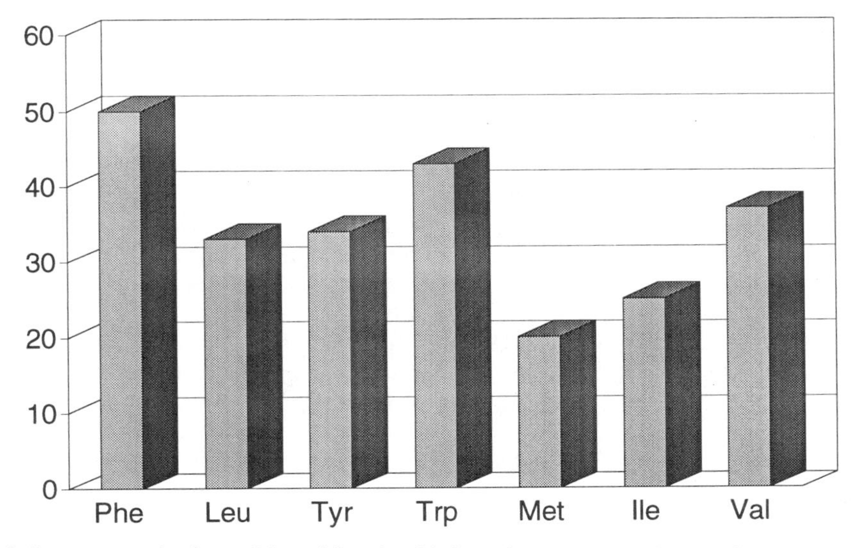

Figure 2. Percentage reductions of the unidirectional influx of LNAA across the BBB in response to intracarotid injection of AVP as calculated from the respective kinetic parameters[4].

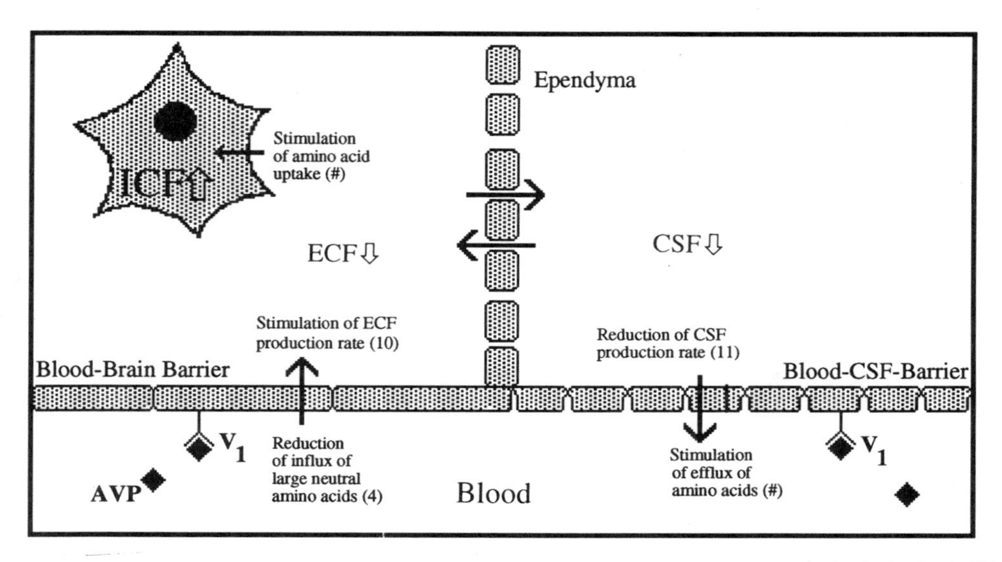

Figure 3. Compartment model of the brain illustrating effects of AVP on amino acid levels in the brain. Solid arrows indicate amino acid fluxes between the compartments, open arrows indicate increase or decrease of amino acid levels in the respective compartments in response to AVP perfusion[6]. The figure compiles reported[4,10,11] as well as postulated mechanisms (#) which in combination may lead to changes in amino acid levels in the brain in response to elevated AVP levels in the blood.

about 32 % in this volume the final reduction of LNAA levels in the total CSF volume would be approximately 4 - 5 %. This value is much smaller than the measured reduction of about 25 %.

This discrepancy suggests that there may be additional events involved also triggered by AVP which, in combination with a reduced amino acid transfer, result in the reduced LNAA levels measured in CSF[8]. Both an increased extrachoroidal contribution or a decreased production of CSF by the choroid plexus, for example, would bring the estimated value closer to the measured result. In fact, there is experimental evidence for such effects of the peptide. AVP has been reported to increase the production rate of brain interstitial fluid[10] as well as to decrease the CSF production rate by the choroid plexus[11] in the rat. Both these effects would increase the extrachoroidal contribution of the CSF production which would further reduce the final LNAA concentrations in the CSF. In order to explain the CSF levels of the LNAA by this mechanism completely, the extrachoroidal contribution needs to become as high as 50 - 60 % of total CSF production. There is however no evidence that AVP can induce changes in CSF production of this magnitude.

Furthermore, it would be reasonable to assume that AVP has a similar effect on the transport of amino acids across the blood-CSF barrier formed by the choroid plexus epithelium, which is also known to express V_1 receptors[12] as it does at the BBB. This argument is supported by findings showing that blood-borne AVP also reduces CSF levels of amino acids which are not significantly taken up at BBB sites, e.g. acidic and small neutral amino acids[8]. An alternative explanation might be that AVP has a stimulatory effect of AVP on amino acid efflux out of the CSF by the choroid plexus epithelium.

Additionally, the intravenous infusion of AVP has been shown to increase intracellular levels of LNAA in hippocampus tissue[8]. An increased accumulation of LNAA by brain cells generally in response to elevated plasma levels of AVP would further contribute to an enhanced reduction of amino acid levels in the brain ECF and be reflected in the CSF. It must be emphasised that the actual mechanism for this effect is not known. It cannot be excluded that circumventricular organs which are known to express AVP receptors[7] respond to plasma AVP levels (e.g. the area postrema) and then transduce and propagate hormonal messages into other areas of the nervous system. Alternatively a second messenger inside the brain may be released or significant amounts of AVP may cross the BBB to influence the hippocampal tissue directly.

The physiological significance of the changes in amino acid concentrations within the brain as a result of increased levels of blood-borne AVP remains uncertain. Plasma levels of AVP increase as a physiological response to a variety of stimuli which are often related to stress[13]. During situations of stress plasma levels of amino acids can differ remarkably from normal, (e.g. as a response to increased sympathetic nervous activity[14]. The therefore altered brain entry rates of amino acids may cause disturbances of brain osmoregualtion due to their potent osmolytic activity. The reduction of the blood-to-brain transfer of LNAA induced by AVP could help to prevent osmotic disturbance in brain ECF when the homeostasis of body fluids, e.g. the plasma, is disturbed.

We conclude that the reduced levels of LNAA in CSF as a response to elevated plasma levels of AVP are likely to be a result of multiple effects induced by the peptide. The reduced blood-to-brain transfer of LNAA across the BBB seems to be only one mechanism, others such as altered ECF and CSF productions rates and/or changes in amino acid uptake by brain cells may also contribute to the effect of the peptide on intracerebral amino acid levels. A stimulatory effect of AVP on the amino acid efflux out of the CSF by the choroid plexus is postulated.

ACKNOWLEDGMENT

The authors would like to thank the British Council for a British-German Academic Research Collaboration Grant (ARC), which has made this collaborative project possible. Andreas Reichel was holding a grant by the German Academic Exchange Service (grant 312/3324 00 075).

REFERENCES

1. Brust, P. Changes in regional blood-brain transfer of L-leucine elicited by arginine vasopressin. *J. Neurochem.*, 46 (1986) 534-547
2. Ermisch, A., Reichel, A. and Brust, P. Changes in the blood-brain barrier transfer of L-phenylalanine elicited by arginine vasopressin. *Endocr. Regul.*, 26 (1992) 11-16
3. Brust, P., Shaya, E.K., Jeffries, K.J., Dannals, R.F., Ravert, H.T., Wilson, A.A., Conti, P.S., Wagner, H.N., Gjedde, A., Ermisch, A. and Wong, D.F. Effects of vasopressin on blood-brain transfer of methionine in dogs. *J. Neurochem.*, 59 (1992) 1421-1429
4. Reichel, A., Begley, D.J. and Ermisch, A. Arginine Vasopressin reduces the blood-brain transfer of L-tyrosine and L-valine: further evidence of the effect of the peptide on the *L* system transporter at the blood-brain barrier. *(submitted)*
5. Oldendorf, W.H. Measurement of brain uptake of radiolabelled substances using a tritiated water internal standard. *Brain. Res.*, 24 (1970) 372-376
6. Gjedde, A., Hansen, A.J. and Siemkowicz, E. Rapid simultaneous determination of regional blood flow and blood-brain glucose transfer in brain of rat *Acta Physiol. Scand.*, 108 (1980) 321-330
7. Ermisch, A., Brust, P., Kretzschmar, R. and Rühle H.-J. Peptides and blood-brain barrier. *Physiol. Rev.*, 73 (1993) 489-527
8. Reichel, A., Begley, D.J. and Ermisch, A. Changes in amino acid levels in rat plasma, cisternal cerebrospinal fluid and brain tissue induced by intravenously infused arginine vasopressin. *Peptides*, 16 (1995) 965-971

9. Davson, H., Welch, K., Segal, M.B. *The Physiology and Pathophysiology of the Cerebrospinal Fluid.* New York: Churchill Livingstone, 1987
10. DePasquale, M., Patlak, C.S. and Cserr, H.F. Brain ion and volume regulation during acute hypernatremia in Brattleboro rats. *Am. J. Physiol.,* 256 (1989) F1059-F1066
11. Faraci, F.M., Mayhan, W.G. and Heistad, D.D. Effect of vasopressin on production of cerebrospinal fluid: possible role of vasopressin (V1)-receptors. *Am. J. Physiol.,* 258 (1990) R94-R98
12. Zlokovic, B.V., Segal, M.B., McComb, J.G., Hyman, S., Weiss, M.H. and Davson, H. Kinetics of circulating vasopressin uptake by choroid plexus. *Am. J. Physiol.,* 260 (1991) F216-F224
13. Bohus B. Physiological functions of vasopressin in behavioural and autonomic responses to stress. In: J.P.H. Burbach & D. DeWied Brain (Eds.) *Functions of Neuropeptides.* pp. 15-40, Parthenon Publ. Group: New York, 1993
14. Eriksson T. and Carlsson A. Adrenergic influence on rat plasma concentrations of tyrosine and tryptophan. *Life Sci.,* 40 (1982) 1465-1472

6

DIFFERENTIAL AMINO ACID UPTAKE INTO CEREBRAL PARENCHYMA AND CAPILLARY CELLS DURING DEVELOPMENT

Hameed Al-Sarraf,[1] Kevin A. Smart,[1] Malcolm B. Segal,[1] and Jane E. Preston[2]

[1] Sherrington School of Physiology
UMDS, St Thomas' Campus
London SE1 7EH, United Kingdom
[2] Department of Gerontology
King's College London, Cornwall House Annex
London SE1 8WA, United Kingdom

1. INTRODUCTION

The amino acids aspartate, glutamate and glycine are neurotransmitters within the central nervous system[1]. High, uncontrolled brain levels of these amino acids is potentially harmful[2,3] therefore entry into the brain must be carefully controlled in the face of plasma elevations. This is accomplished by the blood-brain barrier (BBB). In general, neutral amino acid uptake into brain is high in the immature animal[4] due to the a greater demand for protein synthesis of the developing brain[5]. In our recent studies, we have demonstrated that the neurotransmitter amino acids also show greater uptake into the neonatal brain compared to the adult[6,7]. It is possible that this high brain uptake is due to endothelial cell trapping of amino acid since the neonatal BBB contains many vesicles which are not seen in adults[8]. To investigate the contribution of endothelial cell sequestration to neonatal brain amino acid, we have combined the *in situ* brain perfusion technique with dextran density capillary depletion. This combination allows the differential uptake of amino acid into brain and capillary endothelium to be measured.

2. METHODS

Wistar rats of 1 week, 2 weeks and 3 weeks postnatal age and adults (7-10 weeks) of either sex were used. The *in situ* brain perfusion technique was used to study the entry of ^{14}C radiolabelled amino acids into the brain. This has been previously described[7]. Briefly, the brain was perfused via both carotid arteries in the absence of systemic circulation, using a modified Krebs-Henseleit solution which contained (in mM): Na^+ 142, K^+ 5.7, Cl^- 127, Ca^{2+} 2.5,

Biology and Physiology of the Blood–Brain Barrier, edited by Couraud and Scherman
Plenum Press, New York, 1996

Mg^{2+} 1.21, HCO_3^- 25, $H_2PO_4^-$ 1.2, SO_4^{2-} 1.21, glucose 10 and 4 % bovine serum albumin. The perfusion fluid was saturated with 5% CO_2 in O_2 and kept at 37 °C. At the start of perfusion both jugular veins were sectioned to allow perfusate outflow. ^{14}C labelled aspartate, glutamate, glycine or mannitol were introduce into perfusion circuit via a slow-drive syringe. ^{14}C-Mannitol was used to estimate vascular space. At the end of perfusion time (20 min), decapitation was followed by removal of the brain. The brain was homogenised at 4 °C with physiological buffer (HEPES) in a ratio of 1:5 brain : buffer. A dextran solution (70,000) was then added to a final concentration of 13% and further briefly homogenised. An aliquot of homogenate was taken and the reminder was centrifuged (5,300xg for 15 min). The supernatant was then carefully separated from the pellet and samples of each, and the homogenate and Ringer, were taken for liquid scintillation counting. The supernatant consisted of devascularised brain parenchyma, and the pellet of brain vasculature and brain nuclei.[9]

The radioactivity in each compartment was expressed relative to the activity in the Ringer per unit volume, where:

$$\text{Volume of distribution, } R \text{ ml/g} = (\text{dpm/g tissue})/(\text{dpm/ml Ringer}) \quad (1)$$

The transfer coefficient, K_{in}, into each compartment was then calculated.

$$K_{in\ \text{ml/min/g}} = (R \text{ amino acid} - R \text{ mannitol})/\text{Time} \quad (2)$$

The amino acid K_{in} values for homogenate, supernatant, and pellet at each age were established.

To establish the degree of supernatant contamination with vasculature after capillary depletion, the activity of the vascular enzyme alkaline phosphatase was measured in the supernatant (Sigma Kit 104-LS).

The ATP content of perfused brains was routinely compared with that of non-perfused sham operated controls, using the luciferin bioluminescent method[7].

3. RESULTS

3.1. ATP

The ATP content of whole brain after 20 min *in situ* perfusion was compared to that in the non-perfused sham operated (Fig 1). There was no significant difference between the perfused and the control within each age group ($p>0.05$), however, the adult levels were consistently lower than those for the neonate, indicating the reduced hypoxic tolerance of the adult tissue.

3.2. Alkaline Phosphatase

The activity of vascular enzyme, alkaline phosphatase, was measured in whole brain homogenate, supernatant and pellet. In excess of 94% of total enzyme activity was present in the pellet (Fig 2), which indicates the efficiency of the capillary depletion technique.

3.3. Volume of Distribution

The ^{14}C-mannitol volume of distribution (R, ml/g) in homogenate, supernatant, and pellet is shown in Fig 3. There was a reduction of R in brain homogenate and pellet with age ($p<0.05$) which may reflect endothelial sequestration in vesicles in the younger age groups.

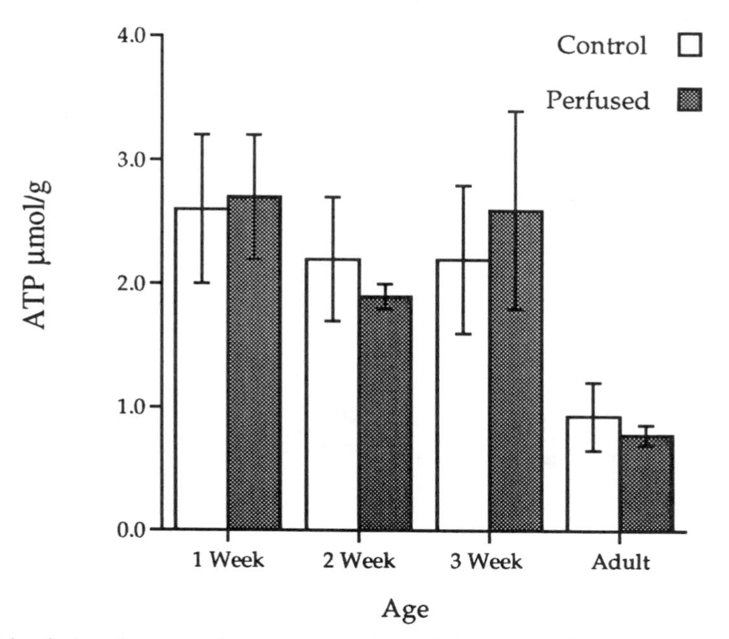

Figure 1. ATP levels (µmol/g) in the brain after 20 min perfusion compared to sham operated controls at different ages. Values are mean±SEM, n=4.

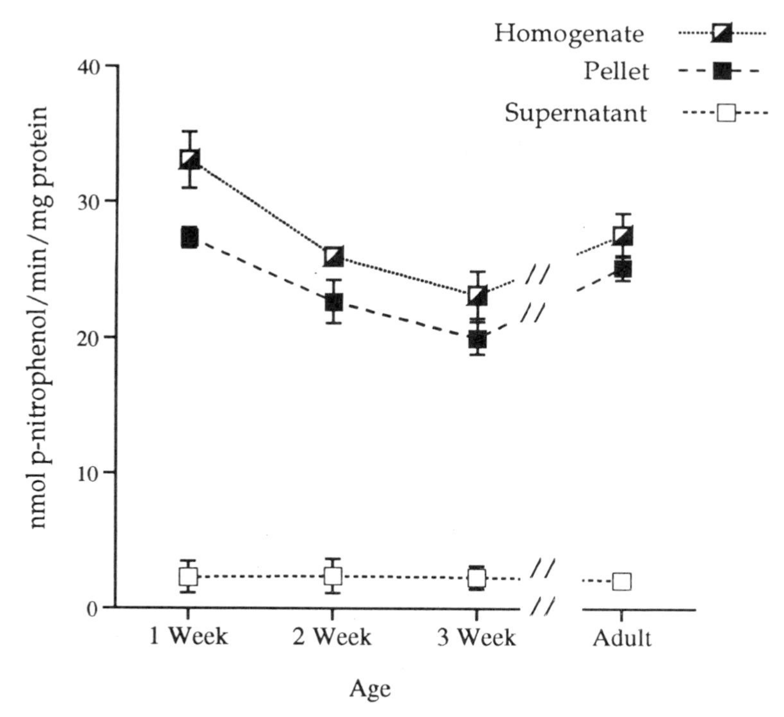

Figure 2. Alkaline phosphatase activity in the brain homogenate, capillary depleted supernatant, and endothelium containing pellet. 94–97% of enzyme activity was found in the pellet at all ages.

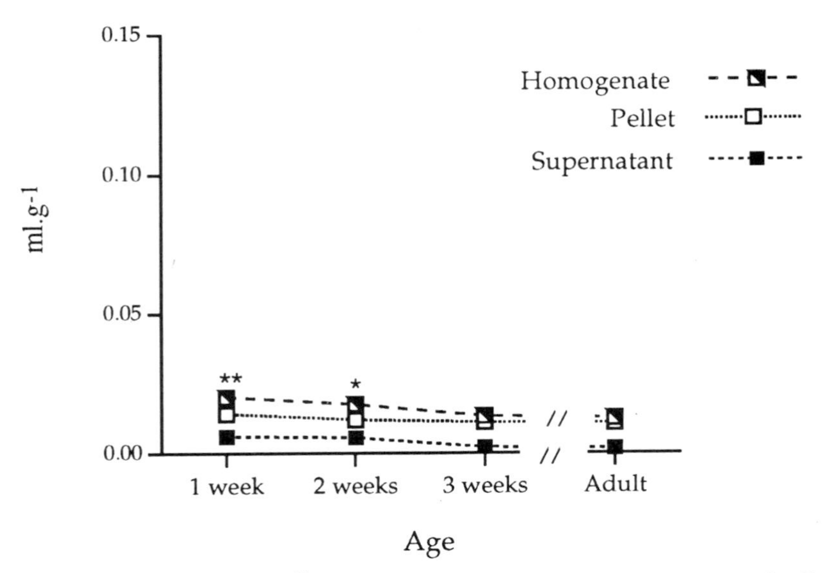

Figure 3. Volume of distribution, *R*, for ^{14}C-mannitol into brain homogenate supernatant and pellet. Values are mean±SEM, n=3. *p<0.05, **p<0.01 difference from adult.

By comparison *R* for ^{14}C-aspartate and ^{14}C-glutamate was much greater than that for mannitol in the supernatant and homogenate (Fig 4).

The supernatant *R* reduced significantly with age, whereas the small pellet *R* did not change with maturity ($p>0.05$). Like the acidic amino acids, ^{14}C-glycine *R* was greater than that for mannitol in supernatant and homogenate and both values reduced with age. However, pellet *R* was greater than that for the acidic amino acids, and demonstrated developmental change (Fig 5).

3.4. Transfer Coefficient, K_{in}

The amino acid K_{in} values, which take account of the vascular space, are shown in Table 1. The whole brain homogenate K_{in} for all tested amino acids was small compared to neutral amino acids, and fell with maturity. This was also reflected in the decline of supernatant K_{in}, indicating less brain parenchyma uptake of amino acids with age.

By comparison, only glycine pellet K_{in} showed a similar maturational change. The acidic amino acid K_{in} for this compartment was very small and did not exhibit an age related change.

4. CONCLUSION

The amino acid entry into brain parenchyma was 4 to 16 times that of the capillary endothelial cells at all ages. However, the uptake of all amino acids reduced by 50% with maturity. The endothelial sequestration of mannitol and glycine in the 1 week old rat exceeded that of the adult, possibly due to the greater number of vesicles in the neonatal

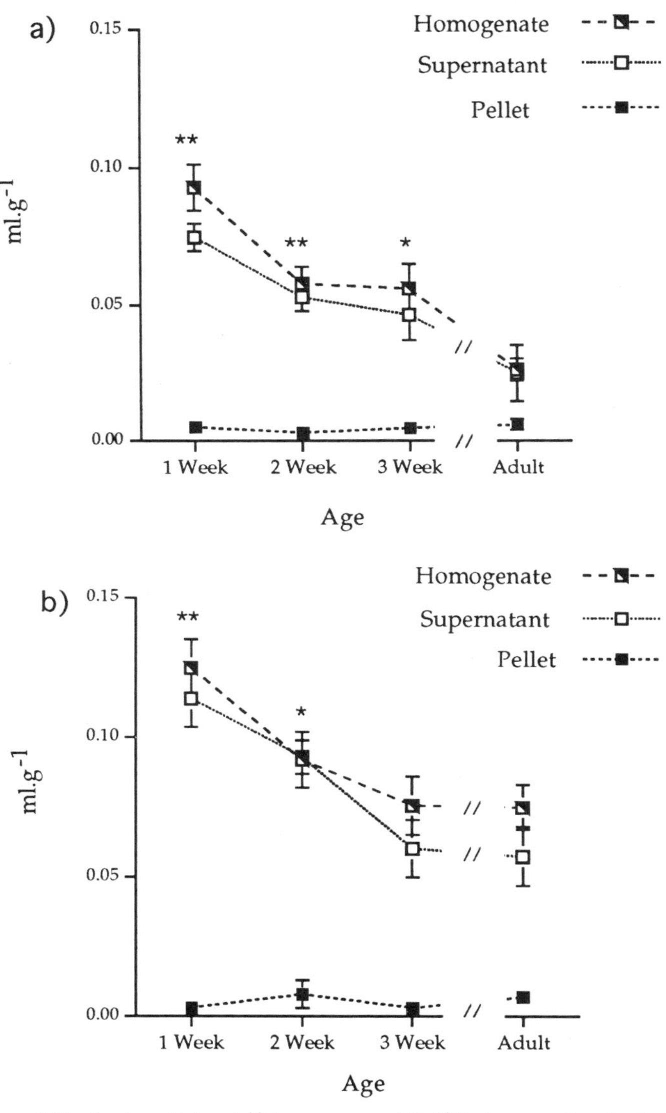

Figure 4. Volume of distribution, R, for a) ^{14}C-aspartate and b) ^{14}C-glutamate into brain homogenate supernatant and pellet. Values are mean±SEM, n=4–7. *p<0.01, **p<0.001 difference from adult.

capillary cells. In contrast, endothelial uptake of aspartate and glutamate did not change with age. This may be due to the different charges found on the two groups of amino acids.

The parenchymal content of the non-transported vascular marker, mannitol, remained constant from 1 week postnatal to adulthood, suggesting that reduced amino acid uptake with age is a function of changes to specific transport systems rather than a non-specific decrease in blood-brain barrier permeability.

This work was supported by The Wellcome Trust.

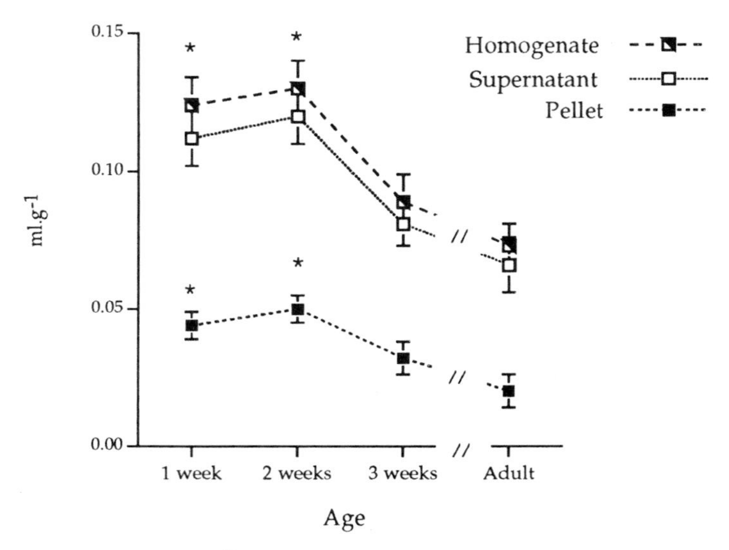

Figure 5. Volume of distribution, R, for ^{14}C-glycine into brain homogenate, supernatant and pellet. Values are mean±SEM, n=5-8. *p<0.05 difference from adult

Table 1. The transfer coefficient, K_{in} for ^{14}C-labelled solutes into brain homogenate, supernatant and pellet for all ages. Mannitol content was subtracted to take account of the vascular space

		^{14}C-amino acid K_{in} (μl/min/g)		
Compartment	Age	Aspartate	Glutamate	Glycine
Homogenate	1 week	3.55±0.25	5.01±0.50	3.47±0.30
	2 weeks	2.26±0.26	3.74±0.30	3.83±0.15
	3 weeks	2.49±0.33	3.27±0.49	1.97±0.49
	Adult	0.95±0.10	3.21±0.40	1.96±0.27
Supernatant	1 week	3.12±0.20	4.86±0.21	3.23±0.31
	2 weeks	2.25±0.24	4.01±0.09	3.90±0.15
	3 weeks	1.28±0.11	2.65±0.23	2.27±0.34
	Adult	0.83±0.32	2.58±0.19	1.80±0.25
Pellet	1 week	0.25±0.11	0.22±0.15	1.31±0.17
	2 weeks	0.21±0.13	0.45±0.30	1.47±0.19
	3 weeks	0.23±0.18	0.18±0.11	0.98±0.15
	Adult	0.28±0.12	0.32±0.21	0.60±0.14

5. REFERENCES

1. Logan WJ & Snyder SH (1971) Unique high affinity uptake systems for glycine, glutamic and aspartic acid in the central nervous tissue of the rat. Nature 234, 297-299
2. Olney JW (1969) Brain lesions, obesity and other disturbances in mice treated with monosodium glutamate. Science 164, 719-721.
3. Schoepp DD, Gamble AY, Salhoff CR, Johnson BG & Ornstein PL (1990) Excitatory amino acid-induced convulsions in neonatal rats mediated by distinct receptor sub-types.
4. Seta K, Sershen H & Lajtha A (1972) Cerebral amino acid uptake in vivo in newborn mice. Brain Res. 47, 415-425.
5. Pardridge WM & Mietus LJ (1982) Kinetics of neutral amino acid transport through the blood-brain barrier of the newborn rabbit. J. Neurochem. 38, 955-962.
6. Preston JE, Al-Sarraf H & Segal MB (1995) Permeability of the developing blood-brain barrier to ^{14}C-mannitol using the rat in situ brain perfusion technique. Dev. Brain Res. 87, 69-76.
7. Al-Sarraf H, Preston JE & Segal MB (1996) The entry of acidic amino acids into brain and CSF during development using in situ perfusion in the rat. Dev. Brain Res. (in press).
8. Xu J & Ling E (1994) Studies of the ultrastructure and permeability of the blood-brain barrier in the developing corpus callosum in postnatal rat brain using electron dense tracers. J. Anat. 184, 227-237.
9. Triguero D, Buciak J & Pardridge WM (1990) Capillary depletion method for the quantification of blood-brain barrier transport of circulating peptides and plasma proteins. J. Neurochem. 54, 1882-1888.

7

TRANSPORT OF ^3H L-ALANINE ACROSS THE BLOOD-BRAIN BARRIER OF *IN SITU* PERFUSED GUINEA PIG BRAIN

Ivanka D. Markovic,[1] Zoran B. Redzic,[2] Suzana S. Jovanovic,[1] Dusan M. Mitrovic,[3] and Ljubisa M. Rakic[1]

[1] Biomedical Research Department
ICN Galenika Institute
Belgrade, Yugoslavia
[2] Institute of Biochemistry
School of Medicine
Belgrade, Yugoslavia
[3] Institute of Physiology
School of Medicine
Belgrade, Yugoslavia

SUMMARY

Transport of ^3H L-alanine through the blood-brain barrier (BBB) was studied using brain vascular perfusion method in guinea pig. Our results indicate that L-alanine passes across the luminal side of the BBB. Unidirectional transport constant K_{in} ranged from 4.871±0.622 μlmin^{-1}g^{-1} in hippocampus to 5.608±0.902 μlmin^{-1}g^{-1} in parietal cortex, which is comparable with the values obtained for other small neutral non-essential amino acids. Addition of unlabelled L-alanine to perfusing medium caused the decrease in L-alanine transport, indicating the importance of saturable component for L-alanine transport. However, presence of high concentrations of unlabelled L-alanine in perfusing medium (up to 12 mmol/l), did not result in complete inhibition of ^3H L-alanine transport through the BBB. Therefore, it seems that another mechanism is also involved in ^3H L-alanine transport across the endothelial cells' luminal membrane. Values for Michaelis-Menten constant for L-alanine transport from blood into brain point out that the affinity of this molecule to its carrier(s) is rather small (K_m >1 mmol/l). Capacity of ^3H L-alanine blood-to-brain transport is very small as well (V_{max} <20 nmol/min/g).

The addition of BCH (4 mmol/l), in order to eliminate the contribution of the L-transport system, did not cause significant decrease in L-alanine blood-to-brain transport. Still, presence of unlabelled L-serine in the perfusing medium (4 mmol/l) resulted in reduction of L-alanine uptake ($p<0.05$). It appears that L transport system may not be of the

Biology and Physiology of the Blood–Brain Barrier, edited by Couraud and Scherman
Plenum Press, New York, 1996

greatest importance in L-alanine transport through the BBB, and that L-serine and L-alanine compete for the same transport systems.

These results indicate that transport of L-alanine across the BBB consist of both saturable and non-saturable component. Saturable uptake of L-alanine is probably mediated by more than one transport system.

RÉSUMÉ

Nous avons étudié, chez le cobaye, par la méthode de perfusion vasculaire cérébrale, le transport de la ^{3}H-L-alanine à travers la barrière hémato-encéphalique. La constante Km de transport unidirectionnel varie de 4,871 ± 0,622 μl min^{-1} g^{-1} dans l'hippocampe à 5,608 ± 0,902 μlmin^{-1} g^{-1} dans le cortex pariétal, valeurs comparables à celles obtenues pour d'autres petits acides aminés neutres non-essentiels. L'addition de L-alanine non marquée dans le milieu de perfusion provoque une réduction du transport de la L-alanine, ce qui montre l'importance d'un composé saturable dans ce transport. Cependant, même de fortes concentrations en L-alanine non marquée (jusqu'à 12 mmol/l) dans le milieu de perfusion n'entraînent pas une inhibition totale du transport de ^{3}H-L-alanine à travers la BHE. Il semble donc qu'un autre mécanisme soit mis en jeu dans ce transport à travers la membrane luminale des cellules endothéliales. Les valeurs de la constante de Michaelis-Menten pour le transport de la L-alanine du sang vers le parenchyme cérébral démontrent que l'affinité de cette molécule pour son ou ses transporteurs est plutôt faible (Km > 1 mmol/l). L'efficacité du transport de la ^{3}H-L-alanine du sang vers le cerveau est également très faible (Vmax < 20 nmol/min/g). L'addition de BCH (4 nmol/l) pour éliminer la contribution du système de L-transport ne provoquait pas de baisse significative du transport de L-alanine. Pourtant, la présence de L-sérine non marquée (4 nmol/l) dans le milieu de perfusion a provoqué une réduction de la capture de L-alanine ($P < 0,05$). Il semble que le système de L-transport n'est peut-être pas essentiel pour le transport de la L-alanine à travers la BHE, et que la L-sérine et la L-alanine se partagent le même système de transport. Ces résultats montrent que le transport de la L-alanine à travers la BHE a des composants saturables et non saturables. La capture saturable de L-alanine est probablement médiée par plus d'un système de transport.

1. INTRODUCTION

Transport of nutrients from blood into brain is essential for maintaining the brain homeostasis. Many research methods have been applied to enlighten amino acids' blood-to-brain transport mechanisms, but transport of small neutral amino-acids was not studied as much as that of the essential ones. First studies performed by Yudilevich et al.[1] showed that transport of L-serine and L-alanine from blood into brain can be considered irrelevant. Later on, other methods have revealed that uptake of these molecules, although significantly smaller in comparison with essential amino acids, is still greater than transport of the inert polar molecules (so called vascular space tracers - mannitol, sucrose...) and those amino acids that take part in neurotransmission[2]. Self-inhibition and cross-inhibition results suggested that saturable mechanism was most likely responsible for transport of small neutral amino acids (except L-alanine) from plasma into brain, although probably not all of them shared the same transport system. Michaelis-Menten constant for transport of L-alanine from blood into rat brain was estimated to be very high, indicating extremely low affinity of these amino acids for their transport system. In order to determine accurately the mechanism as well as the kinetic parameters for L-alanine blood-to-brain transport, we have applied the in situ brain vascular perfusion method developed by Zlokovic et al.[3]

2. MATERIALS AND METHODS

The details of this method have been previously reported [3]. Adult guinea pigs weighing 250-400 g were anesthetized with thiopentone sodium (30-35 mg/kg); the neck vessels exposed and the right common carotid artery cannulated with polyethene tubing connected to the perfusion circuit. Perfusion fluid consisted of washed sheep erythrocytes (hematocrit ~20%) suspended in a saline medium. Immediately after the start of perfusion, the contralateral carotid artery was tied, and both external and internal jugular veins severed to allow drainage of the perfusate. The perfusion medium was pumped from a reservoir (provided with 96% O_2 and 4% CO_2 gas mixture) through the water bath using a peristaltic pump. Perfusion pressure was continuously monitored and kept slightly above the animal's blood pressure, to eliminate any possible ingress from the systemic circulation. At the appropriate times the perfusion was terminated by decapitation of the animal, and the brain removed and prepared for scintillation counting after removal of the chorioid plexus. Isotopically labelled ^{3}H L-alanine (alone or together with the unlabelled L-alanine, BCH or L-serine) was introduced into a perfusion circuit by a slow-drive syringe at a rate of 0.2 ml/min. Radioactivity was determined in a LKB Spectral 1219 counter.

Two compartment analysis of test-solute entry into the ipsilateral parietal cortex and hippocampus was performed in this study. The equation for unidirectional blood to brain flux from multiple time uptake data was developed by Gjedde [4] and Patlak et al. [5] and includes the possibility of initial distribution of the test solute in the rapidly reversible compartments (mostly vascular compartment):

$$\frac{C_{br}(T)}{C_{pl}(t)} = K_{in}\frac{C_{br}(T)dt}{C_{pl}(T)} + V_i \tag{1}$$

where C_{br} is the total amount of test solute measured per unit mass of brain tissue at time t, C_{pl} is the concentration of test solute in the artificial plasma at time t and V_i is the initial volume of distribution of the test solute. In the present experimental conditions, concentration in the influx (C_{pl}) is constant, so the equation (1) becomes

$$\frac{C_{br}(T)}{C_{pl}(T)} = K_{in}\cdot T + V_i. \tag{2}$$

Equation (2) defines a straight line with slope K_{in} (the unidirectional transfer constant - the equivalent of ps product) expressed in ml/min/g of brain tissue (it has a concept of clearance) and ordinate intercept V_i, expressed in ml/g.

The kinetic parameters of ^{3}H L-alanine brain uptake were calculated using the equation

$$K_{in} = \frac{V_{max}}{(K_m + C_{pl})} + K_d \tag{3}$$

where V_{max} equals the maximal influx rate of the saturable component, K_m equals the half-saturation concentration of the saturable component, and K_d is the constant of non-saturable diffusion.

Differences between mean-values were analyzed for statistical significance using analysis of variance. In all cases the criterion for statistical significance was $p<0.05$.

Table 1. Parameters of ^{3}H L-alanine transport from blood into the guinea pig brain

	K_{in} (μl min^{-1}g^{-1})	V_i (ml/1000g)	P (cm/s x 10^{-6})
hippocampus	4.871 ± 0.622	5.272 ± 5.154	0.812 ± 0.104
nc. caudate	5.604 ± 0.836	6.992 ± 6.573	0.934 ± 0.139
cortex	5.608 ± 0.902	8.253 ± 7.674	0.935 ± 0.150

Values are mean ± SE. Number of experiments was n=12 for all examined regions. Vascular permeability constant P was determined if brain capillaries was surface S=100 cm^2/g tissue.

3. RESULTS

Brain uptake of L-alanine, although rather small, is still significantly higher than the uptake of the inert polar molecules. Parameters of this uptake (K_{in}, V_i) are of same order of magnitude as for other small neutral nonessential amino acids (see Table 1).

Addition of unlabelled L-alanine to the perfusing medium caused significant decrease of the brain uptake of ^{3}H labelled L-alanine. The diffusion constant for L-alanine is not significantly different from zero, except in parietal cortex, so free diffusion of L-alanine across the luminal side of the BBB can be considered irrelevant. Values for Michaelis-Menten constant for L-alanine transport from blood into brain point out that the affinity of this molecule to its carrier(s) is rather small (see Table 2). Capacity of ^{3}H L-alanine blood-to-brain transport is very small as well (see Fig. 1).

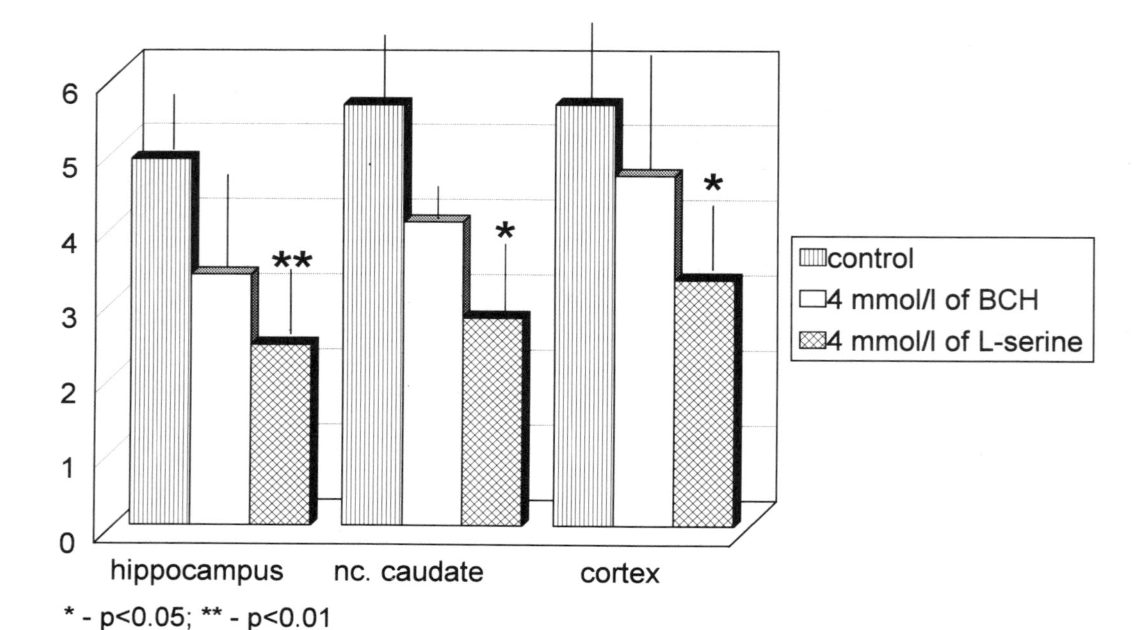

Figure 1. Transport of ^{3}H L-alanine across the blood-brain barrier in the control group, and in presence of BCH (4 mmol/l) and L-serine (4 mmol/l).

Table 2. Values of maximal transport velocity, Michaelis-Menten constant and constant of nonspecific diffusion for ^{3}H L-alanine blood-to-brain transport

	V_{max}(nmol min^{-1} g^{-1})	K_m(mmol l^{-1})	K_d (μl min^{-1}g^{-1})
hippocampus	18.07 ± 8.48	3.55 ± 1.35	0.407± 0.635
cortex	13.79 ± 4.65	2.47 ± 0.71	0.846±0.406

Values are mean ± SE.

Presence of 4 mmol/l of BCH (2-aminobicycloheptane-2-carboxylic acid, a competitive inhibitor of the L transport system) caused a decrease in ^{3}H L-alanine brain uptake. Still, the decrease was below the criterion for the statistical significance, which indicates that L-transport system, although important, plays a minor role in L-alanine brain uptake. Addition of L-serine to perfusing medium caused significant decrease in L-alanine blood-to-brain transport, suggesting that L-alanine and L-serine share the same transport system(s).

4. CONCLUSIONS

These results indicate that transport of L-alanine across the BBB is a saturable process. It is most likely mediated by more than one transport system.

REFERENCES

1. Yudilevich, D., de Rose, N. and Sepulveda, F. (1972). Facilitated transport of amino acids through the blood-brain barrier of the dog studied in a single capillary circulation. Brain Res., 44, 569-578.
2. Smith Q.R., Momma S., Aoyagi M., Rapoport S.I. (1987). Kinetics of Neutral Amino Acid Transport Across the Blood-Brain Barrier. J Neurochem, 49 (5); 1651-1658
3. Zlokoviæ B., Begley D., Durièiæ B. and Mitroviæ D. (1986). Measurement of solute transport across the blood-brain barrier in the perfused guinea pig brain; method and application to N-methyl- α-amino isobutyric acid. J. Neurochem., 46: 1444-1459.
4. Gjedde, A. (1983). Modulation of substrate transport to the brain. Acta Neurol. Scand., 67, 3-25.
5. Patlak, C., Fenstermacher, J. and Blasberg, R. (1983). Graphical evaluation of blood to brain transfer constants from multiple time uptake data. J. Cereb. Blood Flow Metab., 3, 1-7.

8

DISTRIBUTION OF SMALL NEUTRAL AMINO ACIDS AFTER PENETRATING THE LUMINAL SIDE OF THE GUINEA PIG BLOOD-BRAIN BARRIER

S. S. Jovanovic,[1] Z. B. Redzic,[2] I. D. Markovic,[1] D. M. Mitrovic,[2] and Lj. M. Rakic[2]

[1] ICN Galenika Institute
Biomedical Research Department
[2] School of Medicine
Institute of Biochemistry
Yugoslavia

SUMMARY

Distribution of radio labelled small neutral amino acids between endothelial cells and brain parenchyma after transport across the luminal side of the blood-brain barrier was studied in guinea pig. After in situ brain vascular perfusion, the capillary depletion method was applied (Triguero et al. 1990). Endothelial cells were separated from brain parenchyma, using centrifugation (5400 g) in 13% dextran solution. The concentration ratio between endothelial cells (pellet) and brain parenchyma (supernatant) was determined for each amino acid studied (^{3}H L-serine, ^{3}H L-alanine and ^{14}C L-proline) after different perfusion times (1, 3 and 6 minutes).

Our results show the significant increase of pellet/supernatant ratio in time for all three amino acids ($p<0.05$ between 1 and 6 min for both L-alanine and L-serine and $p<0.05$ between 1 and 3 min for L-proline). The increase was due to very slow increase of volume of distribution in postvascular compartment (brain parenchyma), in comparison with the vascular compartment (pellet).

These results indicate that small neutral amino acids, after penetrating the luminal side of the blood-brain barrier, are probably accumulated in brain endothelial cell. The accumulation of these amino acids could reflect their involvement in local metabolic pathways.

RÉSUMÉ

Nous avons étudié, chez le cobaye, par la méthode de perfusion vasculaire cérébrale, le transport de la ^{3}H-L-alanine à travers la barrière hémato-encéphalique. La constante Km

Biology and Physiology of the Blood–Brain Barrier, edited by Couraud and Scherman
Plenum Press, New York, 1996

de transport unidirectionnel varie de 4,871 ± 0,622 μl min^{-1} g^{-1} dans l'hippocampe à 5,608 ± 0,902 μlmin^{-1} g^{-1} dans le cortex pariétal, valeurs comparables à celles obtenues pour d'autres petits acides aminés neutres non-essentiels. L'addition de L-alanine non marquée dans le milieu de perfusion provoque une réduction du transport de la L-alanine, ce qui montre l'importance d'un composé saturable dans ce transport. Cependant, même de fortes concentrations en L-alanine non marquée (jusqu'à 12 mmol/l) dans le milieu de perfusion n'entraînent pas une inhibition totale du transport de ^{3}H-L-alanine à travers la BHE. Il semble donc qu'un autre mécanisme soit mis en jeu dans ce transport à travers la membrane luminale des cellules endothéliales. Les valeurs de la constante de Michaelis-Menten pour le transport de la L-alanine du sang vers le parenchyme cérébral démontrent que l'affinité de cette molécule pour son ou ses transporteurs est plutôt faible (Km > 1 mmol/l). L'efficacité du transport de la ^{3}H-L-alanine du sang vers le cerveau est également très faible (Vmax < 20 nmol/min/g). L'addition de BCH (4 nmol/l) pour éliminer la contribution du système de L-transport ne provoquait pas de baisse significative du transport de L-alanine. Pourtant, la présence de L-sérine non marquée (4 nmol/l) dans le milieu de perfusion a provoqué une réduction de la capture de L-alanine (P < 0,05). Il semble que le système de L-transport n'est peut-être pas essentiel pour le transport de la L-alanine à travers la BHE, et que la L-sérine et la L-alanine se partagent le même système de transport. Ces résultats montrent que le transport de la L-alanine à travers la BHE a des composants saturables et non saturables. La capture saturable de L-alanine est probablement médiée par plus d'un système de transport.

1. INTRODUCTION

Transport of endogenous and exogenous substances through the blood-brain barrier is a complex process. It involves changes of the volume of distribution in the brain tissue after particular time intervals of vascular perfusion. When studying the transport across the blood-brain barrier, it is important to consider the existence of so called "biochemical barrier" between the blood and the brain. The enzymatic activity in the brain capillaries' endothelial cells can result in metabolic changes of the examined substance, with its potential trapping in the vascular compartment. This trapping could change the net influx of the substance from blood into the brain tissue[1] .

It has been shown that the metabolic changes of the test substance could influence its further distribution in the brain tissue[2] . If the enzymes involved in these changes had K_m values of the same order of magnitude as the K_m of the transport systems at the luminal side of the blood-brain barrier, the question arises whether the obtained K_{in} values describe the actual transport across the BBB, or both the transport and metabolism.

2. AIM OF THE STUDY

In order to determine the influence of metabolic changes of the test substance (in brain vascular endothelium as well as in the brain parenchyma) on its blood-to-brain transport, the aim of this study was to determine the distribution of certain essential (^{3}H L-leucine) and small neutral non-essential amino acids (^{3}H L-alanine, ^{3}H L-serine and ^{14}C L-proline) between brain vascular and postvascular compartment.

Methods used in this study were the *in situ* guinea pig brain vascular perfusion and the capillary depletion.

3. METHODOLOGY

The *brain vascular perfusion* was performed by the method of Zlokoviæ et al (1986)[3]. Adult guinea pigs were anesthetized with thiopentone sodium (30-35mg/kg). The neck vessels were exposed and right carotid artery cannulated with polyethylene tubing connected to the perfusion circuit. The perfusing fluid consisted of washed sheep erythrocytes suspended in a saline medium (hematocrit ~20%). The perfusing medium was pumped from a reservoir through the water bath using a peristaltic pump. The radiolabelled test substance (^{3}H L-serine, ^{3}H L- alanine, ^{14}C L-proline and ^{3}H L-leucine) was introduced into perfusing medium by slow drive syringe at a rate of 0.2ml/min. After appropriate times (1-15 min) the perfusion was terminated, the brain removed and the right cortex was used for *capillary depletion* (Triguero et al, 1990)[4]. Cortex was weighed and homogenized in physiological buffer. The 26% dextran solution was added to the homogenate (dextran final concentration was 13%) and homogenized again. The homogenate was then centrifuged at 5,400g for 15min. All procedures were performed at 4°C. The supernatant (postvascular compartment - brain parenchyma) and pellet (vascular compartment - brain endothelial cells) were carefully separated, weighed and solubilized in a tissue solubilizer before scintillation counting.

Volume of distribution (V_d) values were calculated for each substance tested, for the pellet and supernatant:

$$V_d(t) = V_{d\ tiss.}(t) - V_{d(mannitol)} \tag{1}$$

where $V_{d\ tiss.}(t)$ was the volume of distribution of test substance (ml / 100 mg of proteins) in the brain tissue (pellet or supernatant) in time t, and $V_{d(mannitol)}$ was the volume of distribution of ^{14}C d-mannitol in the same tissue. Subtraction of the V_d for d-mannitol (vascular space marker) was necessary to make correction for all the molecules of ^{3}H amino acid tested, "leaking" from the vasculature following homogenization and disruption of the brain microvessels. Since molecular weight of d-mannitol (180) is within same order of magnitude as tested amino acids, it has been chosen as an appropriate vascular space marker.

Values for volume of distribution of ^{3}H L-serine, ^{14}C L-proline ^{3}H L-leucine and ^{14}C mannitol were calculated as

$$V_d(t) = \frac{C_{br}(t)}{C_{pl}(t)} \tag{2}$$

where C_{br} (t) was the concentration of the test substance (expressed as dpm/mg proteins x 10^{-2}) in the brain tissue (pellet or supernatant) in time t, and $C_{pl.}$(t) was the concentration of this substance in the perfusing medium (dpm/µl).

4. RESULTS

Our results show the significant increase of pellet/supernatant ratio in time for all three small neutral amino acids tested ($p<0.05$ between 1 and 6 min for both L-alanine and L-serine and $p<0.05$ between 1 and 3 min for L-proline - Fig. 1). This increase was due to very slow increase of volume of distribution in postvascular compartment (brain parenchyma), in comparison to the vascular compartment (pellet).

However, the pellet/supernatant ratio for [^{3}H]-leucine, (Fig. 2) was different compared to the one for small neutral nonessential amino acids. The obtained values indicate the constant increase of volume of distribution for [^{3}H]-leucin in the brain parenchyma in time.

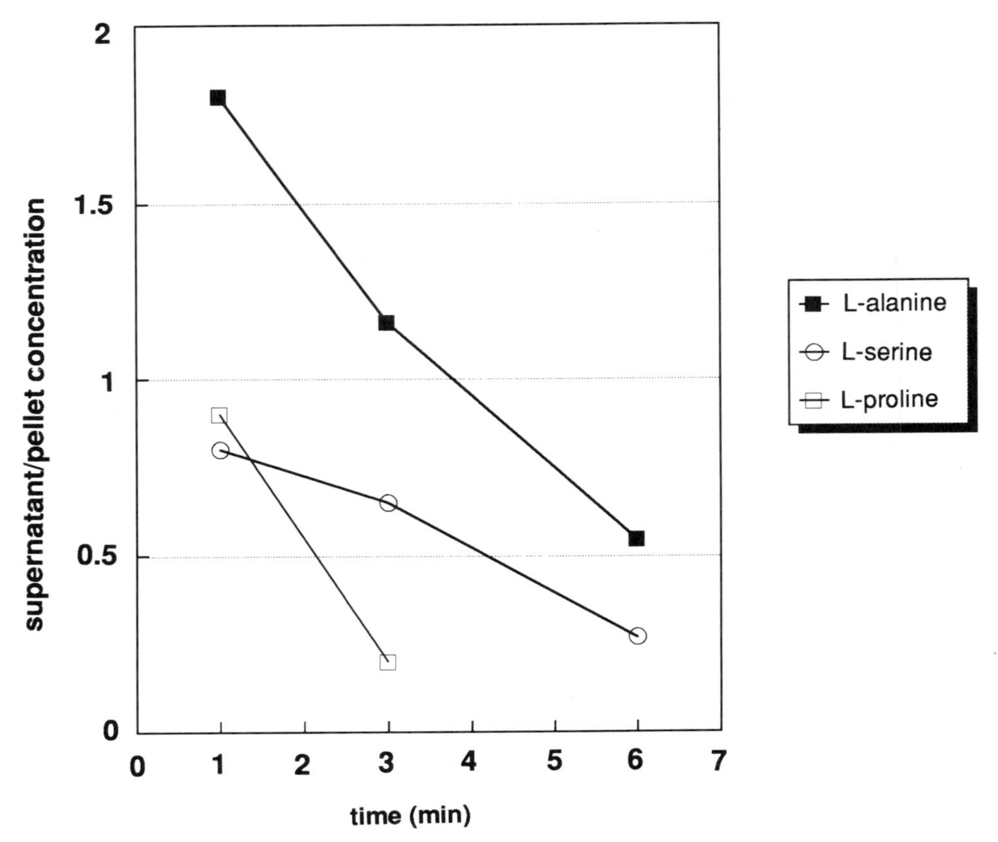

Figure 1. The changes of supernatant/pellet concentration ratio in time for L-alanine, L-serine and L-proline. Each point represents V_d values (mean ± SD) obtained after appropriate time of perfusion. Sign (**) denotes significant difference between the groups (p<0.01).

5. CONCLUSIONS

Our results suggest that the capillary depletion method could be used to determine the distribution ratio between the endothelial compartment and brain parenchyma for endogenous substances.

The brain parenchyma / endothelium ratio for- ^{3}H L-leucine significantly increased in time, indicating that this molecule penetrates from blood into the brain parenchyma, probably without significant accumulation in the endothelial cells.

However, there was a significant decrease of brain parenchyma / endothelium concentrations ratio in time for all three neutral non-essential amino acids studied. It can be assumed that this accumulation in brain endothelial cells is due to metabolic changes of those molecules.

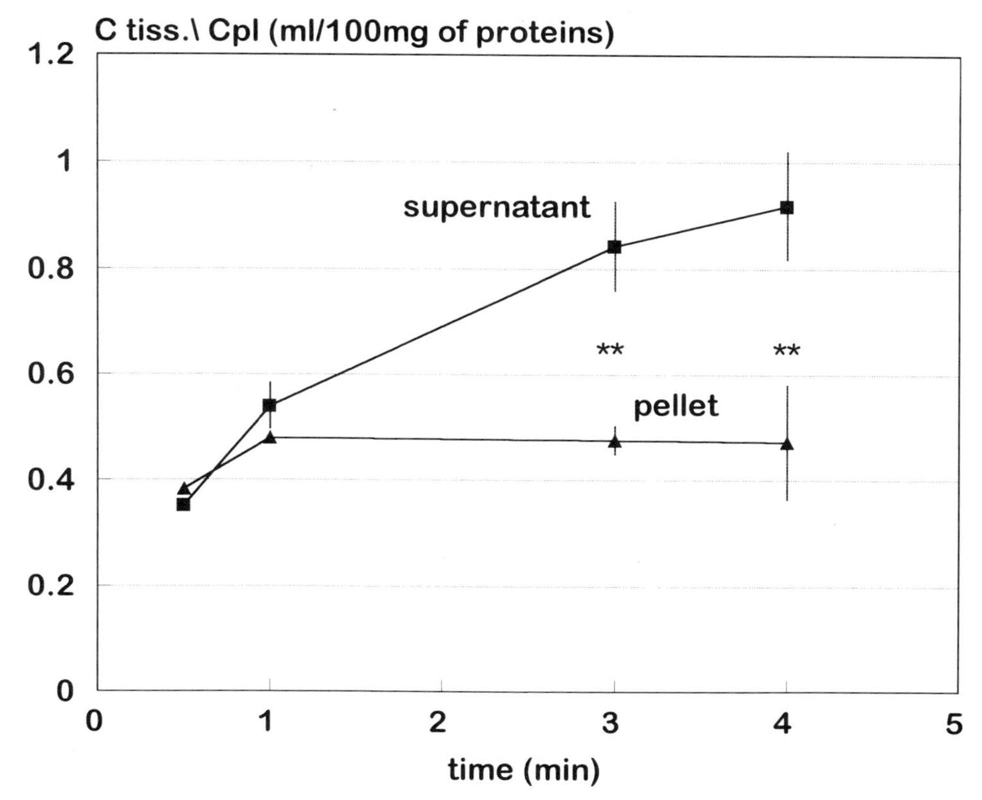

Figure 2. The distribution of [^{3}H] L-leucin in endothelial cells (pellet) and postvascular compartment (supernatant).

REFERENCES

1. Pardridge W.M.: *The Blood-Brain Barrier: Cellular and molecular biology.* Raven Press, New York 1993.
2. Redzic Z., Markovic D., Jovanovic S., Mitrovic D., Zlokovic B., and Rakic Lj. Slow penetration of [^{3}H] tiazofurin into guinea pig brain by a saturable mechanism. *Meth Find Exp Clin Pharmacol* 17 (6) In press
3. Zlokovic BV, Begley D, Duricic B and Mitrovic D. M. Measurement of solute transport across the blood brain barrier in the perfused guinea pig brain; method and application to N-methyl-α-amino isobutyric acid. *J Neurochem (1986)* 46:1444-1451.
4. Triguero D., Buciak J. and Pardridge W.M. Capillary depletion method for quantification of blood-brain barrier transport of circulating peptides and plasma proteins. *J Neurochem* (1990) 54 (6): 1882-1888.

THE BLOOD-BRAIN BARRIER, POTASSIUM, AND BRAIN GROWTH

Richard F. Keep,[1] Jianming Xiang,[1] and A. Lorris Betz[1,2]

[1] Department of Surgery (Neurosurgery)
[2] Departments of Pediatrics and Neurology
University of Michigan
Ann Arbor, Michigan 48105-0532

SUMMARY

Blood-brain barrier (BBB) permeability to small polar molecules, including potassium, is increased early in rat development and this may reflect the need of the growing brain for potassium. The latter was tested by examining the effect of dexamethasone on BBB ^{86}Rb (potassium) permeability and brain growth. This drug substantially reduced BBB ^{86}Rb permeability in 1 day and 1 week old rats (38 and 55%) but had no significant effect in 3 and 7 week old rats. The reduction in BBB permeability in the younger age groups was accompanied by a reduction in brain growth.

RÉSUMÉ

La perméabilité de la barrière hémato-encéphalique (BHE) pour les ions tels que le potassium est plus élevée lors des premiers stades de développement chez le rat. La pénétration cérébrale accrue du potassium pourrait refléter les besoins du cerveau au cours de son développement. Cette hypothèse a été vérifiée en étudiant l'effet de la dexamethasone sur la perméablité de la BHE pour le ^{86}Rb (potassium) et sur la croissance cérébrale. Cette substance réduit significativement la pénétration cérébrale du ^{86}Rb chez des rats de un jour et d'une semaine (respectivement 38 et 55%) mais n'a pas d'effet significatif chez des rats de 3 ou 7 semaines. La réduction de la perméabilité de la BHE pour le potassium par la dexamethasone, observée chez les groupes d'animaux les plus jeunes, s'accompagne d'une réduction de la croissance cérébrale.

INTRODUCTION

A number of studies have shown an increased blood-brain barrier (BBB) permeability to small polar molecules early in development (1). The reason for this increased permeability

Biology and Physiology of the Blood–Brain Barrier, edited by Couraud and Scherman
Plenum Press, New York, 1996

is uncertain. Are proliferating vessels inherently more leaky or does the increased permeability meet some functional need of the developing brain? An insight into these questions may aid in the discovery of compounds that naturally regulate BBB permeability.

Recently we have found that the rate of potassium influx across an adult rat BBB would not be sufficient to meet the rate of potassium accumulation in the rat brain early in development (2). This raises the possibility that the increased BBB permeability early in development is necessary to meet the potassium requirement for brain growth. In this study we examined this hypothesis by studying the effect on brain growth of reducing BBB permeability using the steroid dexamethasone.

METHODS

We examined the effect of dexamethasone on (A) BBB permeability to ^{86}Rb (a marker for potassium) and ^{14}C-urea (a passive permeability marker) and (B) brain and body growth at different ages in the Sprague Dawley rat. For the permeability measurements, two doses of dexamethasone (2 mg/kg, i.p.) or vehicle were given 24 and 4 hours prior to sacrifice. For growth measurements, four doses were given at 72, 48, 24 and 4 hours prior to sacrifice. Four ages groups were examined, with the initial dose of dexamethasone for each type of experiment being given at 1 day or 1, 3 or 7 weeks.

The influx rate constants for ^{86}Rb or ^{14}C-urea were determined in separate groups of rats under pentobarbital anesthesia (50 mg/kg i.p.) using graphical analysis (2). For measurements of brain growth, brain dry weights were determined since dexamethasone can effect brain water content. Brain potassium contents were determined by flame photometry. For body growth, body weights were determined prior to the initial injection of dexamethasone or vehicle and at sacrifice and the growth expressed as a % change.

Statistical comparisons were by analysis of covariance (permeabilities) and analysis of variance (growth). All results are presented as means ± S.E.

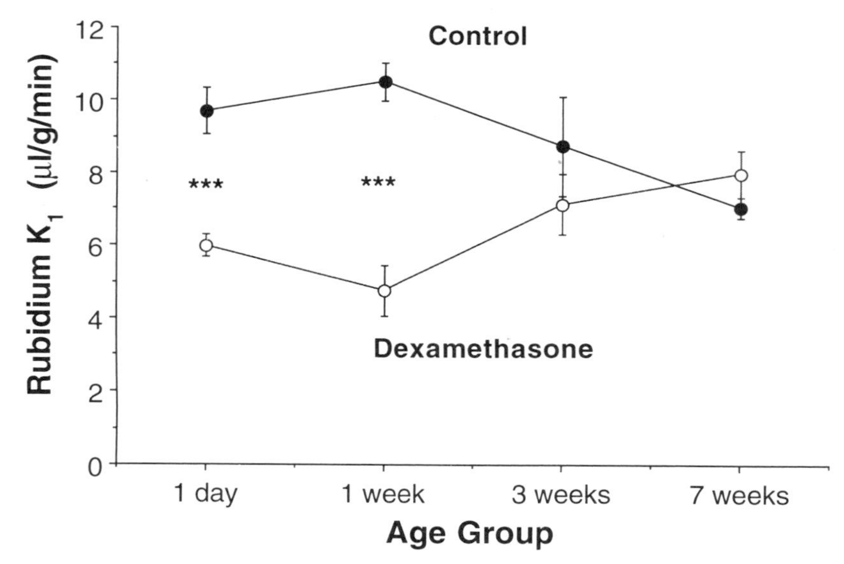

Figure 1. Effect of dexamethasone on the cortical influx rate constant for ^{86}Rb at different ages. Values are means ± S.E., n = 8-21, * indicates a significant difference ($p<0.001$) between control and dexamethasone groups.

Table 1. Effect of dexamethasone on brain and body growth during development

	Brain dry wt (mg)		Δ Body wt (%)	
Age	Vehicle	Dexamethasone	Vehicle	Dexamethasone
1 day	53 ± 1	45 ± 1**	57.0 ± 4.0	2.5 ± 2.2**
1 week	125 ± 3	109 ± 2*	51.4 ± 4.0	8.0 ± 2.0**
3 weeks	266 ± 6	274 ± 4	29.3 ± 1.9	11.1 ± 1.5**
7 weeks	387 ± 6	398 ± 4	7.8 ± 1.0	-10.0 ± 0.4**

Animals received 4 daily doses of dexamethasone (2mg/kg) prior to sacrifice. Controls received vehicle alone. All values are means ± S.E.; n=4-9; * and ** indicate significant differences from control at the $p<0.01$ and $p<0.001$ levels.

RESULTS

Dexamethasone treatment (2 x 2mg/kg) had no significant effect on the influx rate constant (K_1) for ^{86}Rb in 3 and 7 week old rats, but caused a marked reduction in K_1 in 1 day and 1 week old animals ($p<0.001$; Fig. 1). The reduction in ^{86}Rb permeability in young animals does not appear to represent a specific effect on potassium transport since dexamethasone also reduced ^{14}C-urea cortical influx rate constant in 1 week old rats (9.2 ± 1.5 and 5.4 ± 0.4 μl/g/min in vehicle and dexamethasone treated rats respectively; n= 10-11; $p<0.05$).

In the two youngest age groups, dexamethasone treatment (4 x 2 mg) also resulted in a reduction in brain dry weight compared to vehicle-treated controls (Table 1). Mirroring the effect on BBB permeability, however, dexamethasone had no effect on brain weight in the two older age groups. The effect of dexamethasone on total brain potassium content was similar to that on brain dry weight. There was a reduction (12-14%) in the two youngest age groups ($p<0.01$) and no effect in the two oldest groups. Dexamethasone also affected total body growth (Table 1). However, unlike the effect on brain weight, a reduction in body growth was found in all age groups with dexamethasone treatment.

DISCUSSION

Early in development, the permeability of ^{86}Rb (potassium) at the rat BBB is markedly elevated being 12 x 10^{-6} and 4 x 10^{-6} cm/sec in the 21 day gestation and 2 day postnatal rat compared to 1 x 10^{-6} cm/sec in the 50 day old rat (calculated from (2) and (3)). This increased permeability does not appear to reflect a greater active transport since there is a similar developmental change in the permeability of ^{14}C-urea (2). Several structural studies have demonstrated changes in the paracellular pathway with development (e.g. (5)) suggesting that this is the high permeability pathway early in development.

In the present study we found that dexamethasone induces a greater reduction in BBB ^{86}Rb permeability early in development and that this reduction is accompanied by a decrease in ^{14}C-urea permeability, i.e. the effect appears to be on a non-specific, probably paracellular, pathway. The greater effect of dexamethasone in young animals is probably because that paracellular pathway is larger in young animals.

The reduction in BBB permeability in the young rats was accompanied by a reduction in brain growth. This is in accord with our hypothesis that increased BBB permeability to small polar compounds early in development might be related to the potassium requirement for the brain to grow. It should be noted, however, that these experiments do not prove a causative relationship between the reduction in BBB potassium entry and the decline in brain growth. Dexamethasone may alter brain growth by other mechanisms. Indeed, dexamethasone can inhibit growth hormone release (4) and this probably accounts for the reduction in body growth found at all ages in this study. However, unlike with body weight, an effect of dexamethasone on brain weight was only found in younger (1 day and 1 week age groups) rats where dexamethasone also effected BBB permeability. No effect was found in 3 and 7 week age groups, where ^{86}Rb BBB permeability is unaffected, even though, at least in the former, the brain was still growing.

REFERENCES

1. Johanson C. E., 1989, Ontogeny and phylogeny of the blood-brain barrier, *in:* "Implications of the Blood-Brain Barrier and Its Manipulation," E. A. Neuwelt, eds., Plenum Publishing Corp, pp. 157-198.
2. Keep R. F., Ennis S. R., Beer M. E., and Betz A. L., (in press), Developmental changes in blood-brain barrier potassium permeability in the rat: relation to brain growth., *J. Physiol.*
3. Keep R. F. and Jones H. C., 1990, Cortical microvessels during brain development. A morphometric study in the rat, *Microvasc Res* 40:412-426.
4. Nakagawa K., Ishizuka T., Obara T., Matsubara M., and Akikawa K., 1987, Dichotomic action of glucocorticoids on growth hormone secretion., *Acta Endocrinol.* 116:165-171.
5. Stewart P. A. and Hayakawa E. M., 1987, Interendothelial junctional changes underlie the developmental "tightening" of the blood-brain barrier, *Dev Brain Res 32:271-281.*

RECEPTOR-MEDIATED TRANSCYTOSIS OF TRANSFERRIN THROUGH BLOOD-BRAIN BARRIER ENDOTHELIAL CELLS

Laurence Descamps,[1] Marie-Pierre Dehouck,[1]
Gérard Torpier,[1] and Roméo Cecchelli[1, 2]

[1] INSERM U325-SERLIA
Institut Pasteur
Lille, France
[2] Université des Sciences et Technologies de Lille I
Villeneuve d'Ascq, France

1. SUMMARY

A cell culture model of the blood-brain barrier consisting of a coculture of bovine brain capillary endothelial cells (ECs) and astrocytes[1] have been used to examine the mechanism of iron transport to the brain. In contrast to apoTf, we observed a specific transport of holoTf across ECs. This transport was completely inhibited at low temperature. Moreover, the anti-Tf receptor antibody (OX-26) competitively inhibited holoTf uptake by ECs. Pulse chase experiments demonstrated that only 10 % of Tf was recycled to the luminal side of the cells, whereas the majority of Tf was transcytosed to the abluminal side; double labeling experiments clearly demonstrated that iron crosses ECs bound to Tf. No intraendothelial degradation of Tf was observed, suggesting that the intraendothelial pathway through ECs bypasses the lysosomal compartment. These results clearly show that the iron-Tf complex is transcytosed across brain capillary endothelial cells by a receptor-mediated pathway without any degradation.

1. RÉSUMÉ

Afin d'élucider les mécanismes de transport du fer vers le cerveau, nous avons utilisé un modèle "in vitro" de barrière hémato-encéphalique. Ce modèle consiste en une coculture de cellules endothéliales de capillaires cérébraux et d'astrocytes[1]. Contrairement à l'apoTf, il existe un transport spécifique et unidirectionnel de l'holoTf au niveau des cellules endothéliales. En effet, après accumulation de la Tf dans les cellules endothéliales, seulement 10% de la Tf est recyclée face luminale, alors que la majorité (75%) est retrouvée face

Biology and Physiology of the Blood–Brain Barrier, edited by Couraud and Scherman
Plenum Press, New York, 1996

abluminale. De plus, l'endocytose de l'holoTf est complètement inhibée par l'OX-26. Des expériences de double marquage ont clairement démontré que le fer est transporté au niveau des cellules endothéliales lié à la Tf. Aucune dégradation n'est observée lors de la transcytose. Ces résultats montrent qu'il existe une transcytose du complexe fer-Tf au niveau des cellules endothéliales de capillaires cérébraux, et que ce transport est médié par le récepteur de la Tf.

2. RESULTS

2.1. Apical to Basolateral Transport of Tf across Bovine Brain Capillary EC Monolayers

The transport of labeled holoTf from the luminal to the abluminal compartment was reduced severely by an excess of unlabeled Tf (Fig. 1a), suggesting that the Tf transport from the apical side to the basal side of the cells was specific.

The same experiments with ^{125}I-apoTf introduced to the luminal surfaces of ECs were performed. Fig. 1b shows that in contrast to holoTf no specific accumulation of ^{125}I-apoTf reached the abluminal compartment, suggesting that Tf receptor is involved in Tf transport, since apoTf has a low affinity for the Tf receptor. Moreover, experiments using the anti-Tf receptor antibody OX-26, have shown that this antibody completely inhibited Tf endocytosis in endothelial cells (not shown).

The effect of temperature on the transport of sucrose and holoTf from the apical to the basal compartment showed that a decrease in the incubation temperature from 37°C to 4°C slightly affected the passage of sucrose, whereas a dramatic decrease in holoTf transport through the monolayer was observed.

A pulse-chase experiment showed that ten percent of the endocytosed ^{125}I-holoTf at 37 °C was recycled to the luminal side, whereas 75% was transcytosed to the abluminal compartment (Fig. 2). In the control condition (4°C), the major part of the transferrin remained in the cells. No degradation of Tf was observed during transcytosis.

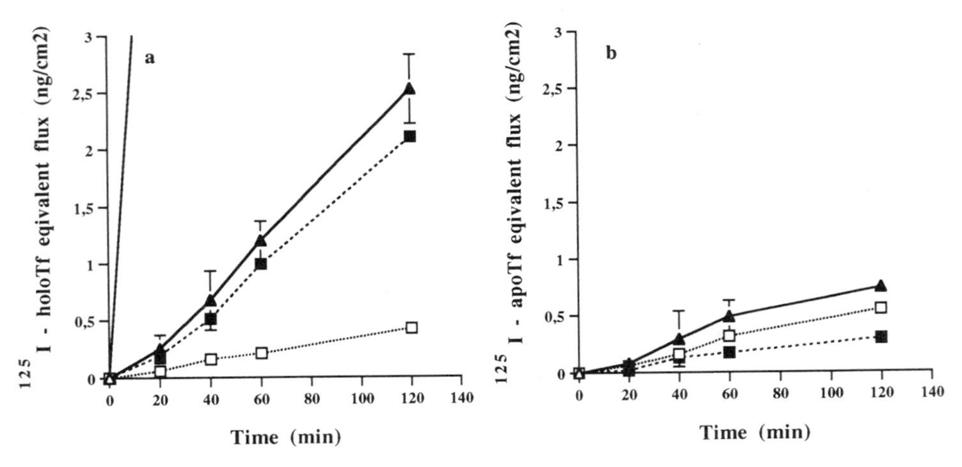

Figure 1. Apical to basolateral transport of ^{125}I-holoTf (a) and ^{125}I-apoTf (b) across bovine brain capillary EC monolayers grown on a porous filter. 1400 ng/ml of ^{125}I-Tf was added to the luminal side of the cells. All values (mean of triplicate inserts ± SEM (bars) (n=3)) represent radioactivity that was TCA precipitable. Tf flux across filters (—), total Tf flux (▲) was corrected for non specific Tf flux (performed with a 100 fold excess of unlabeled Tf), giving the specific Tf flux (■).

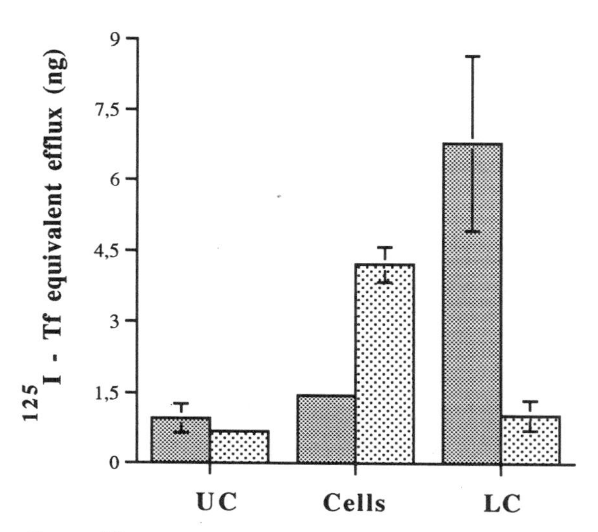

Figure 2. Asymmetric efflux of ^{125}I-Tf from bovine brain capillary EC monolayers grown on a porous filter. Cells were allowed to accumulate ^{125}I-holoTf (1400 ng/ml), from their luminal side for 1 hour at 37°C. Then they were carefully washed, and put in fresh medium at either 37°C (gray) or 4°C (stippled) for 30 minutes. The amount of cell associated radioactivity and TCA precipitable radioactivity in the upper (UC) and lower (LC) compartment were measured. All values are the mean of triplicate inserts ± SEM (bars).

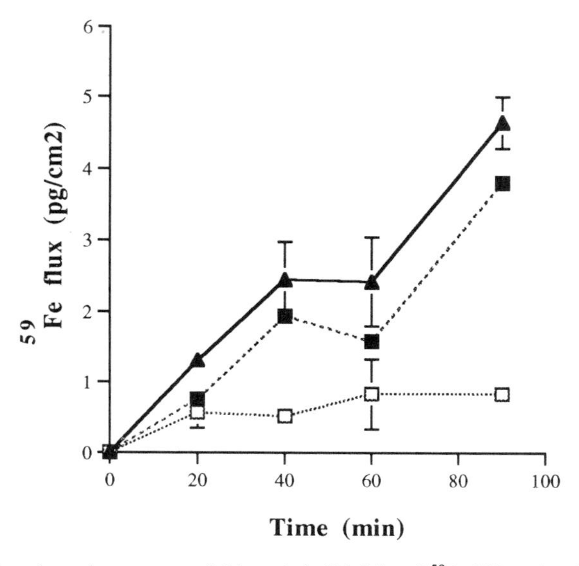

Figure 3. Apical to basolateral transport of 21 μg/ml (26μM) of ^{59}Fe-Tf on bovine brain capillary EC monolayers. Total flux (▲) was corrected for non specific flux (□) giving the specific flux (■) of ^{59}Fe. All values represent the mean of triplicate inserts ± SEM (bars).

Table 1. ^{59}Fe-^{125}I -Tf transport across bovine brain capillary EC monolayers.

	^{59} Fe flux	^{59}Fe-Tf equivalent flux	^{125}I-Tf equivalent flux
Passage ($pmol/cm^2/h$)	0.176	0.088	0.093

All values are the mean of triplicate inserts.

2.2. TRANSENDOTHELIAL IRON TRANSPORT STUDIES USING ^{59}Fe-Tf AND ^{59}Fe-^{125}I-Tf

Experiments using ^{59}Fe-Tf showed that a specific transendothelial transport of ^{59}Fe occurred over 90 minutes (Fig. 3), suggesting that the holoTf receptor could take part in the transport. To examine whether iron was transcytosed through endothelial cells with holoTf, double labeling experiments were carried out. The results are summarized in Table 1. The ^{125}I-holoTf equivalent flux represented TCA precipitable radioactivity. The ^{59}Fe-holoTf calculated equivalent flux (0.088 $pmol/cm^2/h$), which is equal to half the ^{59}Fe flux (0.176 $pmol/cm^2/h$), was within the same range of the observed ^{125}I-holoTf equivalent flux (0.093 $pmol/cm^2/h$).

3. CONCLUSION

Our BBB "in vitro" model has permitted us to demonstrate that the iron-Tf complex is transcytosed across the brain capillary endothelial cells by a receptor-mediated pathway.

REFERENCE

Dehouck B., Dehouck M. P., Fruchart J. C. and Cecchelli R. Upregulation of the low density lipoprotein receptor at the blood-brain barrier : intercommunications between brain capillary endothelial cells and astrocytes. J. Cell. Biol. 126 : 465-473, 1994.

11

INTRACEREBRAL EXPRESSION OF TRANSFERRIN RECEPTORS IN IRON-DEFICIENT RATS

T. Moos and T. M. Hansen

Institute of Medical Anatomy, Section A
The Panum Institute
University of Copenhagen
Denmark

SUMMARY

In iron-deprived animals, transferrin receptors were increasingly expressed in both capillary endothelium and choroid plexus epithelium. Transferrin receptors were also increasingly expressed in neurons. It is very likely that this principal distribution is also seen in other mammals, including man. Prospective physiological studies on iron-homeostasis should account for the ability of neurons to internalize iron-transferrin complexes by receptor-mediated endocytosis. The results are important in the context of current therapies for treatment of brain tumors using conjugation of chemotherapeutics to transferrin or transferrin receptor antibodies, as toxin-conjugated transferrin may have the potential of harming normal neurons expressing the transferrin receptor. Moreover, the fact that neurons hallmark transferrin receptors may open new vistas on the role of an impaired intracerebral iron-homeostasis for formation of neurodegenerative disorders.

RÉSUMÉ

Chez les animaux carencés en fer, les récepteurs à la transferrine sont plus fortement exprimés à la fois dans l'endothélium capillaire et dans l'épithélium du plexus choroïde; cette augmentation est également mise en évidence au niveau des neurones. Il est très vraisemblable que cette répartition est la même chez d'autres mammifères, dont l'Homme.Des études physiologiques prospectives sur l'hémostase du fer devraient montrer la capacité des neurones à internaliser les complexes fer-transferrine par endocytose médiée par des récepteurs. Les résultats sont importants dans le contexte des thérapies courantes pour le traitement des tumeurs cérébrales, en conjuguant des produits de chimiothérapie avec la transferrine, ou avec des anticorps dirigés contre le récepteur de la transferrine, car la transferrine conjuguée à un toxique pourrait avoir un effet néfaste sur les neurones normaux

Biology and Physiology of the Blood–Brain Barrier, edited by Couraud and Scherman
Plenum Press, New York, 1996

qui expriment le récepteur de la transferrine. De plus, le fait que les neurones contrôlent les récepteurs à la transferrine peut ouvrir une autre vision du rôle d'une hémostase intracérébrale anormale du fer sur l'apparition des désordres neuro-dégénératifs.

INTRODUCTION

The brain-barrier system comprises the blood-brain and the blood-CSF barriers, which prevent diffusion of water-soluble macromolecules from blood to brain (Broadwell and Sofroniew, 1993; Moos and Høyer, 1995). Since the brain-barrier system excludes iron-carrying transferrin from gaining access to the brain parenchyma, transferrin receptor is thoroughly distributed on brain capillary endothelial cells (Jeffries et al., 1984) and choroid plexus epithelial cells (Moos, 1995b) in order to permit delivery of the essential metal iron into the brain. Iron is transported through the brain-barriers subsequent to the attachment of liver-derived transferrin ("liver transferrin") carrying iron to the transferrin receptor (Crowe and Morgan, 1992; Morris et al., 1992b; Roberts et al., 1993). Next, iron-free transferrin is recycled to the blood, whereas iron by a so far unknown mechanism is transported into the brain interstitium.

It is now also established that transferrin receptors are present on neuronal cells within the central nervous system (Giometto et al., 1990; Graeber et al., 1989; Moos, 1995a,b). Although still only slight studied in neuroectodermal tissues, the internalization of iron-transferrin in neurons containing transferrin receptors could most likely occur by a mechanism identical to the intracellular trafficking mechanism seen in well-studied cell types derived from other germ layers, e.g. brain endothelial cells, liver cells and different erythropoietic cells (Huebers and Finch, 1987; Morgan, 1995). The intraneuronal transferrin receptor is probably being attached by brain-derived transferrin ("brain transferrin") synthesized in choroid plexus and oligodendrocytes (Bloch et al., 1985, 1987).

Following a period with iron-deficiency, the transferrin receptor is increasingly expressed in different tissues of mesenchymal and endodermal origin (e.g. Huebers and Finch, 1987; Morgan, 1995). There is evidence for an increased transport of iron from blood into the brain in iron-deficient rats (Taylor et al., 1991). The intraneuronal expression during these circumstances has never been investigated. Accordingly, we evaluated the expression of the transferrin receptor protein in the rat brain following iron-depletion using immunohistochemistry.

MATERIALS AND METHODS

Animals

Wistar rats aged 4 weeks were implemented in the study. One ml of blood was tapped from the animals by venous puncture in order to deplete iron-deposits. Next, the animals were subjected to an iron-depleted diet. After 6 weeks, half of the iron-deficient rats received an intraperitoneal injection of 5 mg iron-dextran (Sigma) before returning to a normal, iron-containing diet (group B). The remaining rats were maintained on the iron-free diet (group A). Age-matched control rats fed with a normal diet comprised group C. The animals had free access to water and food and were housed in cages at the Animal Department of the Panum Institute in Copenhagen under constant temperature and humidity conditions with a 12 h light/dark cycle.

For monitoring the iron-status of the rats, samples of blood were achieved immediately before the animals were sacrificed at 10 weeks of age. Hole-blood and serum were used for measurements of hemoglobin-concentration, total iron-binding capacity (TIBC), and iron-content using a commercial available kit (Sigma).

Tissue Processing

For histochemical investigations, animals were deeply anesthetized and transcardially perfused with via the left ventricle with heparinized saline in 0.1M potassium phosphate-buffered saline (KPBS) followed by 4% paraformaldehyde. Brains, livers and small intestine were gently dissected, post-fixed for 4 h at room temperature and immersed in 30% sucrose-KPBS for 48 h. Subsequently, the tissues were cut into serial, coronal 40 μm sections on a cryostat and reacted free-floating.

Histological Investigations

A. Iron Histochemistry. The sections were reacted for non-heme iron using the method described by Moos (1995c). Sections were washed in KPBS for 3 x 10 min and incubated for 20 min in 1% H_2O_2 in KPBS to quench endogenous peroxidase activity. They were then immersed in Perl's solution (1:1, 2% HCl and 2% potassium ferrocyanide) at room temperature for 30 min, and incubated for 10 min in 0.05% 3,3-diaminobenzidine tetrahydrocloride (DAB) in PBS (pH 7.0), and then for 10 min in 0.015% H_2O_2 in DAB/PBS. The free-floating cryostat sections were mounted on lysine-coated glass slides. All slides were dried and embedded in DPX (BDH, UK).

B. Immunohistochemistry. The cryostat sections were treated with 0.1% sodium borohydride (Sigma). Next, they were incubated in 1.0% H_2O_2 in Tris-buffered saline (TBS: 0.05 M Tris, pH 7.4, 0.15 M NaCl) with 0.01% Nonidet P-40 (Sigma) (TBS/Nonidet). For blocking of non-specific binding, the sections were then preincubated with 10% normal goat serum followed by incubation overnight at 4^0C with a mouse monoclonal antibody raised against the rat transferrin receptor protein (CD71) (Serotec, UK) diluted 1:100 in 10% goat serum in Tris/Nonidet (pH 7.4). The antibodies were incubated for 30 minutes at room temperature in biotinylated monoclonal anti-mouse IgG adsorbed with rat immunoglobulins (Sigma) diluted 1:200. The sections were incubated for 30 minutes in streptABComplex/HRP (Dakopatts, DK). The sections were developed for 10 min in DAB followed by 0.015% H_2O_2 in DAB for 10 min.

Since the presence of endogenous biotin in liver and intestine disqualifies the use of the ABC-system, transferrin receptors in these tissues were detected using peroxidase-labeled anti-mouse IgG (Sigma) diluted 1:50. In order to evaluate the extent of non-specific binding of the secondary antibodies, the preincubation agent was substituted for the primary antibody and the immunohistochemical procedures performed as described above. Immunolabeling was not observed in such circumstances.

C. Cytochrome c Oxidase Histochemistry. Sections were reacted for cytochrome c oxidase activity using the protocol of Wong-Riley (1979). A substrate was prepared as 5 mg of cytochrome c and 10 mg DAB dissolved in 20 ml PBS added with 1.8 g of sucrose. The substrate was filtered and incubated with the sections for 1 h at 37 ^{0}C.

Protein-Blotting. Following transcardial perfusion with heparinized saline, brains and livers were dissected and homogenized in a buffer consisting of various inhibitors of proteolytic enzymes. The homogenates were sonicated, purified by centrifugation and stored at -80^0C. Samples were spotted onto nitrocellulose membranes and reacted for transferrin receptor using the protocol described above with the exception that transferrin receptors were visualized using the ECL-detection system (Amersham) instead of DAB.

RESULTS

In group C (normal rats), iron was observed in choroid plexus epithelial cells, oligodendrocytes, microglia-like cells, and ependymal cells. Following iron-deficiency, the iron content was clearly lowered in these cells. After 10 weeks, iron was almost undetectable in group A rats. In group B, iron was detectable in ependymal cells and scattered in glial cells, especially in microglial cells situated in the corpus callosum. In intestine and liver, iron was likewise low in group A animals, whereas the iron-content was clearly raised in group B. Blood-samples from animals of each group confirmed the differences in hemoglobin-concentration, TIBC, and iron-content.

The cytochrome c oxidase (CCO) enzyme-histochemistry reaction revealed enzymatic activity in neurons throughout the brain. Moreover, CCO was seen in non-neuronal cells, liver cells and cells of the small intestines. The pattern of CCO reactivity was in principal equal to that of iron. Thus, the reactivity was high in groups B and C, and low in group A.

Transferrin receptor immunoreactivity (Trf-R-IR) was observed in brain capillary endothelial cells and choroid plexus epithelial cells. The immunoreactive brain capillary endothelial cells comprised capillaries throughout the CNS, with the exception of those situated within circumventricular organs. In choroid plexus epithelial cells, Trf-R-IR was observed in the cytoplasm as densely labeled dots in an otherwise unstained cytoplasm. Trf-R-IR was never observed in astrocytes, oligodendrocytes or microglial cells. Following a period of iron-deficiency, the transferrin receptor expression was clearly raised in endothelial cells and choroid plexus epithelial cells (Fig. 1).

Neuronal elements exhibiting Trf-R-IR were distributed widely in different nuclei and regions of the cerebral cortex, basal ganglia, basal forebrain, olfactory and limbic system, diencephalon, mesencephalon, pons, medulla oblongata, reticular formation, spinal cord, and cerebellum (Fig. 1). The transferrin receptor was consistently higher expressed in iron-deficient rats (Fig. 2).

When reversed to a normal diet, the intracerebral transferrin receptor expression was lower than in group A (Fig. 2). This pattern of immunoreactivity was confirmed in sections of liver and small intestine. The blotting analyses confirmed the findings from the histological examinations. The increasing expression of transferrin receptor in group A in the blotting reflects the increased expression in total brain cells within each individual sample.

Table I summarizes the occurrence of iron, cytochrome c oxidase (CCO), and transferrin receptor expression in rats during the investigation.

DISCUSSION

Being a co-factor for several enzymes in the brain, iron is essential for neuronal function (Wrigglesworth and Baum, 1988; Youdim et al., 1980). Accordingly, the transfer of iron from blood to brain is vital for normal brain function. Physiological studies have revealed an elevated transport of iron from blood to brain in the iron-deficient rat, which likely reflects an elevated transferrin receptor expression at the brain-barrier sites following iron-deficiency (Taylor et al., 1991). Our study provides the first morphological evidence that transferrin receptors are increasingly expressed at brain-barrier sites following iron-de-

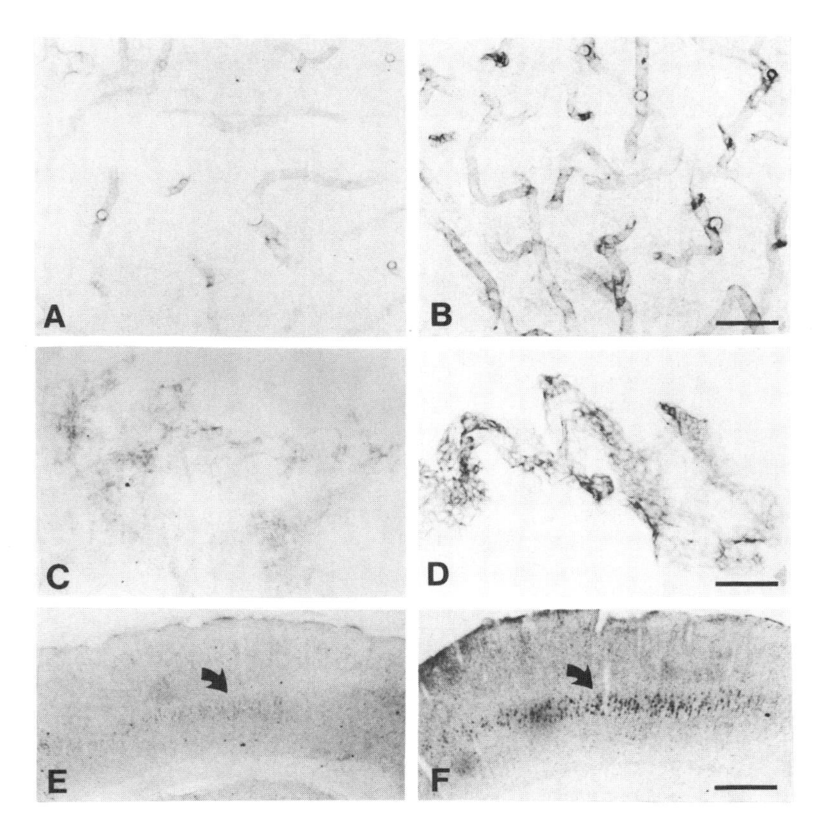

Figure 1. Distribution of transferrin receptor in normal (A,C,E) and iron-deficient (B,D,F) rats. Sections showing the difference in transferrin receptor immunoreactivity in brain capillary endothelial cells (A,B), choroid plexus epithelial cells (C,D), and neurons (E,F). The curved arrows in E,F identify the neocortical layer V in which the difference in intraneuronal immunoreactivity is particularly pronounced. Scale bars: A-B = 40 μm, C-D = 300 μm, E-F = 1 mm.

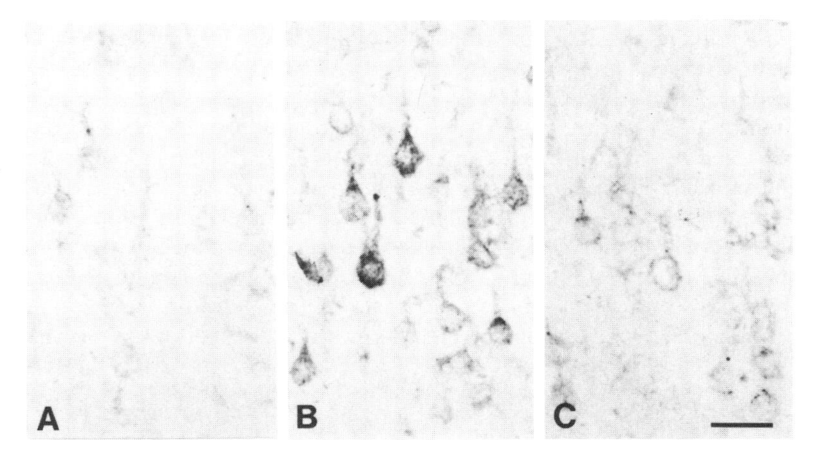

Figure 2. High-power magnification of neocortical neurons of normal rats (A), iron-deficient rats (B), and iron-deficient rats reversed to normal diet (C). The higher immunoreactivity in iron-deficient rat is clearly seen. Scale bars = 20 μm.

Table 1. Diagrammatic representation of the temporal profile of iron, cytochrome c oxidase (CCO), and transferrin receptor expression in rats following a period of iron-deficiency[a]

	C[b]	A	B
Iron[c]			
Endothelial cells	–	–	–
Choroid plexus epithelium	++	–	+
Neurons	–	–	–
Glial cells	++	–	+
Liver, small intestine epithelium	+++	+	++
CCO			
Endothelial cells	++	+	++
Choroid plexus epithelium	++	+	++
Neurons	+++	+	++
Glial cells	++	+	++
Liver, small intestine epithelium	+++	+	++
Trf-R-IR			
Endothelial cells	+	+++	+
Choroid plexus epithelium	+	+++	+
Neurons	+	+++	++
Glial cells	–	–	–
Liver, small intestine epithelium	+	+++	+

[a]Symbols are related to the density of cells exhibiting reactivity, which was semi-quantitatively assessed in four degrees: absent -, weak +, moderate ++, strong +++.
[b]The terms C,A,B refer to rats subjected to a normal (C), iron-depleted (A), and normal diet after a period of iron-depletion (B).
[c]The iron-histochemistry technique only detects non-heme iron.

pletion. We also demonstrate that intraneuronal transferrin receptors respond very promptly to iron-depletion. The function of transferrin receptor in the non-dividing neuronal cells is thought to reflect the need for iron in metabolically active neurons in which oxidative phosphorylation in the mitochondrial respiratory chain is provided by enzymes having iron as part of their prosthetic group (Morris et al., 1992a).

Transferrin and its receptor have also attracted increasing interest in terms of drug delivery of different pharmacological compounds to the brain. One method has been intracerebral injection of transferrin conjugated with chemotherapeutics for treatment of cerebral cancer. The rationale for this mode of therapy should be that proliferating cancer cells would be the sole cells within the CNS to express transferrin receptors (Martell et al., 1993; Laske et al., 1994). Thereby, a highly selective cancer treatment with rare side effects should be possible. However, the possibility for binding of toxin-conjugated transferrin to transferrin receptors present on subsets of normal brain cells, leading to the impairment of normal brain function, is a dramatic potential that warrants further investigation.

ACKNOWLEDGMENTS

This work was supported by Læge Sofus Carl Emil Friis og Hustru Olga Doris Friis' Legat, Beckett-Fonden, and Novo Nordisk Fonden.

REFERENCES

Bloch B, Popovici T, Chouham S, Levin MJ, Tuil D, Kahn A (1987) Transferrin gene expression in choroid plexus of the adult rat brain. Brain Res Bull 18:573-576

Bloch B, Popovici T, Levin MJ, Tuil D, Kahn A (1985) Transferrin gene expression visualized in oligodendrocytes of the rat brain by using in situ hybridization and immunocytochemistry. Proc Natl Acad Sci, USA 82:6706-6710

Crowe A, Morgan EH (1992) Iron and transferrin uptake by brain and cerebrospinal fluid in the rat. Brain Res 592:8-16

Giometto B, Bozza F, Argentiero V, Gallo P, Pagni S, Piccinno MG, Tavolato B (1990) Transferrin receptors in the rat central nervous system. J Neurol Sci 98:81-90

Graeber MB, Raivich G, Kreutzberg GW (1989) Increase of transferrin receptors and iron uptake in regenerating motor neurons. J Neurosci Res 23:342-345

Huebers HA, Finch CA (1987) The physiology of transferrin and transferrin receptors. Physiol Rev 67:520-582

Jeffries WA, Brandon MR, Hunt SV, Williams AF, Gatter KC, Mason DY (1984) Transferrin receptor on endothelium of brain capillaries. Nature 312:162-163

Laske DW, Ilercil O, Akbasak A, Youle RJ, Oldfield EH (1994) Efficacy of direct intratumoral therapy with targeted protein toxins for solid human gliomas in nude mice. J Neurosurg 80:520-526

Martell LA, Agrawal A, Ross DA, Muraszko KM (1993) Efficacy of transferrin receptor-targeted immunotoxins in brain tumor cell lines and pediatric brain tumors. Cancer Res 53:1348-1353

Moos T (1995a) Age-dependent retrograde axonal transport of exogenous albumin and transferrin in rat motor neurons. Brain Res 672:14-23

Moos T (1995b) Immunohistochemical Localization of Intraneuronal Transferrin Receptor Immunoreactivity in the Adult Mouse Central Nervous System. Submitted

Moos T (1995c) Developmental profile of non-heme iron distribution in the rat during ontogenesis. Dev Brain Res 87:203-213

Moos T, Høyer PE (1995) Detection of plasma proteins in CNS neurons: Conspicuous influence of tissue-processing parameters and the utilization of serum for blocking non-specific reactions. Submitted

Morgan EH (1995) Iron metabolism and transport. In: D. Zakim and T. Bayer (eds): Hepatology. A textbook of liver disease, 3rd edition. Philadelphia: Saunders

Morris CM, Candy JM, Bloxham CA, Edwardson JA (1992a) Distribution of transferrin receptors in relation to cytochrome oxidase activity in the human spinal cord, lower brain stem and cerebellum. J Neurol Sci 111:158-172

Morris CM, Keith AB, Edwardson JA, Pullen RGL (1992b) Uptake and distribution of iron and transferrin in the adult rat brain. J Neurochem 59:300-306

Roberts RL, Fine RE, Sandra A (1993) Receptor-mediated endocytosis of transferrin at the blood-brain barrier. J Cell Sci 104:521-532

Taylor EM, Crowe A, Morgan EH (1991) Transferrin and iron uptake by the brain: Effects of altered iron status. J Neurochem 57:1584-1592

Wong-Riley M (1979) Changes in the visual system of monocularly sutured or enucleated cats demonstrable with cytochrome oxidase histochemistry. Brain Res 171:11-28

12

VASCULAR PERMEABILITY TO HEMORPHINS IN THE CENTRAL NERVOUS SYSTEM

An Experimental Study Using ^{125}I-Hemorphin-7 in the Rat

H. S. Sharma,[1,2] K. Sanderson,[1] E.-L. Glämsta,[1] Y. Olsson,[2] and F. Nyberg[1]

[1] Department of Pharmaceutical Biosciences
Biomedical Centre, Uppsala University
[2] Laboratory of Neuropathology
University Hospital
S-751 85 Uppsala, Sweden

SUMMARY

Blood-brain barrier (BBB) permeability to hemorphin-7, an endogenous morphine like peptide derived from haemoglobin, is unknown. The present investigation was undertaken to examine the microvascular permeability to ^{125}I-labelled hemorphin-7 in different regions of the brain and spinal cord in normal male rats. In addition, the influence of hemorphin on brain water content and morphology of cerebral microvessels was examined at ultrastructural level. Rats received ^{125}I-sodium instead of hemorphin served as controls. The BBB permeability to ^{125}I-labelled hemorphin-7 was significantly higher in various brain and spinal cord regions at 3 min (35-70 %) and 15 min (150-170 %) after administration compared to ^{125}I-sodium. On the other hand, the microvascular permeability to another form of hemorphin, Leu-Val-Val-hemorphin-7 was very close to that of radioactive iodine at both 3 or 15 min circulation period. Infusion of hemorphin-7 however, did not influence the regional brain water content or ultrastructure of the cerebral microvessels compared to either radioactive iodine or Leu-Val-Val-hemorphin-7. These results suggest that hemorphin-7 has the capacity to cross the BBB of normal rats without affecting brain edema formation or cerebrovascular ultrastructure.

RÉSUMÉ

L'hémorphine 7 est un peptide dérivé de l'hémoglobine, de la famille de la morphine. La perméabilité de la barrière hémato-encéphalique (BHE) de ce peptide n'est pas connue et cette étude avait pour but d'évaluer la perméabilité microvasculaire de l'hémorphine 7

Biology and Physiology of the Blood–Brain Barrier, edited by Couraud and Scherman
Plenum Press, New York, 1996

marquée à l'iode 125 dans différentes régions du cerveau et de la moëlle épinière chez des rats normaux males. De plus, les effets de l'hémorphine sur la régulation hydrique du cerveau ainsi que sur la morphologie des microvaisseaux a été étudiée. La perméabilité de la BHE du peptide marqué est supérieure dans de nombreuses régions du cerveau et de la moëlle épinière en comparaison d'iode 125 utilisé comme temoin : 35 à 70%, 3 min après l'administration du peptide et 150 a 170% après 15 min. Par contre, la perméabilité microvasculaire du peptide leu-val-val-hémorphine 7 est très proche de celle de l'iode 125 témoin après les mêmes temps de circulation. Toutefois, l'administration de l'hémorphine 7 ne semble pas provoquer de changements de l'équilibre hydrique du cerveau ni de modifications de l'ultrastructure des microvaisseaux cérébraux par rapport aux témoins, iode 125 ou leu-val-val-hémorphine 7. Ces résultats suggèrent que l'hémorphine peut traverser la BHE de rats normaux sans provoquer d'oedème cérébral ni de modification de l'ultrastructure microvasculaire

INTRODUCTION

The hemorphins are endogenous opioid peptides derived from haemoglobin[2]. The physiological function of these peptides is not known in all details. An increased plasma level of hemorphin-7 is found in marathon runners[3]. Patients with cerebrovascular bleeding show an increased level of the peptide in their cerebrospinal fluid[4]. The functional significance of such findings are still unclear. Since binding affinity of hemorphin-7 for brain opioid receptors has recently been demonstrated[2,4], a possibility exists that the peptide has some central effects. However, the passage of hemorphin across the blood-brain barrier (BBB) in normal conditions is still unknown[1]. The present investigation was undertaken to investigate the vascular permeability of hemorphin in normal anaesthetised rats. In addition, the influence of hemorphin on brain water content and ultrastructure was also examined.

MATERIALS AND METHODS

Animals

Experiments were carried out on inbred Sprague Dawley male rats (Alab, Stockholm) housed at controlled ambient temperature (21±1° C) with a 12 h light and 12 h dark schedule. Food and tap water were provided *ad libitum*.

Blood-Brain Barrier Permeability to Hemorphin

The blood-brain barrier (BBB) permeability to hemorphin was examined using radiolabelled hemorphin-7. For this purpose, radio iodinated hemorphin-7 (^{125}I-hemorphin) (about 10^6 CPM) was administered into the right femoral artery[5,7]. The tracer was allowed to circulate for 3 min (n = 5) or 15 min (n = 5). To explore the influence of its physicochemical properties at the site of the BBB, ^{125}I-Leu-Val-Val-hemorphin-7 (LVV-hemorphin) was injected as a tracer in a separate group of rats (n = 10). Immediately before sacrifice, intravascular tracer was washed out by a brief saline rinse followed by perfusion with 4 % paraformaldehyde, 0.5 % glutaraldehyde in 0.1 M sodium-potassium phosphate buffer (pH 7.4) containing lanthanum (2.5 %) at room temperature[6]. Radioactivity in selected tissue pieces from the brain and spinal cord was determined in a gamma counter. After counting the radioactivity, samples were placed in an oven maintained at 90° C to determine their dry weight. Some pieces were embedded in epon and processed for transmission electron microscopy using standard procedures[5,6]. Control rats (n = 10) received ^{125}I-sodium as tracer instead of hemorphin.

Statistical Evaluation

ANOVA followed by Dunnet's test for multiple group comparison was used to evaluate statistical significance of the data obtained. A p- value less than 0.05 was considered to be significant.

RESULTS

Blood-Brain Barrier Permeability to Hemorphin

The BBB permeability to radioactive iodine is very similar in the brain and spinal cord samples taken from the rats either at 3 min or 15 min after the injection. However, a significant progressive increase in the extravasation of hemorphin was noted in various regions of the brain and spinal cord when the tracer was allowed to circulate for 3 min (35-50 %) and 15 min (150-170 %) compared to iodine (Fig 1a). On the other hand, passage of LVV-hemorphin was similar to iodine at both 3 min and 15 min after injection into the blood stream.

Brain Water Content

There was no significant difference in the regional water content of the brain and spinal cord in samples taken from either iodine, hemorphin or LVV-hemorphin infused rats at 3 min or 15 min after injection. Regional brain water content of iodine, hemorphin and LVV-hemorphin infused rat with 15 min survival period are shown in Fig 1b. The regional brain water content did not differ in either group compared to control.

Ultrastructural Studies

Morphological investigation showed normal brain and perivascular structures in hemorphin, LVV-hemorphin or iodine infused rats. Thus no signs of edema, membrane damage or neuronal distortion can be seen in either groups of rats. A representative example of one microvessel from hemorphin-7 infused rat is shown in Fig 2. The microvessel and surrounding neuropil is quite normal in appearance.

DISCUSSION

The salient new findings of our study is that hemorphin-7 can cross the normal BBB when allowed to circulate for 3-15 min. This increased passage of hemorphin across the BBB is time dependent. The detailed molecular mechanisms of hemorphin transport across the BBB is not clear from this study. However, the passage of LVVV-hemorphin at the BBB is quite similar to that of iodine. This indicates that molecular size and structure appears to play an important role in tracer transport across the BBB[5,7].

This study further show that this increase in hemorphin permeability, however, was not associated with alteration in cerebrovascular ultrastructure or water content. This

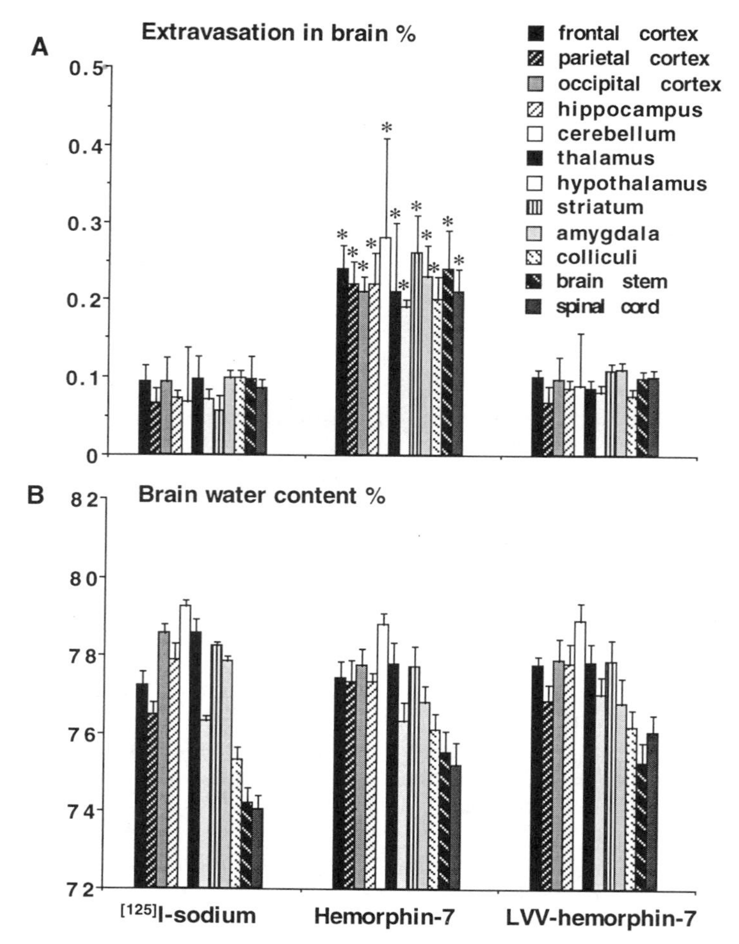

Figure 1. Regional blood-brain barrier permeability (A) and brain water content (B) in rats infused with either radioactive iodine (n = 5), hemorphin-7 (n = 6) or Leu-Vel-Val-hemorphin-7 (n = 6). The parameters were measured 15 min after injection of either tracer. * = P <0.01, ANOVA followed by Dunnet's test for multiple group comparison. Each column represents mean and bar over the column depicts standard deviation.

indicates that increased permeability to hemorphin-7 is not influencing brain structure and function.

Previous investigations from our laboratory suggests that specific binding sites of hemorphin in the CNS are present which is very closely associated with μ opioid receptors[3]. In present study the magnitude of permeability increase of hemorphin does not correlate with the density of binding sites to μ opioid receptors distribution in different brain regions. This indicates that passage of hemorphin across the BBB is not dependent on the hemorphin binding sites in the brain and spinal cord. It may be that hemorphin binding sites are present on the cerebral endothelium. Thus, a possibility exists that hemorphin-7 can across the cerebral endothelium in certain brain regions via a receptor mediated mechanisms, a feature which requires additional study.

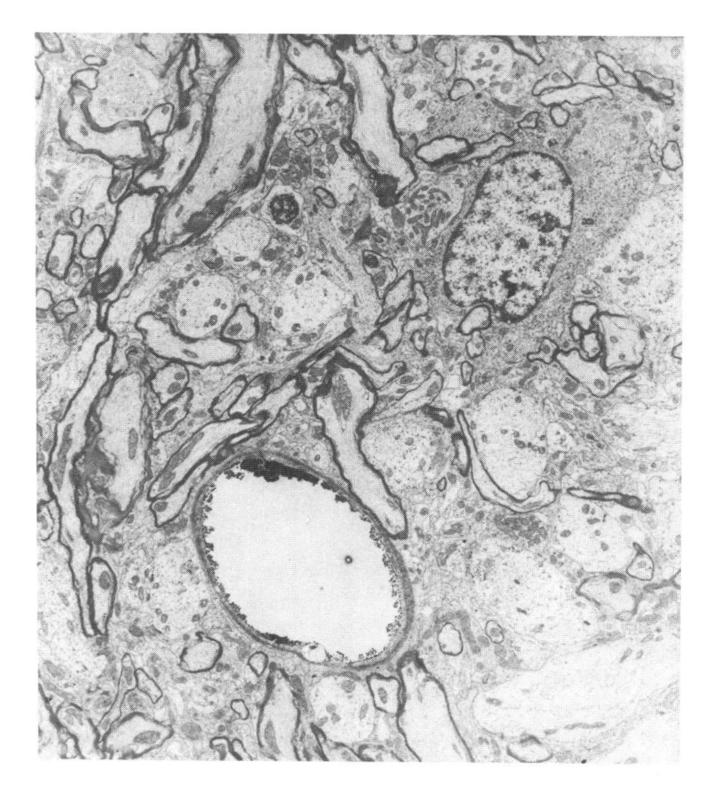

Figure 2. Low power electron micrograph of one microvessel from dorsal horn of the cervical spinal cord in a rat received hemorphin-7. The rat was killed 15 min after hemorphin injection. The microvessel and surrounding neuropil appears normal in structure and no sign of perivascular edema or cell changes can be seen. Bar = 1 μm.

ACKNOWLEDGMENTS

This study is supported by grants from Swedish Medical Research Council project nos. 2710, 9459, 9710 and 10 523, and Göran Gustafsson Foundation, Sweden. The skilful technical assistance of Kärstin Flink, Ingmarie Olsson, Madeleine Thörnwall Madeleine Jarild and Gunilla Tibling is highly appreciated.

REFERENCES

1. Black K L (1995) Biochemical opening of the blood-brain barrier. Adv Drug Del Rev 15: 37-52.
2. Glämsta, E-L, Marklund A, Hellman U, Wernstedt C, Terenius L, Nyberg F (1991) Isolation and characterization of haemoglobin-derived opioid peptide from the human pituitary gland. Reg Pept 34: 169-179.
3. Glämsta, E-L, Mørkrid L, Lantz I, Nyberg F (1993) Concomitant increase in blood plasma levels of immunoreactive hemorphin-7 and β-endorphin following long distance running. Reg Pept 49: 9-18.
4. Glämsta E-L, Meyerson B, Silberring J, Terenius L, Nyberg F (1992) Isolation of a haemoglobin-derived peptide from cerebrospinal fluid of patients with cerebrovascular bleeding. Biochem Biophys Res Commun 184: 1060-1066.

5. Sharma H S, Olsson Y, Persson S, Nyberg F (1995) Trauma induced opening of the blood-spinal cord barrier is reduced by indomethacin, an inhibitor of prostaglandin biosynthesis. An experimental study in the rat using Evans blue, ^{131}I-sodium and lanthanum as tracers. Restor Neurol Neurosci 7: 207-215.
6. Sharma H S, Cervós-Navarro J (1990) Brain oedema and cellular changes induced by acute heat stress in young rats. Acta Neurochir (Wien) Suppl 51: 383-386.
7. Sharma H S, Olsson Y, Dey P K (1990) Blood-brain barrier permeability and cerebral blood flow following elevation of circulating serotonin level in the anaesthetized rats. Brain Res 517: 215-223.

13

THE EFFECT OF GLYCOSYLATION ON THE UPTAKE OF AN ENKEPHALIN ANALOGUE INTO THE CENTRAL NERVOUS SYSTEM

Sarah A. Williams,[1] Thomas J. Abbruscato,[1] Lajos Szabo,[2] Robin Polt,[2] Victor Hruby,[2] and Thomas P. Davis[1]

[1] Department of Pharmacology
[2] Department of Chemistry
University of Arizona
Tucson, Arizona 85724

1. SUMMARY

In contrast to unglycosylated controls, glycosylated [D-Cys^{2,5}]enkephalin-ser-gly (glycosylated DCDCE-ser-gly) elicits analgesia after intraperitoneal administration. This was postulated to be due to the presence of the glucose moiety allowing the analogue to cross the BBB via the glucose carrier. To test this hypothesis, the present study investigated the biological stability and the CNS uptake of unglycosylated and glycosylated DCDCE-ser-gly. Interestingly, the metabolic half-lives and ability to cross the *in vitro* BBB was found to be similar for both analogues. *In situ* brain perfusion indicated that the brain uptake of glycosylated DCDCE-ser-gly was greater than that for the vascular marker, [^{14}C]sucrose, but similar to the CSF uptake of the peptide. CNS uptake of glycosylated DCDCE-ser-gly was not affected by replacing D- with L- glucose, nor with the addition of 10 μM unlabelled glycosylated DCDCE-ser-gly. In summary, the difference in analgesic response of glycosylated compared to unglycosylated DCDCE-ser-gly, is not related to either differing metabolic profiles, nor the ability of the glycosylated analogue to use the glucose carrier to enter the CNS. However, this study does not eliminate the involvement of a different low affinity, saturable uptake system taking the glycosylated, but not the unglycosylated form.

1. RÉSUMÉ

Après administration par voie intrapéritonéale, le peptide glycosylé (D-cys 2,5)enképhaline-ser-gly (DCDCE-ser-gly) montre une activité analgésique, contrairement au même peptide non-glycosylé. L'effet pharmacologique spécifique de l'enképhaline glycosylée pourrait être expliqué par l'addition du radical glycosylé qui permettrait àl'enképhaline de traverser la barrière hémato-encéphalique (BHE) en utilisant le transporteur du glucose. Afin de vérifier cette hypothèse, la stabilité ainsi que la pénétration cérébrale des molécules glycosylées et non glycosylées ont été étudiées. La demi-vie métabolique et la perméabilité de la BHE pour ces deux composés est semblable. Des études de microdialyse dans le cerveau montrent que la capture cérébrale du dérivé glycosylé est superieure à celle du (14 C)sucrose, mais similaire àla capture du peptide par le liquide cérébro-spinal. De plus, la capture cérébrale de la molécule glycosylée marquée n'est pas inhibée par le remplacement du D-glucose par du L-glucose ni par la dilution de la molécule marquée avec 10mM d'enképhaline glycosylée non marquée. En conséquence, les différences de réponse analgèsique observées entre la DCDCE glycosylée et non glycosylée ne sont pas liées à une plus grande stabilité de la molécule glycosylée, ni au transport spécifique de ce composé par le transporteur du glucose de la BHE. Cependant cette étude n'écarte pas la possibilité du transport specifique du dérivé glycosylé par un autre système de transport saturable à faible affinité.

2. INTRODUCTION

The L-serinyl β-D-glucoside analogues of [Met^5]enkephalin were developed by our research group, in an attempt to design opioid-receptor selective drugs, which elicited favourable pharmacological effects such as analgesia [11]. As shown in Table 1, when glycosylated [D-Cys^{2,5}]enkephalin-ser-gly (glycosylated DCDCE-ser-gly) was administered intraperitoneally, a centrally-mediated analgesia was produced in contrast to the unglycosylated control. Since both analogues were found to bind to the μ- and δ- opioid receptors, Polt et al. [11] postulated that this difference was related to the presence of the glucose moiety allowing the glycosylated analogue to cross the blood-brain barrier (BBB) via the glucose carrier. Previous studies have indicated that attachment of a D-glucose does increase the ability of proteins and peptides to cross the blood-brain and blood-nerve barriers [10].

Glucose is the primary energy substrate of the brain, so it is not surprising that it is rapidly transported across the cerebral capillary endothelium by a carrier-mediated process [9]. Seven different isoforms (GLUT 1-7) of the hexose transporter have been characterised in mammalian cells, with GLUTs 1, 3 and 5 being identified in brain [1]. GLUT-1 is detected in two forms, either as a 55 kDa protein, which attains one of its highest expressions in brain

Table 1. Table summarizing the peptide structure and results obtained from the antinociception studies for unglycosylated and glycosylated DCDCE-ser-gly.

PEPTIDE	**Analgesic activity after i.p. administration** ***(Polt et al., 1994)***	**PEPTIDE STRUCTURE**
DCDCE-ser-gly	No significant analgesic activity	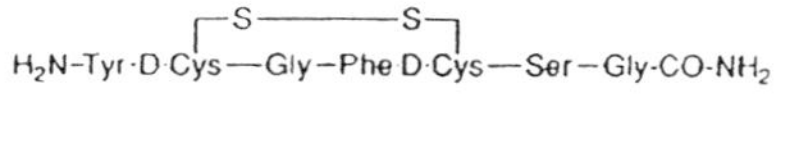 H_2N-Tyr·D·Cys—Gly—Phe D·Cys—Ser—Gly·CO·NH_2 (S—S bridge between the two Cys)
glycosylated DCDCE-ser-gly	Long-lasting analgesia	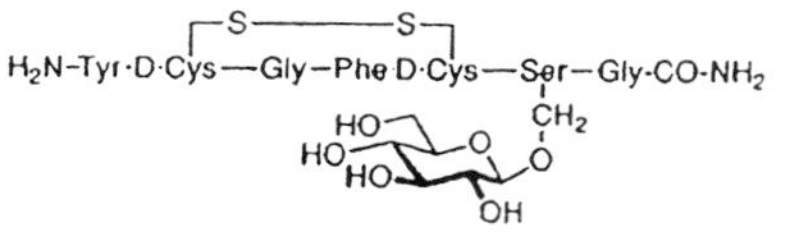 H_2N-Tyr·D·Cys—Gly—Phe D·Cys—Ser—Gly·CO·NH_2 (S—S bridge between the two Cys; Ser CH_2–O– linked to glucose: HO, HO, HO, O, OH)

capillary endothelium, or as a 45 kDa protein, which is thought to be the transporter in the neuronal/glial membranes [6]. GLUT-1 is also found at a high density in the basolateral membrane of choroidal epithelium [5]. GLUT-3 appears to be expressed in neurones [6], as well as the cerebral capillary endothelium of certain species [4]. GLUT-5 is thought to be primarily a fructose transporter [1] and has been detected in both human cerebrovascular endothelium [7] and cerebral microglial [6].

The aim of the present study was to elucidate if glycosylation of DCDCE improved CNS entry, because it allowed the drug to use these glucose carriers present at the blood-brain and blood-cerebrospinal fluid (CSF) barriers. To ensure that the elicitation of analgesia by glycosylated DCDCE-ser-gly was not due to an enhanced biological half-life, the initial study investigated the stability of both glycosylated and unglycosylated DCDCE-ser-gly in serum and brain. The uptake of these analogues across the blood-brain and blood-CSF barriers was then examined.

3. METHOD

3.1. *In Vitro* Stability Incubations

The stability of unglycosylated and glycosylated DCDCE-ser-gly in mouse brain homogenate and serum was examined over a period of 4 hours, as previously described [11].

3.2. *In Vitro* BBB

The *in vitro* BBB model employs primary cultures of bovine brain microvessel endothelial cells (BMEC) and has been described by Weber et al. [14]. Passage of the test solute across the confluent BMEC monolayers was determined by HPLC analysis of the samples (as described below) and was expressed in the form of a permeability coefficient (PC).

$$PC = X/(A \times t \times CD)$$

where PC is the apparent permeability coefficient in cm/min, X is the amount of substance in moles in the receptor chamber after correction for sampling and paracellular passage based on sucrose levels at time t in minutes, A is the diffusion area (i.e. 0.636 cm^2) and CD is the concentration of substance in the donor chamber in $mol.cm^{-3}$ (CD remains >90% of the initial value over the time of the experiments).

3.3. *In Situ* Brain Perfusion

The *in situ* brain perfusion method was used to characterise the CNS uptake of glycosylated DCDCE-ser-gly [12,17,18]. Adult rats were anaesthetised (sodium pentobarbital; 64.8 $mg.kg^{-1}$) and heparinized (10,000 $U.kg^{-1}$). The carotid arteries were then cannulated with silicone tubing and the jugular veins sectioned. The perfusion medium consisted of a mammalian Ringer (NaCl 117.0 mM; KCl 4.7 mM; $MgSO_4$ ($3H_20$) 0.8 mM; $NaHCO_3$ 24.8 mM; KH_2PO_4 1.2 mM; $CaCl_2$ ($6H_20$) 2.5 mM; D-glucose 10 mM; dextran (M.W. 70,000) 39 $g.l^{-1}$ and bovine serum albumin 1 $g.l^{-1}$), which had been thoroughly oxygenated with 95% O_2, 5% CO_2. It was passed by means of a peristaltic pump (6.2 $ml.min^{-1}$) through a heating coil (37°C), and then filtered and debubbled before entering into the animal. The [^{14}C glycine]glycosylated DCDCE-ser-gly (specific activity 30.4 μCi/μmol; ~247 nM/animal) and sucrose were then introduced into the perfusate for 20 minutes. After the set perfusion period, a cisterna magna CSF sample was taken and the animal decapitated. The brain was then taken for capillary depletion analysis [12,17]. Brain and CSF uptake was expressed as a percentage ratio of tissue to plasma radioactivities and termed the R_{Tissue} % (ml.g^{-1} or ml.ml^{-1}). In a group of experiments, the 10 mM D-glucose in the perfusion medium was replaced with 10 mM L-glucose. The stability and integrity of the label to glycosylated DCDCE-ser-gly in the perfusion medium was analysed by HPLC.

3.4. HPLC Analysis

Samples from both the *in vitro* and *in situ* experiments were analysed using a series 410 HPLC gradient system (Perkin-Elmer, Norwalk, CT). All samples were eluted from a 4.6 x 250 mm Vydac 218TP54 column (Vydac, Hesperia, CA) with a linear gradient of 5-35 % 0.1% trifluoroacetic acid (TFA) in acetonitrile versus 0.1% aqueous TFA over 30 minutes at 1.5 ml/min, 37°C. The *in situ* outflow was then directed to an A200 Flo-One Radioactive Detector (Packard, Tampa Bay, FL), where it was mixed with Flo-Scint 111 (Packard) before passing through the 0.5 ml flow-cell for real time analysis of the radioactive sample.

4. RESULTS

4.1. *In Vitro* Stability

The stability incubations (Table 2) indicated that glycosylated and unglycosylated DCDCE-ser-gly had half-lives in the serum and brain homogenate of approximately 280 and > 500 minutes, respectively.

4.2. *In Vitro* BBB

The permeability coefficients (PC) determined for the passage of unglycosylated and glycosylated DCDCE-ser-gly across the *in vitro* BBB were found to be not statistically different from each other (Table 2). However, while there was no significant difference between the PC values for glycosylated DCDCE-ser-gly and the membrane impermeant marker, [^{14}C]sucrose, the PC value for DCDCE-ser-gly was significantly higher than that for [^{14}C]sucrose.

Table 2. The stability and BBB passage of unglycosylated and glycosylated DCDCE-ser-gly expressed in the form of half-lives and permeability coefficients (PC), respectively

Compound	Half-Lives (min)		PC±S.E.M. (cm/min x 10^{-4}) (n = 4)	Students' t-test Comparison with [^{14}C]Sucrose PC
	Brain	Serum		
[^{14}C]Sucrose (M.W. 342)	–	–	9.68±0.91	–
DCDCE-ser-gly (M.W. 847)	>500	276	15.58 ±0.24	P < 0.05
Glycosylated DCDCE-ser-gly (M.W. 1029)	>500	286	14.34±1.94*	P > 0.05

*P>0.05; the PC value for glycosylated DCDCE-ser-gly is not significantly different than that for DCDCE-ser-gly.

4.3. *In Situ* Brain and CSF Uptake

Figure 1 shows the uptake of radiolabelled glycosylated DCDCE-ser-gly and sucrose into the CNS measured using the *in situ* brain perfusion technique. As can be seen the uptake of glycosylated DCDCE-ser-gly into the brain was significantly greater ($P<0.05$) than that for the vascular marker, sucrose. In addition, after considering vascular space, brain uptake of glycosylated DCDCE-ser-gly was not significantly higher than that for CSF uptake. Furthermore, [^{14}C]glycosylated DCDCE-ser-gly eluted as a single peak in the perfusion medium (Figure 2). Retention times matching each other, as well as the radioactive standard (not shown).

The results obtained after capillary depletion analysis of these *in situ* perfused rat brains show there was no significant difference between the values obtained for $R_{Homogenate}$ and $R_{Supernatant}$ (Table 3). In addition, R_{pellet} values were significantly lower than both $R_{Homogenate}$ and $R_{Supernatant}$.

Furthermore, the uptake of the radiolabelled glycosylated peptide was not significantly affected by replacing D- with L-glucose, nor with the addition of 10 μM glycosylated DCDCE-ser-gly in the perfusion medium (Figure 1). BBB integrity (sucrose space) was not affected by this manipulation of the perfusion medium (data not shown).

5. DISCUSSION

The initial investigation indicated that glycosylated and unglycosylated DCDCE-ser-gly had similar half-lives in brain and serum. (Table 2). A metabolic half-life of > 7 hours

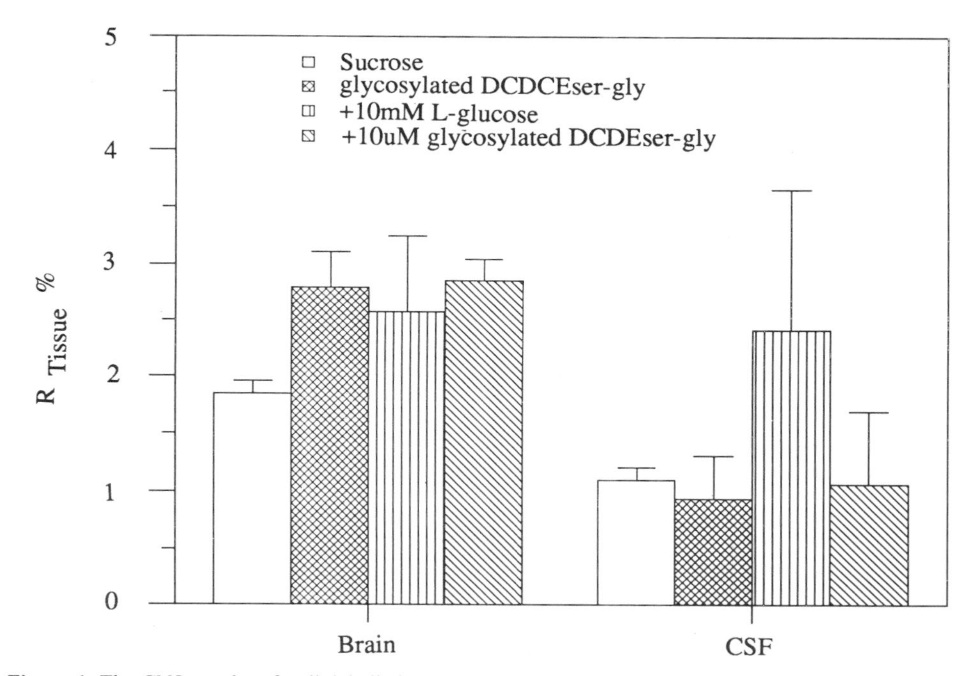

Figure 1. The CNS uptake of radiolabelled sucrose and glycosylated DCDCE-ser-gly under varying conditions. Uptake is expressed as a percentage ratio of tissue to perfusion medium radioactivities (R_{Tissue} % ± S.E.M. for 3-7 animals; $ml.g^{-1}$ or $ml.ml^{-1}$). Perfusion time 20 minutes.

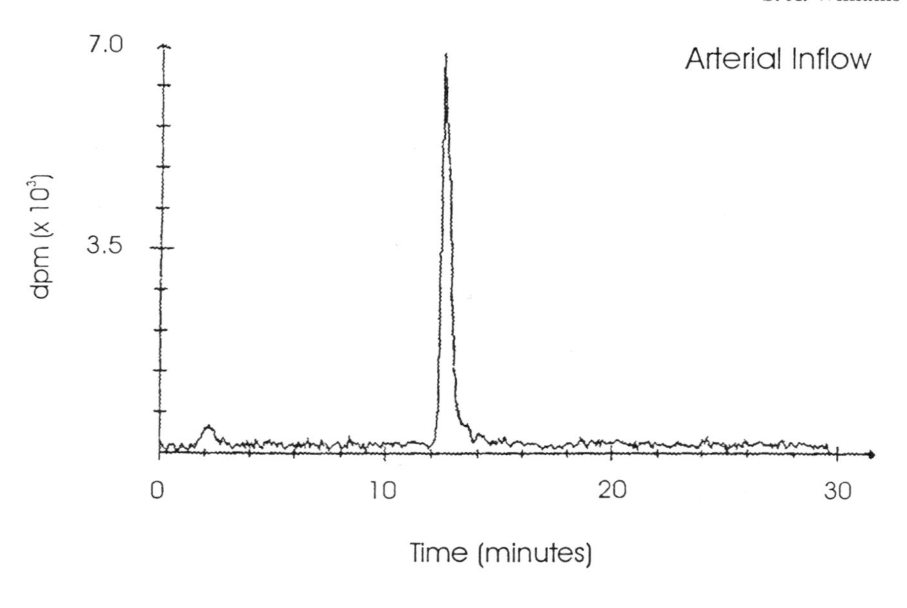

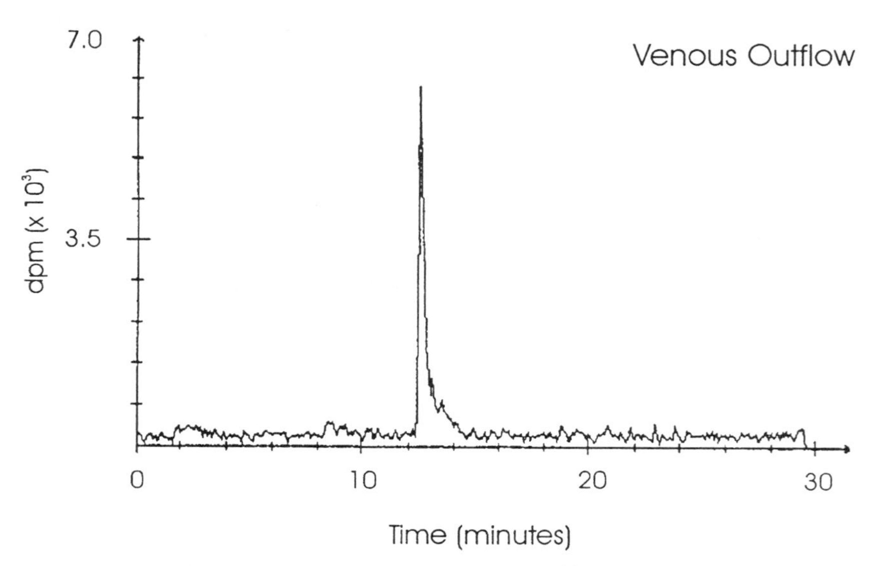

Figure 2. HPLC/Flo-One radioactive detector chromatograms of [^{14}C]glycosylated DCDCE-ser-gly in the perfusion medium before (arterial inflow) and after (venous outflow) it had passed through the cerebral circulation.

has previously been reported for glycosylated DCDCE-ser-gly in brain [11]. These results confirm that differing metabolic profiles were not responsible for the contrasting analgesic effects of glycosylated DCDCE-ser-gly compared to unglycosylated DCDCE-ser-gly (Table 1). Furthermore, both analogues have long metabolic half-lives in comparison to endogenous enkephalins, which have serum half-lives of a few minutes. This resistance to enzymatic degradation can be related to the disulphide bridge [15] and the presence of D-cysteine, instead of L-glycine and L-methionine at the 2 and 5 position, respectively [3].

Table 3. Table showing the amount of radiolabelled glycosylated DCDCE-ser-gly found accumulated in the capillary endothelial cells (R_{Pellet}) after the capillary depletion technique was applied to the *in situ* perfused brains

R_{Tissue}	$[^{14}C]$Glycosylated DCDCE-ser-gly % ± S.E.M. (n = 4)	Students t-test Comparison with $R_{Homogenate}$
$R_{Homogenate}$	2.60±0.34	–
$R_{Supernatant}$	2.46±0.34	P>0.05
R_{Pellet}	0.81±0.47	P<0.01

The passage of DCDCE-ser-gly and glycosylated DCDCE-ser-gly into the CNS, was compared by means of the *in vitro* BBB model. There was no significant difference between the PC values for the glycosylated analogue and the membrane impermeant marker molecule, $[^{14}C]$sucrose (Table 2). Although, the unglycosylated form had a PC value which was significantly greater than that for $[^{14}C]$sucrose, further statistical analysis revealed that the PC value for DCDCE-ser-gly was not significantly different to that obtained for glycosylated DCDCE-ser-gly. Thus, these data suggest that enhanced passage across the BBB of the glycosylated analogue is not responsible for its analgesic effect (Table 1). However, these results may be due to the absence of astrocytic factors in this BBB model, which induce glucose transporter gene expression [8,16]. Therefore, if glycosylated DCDCE-ser-gly does use the glucose carrier, its PC may have been decreased and similar to DCDCE-ser-gly, due to the diminished presence of this transporter and the decrease in lipophilicity [11], increase in molecular size and number of hydrogen bonds, associated with the attachment of glucose.

The CNS entry of glycosylated DCDCE-ser-gly was further characterised by means of the *in situ* brain perfusion technique. As can be seen in Figure 1, the brain uptake of the glycosylated enkephalin is significantly greater than the uptake of the vascular marker, sucrose. In addition, after considering vascular space the brain uptake of $[^{14}C]$glycosylated DCDCE-ser-gly is significantly greater than its CSF uptake. Capillary depletion analysis revealed that there was little accumulation of $[^{14}C]$glycosylated DCDCE-ser-gly within the endothelium-enriched pellet and that the levels of radioactivity that were detected in the homogenate and supernatant were similar (Table 3). This indicates that the bulk of $[^{14}C]$glycosylated DCDCE-ser-gly had actually traversed the BBB and entered the CNS. Figure 2 confirms the inherent stability of glycosylated DCDCE-ser-gly (Table 2), as well as the integrity of the $[^{14}C]$label to this peptide analogue in the perfusion medium, before and after it had passed through the cerebral circulation. Together, these results suggest that there is a significant CNS uptake of $[^{14}C]$glycosylated DCDCE-ser-gly in the rat. Considering the smaller surface area of the choroid plexuses compared to the BBB, as well as the fact that the CSF is more likely to act as a sink to the brain, than the brain acting as a sink to the CSF [2], then these results would suggest that $[^{14}C]$glycosylated DCDCE-ser-gly enters the brain predominately through the BBB and the blood-CSF barrier plays only a minor role.

A permeability constant (cm/min) can be determined from this *in situ* brain perfusion data, by assuming a cerebrovascular surface area in the rat of $100\ cm^2.g^{-1}$ and that influx of $[^{14}C]$glycosylated DCDCE-ser-gly into the brain from the perfusion medium is greater than backflux [18]. Correlations between this *in situ* ($1.40 \pm 0.16 \times 10^{-5}$) and *in vitro* ($1.43 \pm 0.19 \times 10^{-3}$; Table 2) data revealed a 102-fold difference, which is similar to that previously observed by Pardridge et al. ([8]; 150-fold). This greater permeability of the *in vitro* model is thought to be partly explained by the absence of astrocytic factors attributing to the dedifferentiation observed in cultured brain endothelial cells [8,13].

It is generally accepted that the glucose carrier is stereospecific, carrying D-, but not L- glucose [9]. By replacing the D- with L- glucose in the *in situ* perfusion medium, any competition against [^{14}C]glycosylated DCDCE-ser-gly for the glucose carrier has been removed. Figure 1 shows that there was no significant difference in the brain uptake of [^{14}C]glycosylated DCDCE-ser-gly in the presence of 10 mM D- or L- glucose. Although, the CSF uptake of [^{14}C]glycosylated DCDCE-ser-gly was higher in the presence of the L-enantiomer, this was found not to be statistically significant and may be related to the difficulties in obtaining clear CSF samples. These results indicate that [^{14}C]glycosylated DCDCE-ser-gly does not use the glucose carrier to enter the CNS of the anaesthetised rat.

If one considers the approximately 100-fold greater permeability of the *in vitro* versus the *in situ* BBB, it is unusual that glycosylated DCDCE-ser-gly crossed the *in situ*, but not the *in vitro* BBB. Initially, it was hypothesized that glycosylated DCDCE-ser-gly failed to cross the *in vitro* BBB because of the absence of specific astrocytic factors, which induce glucose transporter gene expression [16]. Figure 1 has now shown that the absence of the glucose transporter at the *in vitro* BBB is not responsible for the results in Table 2. However, it still may be related to the loss of other BBB-specific proteins, associated with primary cultures of endothelial cells [8], even though self-inhibition experiments indicated that a high affinity saturable uptake system is not involved (Figure 1).

In summary, the difference in analgesic response of glycosylated compared to unglycosylated DCDCE-ser-gly, is not related to either differing metabolic profiles, nor the ability of the glycosylated analogue to use a glucose carrier to enter the CNS. However, this study does not eliminate the involvement of a different low affinity, saturable uptake system taking the glycosylated, but not the unglycosylated form into the CNS.

6. ACKNOWLEDGMENTS

This work was supported by N.I.H. grant DA 06284.

7. REFERENCES

1. Bell G. I., Burant C. F., Takeda J. and Gould G. W. (1993) Structure and function of mammalian facilitative sugar transporters. *J. Biol. Chem.* **268,** 19161-19164.
2. Davson H., Kleeman C. R. and Levin E. (1961) Blood-brain barrier and extracellular space. *J. Physiol.* **159,** 67P-68P.
3. Dooley C. T., Chung N. N., Wilkes B. C., Schiller P. W., Bidlack J. M., Pasternak G. W. and Houghten R. A. (1994) An all D-amino acid opioid peptide with central analgesic activity from a combinatorial library. *Science.* **266,** 2019-2022.
4. Gerhart D. Z., Broderius M. A., Borson N. D. and Drewes L. R. (1992) Neurons and microvessels express the brain glucose transporter protein GLUT-3. *Proc. Natl. Acad. Sci.* **89,** 733-737.
5. Harik S. I., Hall A. K. and Perry G. (1992) Ontogeny of glucose transporter in the rat central nervous system. *Neurology.* **42,** 407.
6. Maher F., Vannucci S. J. and Simpson I. A. (1994) Glucose transporter proteins in brain. *FASEB J.* **8,** 1003-1011.
7. Mantych G. J., James D. E. and Devaskar S. U. (1993) Jejunal/kidney glucose transporter isoform (Glut-5) is expressed in the human blood-brain barrier. *Endocrinology.* **132,** 35-40.
8. Pardridge W. M., Triguero D., Yang J. and Cancilla P. A. (1990) Comparison of *in vitro* and *in vivo* models of drug transcytosis through the blood-brain barrier. *J. Pharmacol. Exp. Ther.* **253,** 884-891.
9. Pelligrino D. A., LaManna J. C., Duckrow R. B., Bryan Jr. R. M. and Harik S. I. (1992) Hyperglycemia and blood-brain barrier glucose transport. *J. Cerebr. Blood Flow and Metab.* **12,** 887-899.
10. Poduslo J. F. and Curran G. L. (1994) Glycation increases the permeability of proteins across the blood-nerve and blood-brain barriers. *Mol. Brain Res.* **23,** 157-162.

11. Polt, R., Porreca, F., Szabò, L. Z., Bilsky E. J., Davis, P., Abbruscato, T. J., Davis, T. P., Horvath, R., Yamamura, H.I. and Hruby, V.J. (1994). Glycopeptide enkephalin analogues produce analgesia in mice: Evidence for penetration of the blood-brain barrier. *Proc. Natl. Acad. Sci. USA*. **91,** 7114-7118.
12. Preston J. E., Al-Sarraf H. and Segal M. B. (1995) Permeability of the developing blood-brain barrier to ^{14}C-mannitol using the rat *in situ* brain perfusion technique. *Dev. Brain Res.* **87,** 69-76.
13. Raub T. J., Kuentzel S. L. and Sawada G. A. (1992) Permeability of bovine brain microvessel endothelial cells *in vitro*: barrier tightening by a factor released from astroglioma cells. *Exp. Cell Res.* **199,** 330-340.
14. Weber S. J., Abbruscato T. J., Brownson E. A., Lipkowski A. W., Polt R., Misicka A., Haaseth R. C., Bartosz H., Hruby V. J. and Davis T. P. (1993) Assessment of an *in vitro* blood-brain barrier model using several [Met^5]enkephalin opioid analogs. *J. Pharmacol. Exp. Ther.* **266,** 1649-1655.
15. Weber S. J., Greene D. L., Hruby V. J., Yamamura H. I., Porreca F. and Davis T. P. (1992) Whole body and brain distribution of [^{3}H]Cyclic[D-Pen^2,D-Pen^5]enkephalin after intraperitoneal, intravenous, oral and subcutaneous administration. *J. Pharmacol. Exp. Ther.* **263,** 1308-1316.
16. Weiler-Güttler H., Zinke H., Möckel B., Frey A. and Gassen H. G. (1989) cDNA cloning and sequence analysis of the glucose transporter from porcine blood-brain barrier. *Biol. Chem. Hoppe-Seyler* **370,** 467-473.
17. Williams S. A., Abbruscato T. J., Hruby V. J. and Davis T. P. (1995) The passage of a *delta*-opioid receptor selective enkephalin, DPDPE, across the blood-brain and blood-cerebrospinal fluid barriers. *J. Neurochem.* (in press).
18. Zlokovic B. V., Begley D. J., Djuricic B. M. and Mitrovic D. M. (1986) Measurement of solute transport across the blood-brain barrier in the perfused guinea-pig brain: method and application to N-methylaminoisobutyric acid. *J. Neurochem.* **46,** 1444-1451.

14

BLOOD-RETINA BARRIER PROPERTIES OF THE PECTEN OCULI OF THE CHICKEN

Holger Gerhardt, Stefan Liebner, and Hartwig Wolburg

Institute of Pathology
University of Tübingen
D-72076 Tübingen, Germany

The site of the blood-brain or blood-retina barrier is the tight junction between endothelial and pigment epithelial cells, respectively. Considerable effort has been provided for understanding the nature of tight junctions and their regulation. It has become clear that morphology and physiology of blood brain barrier endothelium heavily alters if cultured in vitro[1] and that, as a result, blood brain barrier properties are highly dependent on the brain microenvironment. Therefore, we looked for a new model-system, preferably posessing the combined validity of the in vivo situation and the simplicity of an in vitro system. We found it in the pecten oculi of the avian eye, which is a pleated vascular structure, protruding from the optic nerve head into the vitreous (for a review, see [2]). Although the pecten is well-known for a long time, its function is still a matter of controversal speculation. Regulation of the pH-value, of the temperature and of the inner bulb pressure of the eye, as well as support in movement detection and nutrition of the retina have been considered.

The pecten contains mainly two types of cells: pigmented glial cells and endothelial cells. Originating in the retinal pigment epithelium (RPE), the pigmented cells migrate during development into the cleft between the verges of the optic fissure, giving rise to a highly proliferative accumulation of undifferentiated glial cells, i.e. the pecten primordium. Throughout lifetime the basal portions of the pecten seem to be continously fed by migrating RPE cells. Prior to the migration, in terms of the RPE, these cells already express barrier properties such as complex tight junctions associated to the protoplasmic fracture face (P-face), the glucose transporter isoform GluT-1, the barrier-specific antigen HT7 and the impermeability to lanthanum. But at the site of their destination, i.e.the pecten, they lose these properties step by step. Primarily, the pigmented cells form a border between vitreous and blood vessels; later on, they retract, assuming a position between the blood vessels which - as a result - gain direct contact to the vitreous. The endothelial cells invade the pecten primordium from the choroidal plexus early in development. Whereas the choroidal vessels later on become fenestrated, the endothelium of the pecten gains barrier properties. This furthermore supports the idea that environmental factors determine barrier properties of endothelial cells. However, the pecten endothelial cells differ from brain endothelium cells in terms of their shape and arrangement. Whereas in brain capillaries only one endothelial cell forms the whole circumference of the vessel, in the pecten frequently more than four

Biology and Physiology of the Blood–Brain Barrier, edited by Couraud and Scherman
Plenum Press, New York, 1996

cells line the capillary lumen. Whereas in brain microvessels the endothelial cells are extremely flat and the contact zones run obliquely to the luminal membrane, the pecten endothelial cells are cuboid cells with a basal labyrinth, lateral interdigitations and apical microfolds. In freeze-fracture replicas, the tight junctions were found to be highly complex and mainly associated with the P-face.

Immunofluorescence- and Immunogold-labeling experiments revealed highly specific immunoreactivities for both anti-GluT-1 and anti-HT7 at all endothelial membranes. Here we found another difference between capillary endothelial cells of brain and pecten: in brain microvessels the GluT-1 glucose transporter is distributed asymmetrically at abluminal and luminal membranes [3,4]. In pecten endothelium we were not able to detect any asymmetry of labeling comparing the apical and basal membranes. In contrast to brain endothelium, pecten endothelium did not show any labeling in the cytoplasmic compartment. Double labeling experiments using a haptenized form of the anti-GluT-1 and different sizes of immunogold particles showed about 24% of the HT7 antigen directly colocalized to GluT-1. This could be interpreted as a first hint for functional interactions of both molecules. Pigmented cells were devoid of any labeling.

The unusual structure of the pecten endothelial cells could possibly elucidate the complex requirements which have to be satisfied by the pecten. Nevertheless, our description of the blood-brain barrier properties does not exhibit any requirements, which could explain the structural peculiarity of the pecten endothelial cells. Assuming that the development of cell polarization is highly dependent on the extracellular milieu, the development of a basal labyrinth and apical microvilli in many epithelia is most likely related to exogenous environmental cues. This development is accompanied with the acquisition of tight junctions expressing a PFA beyond 90%. These two characteristics could be related, as they are both found in the endothelial cells of the pecten, which are undoubtedly derived from endothelial cell lineage. As there are two different types of blood vessels found in the Papilla nervi optici, the described pectinate endothelia and normal brain-type endothelia, surrounded by PCs and astrocytes respectively, the induction of the above mentioned properties is most likely dependent on the vincinity of the PCs. Perhaps, the metabolic activities of the pecten allow supply of the retina with oxygen at a rate which would explain the lack of blood vessels in the avian retina. Recently, many studies pointed out the importance of the vascular endothelial growth factor (VEGF) for angiogenesis, in particular the up- (down-) regulation of VEGF in retinal glial cells by hypoxia (hyperoxia)[5]. Indeed, it is conspicuous that the thick and complex avian retina is avascular whereas avascular retinae in mammalian species are very thin and supplied with oxygen from the choroid vasculature. On the other hand, it is unlikely that the oxygen partial pressure alone determines whether or not blood vessels sprout into the retina. Another important factor might be the endowing of growing endothelial cells with adhesion molecules. Preliminary results showed that the expression of different cadherins on the surface of endothelial and pigmented cells altered with developmental stages. This could indicate that the expression pattern of cadherins correlates to the acquisition or loss of barrier properties and the capability to migrate, respectively.

REFERENCES

1. Wolburg H, Neuhaus J, Kniesel U, Krauß B, Schmid E, Öcalan M, Farrell C, Risau W (1994) Modulation of tight junction structure in blood-brain barrier endothelial cells; Effects of tissue culture, second messengers and cocultured astrocytes. J Cell Sci 107:1347-1357
2. Romanoff AL (1960) The organ of special sense. In: Romanoff AL (ed) The Avian Embryo. Macmillan Company, New York. pp 396-404

3. Farrell CL, Pardridge WM (1991) Blood-brain barrier glucose transporter is asymmetrically distributed on brain capillary endothelial lumenal and ablumenal membranes: An electron microscopic immunogold study. Proc Natl Acad Sci USA 88:5779-5783
4. Bolz et al.(1995), in press.
5. Pierce EA, Avery RL, Foley ED, Aiello LP, Smith LEH (1995) Vascular endothelial growth factor/vascular permeability factor expression in a mouse model of retinal neovascularization. Proc Natl Acad Sci USA 92:905-909

THE MICROARCHITECTURE OF CEREBRAL VESSELS

Kenneth R. Davies, Grace Richardson, Wendy Akmentin, Virgil Acuff, and Joseph D. Fenstermacher

Department of Anesthesia
Henry Ford Hospital
Detroit, Michigan 48202
Department of Neurosurgery
State University of New York
Stony Brook, New York 11794

1. INTRODUCTION

The morphology of microvascular systems within the brain is routinely studied in two dimensions using light microscopy or transmission electron microscopy. Limited to height and width, the flat images generated by these techniques lack the depth necessary to visualize the complete architecture of cerebral microvascular beds. To sculpture a "truer" representation, we integrated multiple-sequential, two-dimensional microscopic pictures to reconstruct a three-dimensional image of the microvascular network of the cerebral cortex and hippocampus.

2. BACKGROUND

Examination of cerebral capillary structure was attempted as early as 1939 by Colin, who injected solutions of India ink into cats at pressures as high as 180 mm Hg in order to distend and stain capillary beds. Images of twenty micron thick sections were projected onto blank paper at a magnification of 600x and the length of each capillary was traced. No attempt to determine vessel diameter was made, rather the length of each individual tracing was summated to produce an overall measurement of vascularity. A considerably degree of tortuosity in capillary structure was noted but attributed to an artifact of tissue shrinkage during preparation (Colin, 1939). The least vascular structure found in this early study was the hippocampus which contained only one-third the number of capillaries as the lateral geniculate body.

Fifty years later, the most commonly used method for studying brain microvessels still involves injection of India ink with rapid tissue fixation. Transparent blocks of tissue

Biology and Physiology of the Blood–Brain Barrier, edited by Couraud and Scherman
Plenum Press, New York, 1996

are now rotated under a microscope and viewed on their side. Despite stereoscopic microscopes and rotation of tissue blocks, Duvernoy et. al. (1981) complained that it was impossible to obtain good photographs of vascular structural features due to the great degree of vessel branching. In a new approach aimed at avoiding dehydration and fixation, vessel were injected with methyl methacrylate to obtain casts of cerebral vessels (Duvernoy et al., 1981). The surrounding tissue was then removed by digestion with potassium hydroxide and the vascular casts were viewed under either a stereoscopic microscope or scanning electron microscope. The problem with this method is that after removal of the surrounding supporting tissue, small microvessels become very fragile and subject to easy destruction. Dynamic examination of capillaries using microtransillumination, microcinephotography and stereological analysis of the cerebral cortex was pioneered by Pawlik et. al. in 1981. His methods produced some of the first measurements of capillary length, surface areas and volumes; his comparisons confirmed significant interregional differences.

Nopanitaya, a corrosion-casting compound developed in the middle 1980s, allowed Motti et.al. (1986) to study detailed models of vascular casts under a scanning electron microscope (SEM). Unfortunately, SEM inspection was limited to the surface microvasculature of the cortex, that is, a depth of 300 to 400 μm. Attempts to cut the sections caused obvious disruption of the vascular casts. A similar approach was used by Yoshida and Ikuta (1984) to examine cerebrovascular structure in fetal and adult rats, but no data on parameters such as capillary length, arteriole-capillary branching, and capillary anastomosis were produced.

Capillary circulation has been studied with fluorescent dye labeled erythrocytes (Hudetz et al., 1993). Structures up to 70 μm deep are viewed with an epifluorescence microscopy. Done in living animals, these studies are essential limited to two dimensions since only the surface of the cortex can be studied and overlying large vessels partially obscure the capillaries underneath. Their analysis of the superficial cortex revealed large anastomosing loops of capillaries. A broad variation in the velocities of erythrocytes within these loops was reported, but, as stated by the authors, this may be partly the result of limited depth of field. In later work from this laboratory (Hudetz et al., 1995), it was noted that red cell flow velocity is maintained fairly constant when cerebral perfusion pressure is lowered but red cell flux through the visualized microvessels appeared to be lowered possibly because of a reduced hematocrit.

Heterogeneity of transit-times through cerebral capillaries and related larger vessels was first hypothesized by Hertz and Paulson (1980) to explain the differences in extractional loss of various test substances passing through the cerebral circulation. These indicator dilution measured differences in transit times of various test substances indicate a range of blood flow velocities and/or capillary lengths among local cerebral microvascular systems.

Capillary bed structure has been studied in both the Wistar-Kyoto rat (WKY) and the spontaneously hypertensive rat (SHR) by Lin et al.(1990) and Gesztelyi et al.(1993). The documented cerebral atrophy and hydrocephalus of SHR animals provided a possible basis for differences in vascularity between SHR and normotensive WKY. The two-dimensional morphology of capillaries was evaluated from light micrographs of two μm thick sections by Lin et al. (1990) and from slides directly by a video camera coupled to a computer-driven image analyzing system by Gesztelyi et al.(1993). Variables such as the mean number of capillary profiles, capillary surface area, and luminal (blood) volume were determined in both studies. No appreciable difference existed between either the frequency or size of microscopic (<50 μm) blood vessels when similar structures were compared in WKY and SHR rats. In contrast, large variations in the size and number of capillaries were observed between different structures within the same animal. The frequency of small microvessels (<7.5 μm) varied ten-fold among 21 different brain areas, with white matter containing one-sixth the blood volume of some highly vascular hypothalamic nuclei. Two-dimensional visual analysis could not determine the reason for these dissimilarities in vascularity. In addition to the observations just noted, Lin et al. (1990) reported that capillary diameter was seemingly inversely related to capillary profile frequency, for instance, mean capillary diameter was less in a highly vascular areas than in an areas of low microvascularity. To understand the implications of this finding on the flow of blood cells and plasma through small diameter capillaries *vis-a-vis* larger ones, knowledge of such variables as capillary length and capillary branching, which can only be done with three-dimensional structural analysis, would be invaluable.

3. METHODS

3.1. Animal Preparation for Morphometry

Animal care, surgical preparation and experimental procedures were designed in accordance with federal guidelines developed by the American Association for Accreditation of Laboratory Animal Care (AAALAC). The general procedures have been previously reported (Gesztelyi et al. 1993, Lin et al. 1990, Wei et al. 1992). As described in Gesztelyi et. al. (1993), rats are anesthetized with sodium pentobarbital (40 mg/kg i.p.), rapidly thoracotomized, and immediately heparinized with 500 IU heparin injected into left ventricle. Immediately thereafter, the vascular system is perfused with warm, buffered saline solution via a catheter passed through the left ventricle and into the aorta. The rate of the saline solution infusion is 35 ml/min and the duration of the perfusion is 45 seconds. Past experiments have indicated that this procedure cleared virtually all blood from the vascular system without raising femoral artery pressure above 30 mm Hg. Subsequently, the rats are perfusion fixed by 2% paraformaldehyde and 2% glutaraldehyde in 0.1 M phosphate buffer at the same rate for the next 30 minutes. During the course of the fixative infusion, arterial pressure is monitored and not allowed to rise above 30 mm Hg.

Fixed animals are placed in a refrigerator at 4°C for three hours. The brains are then stored in cold fixative, and cut the next day into a series of 0.2 mm thick coronal blocks by a vibratome. Each sample is washed in four changes of 0.1 M phosphate buffer, postfixed in buffered 1% osmium tetroxide (w/v 0.1 M phosphate buffer), dehydrated in a graded ethanol series and embedded in electron microscopy resin. From each embedded tissue block, a series of 1 μm thick sections are cut with a glass knife, placed on a glass slide, and stained with 1% Toluidine blue.

3.2. Three-Dimensional Morphometry

These tissue sections are viewed under a Nikon Optiphot light microscope. Capillaries appear as white circles if intersected perpendicular to the plane of the tissue, elongated cylinders if transecting the plane of the slide, and other shapes such as ellipses and the letters L and Y. In all cases, the lumen of each is clear and recognizable. The 625x magnification currently in use produces an image with the dimensions of 320 μm by 240 μm; a thickness of one micron per section is determined by microtome sectioning. Each image is scanned by a video camera and saved as a computer file (Fig 1 & 2). Following scanning, a ghostly-image of many shadows is produced for each section and retained as an overlay in order that subsequent pictures may be appropriately aligned for processing. Alignment is necessary since the orientation of each tissue section on the original slide is different. The final output is a series of computer files, each representing a cross-sectional image of the rat's cerebral tissue in a one micron-thick plane. If alignment is correctly done, then the integration of all the computer files will produce an image of an uninterrupted vertical column of brain tissue.

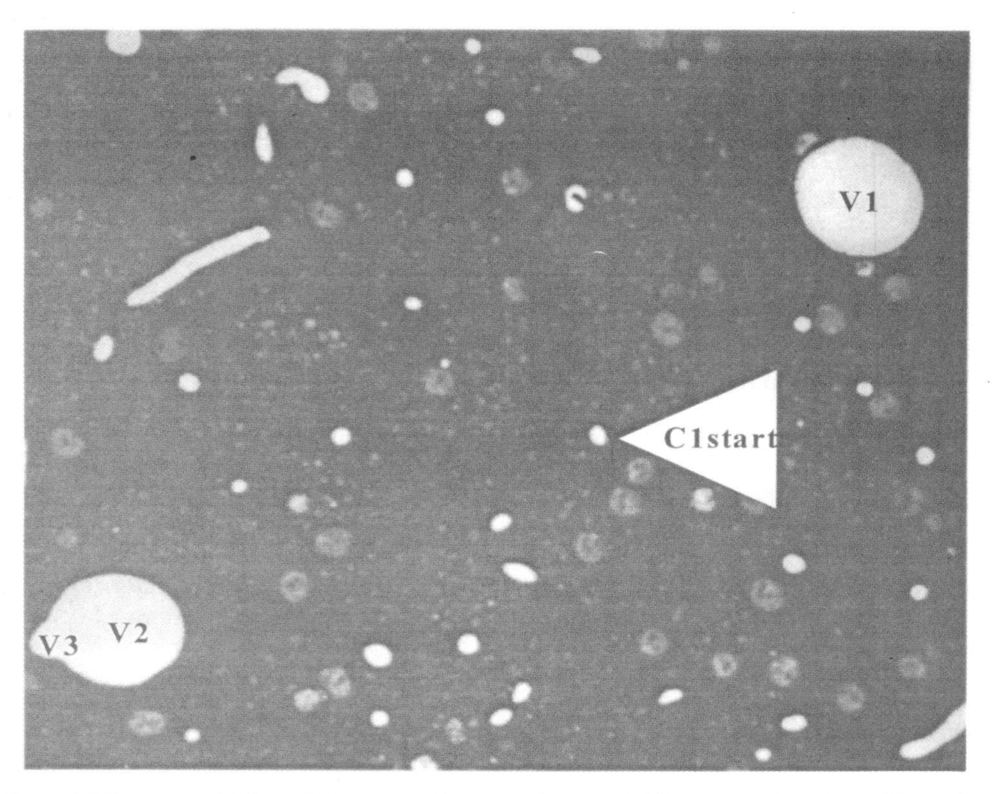

Figure 1. Micrograph of 1.0 μm thick section of rat cerebral cortex at 625x. Image dimension is 320 μm by 240 μm. Two small venules (V1&V2) are located in opposite corners. A third venule (V3) is forming on the left side of V2. The remaining transparent spaces are capillaries transected either lengthwise or on end. The arrow marks the start of an example capillary (C1).

The image processing system developed by *Imaging Research Inc., St. Catharines, Ontario, Canada*, consists of a video camera for scanning micrographs, and a 486DX2 computer with two separate SVGA video boards. One video card is used for displaying and aligning micrographs the other for displaying a composite three-dimensional image. The system can integrate and stack these images and produce a single three-dimensional representation of the sample tissue volume. Full three dimensional structural analysis of the tissue on a two dimensional computer screen requires that three separate views (X-Y plane, X-Z plane and Y-Z plane) are displayed simultaneously (Fig. 3). The system operator must then outline each blood vessel by tracing it's course within the tissue volume with a variable diameter, test circle; movement in any one plane must be confirmed in the other two planes. Analysis of the entire tissue block at one time is not possible due to limitations in current processor memory, necessitating the division of the sample into twelve separate images. Once fully outlined in all three planes, the size and spacial orientation of the vessel is simulated on a graphical display (Fig. 4). This labor intensive procedure is repeated for each vessel within each of the twelve sections until all of the circulatory system is identified and reproduced on the graphical display. Full appreciation of the three dimensional vascular structure is achieved by rotating the graphical image to produce views from each plane.

The number of vessel fragments or segments is counted by the operator; vessel fragment diameters and lengths are automatically determined from each outlined microves-

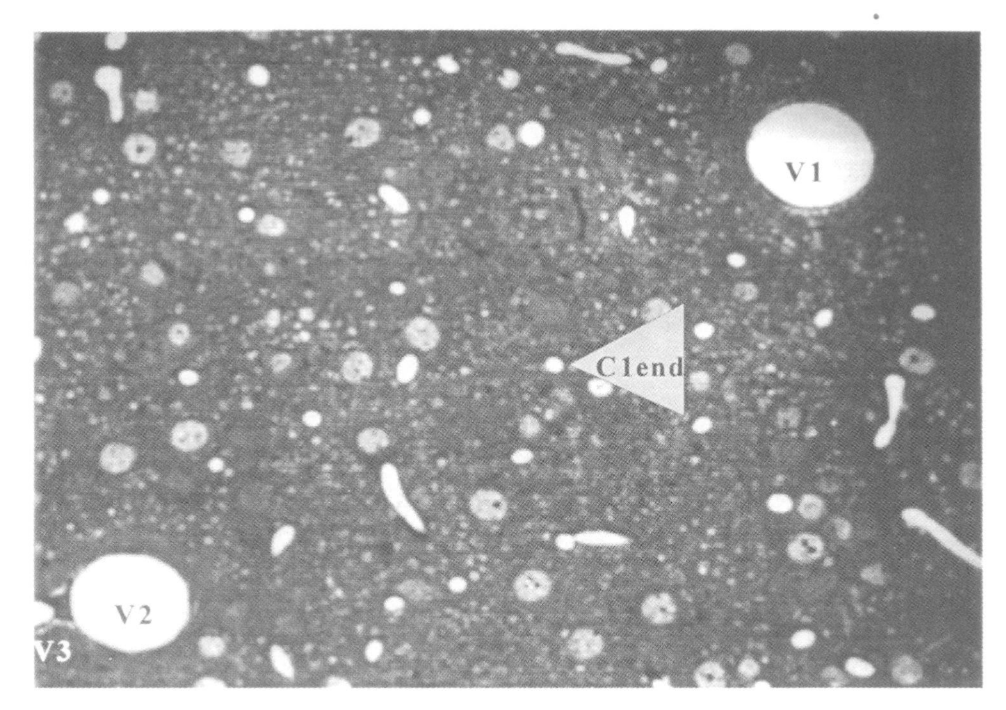

Figure 2. Micrograph of 1.0 μm thick section located 12μm directly below Fig. 1. V1 and V2 are still in approximately the same position. V3 has completely separated from V2 and is moving off the left corner of the picture. C1 has now moved up and to the left.

sel. The product of the latter two values is the vascular volume of an individual vessel. Assuming all blood containing structures are identified in the tissue sample, then total blood volume is measured and can be expressed as an absolute volume in μm^3 and as a percentage of the tissue volume. Vessel fragment lengths, volumes, and surface areas in one tissue volume element (voxel) can be added to those of the same vessel extended into an adjacent voxel.

4. RESULTS

Initial studies included five tissue blocks reconstructed over the last four months. All specimens were obtained from a single Sprague-Dawley (SpD) rat. An entire cerebral cortex was cut into 1500 one micron thick sections, in addition several hundred sections were cut from the hippocampus of the same animal. The first tissue block for vascular reconstruction was the superficial 70 μm of the cerebral cortex. Prior to alignment the tissue is examined under a light microscope in order to find a suitable area for imaging. One or several vessels that appear large on the slide (in actuality a small arteriole or venule) are used as an "anchor" to position and align future sections. In attempting to find and identify such a vessel, the sparsity of arterioles was first noted. The majority of microscopic views at 625x contained only capillaries; venules appeared in the minority of images, and an arteriole was rare. One section of tissue with an average size venule (40 μm diameter) was chosen . This venule was subsequently identified on the next 69 tissue sections and each one positioned under the microscope in the exact same orientation as the first slide. Subsequent integration, manual

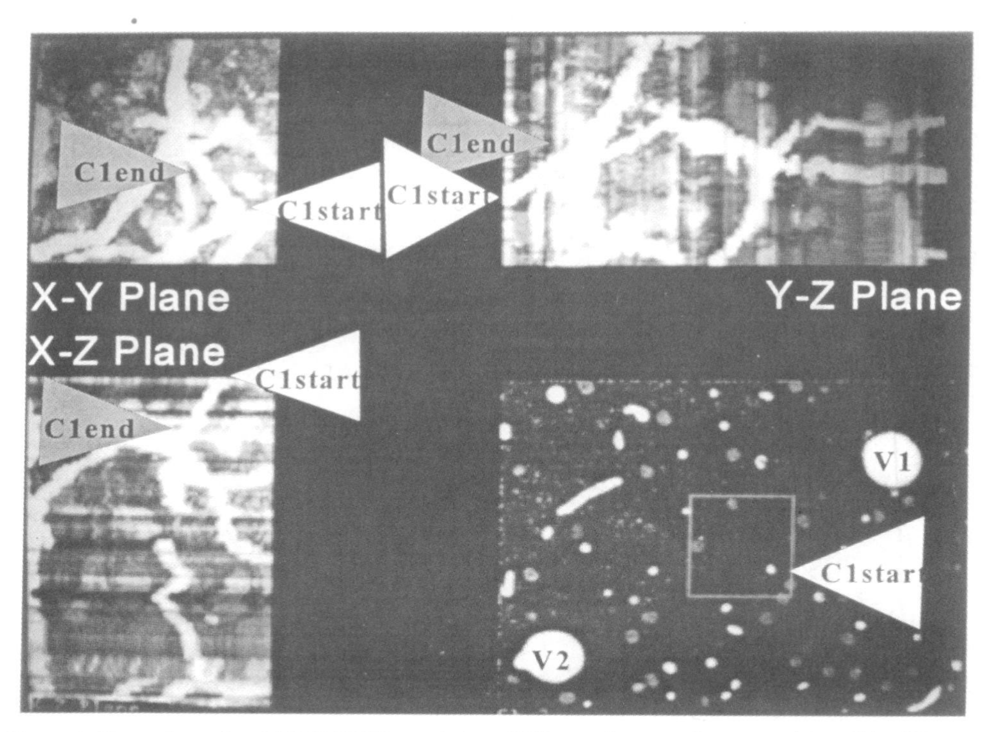

Figure 3. Three plane view (X-Y,X-Z,Y-Z) consisting of 100 μm of consecutive cortex tissue slides. The most cephalad section (Fig 1) is displayed in the lower right for reference during the outlining of vessels. White arrows represent the start of the capillary. Gray arrows, which can be seen only on the three dimensional projections, indicate the direction of the vessel. All vessels must be manually traced from start to finish by the operator.

vessel outlining, and three dimensional imaging were done as described above in the methods section. Initial examination of the vascular structure revealed the following observations: 1) capillary distribution was extremely random and seemed to be patternless; 2) capillaries branched often and unpredictably with many capillary-capillary junctions appearing as trifurcations; and 3) the distance between capillaries varied broadly, producing voids with no microvessels and regions with an abundance of capillaries.

The cortex tissue block described above measured 320x240x70 μm; the total tissue volume was therefore $5.4\times10^{6}\mu m^{3}$ ($5.4\times10^{-3}mm^{3}$). By measuring both the length and diameter of each reconstructed vessel, it is possible to determine the volume of blood in each one. Summation of these values yields the blood volume for the entire tissue block which equals $7.7\times10^{4}\mu m^{3}$. It is worth noting that $7.7\times10^{4}\mu m^{3}$ of blood equals 77 picoliters of blood [$1000\mu m^{3}=1.0\times10^{-12}$l(one picoliter)]. Dividing this figure by the total tissue volume previously calculated, yields a *very localized* calculation of 1.4% for this specimen's blood volume. A value that is similar to the 1.7% that Bereczki et. al. (1992) measured in cortical tissue using radiolabeled red blood cells and albumen. The same measurements that yielded the blood volume for the tissue block can produce the surface area of each vessel within the specimen. For this tissue sample the surface area value was 7.4 mm^{2}/mm^{3} of tissue. Since the blood brain barrier is the endothelial cells of these capillaries, the surface area is actually a measurement of the surface area of the blood side, sometimes called the luminal side, of the barrier.

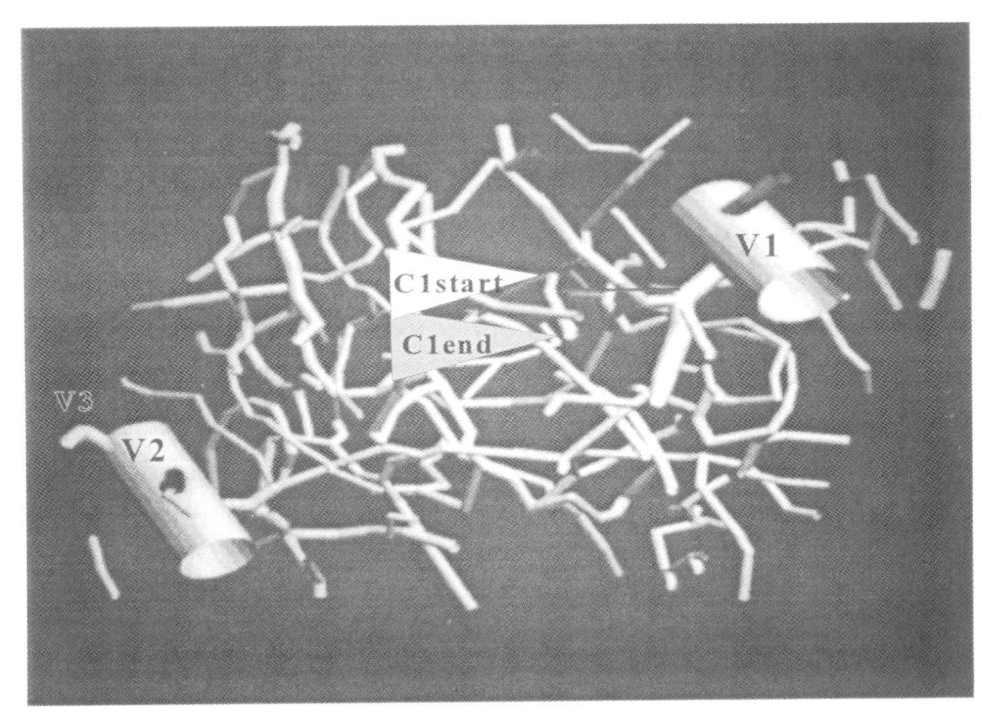

Figure 4. Three dimensional reconstruction of the microvessels of rat cerebral cortex based on 100 serial sections. Image measures 320 μm x 240 μm x 100 μm. White and gray arrows correspond to the beginning and end of the capillary described in Fig 1, 2 and 3. The complete genesis of vein V3 from vein V2 is visible (compare to Fig 1).

The next tissue block studied was a 40 μm thick section located approximately 100μm below the cortical surface. This small area was chosen for study because of the very unusual characteristic of containing an arteriole as well as a venule and capillaries. The only branching apparent in the reconstruction was between capillary and capillary. Tributaries arise from neither the postcapillary venule nor the precapillary arteriole. The rarity of arteriolar branching has been confirmed in subsequent larger tissue specimens as have the infrequent occurrence of capillaries converging with small postcapillary venules or of venules with other venules. Around both vessels there existed an avascular void, with no capillaries seen in proximity to either the arteriole or the venule. Despite a lack of capillary vascular supply, these voids contained large neurons. Similar spaces have been described around larger vessels (Pfeiffer spaces) as early as 1939 and confirmed by casting models in 1981 by Duvernoy et. al. The functional significance of these spaces is unknown. Total tissue volume in this specimen is $3.1x10^{6}\mu m^{3}$ ($3.1x10^{-3}mm^{3}$), the blood volume within all vessels is $4.6x10^{4}\mu m^{3}$. The resulting 1.5% blood volume for this piece of cortical tissue is nearly identical with the first tissue block. The blood brain barrier (capillary) surface area for this tissue block is 8.4 mm^2 / mm^3 of tissue.

Next a 37 μm thick cortical specimen located 200 μm below the surface was examined. The interesting feature of this reconstruction was finding a long continuous segment of capillary. This capillary was over 200 μm long with only one branch in the middle. A long uninterrupted capillary raises questions about the differences in blood cell and platelet flow between this type of capillary and the more typical, curving, anastomosing ones.

The final specimen was a 102 μm thick hippocampal tissue block. The blood volume for this small localized piece of hippocampal tissue is $6.9 \times 10^4 \mu m^3$, 0.9% of total tissue volume, which is 35% less than our measured cortical blood volume. Radiolabeled red blood cells and albumin have indicated that the hippocampal blood volume is less than cortical blood volume by approximately 25%. The blood brain surface area is 7.8 $mm^2 \setminus mm^3$ of tissue, which is similar to values obtained in the cortical tissue specimen. Overall comparison of this hippocampal tissue to previous cortical tissue shows a larger surface area to blood volume ratio.

5. CONCLUSIONS

The tortuous, complex pattern of cerebral capillaries has been previously suggested from scanning electron micrographs of microvascular casts and by epifluorescence. Our preliminary observations confirm and extend this concept of very tortuous, very irregular microvascular beds. We observed that in some brain areas there is: 1) an abundance of capillary-capillary junctions; 2) great local variations in capillary density within a relatively small volume of tissue; 3) feeding arterioles are few and far between; 4) many capillary segments are remote from their feeding arteriole and can be perfused by two or more arterioles by way of the intercapillary network; and 5) postcapillary venules are more abundant than precapillary arterioles.

A capillary surface area to blood volume ratio can be calculated for each structure that was studied. The cortical ratio is 530 μm^2/picoliters of blood, and the hippocampal ratio is 870 μm^2/picoliters of blood. Hippocampal capillaries are highly organized into regular lattice patterns, in contrast the cortical capillaries appear in a disorganized, random, and tortuous disarray. The highly structured vascular pattern of the hippocampus may allow the surface area of the BBB to be slightly larger than in the cortex with a 35% smaller blood volume. This larger surface area/ blood volume ratio of the hippocampus may predispose it to injury during ischemia.

The many large spaces or voids between capillaries and the even greater distance to the nearest arterioles raise the question of how neurons, astrocytes, and other glial cells signal their metabolic requirements and regulate their blood supply. Simple passive diffusion of metabolites and messengers over distances of several hundred microns would not allow the rapid communication between brain cells and arterioles necessary for continuous, virtually immediate regulation of local capillary blood flow. In view of the limitation, it may be that blood flow proceeds at some relatively brisk rate, generally delivering an abundance of essential material to the capillaries, and that regulation of the distribution of many materials between blood and brain is set by the blood-brain barrier and the transporters sited in it. This hypothesis has been offered by us before and was based on the findings that changes in cerebral blood flow and influx of glucose do not seem to be linked. The complex, irregular patterns of capillaries within the parenchyma also suggests that anatomical distance is not an important factor in their formation and that other variables such as neuronal function and cognitive development may be influential in microangiogenesis.

6. REFERENCES

1. Bereczki D, Wei L, Acuff V, Gruber K, Tajima A, Patlak C, Fenstermacher J (1992) Technique-dependent variations in cerebral microvessel blood volumes and hematocrits in the rat. *Journal of Applied Physiology* 73(3):918-924

2. Colin A (1939) Variation in Vascularity and Oxidase Content in Differnet Regions of the Brain of the Cat. *Archives of Neurology and Psychiatry* 41(2):223-242
3. Duvernoy H, Delon S, Vannson J (1981) Cortical Blood Vessels of the Human Brain. *Brain Research Bulletin* 7:519-579
4. Gesztelyi G, Finnegan W, DeMaro JA, Wang JY, Chen JL, Fenstermacher J (1993) Parenchymal microvascular systems and cerebral atrophy in spontaneously hypertensive rats. *Brain Research* 611:249-257
5. Hertz M, Paulson O (1980) Heterogeneity of Cerebral Capillary Flow in Man and Its Consequences for Estimation of Blood-Brain Permeability. *J. Clin. Invest.* 65:1145-1151
6. Hudetz A, Greene A, Feher G, Knuese D, Cowley A (1993) Imaging System for Three-Dimensional Mapping of Cerebrocortical Capillary Networks in Vivo. *Microvascular Research* 46:293-309
7. Hudetz A, Feher G, Weigle G, Knuese D, Kampine J (1995) Video microscopy of cerebrocortical capillary flow: response to hypotension and intracranial hypertension. *American Journal of Physiology* 268:H2202-2210
8. Lin S, Sposito N, Pettersen S, Rybacki L, McKenna E, Pettigrew K, Fenstermacher J (1990) Cerebral Capillary Bed Structure of Normotensive and Chronically Hypertensive Rats. *Microvascular Research* 40:341-357
9. Motti E, Imhof H, Yasargil M (1986) The terminal vascular bed in the superfacial cortex of the rat. An SEM study of corrosion casts. *Journal of Neurosurgery* 65:834-846
10. Pawlik G, Rackl A, Bing RJ (1981) Quantitative Capillary Topography and Blood Flow in the cerebral Cortex of Cats: An in vivo Microscopy Study. *Brain Research* 208:35-58
11. Wei L, Lin S, Tajima A, Nakata H, Acuff V, Patlak C, Pettigrew K, Fenstermacher J (1992) Cerebral Glucose Utilization and Blood Flow in Adult Spontaneously Hypertensive Rats. *Hypertension* 20(4):501-510
12. Yoshida Y, Ikuta F (1984) Three-Dimensional Architecture of Cerebral Microvessels with a Scanning Electron Microscope: A Cerebrovascular Casting Method for Fetal and Adult Rats. *Journal of Cerebral Blood Flow and Metabolism* 4:290-296

16

DEVELOPMENT OF BLOOD-BRAIN BARRIER TIGHT JUNCTIONS

Uwe Kniesel,[1,2] Werner Risau,[1] and Hartwig Wolburg[2]

[1] MPI für Physiologische und Klinische Forschung
W.G. Kerckhoff-Institut
D-61231 Bad Nauheim, Germany
[2] Institut für Pathologie
Universität Tübingen
D-72076 Tübingen, Germany

SUMMARY

The development of blood-brain barrier tight junctions (TJs) *in situ* was investigated on the electron microscopic level by means of quantitative freeze fracture techniques. The structural parameters for morphological analysis were the complexity of the TJ-network as characterized by fractal dimension, the overall particle density and the degree of association of TJ-particles with the protoplasmic fracture face (PFA). Capillaries from rat cortices of embryonic day (E) 13, E15, E17, postnatal day (P) 1 and from adult brains were observed. Additionally, structural properties of cultured TJs from bovine brain endothelial cells, bovine brain capillary fragments and rat brain capillary fragments were compared with the *in situ* data.

Between freshly isolated capillary fragments and TJs *in situ* no significant differences could be observed. The complexity of Tjs *in situ* increases steadily from E13 with a most dramatic enhancement between stages E18 and P1. Particle insertion starts between E13 and E15 and shows predominant association to the extracellular fracture face until stage E18. At E13 almost no particles can be observed, neither on P-, nor on E-faces. Analogue results were obtained from Ca^{2+}-depleted cultures of bovine brain endothelial cells, which may give an additional hint on the involvement of cadherins in TJ-formation and maintainance. From E18 particle insertion is completed and particle densities between 95% and 100% were obtained. The development of particle distribution shows a major alteration between stages E18 and P1, where predominance of particle association switches from E- to P-face. Furthermore predominant P-face association of TJ-particles is maintained on a quite low level (approx. 60%) compared to TJs in epithelial cells (approx. 95%). All cultured and peripheral non-barrier endothelial cells show almost complete E-face association of TJ-particles.

Biology and Physiology of the Blood–Brain Barrier, edited by Couraud and Scherman
Plenum Press, New York, 1996

RÉSUMÉ

Nous avons étudié, par microscopie électronique en cryofracture quantitative, le développement des jonctions serrées (JS) de la BHE *in situ*. Les paramètres structuraux de l'analyse morphologique étaient la complexité du réseau de JS, caractérisée par la dimension fractale, la densité particulaire totale, et le degré d'association des particules et des JS avec la face protoplasmique de la fracture (FPF). Nous avons observé des capillaires de cortex de rat embryonnaire (jour E13, E15, E17), postnatal (P1) et adulte. De plus, nous avons comparé les propriétés structurales des JS de cerveau de boeuf en culture, des fragments de capillaires de cerveau de boeuf et de rat, avec les données obtenues *in situ*..

Nous n'avons pas observé de différence significative entre les fragments de capillaires fraîchement isolés et les JS *in situ*. La complexité des JS *in situ* augmente régulièrement à partir de E13, avec un rythme nettement plus élevé entre E18 et P1. L'insertion des oarticules commence entre E13 et E15, et présente une association prédominante à la face extracellulaire de la fracture jusqu'au stade E18. A E13, on n'observe pratiquement pas de particules, ni sur la face P- ni sur la face E. Nous avons obtenu des résultats analogues avec les cellules endothéliales de cerveau de boeuf carencées en calcium, ce qui peut donner plus d'informations sur l'implication des cadhérines dans la formation et la maintenance des JS. A partir de E18, l'insertion des particules est achevée, avec des densités particulaires de 95 et 100%. Le développement de la distribution des particules présente une importante altération entre les stades E18 et P1, où l'association prédominante passe de la face E à la face P. De plus, l'association prédominante à la face P des particules et des JS se maintient à un niveau relativement bas (environ 60%) par rapport à celle des JS et des cellules épithéliales (environ 95%).

Tight junctions (TJs) between endothelial cells of the brain are the morphological correlate to the physiologically defined blood-brain barrier. The molecular mechanisms of TJ formation are unknown and the physiological function of the complex TJ network is so far only adequately described by morphological parameters as visualized by freeze-fracture electron microscopy. Structural properties of TJs which showed significant correlation to physiological data *in vitro* are the overall complexity of the TJ-network and the extent of association of TJ particles to the protoplasmic fracture face (P-face association, PFA). Whereas epithelial TJs are almost completely associated with the P-face, blood-brain barrier endothelial TJs reveal a PFA in the order of 50-60%. In cultured bovine brain endothelial cells, both the complexity and the PFA of TJs strongly decrease; the association of the TJ particles with the external fracture face of the membrane (E-face) dramatically increase. Recently, we were able to show that various treatments of bovine brain endothelial cells *in vitro* are capable to modulate morphological TJ-parameters as well as transendothelial electrical resistance and inulin permeability. TJ-complexity could be increased to values comparable to those found *in vivo*, whereas the PFA-enhancement never achieved the *in vivo* values [1]. We suggest that the PFA of blood-brain barrier endothelial TJs *in situ* is determined by the brain microenvironment, and that the switch of the particles to the E-face is a characteristic parameter of endothelial cells under *in vitro*-conditions.

The correlation of morphological parameters to transendothelial electrical resistance and permeability is one of the key questions concerning the understanding of the regulation of the blood-brain barrier. It is known from studies in pial vessels of the rat brain that transendothelial resistance and permeability to macromolecules are modulated during development [2]. Cortical vessels lose their fenestrations around embryonic day (E) 13 [3]. We therefore focused our analysis on the modulation of morphological TJ parameters in the rat cortex from E13 to adult. The ratio of PFA as TJ complexity as defined by fractal analysis were evaluated by quantitative freeze-fracture techniques [4].

One of the most intriguing aspects of TJ-fine structure during development was the observation of TJ-like membrane specializations devoid of TJ-particles at developmental stage E 13. Starting from particle-free membrane structures resembling the organization of a TJ-network, particles were increasingly inserted in the membrane showing predominant association to the E-face up to developmental stage E 18 when overall particle density reached 95% of TJ length. Between stages E18 and P1 a significant shift from predominant E-face to predominant P-face association occurred. Additionally, the TJ-complexity is strongly increased between stages E18 and P1.

Comparing structural properties of TJs from cultured bovine brain endothelial cells under conditions of Ca^{2+}-depletion with the *in situ*-data, we observed a striking analogy to the TJ-like membrane structures at E 13. As well, the TJ structures are free of TJ-particles suggesting an involvement of Ca^{2+}-dependent adhesion molecules such as cadherins in TJ formation.

Our observation of particle-free TJ-like membrane structures gives a first hint, suggesting that TJ formation proceeds in at least two steps: the first step includes the formation of particle-free TJ-like membrane-specializations, which may represent the sub-membranous organization of TJ-linked cytoskeletal elements and which seems to indicate a completed determination of TJ-structure before particle-insertion takes place; the second step includes the insertion of particles, which most likely are of proteinaceous nature, since Furuse et al. described occludin as a real transmembrane constituent of TJs [5].

In summary, not only a high complexity of TJs, but also the prominent P-face association seems to be crucial for the development of an intact barrier *in situ*. The TJ formation starts with the elaboration of strand-like membrane-specializations, which consecutively are equipped with intramembranous particles. We are currently testing the hypothesis that blood-brain barrier endothelial TJs are characterized - in contrast to epithelial or non-barrier endothelial TJs - by an intercellular adhesion of TJ particles, which might be stronger than the anchorage of the particles to the cytoskeleton.

REFERENCES

1. Wolburg H., Neuhaus J., Kniesel U., Krauss B., Schmid E., Öcalan M. and Risau W. (1994). Modulation of tight junction structure in blood-brain barrier endothelial cells: effects of tissue culture, second messengers and cocultured astrocytes. *J. Cell Sci.*, **107**, 1347-1357.
2. Butt A.M., Jones H.C. and Abbott N.J. (1990). Electrical resistance across the blood-brain barrier in anaesthetized rats: a developmental study. *J. Physiol.* **429**, 47-62.
3. Stewart P.A. and Hayakawa K. (1994). Early ultrastuctural changes in blood-brain barrier vessels of the rat embryo. *Develop. Brain Res.* **78**, 25-34.
4. Kniesel U., Reichenbach A., Risau W. and Wolburg H. (1994). Quantification of tight junction complexity by means of fractal analysis. *Tiss. Cell* **26**, 901-912.
5. Furuse M., Hirase T., Itoh M., Nagafuchi A., Yonemura S., Tsukita S. and Tsukita S. (1993). Occludin: a novel integral membrane protein localizing at tight junctions. *J. Cell Biol.* **123, 1777-1788.**

17

MEMBRANE FRACTIONATION OF BRAIN CAPILLARY ENDOTHELIAL CELLS AND ANALYSIS OF LIPID POLARITY

Bernhard J. Tewes and Hans-Joachim Galla

Department of Biochemistry
University of Muenster
D-48149 Muenster, Germany

SUMMARY

Membrane fractionation of cultured porcine brain capillary endothelial cells (BCEC) was performed and the phospholipid composition of the obtained membrane fractions was analyzed. Strong differences between the apical and the basolateral membrane domains were observed for phosphatidylcholine pointing out that membrane lipids are polar distributed in BCEC.

RÉSUMÉ

Nous avons étudié la composition en phospholipides des membranes de cultures de cellules endothéliales de microvaisseaux cérébraux de porc, après fractionnement membranaire.Nous avons observé que la composition des membranes apicales et basolatérales en phosphatidylcholine était différente, suggérant que la distribution membranaire des lipides est polarisée dans les cellules endothéliales de capillaires cérébraux de porc établies en culture.

INTRODUCTION

Cellular polarity is essential for BBB-function and the polarized distribution of several membrane proteins (e.g. Na-K-ATPase) in BCECs is well investigated (1). However, there is no information available about lipid polarity in this cell type so far.

Epithelial cells, a similar type of polarized cells, show clear differences in lipid composition between the apical and the basolateral membrane domain. The apical membrane is highly enriched in glycosphingolipids at the expense of phosphatidylcholine, whereas the main lipid of the basolateral domain is phosphatidylcholine (2). This polar lipid distribution

Biology and Physiology of the Blood–Brain Barrier, edited by Couraud and Scherman
Plenum Press, New York, 1996

is generated by a polarized sorting process and maintained by tight junctions (TJ) acting as a lateral diffusion barrier. Recently it was shown, that a fluorescent lipid applied at the apical site of MDCK-cells diffuses to neighboring cells but not to the basolateral site of the cells (3).

Since BCECs have a polar organization with complex tight junctions, we determined, whether the two plasma membrane domains possess different lipid compositions. The membrane fractionation method used in this study was developed by Stolz and Jacobsen for fractionation of aortic endothelial cells (4). With electron microscopic and fluorescence microscopic studies we showed that this efficient method is also applicable to brain capillary endothelial cells. A phospholipid based on high performance thin layer chromatography was performed to determine the lipid composition of the fractionated membranes.

METHODS

A. Cell Culture

Porcine BCECs were isolated as described previously (5). For growing homogeneous cell-monolayers without any capillary fragments or contaminating pericytes, we purified the primary cultured cells by fractionated trypsination: trypsination at 20°C was stopped, when most endothelial cells were deattached, while the capillary fragments with sprouting pericytes remained on the culture dish. The obtained cell suspension was filtered trough a 60m-nylon mesh and seeded on collagen-coated dishes. Cell monolayers used for membrane fractionation were one or two days postconfluent.

B. Membrane Fractionation

Membrane fractionation for cultured porcine BCECs was performed using the cationic colloidal silica method (4).

The principles of this new and powerful method are the following (for detailed informations see ref. 4): Cell-monolayers are coated first with cationic colloidal silica microbeads and then incubated with polyacrylic acid. In this step cationic colloidal silica is attached to the negatively charged plasmamembrane and the coating is stabilized with the anionic polyacrylic acid. Then the coated cells are treated with a hypotonic lysis buffer and disrupted by a buffer stream applied with a syringe. Apical membrane sheets and cell organelles are in the supernatant, while the basolateral membranes are still remaining on the culture dish. The basolateral fraction is obtained by scraping off the culture dish. It should be noted, that pericytes growing between culture surface and BCEC-layer strongly contaminate the basolateral fraction, so that a homogenous monolayer without pericytes is absolutely required. For isolation of the apical fraction the membrane fragments from the supernatant are collected in a low-spin centrifugation and separated in a density gradient centrifugation with metrizamide. Because of the silica-coating apical membranes become more dense than the other membranes and are found in the pellet, while the other membranes are collected at the buffer-metrizamide-interface (figure 1).

C. Lipid Analysis

Phospholipids were extracted from membrane fractions with chloroform / methanol 2:1 and separated varying the method of Macala et al. (6). HPTLC-plates (Merck) were first developed in chlorform / methanol / acetic acid / formic acid / water 35:15:6:2:1 and then in n-hexane / isopropylether / acetic acid 65:35:2. Large samples were quantified by

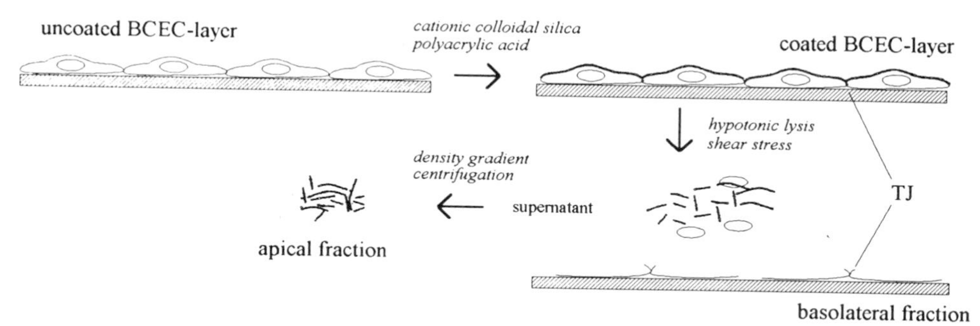

Figure 1. Schema of the mean steps of the cationic colloidal silica microbead membrane fractionation method (4).

phosphorus analysis of the separated lipids. For small samples sizes HPTLC-plates were stained by dipping them in a copperacetate solution (3% in 8% phosphoric acid). The lipid content was quantified with a laser scanning densitometer (Ultroscan XL, Pharmacia).

RESULTS

With electron- and fluorescence microscopy we examined several steps of the membrane fractionation. As shown in fig. 2 a,b the apical membrane of BCECs can be coated with cationic colloidal silica microbeads. Together with the polyacrylic acid the silica forms a dense cover of the whole monolayer. After disruption of the cells large sheets of covered apical membranes are found in the supernatant (fig. 2c). Furthermore, we labelled BCECs with a fluorescent dye, that is able to incorporate into lipid bilayers, to stain both the apical and the basolateral membrane. After membrane fractionation the material attached on the culture surface consists of large areals of intact lipid bilayers, that correspond with the basolateral membranes of the former BCECs (fig 2 d,e). With immunostaining of ZO-1 protein we localized the tight junctions in intact BCEC and in the basolateral fraction obtained by this method (fig 2 f). Since there are fragments of tight junctions still connected to the basolateral membranes, the cell disruption happens directly above or below the tight junction region. These results clearly show, that the method of Stolz and Jacobsen can be transferred to fractionation of BCEC.

Phospholipids of the whole-cell extract and of the extracts of apical and basolateral fractions were separated by HPTLC. For the whole-cell extract lipids were quantified by phosphorus analysis. Due to the sample sizes the quantification of the apical and basolateral fraction lipid extracts was performed only by laser-scanning-densitometry.

Figure 3 shows the whole-cell lipid composition and in comparison the lipid composition of the apical and the basolateral membranes. The extracts from both plasma membrane fractions exhibit clear differences to the whole-cell-membranes extract. Both fractions are enriched in Sphingomyelin and depleted in Phosphatidylethanolamine, wich has to be considered a typical plasma membrane composition.

Furthermore there are significant differences in the lipid composition between the apical and the basolateral membrane: The main phospholipid in the apical membrane is Phosphatidylcholine, whereas the main phospholipid of the basolateral membrane is Sphingomyelin. Despite of these differences the ratio Sph:PS:PE (4:1:1) is the same in both membrane domains.

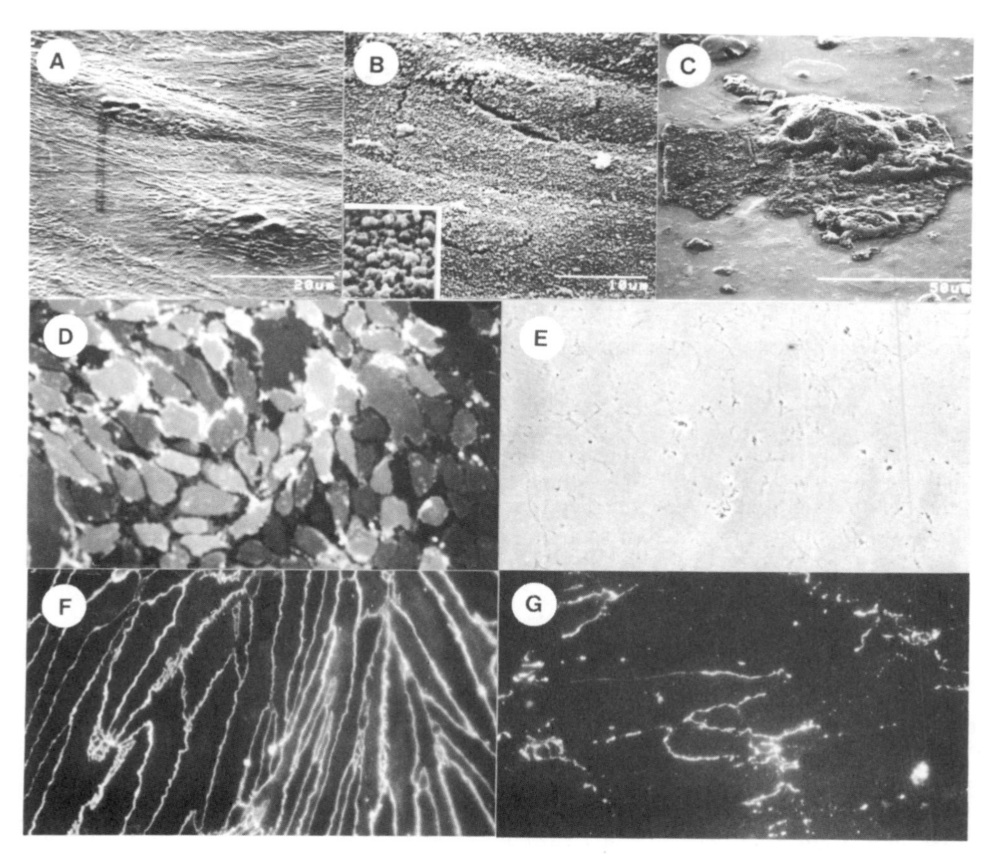

Figure 2. Various steps of the BCEC-membrane fractionation: a) scanning electron micrograph of a confluent BCEC-monolayer; b) scanning electron micrograph of silica-coated BCECs, insert: same sample at higher magnification; c) large sheet of silica coated apical membrane found in the supernatant after cell-disruption d) fluorescence micrograph of the basolateral fraction labelled with a fluorescent dye (1,1'-dioctadecyl-3,3,3',3'-tetraacetyl-indocarbocyanin perchlorate) e) phase contrast micrograph of the same sample as in d); f) fluorescence micrograph of an anti-ZO-1-immunostaining of an intact BCEC-monolayer; g) fluorescence micrograph of an anti ZO-1-immunostaining of the basolateral membrane fraction.

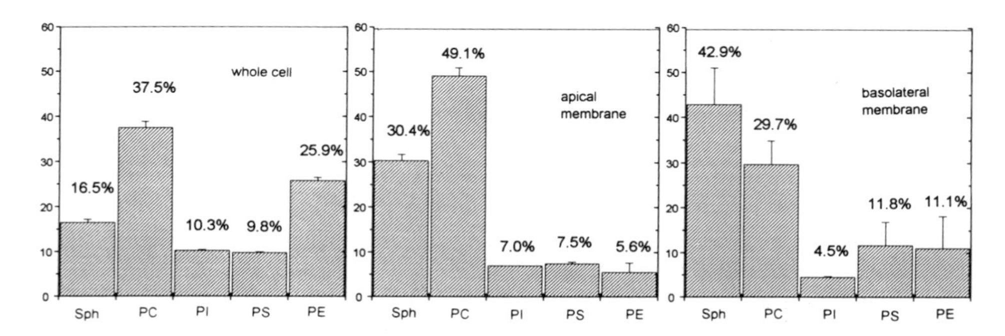

Figure 3. Phospholipid composition of the whole-cell, the apical fraction- and the basolateral fraction extracts. Sph: sphingomyelin, PC: phosphatidylcholine, PI: phosphatidylinositol, PS: phosphatidylserine, PE: phosphatidylethanolamine. error bars: standard deviation

DISCUSSION

Using the cationic-colloidal-silica-microbead technique a membrane fractionation of cultured brain capillary endothelial cells was established. It was shown that the method originally developed for aortic endothelial cells can be transferred to BCECs.

Both membrane fractions show a unique lipid composition with clear differences to each other and to the whole cell extract. PC is polar distributed with 2-fold enrichment in the apical membrane, while Sph, PS and PE are equal distributed in both membranes. Also PI seems to be apically enriched, but due to the small content in both membrane samples the PI-value is not as valid as the values of the other lipids. This lipid-polarity is remarkably different compared to epithelial cells. There PC is also polar distributed, but enriched in the basolateral membrane.

This is the first report showing lipid polarity in BCEC. The following experiments on the distribution of the other lipids and the study of lipid polarity in endothelial cells from other sources could reveal, whether this lipid polarity is a special BCEC-property and may be important for BBB-function.

REFERENCES

1. Vorbrodt A.J. :*Ultrastructural Cytochemistry of Blood-Brain Barrier Endothelia.* G. Fischer Verlag, Stuttgart, New York (1988)
2. Simons, K. and van Meer, G. ,Biochemistry 27, (1988) 6197-6202
3. Grebenkemper, K and Galla, H.J. Chem. & Phys. of Lip. 71, (1994) 133-143
4. Stolz, D.B. and Jacobsen, B. .J. Cell Sci. 103, (1992) 39-51
5. Mischek, U., Meyer, J. and Galla, H. J. Cell Tiss. Res. 256, (1989) 221-226
6. Macala, L.J., Yu, R.K. and Ando, J. Lip. Res. 24, (1983) 1243-1250

18

"BRAIN-TISSUE-SPECIFIC" VERSUS "SERUM-SPECIFIC" POSTTRANSLATIONAL MODIFICATION OF HUMAN CEREBROSPINAL FLUID POLYPEPTIDES WITH N-LINKED CARBOHYDRATES

Andrea Hoffmann, Eckart Grabenhorst, Manfred Nimtz, and Harald S. Conradt

Department of Molecular Biology
Gesellschaft für Biotechnologische Forschung
D-38124 Braunschweig, Germany

SUMMARY

Comparison of the patterns of N-linked oligosaccharides attached to several proteins isolated from human cerebrospinal fluid and serum reveals striking differences ("brain-type" versus "serum-type" glycosylation) which are attributed to the different physiological surroundings of both body fluids. No modification of oligosaccharides was found to occur during passage of proteins across the blood-brain barrier. Glycosylation analysis of CSF-proteins can thus be used to gain information on the site of biosynthesis (i.e. systemic or intrathecal) of specific proteins in question.

RÉSUMÉ

La comparaison de structures des oligosaccharides attachés à plusieurs protéines obtenues du fluide cérébrospinale et du sérum humains a montrée des différences frappantes ("brain-type" versus "serum-type" glycosylation) qui ont été attribuées aux alentours differents de ces deux fluides. On n'a pas rencontré une modification des oligosaccharides pendant le passage au-delà de la barrière hémato-encephalique. La recherche de la glycosylation de protéines cérébrospinales permet alors d'obtenir des informations sur le lieu de biosynthèse (dans le cerveau ou ailleurs dans le corps) des protéines particulières en question.

Biology and Physiology of the Blood–Brain Barrier, edited by Couraud and Scherman
Plenum Press, New York, 1996

1. INTRODUCTION

About 80% of the protein contained in human cerebrospinal fluid (CSF) originates from plasma via passage across the blood-brain barrier (BBB) including albumin and α_1-acid glycoprotein (AGP)[1]. The rest is synthesized locally within the central nervous system (CNS), e.g. β-trace protein (β-TP) and asialotransferrin (β_2-transferrin, τ-globulin) the synthesis of which has been demonstrated, amongst others, by choroid plexus and oligodendrocytes.

The majority of CSF-proteins is glycosylated, and glycosylation has been shown to have such important functions as recognition of ligands or cells, targeting of proteins, and mediating specific recognition processes. It has also been proposed to be involved in the mechanisms underlying transfer of proteins from blood to brain[2,3,4].

In a first approach towards studies of the mechanisms by which serum proteins are transferred into the brain and how they are retained there and into mechanisms of control of the chemical surroundings and homeostasis within the CNS we have investigated the N-glycosylation patterns of four human CSF-proteins: β-TP, transferrin, hemopexin (Hpx), and AGP.

2. METHODS

β-TP and transferrin were isolated from pooled human non-pathological CSF-specimens by anion-exchange chromatography and gel filtration[5,6]. β-TP from human serum as well as Hpx and AGP from CSF were purified by immunoaffinity chromatography. Transferrin, Hpx, and AGP from human serum were purchased (Behring). Oligosaccharides were liberated after tryptic digestion by enzymatic cleavage of the glycopeptides with PNGase F (Boehringer) or by hydrazinolysis using the GlycoPrep™ (Oxford Glycosystems). They were analyzed by sugar component and methylation analysis, high-pH anion exchange chromatography, and fast-atom bombardment or matrix-assisted laser desorption/ ionization mass spectrometry as described[6,7]. After molecular cloning of the β-TP cDNA from a human placenty mRNA-library into eukaryotic expression vectors[5] recombinant protein was produced from BHK-21 and baculovirus-infected Sf21-insect cells and purified by immunoaffinity chromatography.

3. RESULTS

All four proteins investigated only contain N-linked oligosaccharides.

β-TP and the unsialylated transferrin variant contained in CSF are characterized by truncated biantennary complex-type oligosaccharides bearing large amounts of terminal galactose (asialo-oligosaccharides), terminal N-acetylglucosamine (asialo-agalacto-oligosaccharides and bisecting N-acetylglucosamine), complete proximal and partial peripheral fucosylation (Lewis^x- and sialyl-Lewis^x-structures) as well as presence of α2,6- and α2,3-linked neuraminic acid (Figure 1). These features were termed "brain-type" N-glycosylation[6,7] since they strikingly contrast with those of serum proteins: complete sialylation (mainly α2,6-linked neuraminic acid), low amounts of terminal N-acetylglucosamine, no proximal fucose (Figure 2). This is due to the existence of hepatic clearance systems recognizing terminal galactose (asialoglycoprotein- or Ashwell receptor), N-acetylglucosamine, and Lewis^x and explains the very low concentration of β-TP in serum.

From a developmental biological perspective, no such lectin clearance receptors are apparently necessary in the human CNS since the CNS is protected from its surroundings

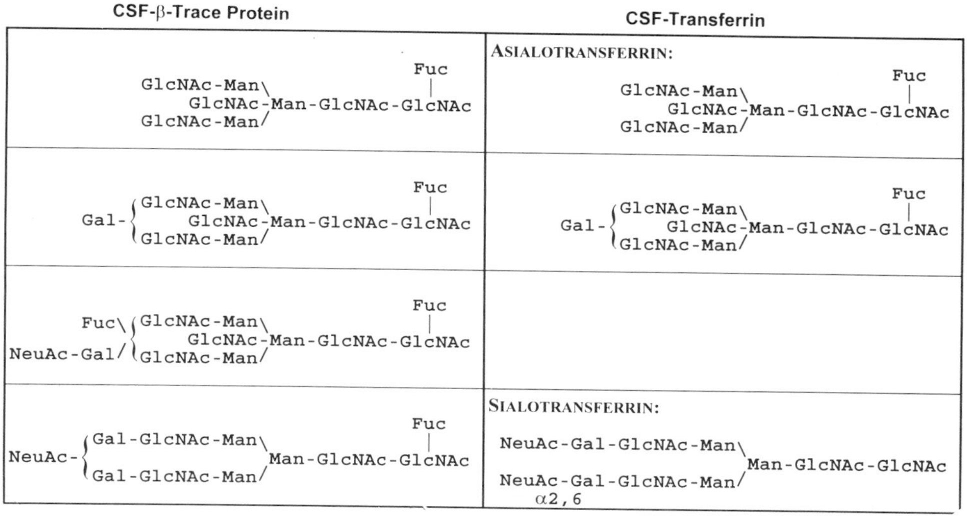

CSF-β-Trace Protein	CSF-Transferrin
Fuc GlcNAc-Man\ | GlcNAc-Man-GlcNAc-GlcNAc GlcNAc-Man/	ASIALOTRANSFERRIN: Fuc GlcNAc-Man\ | GlcNAc-Man-GlcNAc-GlcNAc GlcNAc-Man/
Fuc Gal-{GlcNAc-Man\ | GlcNAc-Man-GlcNAc-GlcNAc GlcNAc-Man/	Fuc Gal-{GlcNAc-Man\ | GlcNAc-Man-GlcNAc-GlcNAc GlcNAc-Man/
Fuc Fuc\ {GlcNAc-Man\ | GlcNAc-Man-GlcNAc-GlcNAc NeuAc-Gal/ GlcNAc-Man/	
Fuc NeuAc-{Gal-GlcNAc-Man\ | Man-GlcNAc-GlcNAc Gal-GlcNAc-Man/	SIALOTRANSFERRIN: NeuAc-Gal-GlcNAc-Man\ Man-GlcNAc-GlcNAc NeuAc-Gal-GlcNAc-Man/ α2,6

Figure 1. Major oligosaccharide structures of β-trace protein and transferrin from human CSF.

by the BBB and blood-CSF barriers which strictly control exchange of substances with the systemic circulation. Thus, CSF-proteins are catabolized after resorption into the venous blood via the arachnoidal granulations and/or the nerve roots and lymphatic pathways in the liver and kidney. "Brain-type" glycosylation may thus contribute to the maintenance of compartmentation and homeostasis within the human CNS and may represent a mechanism for retention of locally synthesized proteins in the CSF.

In order to conclusively demonstrate that the truncated oligosaccharide chains of human CSF-β-TP indeed represent a tissue-specific glycosylation recombinant β-TP was obtained from BHK-21 cells. The recombinant protein turned out to have exclusively fully processed biantennary carbohydrate chains bearing 1 or 2 neuraminic acid residues and proximal fucose but no bisecting N-acetylglucosamine or peripheral fucose (the respective glycosyltransferases are not expressed by BHK-cells: host-specific glycosylation). Thus, the extensive truncation of antennae found with CSF-β-TP and asialotransferrin from human CSF is not encoded by their tertiary structures but reflects a specialized glycosylation capacity of the different brain cell types involved in the biosynthesis of these polypeptides (tissue-specific glycosylation; we measured only low neglibible amounts of glycosidase activities in human CSF). However, β-TP itself determines the biantennarity of the carbohydrate chains since BHK-cells are capable of synthesizing tri- and tetra-antennary chains (protein-specific glycosylation).

In contrast, the sialylated transferrin variant contained in human CSF as well as hemopexin and α_1-acid glycoprotein from CSF exhibit "serum-type" glycosylation characteristics and are probably derived from serum by transport across the blood-brain barrier: absence of proximal fucosylation, low amounts of terminal galactose and no terminal N-acetylglucosamine; neuraminic acids in α2,6-linkage largely prevail over α2,3-linked ones.

β-TP from human serum also contains fully sialylated and galactosylated biantennary oligosaccharide chains. However, as in the case of its CSF-counterpart, these glycans are proximally fucosylated and many of them bear a bisecting GlcNAc. We thus propose that serum β-TP represents glycoforms of the CSF-protein which have escaped the hepatic

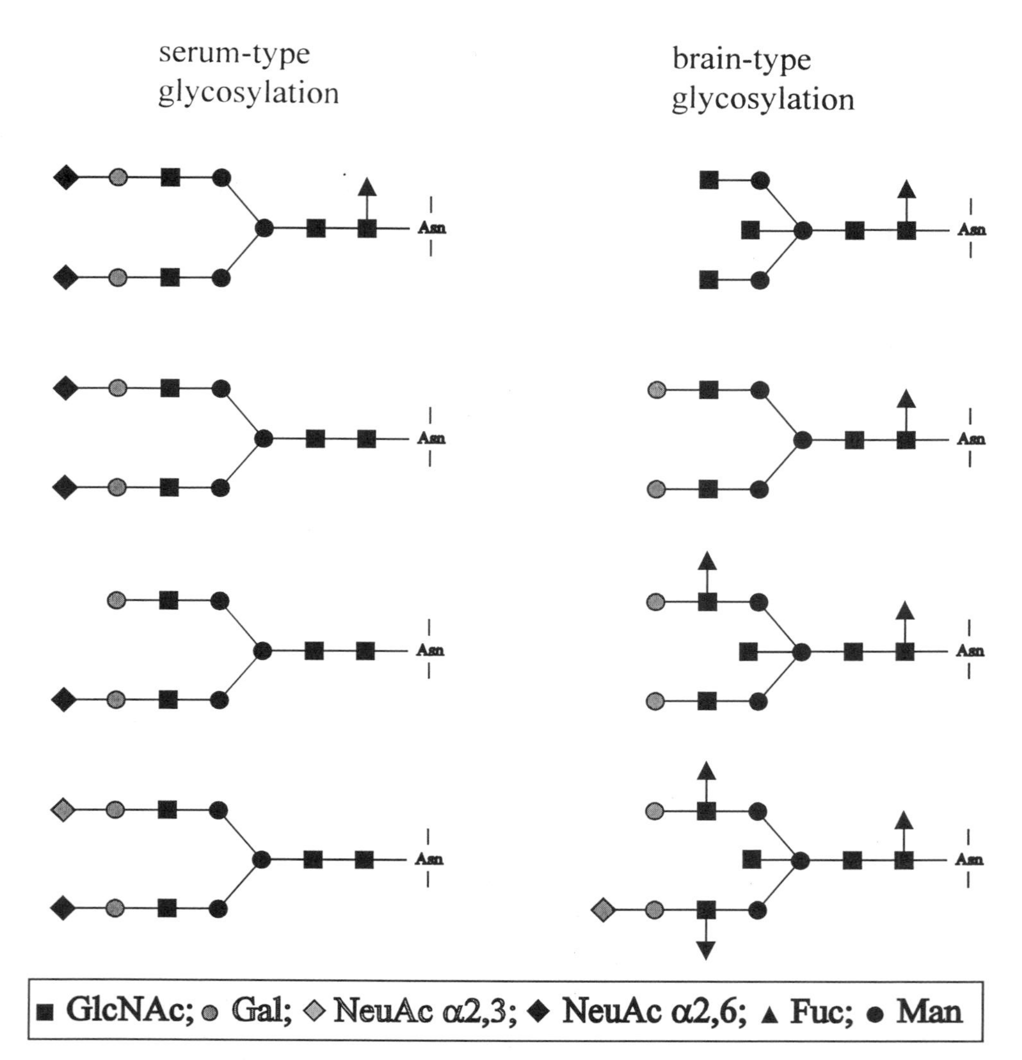

Figure 2. Schematic representation of "brain-type" versus "serum-type" N-glycosylation of human glycoproteins.

clearance receptors after bulk resorption of CSF at the arachnoidal granulations into the venous blood.

4. DISCUSSION AND PERSPECTIVES

In contrast to β-trace and transferrin, Hpx and AGP in human CSF are believed to be exclusively of serum origin. The results of glycosylation analysis presented above support this notion: hemopexin, α_1-acid glycoprotein and the sialylated transferrin variant are, based on their specific carbohydrate structures, derived from serum whereas asialo-transferrin and β-trace protein both are produced intrathecally. Furthermore, our studies demonstrate that no modification of glycosylation patterns is involved in the transport of serum proteins across the BBB. Especially, desialylation of serum proteins does not occur in the course of this process thus showing that "brain-type" N-glycosylation is conferred to proteins by cells located intrathecally but not by cells contributing to the BBB.

The differently glycosylated forms of natural and recombinant β-TP and other glycoproteins will be used in an animal system for elucidating communication pathways between body fluids and clearance mechanisms of glycoproteins and for investigating the molecular processes occurring during the transfer of individual serum protein across the blood-brain and blood-CSF barriers.

5. REFERENCES

1. Thomas, T., Schreiber, G., and Jaworowski, A. (1989). Dev. Biol. *134*, 38-47,
2. Poduslo, J.F. and Curran, G.L. (1994). Molec. Brain Res. *23*, 157-162
3. Irie, S. and Tavassoli, M. (1991). Cell Biol. Rev. *25*, 317-331.
4. Rademacher, T., Parekh, R.B., and Dwek, R.A. (1988). Ann. Rev. Biochem. *57*, 785-838.
5. Hoffmann, A., Conradt, H.S., Gross, G., Nimtz, M., Lottspeich, F., and Wurster, U. (1993). J. Neurochem. *61*, 451-456.
6. Hoffmann, A., Nimtz, M., Getzlaff, R., and Conradt, H.S. (1995). FEBS Lett. *359*, 164-168
7. Hoffmann, A., Nimtz, M., Wurster, U., and Conradt, H.S. (1994). J. Neurochem. *63*, 2185-2196.

19

BLOOD-BRAIN BARRIER PROPERTIES *IN VITRO* AS RELATED TO THE NEUROTRANSMITTER SEROTONIN

P. Brust, S. Matys, J. Wober, A. Friedrich, R. Bergmann, and B. Ahlemeyer

Research Center Rossendorf
Institute of Bioinorganic and Radiopharmaceutic Chemistry
01314 Dresden, Germany

SUMMARY

There is evidence that the neurotransmitter serotonin (5-HT) is involved in the regulation of blood-brain barrier (BBB) functions. Therefore we have started to study the transport and the receptor binding of 5-HT at the BBB of pigs. The 5-HT uptake system was studied using [^{3}H]imipramine and [^{3}H]paroxetine as radiotracers. For both tracers specific binding to isolated brain microvessels was demonstrated indicating the presence of the 5-HT transporter at the BBB. In addition, the specific binding of [^{3}H]paroxetine in freshly prepared brain endothelial cells and cells cultured for various time periods was measured. The finding supports the asssumption that the expression of the 5-HT transporter is not influenced by the conditions of cell culture. Furthermore we obtained evidence for the presence of specific serotonin receptor(s) at the brain endothelium. Binding studies with different types of radiotracers ([^{3}H]ketanserin, [^{3}H]mesulergine, [^{3}H]OH-DPAT) indicate that the 5-HT receptor(s) belong(s) to the 5-HT$_2$ subtype.

RÉSUMÉ

Il est établi que le neurotransmetteur sérotonine (5-HT) est impliqué dans la régulation de la barrière hémato-encéphalique (BHE). Aussi avons- nous étudié, au niveau de la BHE de porc, le transport de la 5-HT ainsi que la pharmacologie de ce neurotransmetteur au moyen de la (^{3}H)imipramine et de la (^{3}H)paroxetine.Ces deux types de ligands des récepteurs de la 5-HT se fixent de manière spécifique sur les microvaisseaux cérébraux isolés, démontrant l'existence d'un recepteur de la sérotonine au niveau de la BHE. De plus, la liaison spécifique de la paroxetine a pu être démontrée sur des cellules endothéliales cérébrales fraîchement isolées ou après differents temps de culture. Ces resultats semblent montrer que l'expression du récepteur de la sérotonine dans des cultures de cellules endothéliales

Biology and Physiology of the Blood–Brain Barrier, edited by Couraud and Scherman
Plenum Press, New York, 1996

cérébrales n'est pas modifiée par les conditions de culture. De plus, nous avons démontré la présence de récepteur(s) spécifique(s) sur l'endothelium cérébral. Des etudes pharmacologiques avec différents types de ligands tritiés ((^{3}H)ketansérine, (^{3}H)mésulergine, (^{3}H)OH-DPAT) semblent montrer que les cellules endothéliales cérébrales expriment le récepteur sérotoninergique de type 5-HT_2.

1. INTRODUCTION

It is known from recent studies that the neurotransmitter serotonin (5-HT) is involved in the regulation of blood-brain barrier (BBB) functions. First evidence for a sensitivity of the brain vasculature to serotonin was obtained by Crone who measured a decrease of the electrical resistance in cerebral postcapillary venules of frogs indicating an increase of vascular permeability[1]. Later Sharma and co-workers have shown that the BBB permeability as measured with Evans blue and [^{131}I]sodium iodide is increased after infusion of serotonin or under conditions associated with increased levels of circulating 5-HT[2,3]. Inhibition of the brain-blood transfer of opioid peptides by 5-HT was described by Banks and Kastin[4] Also, 5-HT elicits endothelium-dependent mechanisms of the regulation of cerebral blood flow[5,6].

Despite this evidence the mechanisms underlying the effects of 5-HT are poorly understood. There is still lack of information with regard to the specific type of 5-HT receptors involved in the mentioned effects on the brain endothelium. Furthermore, the transport of 5-HT into the endothelial cells and the regulation of its metabolism by the enzyme monoamine oxidase needs to be characterized in detail.

2. A SPECIFIC TRANSPORT SYSTEM FOR SEROTONIN IS PRESENT AT THE BRAIN ENDOTHELIUM

2.1. Binding Studies with [^{3}H]Imipramine

First evidence for the involvement of an energy-dependent system in the transport of 5-HT at the brain endothelium was obtained by Maria Spatz more than ten years ago[7]. At that time knowledge about specific serotonin transporter was very incomplete. During the last years it has been improved. Now it is known that there are at least two different systems involved in the transport of 5-HT: a) the vesicular monoamine transporter[8] and b) the sodium-dependent 5-HT transporter[9] which is present in neurons and in a selective group of other tissues[9-13]. To find out whether such type of transporter is also present at the BBB we have started to use isolated microvessels and cultured endothelial cells from porcine brain for *in vitro* investigations.

Recently, it was shown that antidepressants such as imipramine block the uptake of 5-HT[10]. High-affinity [^{3}H]imipramine binding was used to characterize the distribution of 5-HT uptake sites in the mammalian brain[10-12] and on human platelets[13]. Therefore, we have used [^{3}H]imipramine to demonstrate the presence of the serotonin transporter on brain endothelial cells.[14] The specific binding of imipramine was measured at free [^{3}H]imipramine concentrations between 0.04 and 67 nM. The curvilinear type of the Rosenthal-Scatchard plot derived from these data indicated two binding sites. The affinities (K_{d1}= 1.2 nM; K_{d2}= 518 nM) and densities (B_{max1}= 38 fmol/mg; B_{max2}= 3.3 pmol/mg) of the two binding sites differed by more than two orders of magnitude.

In addition, the displacement of [^{3}H]imipramine binding by unlabeled imipramine was studied on both brain tissue membranes and on brain microvessels. The affinities of

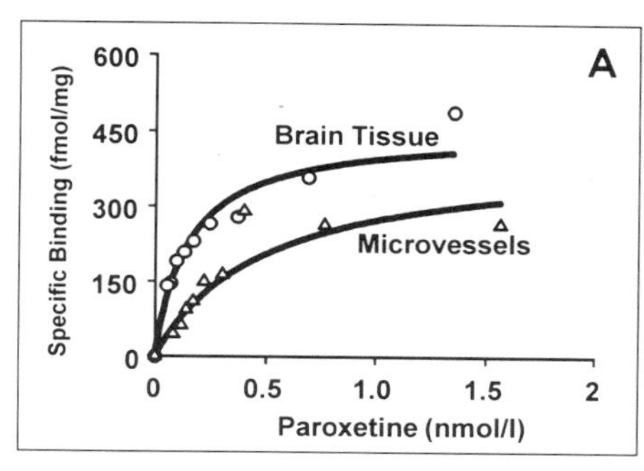

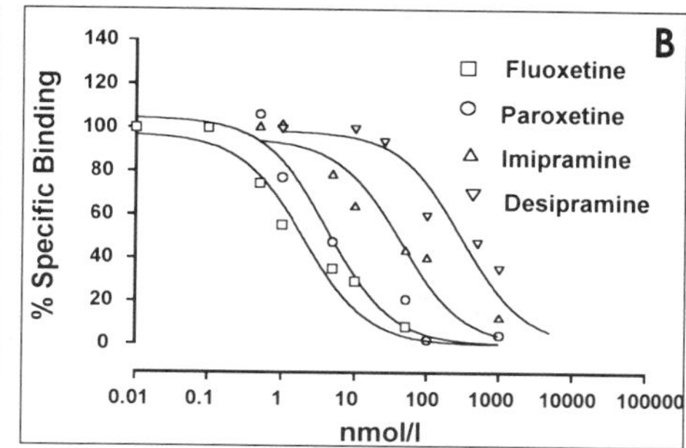

Figure 1. Binding of [^{3}H]paroxetine on brain microvessels and membranes **A.** Saturation analysis of [^{3}H]paroxetine binding at free concentrations between 0.04 nM and 2 nM. The curve was analysed by a non-linear fitting routine (Fig. P, Biosoft, Cambridge, U.K.). The binding parameters for brain tissue and microvessels obtained from these data are: K_d= 0.15 nM and 0.49 nM, respectively, and B_{max}= 450 fmol/mg and B_{max}= 407 fmol/mg, respectively. **B.** Displacement of [^{3}H]paroxetine binding by various uptake inhibitors. The IC_{50} values calculated from these curves are 2.0 ± 1.4 nM (fluoxetine), 3.7 ± 1.3 nM (paroxetine), 95 ± 39 nM (imipramine), and 304 ± 201 nM (desipramine).

[^{3}H]imipramine binding calculated from these data were similar in brain tissue and microvessels. However, differences of the transporter densities were observed. The high-affinity binding site was only slightly enriched in the microvessels (B_{max1}= 32 fmol/mg in brain versus 59 fmol/mg in microvessels). But the low-affinity binding site was enriched by factor 28. Hence the relative parts of these two binding sites has been changed during the microvessel preparation. The nature of the low affinity binding site was never identified even though their involvement in the central nervous actions of antidepressants is expected[11]. It remains still unclear from our studies. However, we have obtained evidence that this may be a protein which is abundant in brain endothelial cells. Imipramine and other tricyclic antidepressants have been found to inhibit the activity of the monoamine oxidase (MAO; E.C. 1.4.3.4)[15] and to reduce the metabolism of 5-HT in microvessels isolated from gerbil's brain[7]. From these data, we expected the low-affinity binding site to be identical with MAO. This was supported by direct measurement of MAO activity in the presence of 10 μM and 100 μM imipramine. Both concentrations of this drug significantly inhibited the activity of MAO in our microvessel preparation by 7.4 ± 2.5 % and 34.1 ± 2.3 %, respectively. Interestingly, also kinase inhibitors has been shown to compete with imipramine for binding and inhibition of serotonin transport[16].

2.2. Binding Studies with [^{3}H]Paroxetine

[^{3}H]paroxetine represents a more specific 5-HT uptake inhibitor without the disadvantage of a complex binding kinetics[17-20]. We have used this radioligand to confirm our previous findings. Saturation studies (0.04 - 2 nM) revealed the presence of a single binding site in both membranes prepared from porcine brain and microvessels (Fig. 1A). The transporter density in both preparations was similar (membranes: 450 ± 39 fmol/mg, microvessels: 407 ± 39 fmol/mg) while the affinity was about three-fold higher in membranes (K_d= 0.15 ± 0.03 nM) than in microvessels (K_d= 0.49 ± 0.17 nM). Various uptake inhibitors were used to displace the binding of [^{3}H]paroxetine (Fig. 1B). Paroxetine, fluoxetine and imipramine are known as potent inhibitors of the serotonin transporter[10,11,18-21]. The pharmacological profile of inhibition of high-affinity [^{3}H]paroxetine binding to platelets or brain membranes by these and other drugs was reported to correlate significantly with the rank order of potency of the drugs to inhibit 5-HT uptake[18-20,22]. Therefore, it is generally accepted that this binding site is associated with the sodium-dependent serotonin transporter[9,17,18]. In our study, the IC_{50} values obtained with the non-tricyclic antidepressants paroxetine and fluoxetine as displacer were lower than the values of the tricyclic antidepressants imipramine and desipramine. This finding is in agreement with studies of the neuronal and placental serotonin transporter[17-20,22]. Therefore, we conclude that in our preparation [^{3}H]paroxetine labels 5-HT uptake sites localized on the brain microvessels.

In addition, we have measured the total and the specific binding of [^{3}H]paroxetine in freshly prepared brain endothelial cells and cells cultured for various time periods. The specific binding was about 50% of total binding at a concentration of 0.61 nM [^{3}H]paroxetine. 1 μM fluoxetine was used to define the unspecific binding. The specific binding in freshly prepared cells (52 ± 9 fmol/mg, n=6) was similar to the binding found in cells after 4 days in culture (69 ± 19 fmol/mg). The finding supports our asssumption that the serotonin transporter is present on brain endothelial cells. Furthermore, it indicates that the expression of this protein is not influenced by the conditions of cell culture.

2.3. Is the Serotonin Transporter Involved in the Regulation of Monoamine Oxidase Activity of the Brain Endothelium?

We expect that the serotonin transporter at the BBB is involved in the metabolism of serotonin by presenting this monoamine to its metabolizing enzyme monoamine oxidase. We were interested to know whether the blockade of the serotonin transporter influences the activity of this enzyme in the brain endothelium. The finding that chronic treatment with fluoxetine resulted in adaptive changes of the 5-HT_2 receptor status in the rat brain[23] supports such a possibility. The MAO activities measured in microvessels (2.88 ± 0.01 nmol/mg/min) and freshly isolated cells (1.31 ± 0.52 nmol/mg/min) were similar to data published previously[24]. But the values measured in cultured cells were about 70-90% lower. Furthermore, we have found a decrease of MAO activity during cell culture (day 4: 0.36 ± 0.08 nmol/mg/min; day 8: 0.27 ± 0.05 nmol/mg/min). However, no influence of 1 μM fluoxetine (present in the serotonin-containing culture medium for several days) was observed. Therefore, substrate availability does not seem to be a major regulator of MAO activity in the brain endothelium.

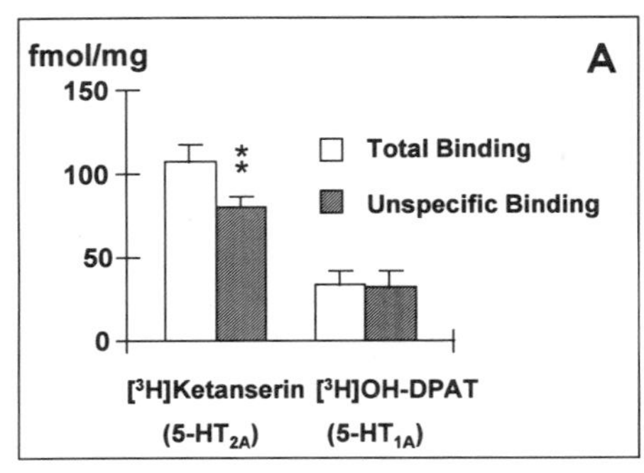

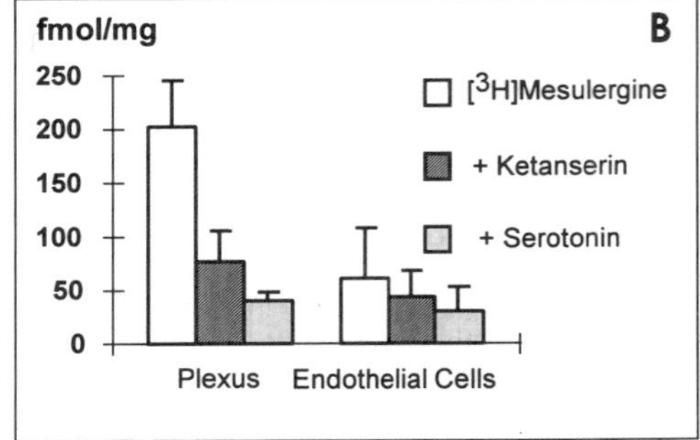

Figure 2. Binding of various 5-HT receptor ligands to brain microvessels and endothelial cells. **A.** Binding of [^{3}H]ketanserine and [^{3}H]8-OH-DPAT on microvessels isolated from porcine brain. 1 μM mianserin (5-HT_{2A}) and 10 μM serotonin (5-HT_{1A}) were used to define unspecific binding. No specific binding of [^{3}H]8-OH-DPAT was observed. The specific binding of [^{3}H]ketanserine was 27 ± 9 fmol/mg (25 ± 7 % of total binding). **B.** Specific binding of [^{3}H]mesulergine (1 nM) on choroid plexus and brain endothelial cells. 50 nM ketanserin or 20 nM serotonin inhibited the binding on choroid plexus by 62 % and 80 %, respectively, and the binding on endothelial cells by 30 % and 51 %, respectively.

3. EVIDENCE FOR THE PRESENCE OF SPECIFIC SEROTONIN RECEPTORS AT THE BRAIN ENDOTHELIUM

Presently, at least 15 different subtypes of serotonin receptor are known. They are classified into seven groups with the 5-HT_1 and the 5-HT_2 receptors as the main classes[25]. We have selected radioligands with specificity for the 5-HT_{1A}, 5-HT_{2A} and the 5-HT_{2C} (the former 5-HT_{1C}) subtype to search for 5-HT receptors at the brain endothelium. 5-HT_{1A} receptors are present with high density in the hippocampus[26]. The preferred radioligand for 5-HT1_A receptors is [^{3}H]OH-DPAT[27]. [^{3}H]ketanserin is a specific radioligand for 5-HT_{2A} receptors which have a high density in the frontal cortex[28]. However, also affinity to 5-HT_{2C} receptors was described. [^{3}H]mesulergine, a specific ligand for the 5-HT_{2C} receptor[29], binds with high density in the plexus epithelium, where only the 5-HT_{2C} receptor subtype exists[30,31].

Using the 5-HT_{1A}-antagonist [^{3}H]OH-DPAT no specific binding in isolated brain microvessels was observed (Fig. 2A). In contrast, the 5-HT_{2A}-antagonist [^{3}H]ketanserin has bound specifically to isolated microvessels and endothelial cells from porcine brain. The amount of specific binding was about 20%. We have compared endothelial cells at different days of culture. No difference of specific binding (at 0.16 nM [^{3}H]ketanserin) was found between freshly prepared cells and cultured cells (4 and 6 days). Because [^{3}H]ketanserin binds also to other 5-HT receptors such as the 5-HT_{2C} receptor we have performed binding studies with [^{3}H]mesulergine. For comparison we have used homogenate of the rat choroid plexus (Fig. 2B). The specific binding of [^{3}H]mesulergine was measured at a concentration of 1 nM. 50 nM ketanserin was able to displace the tracer by about 60% in the plexus and by about 30% in the endothelial cells. Serotonin was found to be a much better displacer. 20 nM inhibited the specific binding of [^{3}H]mesulergine by 80% in the plexus and by 51% in the endothelial cells. This finding is consistent with the binding characteristics of 5-HT_{2C} receptors[32].

4. CONCLUSIONS

We have obtained evidence that the expression of the 5-HT transporter and receptor(s) is a feature of the mature differentiated endothelium which seems to be preserved during cell culture. Therefore, cultured endothelial cells from porcine brain may be used to study in detail the properties of the BBB as related to the neurotransmitter serotonin.

These data demonstrate that a serotonin transporter is present on endothelial cells of the porcine brain. The transport protein has a similar density as in brain tissue. However, the binding affinity of the two uptake inhibitors imipramine and paroxetine on brain microvessels is reduced by 50% and 70%, respectively. Furthermore, there is evidence for the presence of 5-HT_2 receptor(s) at the brain endothelium.

Our findings support the concept of neurogenic vasodilatation and regulation of BBB permeability by serotonergic nerve fibers.

5. ACKNOWLEDGMENT

This work was supported in part by a research grant from the Saxon State Ministry of Science and Arts (Grant 7541.82-FZR/309).

6. REFERENCES

1. Crone, C., and Olesen, S.-P., Autacoids and changes of capillary permeability, *Prog. Appl. Microcirc.* 10:21-31,1986.
2. Sharma, H.S., Olsson, Y., and Dey, P.K., Changes in blood-brain barrier and cerebral blood flow following elevation of circulation serotonin level in anesthetized rats, *Brain Res.* 517:215-223, 1990.
3. Sharma, H.S., Westman, J., Nyberg, F., Cervos-Navarro, J., and Dey, P.K., Role of serotonin and prostaglandins in brain edema induced by heat stress. An experimental study in the young rat, *Acta Neurochir. [Suppl.]* 60:65-70, 1994.
4. Banks, W. A., and Kastin A. J., Effect of neurotransmitters on the system that transports Tyr-MIF-1 and the enkephalins across the blood-brain barrier: a dominant role for serotonin, *Psychopharmacology* 98:380-385, 1989.
5. Marcusson, J., Fowler, C.J., Hall, H., Ross, S.B., and Winblad, B., Specific binding of [^{3}H]imipramine to protease-sensitive and protease-resistant sites, *J. Neurochem.* 44:705-711, 1985.
6. Underwood, M. D., Bakalian, M. J., Arango, V., Smith, R. W., and Mann, J. J., Regulation of cortical blood flow by the dorsal raphe nucleus: topographic organization of cerebrovascular regulatory regions, *J. Cereb. Blood Flow Metab.* 12:664-673, 1992.
7. Spatz, M., Maruki, C., Abe, T., Rausch, W.-D., Abe, K., and Merkel, N., The uptake and fate of the radiolabeled 5-hydroxytryptamine in isolated cerebral microvessels, *Brain Res.* 220:214-219, 1981.
8. Schuldiner, S., A molecular glimpse of vesicular monoamine transporters, *J. Neurochem.* 62:2067-2078, 1994.
9. Blakely, R.D., Defelice, L.J., and Hartzell, H.C., Molecular physiology of norepinephrine and serotonin transporters, *J. Exp. Biol.* 196:263-281, 1994.
10. Langer, S.Z., Moret, C., Raisman, R., Dubocovich, M.L., and Briley, M., High-affinity [^{3}H]imipramine binding in rat hypothalamus: association with uptake of serotonin but not norepinephrine, *Science* 210:1133-1135, 1980.
11. Hrdina, P. D., Differentiation of two components of specific [^{3}H]imipramine binding in rat brain, *Eur. J. Pharmacol.* 102:481-488, 1984.
12. Reith, M.E.A., Sershen, H., Allen, D., and Lajtha, A., High- and low-affinity binding of [^{3}H]imipramine in mouse cerebral cortex, *J. Neurochem.* 40:389-395, 1983.
13. Biessen, E.A.L., Norder, J. A., Horn, A.S., and Robillard, G.T., Evidence for the existence of at least two different binding sites for 5HT-reuptake inhibitors within the 5HT-reuptake system from human platelets, *Biochem. Pharmacol.* 37:3959-3966, 1988.
14. Brust, P., Bergmann, R., and Johannsen, B., Specific binding of [^{3}H]imipramine indicates the presence of a specific serotonin transport system on endothelial cells of porcine brain, *Neurosci. Lett.* 194:1-4, 1995.
15. Strolin Benedetti, M., and Dostert, P., Monoamine oxidase- from physiology and pathophysiology to the design and clinical application of reversible inhibitors, *Adv. Drug Res.* 23: 65-125, 1992.
16. Helmeste, D.M., and Tang, S.W., Kinase inhibitors compete with imipramine for binding and inhibition of serotonin transport, *Eur. J. Pharmacol.* 239-242, 1994.
17. Habert, E., Graham D., Tahraoui, L., Claustre, Y., and Langer, S.Z., Characterization of [^{3}H]paroxetine binding to rat cortical membranes. *Eur. J. Pharmacol.* 118:107-114, 1985.
18. Graham, D., and Langer, S.Z., The neuronal sodium-dependent serotonin transporter: studies with [^{3}H]imipramine and [^{3}H]paroxetine, *in:* Neuronal Serotonin, N.N. Osborne, and M. Hamon, eds., Wiley, New York, pp. 367-391, 1988.
19. Cheetham, S.C., Viggers, J.A., Slater, N.A., Heal, D.J., and Buckett, W.R., [^{3}H]paroxetine binding in rat frontal cortex strongly correlates with [^{3}H]5-HT uptake - effect of administration of various antidepressant treatments, *Neuropharmacology* 32:737-743, 1993.
20. Cool, D.R., Leibach, F.H., and Ganapathy, V., High-affinity paroxetine binding to the human placental serotonin transporter, *Am. J. Physiol.* 259:C196-C204, 1990.
21. Sette, M., Briley M. S., and Langer, S.Z., Complex inhibition of [^{3}H]imipramine binding by serotonin and nontricyclic serotonin uptake blockers, *J. Neurochem.* 40:622-628, 1983.
22. Cool, D.R., Leibach, F.H., and Ganapathy, V., Interaction of fluoxetine with the human placental serotonin transporter, *Biochem. Pharmacol.* 40: 2161-2167, 1990.
23. Hrdina, P.D., and Vu, T.B., Chronic fluoxetine treatment upregulates 5-HT uptake sites and 5-HT2 receptors in rat brain - an autoradiographic study, *Synapse* 14:324-331, 1993
24. Kalaria, R.N., and Harik, S.I., Blood-brain barrier monoamine oxidase: enzyme characterization in cerebral microvessels and other tissues from six mammalian species, including human, *J. Neurochem.* 49:856-867, 1987.

25. Hoyer, D., Clarke, D.E., Fozard, J.R., Hartig, P.R., Martin, G.R., Mylecharane, E.J., Saxena, P.R., and Humphrey, P.P.A., International union of pharmacology classification of receptors for 5-hydroxytryptamine (serotonin), *Pharmacol. Rev.* 46:157-203, 1994.
26. Hoyer, D., Pazos, A., Probst, A., and Palacios, J.M., Serotonin receptors in the human brain. I. Characterization and autoradiographic localization of 5-HT_{1A} recognition sites. Apparent absence of 5-HT_{1B} recognition sites, *Brain Res.* 376:85-96, 1986.
27. Hall, M.D., El Mestikawy, S., Emerit, M.B., Pichat, L., Hamon, M., and Gozlan, H., [^{3}H]8-Hydroxy-2-(di-n-propylamino)tetralin binding to pre- and postsynaptic 5-hydroxytryptamine sites in various regions of the rat brain, *Eur. J. Pharmacol.* 100: 263-276, 1984.
28. Leysen, J.E., Niemeegers, C.J.E., Van Nueten, J.M., Laduron, P.M., [^{3}H]ketanserin (R41468), a selective ligand for serotonin$_2$ receptor binding sites. Binding properties, brain distribution, and functional role, *Mol. Pharmacol.* 21: 301-314, 1982.
29. Hoyer, D., Engel, G., and Kalkman, H.O., Molecular pharmacology of 5-HT_1 and 5-HT_2 recognition sites in rat and pig brain membranes: radioligand binding studies with [^{3}H]5-HT, [^{3}H]8-OH-DPAT, (-)[^{125}I]iodocyanopindolol, [^{3}H]mesulergine and [^{3}H]ketanserin, Eur. J. Pharmacol. 118:13-23, 1985.
30. Yagaloff, K.A., and Hartig, P.R., ^{125}I-Lysergic acid diethylamide binds to a novel serotonergic site on rat choroid plexus epithelial cells. *J. Neurosci.* 5: 3178-3183, 1985.
31. Barker, E.L., and Sanders-Bush, E., 5-Hydroxytryptamine$_{1C}$ receptor density and mRNA levels in choroid plexus epithelial cells after treatment with mianserin and (-)-1-(4-bromo-2,5-dimethoxyphenyl)-2-aminopropane, *Molec. Pharmacol.* 44:725-730, 1993.
32. Havlik, S., and Peroutka, S.J., Differential radioligand binding properties of [^{3}H]5-hydroxytryptamine and [^{3}H]mesulergine in a clonal 5-hydroxytryptamine$_{1C}$ cell line, *Brain Res.* 584:191-196, 1992.

20

A 5-HT_2 RECEPTOR-MEDIATED BREAKDOWN OF THE BLOOD-BRAIN BARRIER PERMEABILITY AND BRAIN PATHOLOGY IN HEAT STRESS

An Experimental Study Using Cyproheptadine and Ketanserin in Young Rats

H. S. Sharma,[1-3] J. Westman,[2] J. Cervós-Navarro,[1] and F. Nyberg[3]

[1] Institute of Neuropathology
Free University
Berlin, Germany
[2] Department of Anatomy
Biomedical Centre
Uppsala University
S-751 23 Uppsala, Sweden
[3] Department of Pharmaceutical Biosciences
Biomedical Centre
Uppsala University
S-751 23 Uppsala, Sweden

SUMMARY

The possibility that 5-HT is involved in the pathophysiology of the blood-brain barrier (BBB) and subsequent brain damage via a receptor mediated mechanism was examined in a rat model of heat stress. Exposure of young rats (age 8-9 weeks) to heat at 38° C in a biological oxygen demand (BOD) incubator for 4 h resulted in a marked increase in the permeability of the blood-brain barrier (BBB) to Evans blue, ^{131}I-sodium and lanthanum in the cerebral cortex, cerebellum, hippocampus, thalamus, hypothalamus and brain stem. These brain regions exhibited 3-4% increase in brain water content and 30-40% reduction in the cerebral blood flow (CBF). Morphological examinations showed upregulation of heat shock protein (HSP) 72 kD immunostaining along with distorted neurones, glial cells, perivascular edema, and myelin disruption. Pretreatment with 5-HT_2 receptor antagonists cyproheptadine and ketanserin markedly reduced the BBB breakdown and brain edema formation. The CBF restored to normal values. Occurrence of HSP response and cell

Biology and Physiology of the Blood–Brain Barrier, edited by Couraud and Scherman
Plenum Press, New York, 1996

reactions were much less evident. This suggests that 5-HT participates in the pathophysiology of BBB breakdown and resulting brain damage in heat stress via 5-HT_2 receptors.

RÉSUMÉ

L'implication possible de la 5-HT, médiée par un récepteur spécifique, dans la pathophysiologie de la BHE conduisant à des lésions cérébrales a été étudiée dans le modèle animal de stress à la chaleur chez le rat. L'exposition de jeunes rats (8-9 semaines) à une température de 38° C pendant 4 heures dans un incubateur régulant le taux d'oxygène physiologique conduit à un accroîssement de la perméablité de la BHE pour le bleu Evans, l'iode 125 et le lanthane dans le cortex cérébral, l'hippocampe, le thalamus, l'hypothalamus et le bulbe cervical.Ces différentes régions présentent une augmentation de 3 a 4% de la teneur en eau du parenchyme cérébral et une diminution de 30 a 40% du flux sanguin cérébral. Des examens histologiques montrent une surexpression de la protéine heat shock (HSP) de 72 kD révelée par immunomarquage ainsi que des modifications morphologiques des neurones et des cellules gliales, ainsiqu'un oedème périvasculaire et des ruptures de la myeline. le prétraitement des animaux avec des antagonistes du récepteurs de la 5-HT_2 tels que la cyproheptadine et la kétansérine réduit les phénomènes de rupture de la BHE, de formation d'oedème et d'augmentation du flux sanguin cérébral. Ces prétraitements ne semblent par contre pas modifier l'expression de la HSP ou les lésions cellulaires d'une manière aussi nette. Ces résultats suggèrent que la 5-HT, via les récepteurs 5-HT_2 des cellules endothéliales cérébrales, est impliquée dans la pathophysiologie de rupture de la BHE conduisant à des lésions cérébrales dans ce modèle de stress a la chaleur.

INTRODUCTION

Heat stress (HS) is a serious clinical problem associated with brain pathology[1]. Post-mortem findings show micro-haemorrhages, edema and tissue softening in many parts of the brain[1]. The clinical symptoms of heat stress include hyperpyrexia, delirium, coma, unconsciousness and eventually death in more than 50% of the victims. The probable mechanisms underlying brain injury following HS is not well known.

It appears that breakdown of the blood-brain barrier (BBB) play an important role in the pathophysiology of brain damage observed in heat stress[2,5,7-9,12,14]. The BBB is a physiological dynamic barrier which maintains a constant composition of the extracellular fluid in which neurons bath[2,5,14]. It appears that a slight alteration in the neuronal microenvironment will lead to abnormal brain function. This is further evident with the fact that a breakdown of the BBB occurs in a wide variety of neurological and psychiatric disorders[5,14]. Such diseases are often associated with brain edema and structural changes in neurons, glial cells and myelin.

Increased permeability of the blood-brain barrier (BBB) occurs in heat stressed rats[12]. These animals also exhibit edema and cell changes in many parts of the brain[7-9]. This suggests that disruption of the BBB could be instrumental in the pathophysiology of brain edema and cell changes. However, the probable mechanisms of increased BBB permeability and brain edema in heat stress is not well characterised. It appears that various neurochemicals may play an important role.

There are several neurochemical mediators of BBB permeability and brain edema such as serotonin (5-HT), prostaglandins, histamine, amino acids, nitric oxide, free radicals, bradykinin and neuropeptides[2,14]. Under *in vivo* conditions, no single chemical compound is found to be responsible for all the observed pathophysiological changes. In fact, various experimental or diseases conditions suggest a wide spectrum of neurochemicals to be involved in the pathogenesis of BBB permeability and brain edema[2,5,14]. The detail mechanisms by which these neurochemicals will exert their actions on cerebral endothelium or brain parenchyma causing BBB disruption and adverse cell reaction is not well understood. One possibility could be that these neurochemicals may act through a receptor mediated mechanism, a hypothesis which require additional investigation.

This study is focused on 5-HT as a potential mediator of BBB permeability and brain edema. We wanted to know whether 5-HT can influence microvascular permeability and brain edema in heat stress through a receptor mediated mechanism. More than seven kinds of serotonergic receptors together with multiple subtypes have so far been identified in the central nervous system (CNS)[3,4]. The precise function of these receptor subtypes are not yet fully characterised. Since vasoconstrictor action of the amine in cerebrovascular bed is largely mediated via 5-HT$_2$ receptors[3,4], we examined the potential therapeutic efficacy of two potent 5-HT$_2$ receptor antagonists ketanserin[4] and cyproheptadine[13] on the BBB disturbances, brain edema formation and cell reaction in a rat model of heat stress.

MATERIALS AND METHODS

Animals

Experiments were carried out on male Sprague Dawley rats (body weight 90-100 g) housed at controlled ambient temperature (21±1 ° C) with 12 h light and 12 h dark schedule. Food and tap water were supplied ad libitum.

Exposure to Heat Stress

All experiments were commenced between 8.00 and 9.00 A.M. Rat were exposed to 4 h heat stress in a biological oxygen demand (BOD) incubator maintained at 38 °C. The relative humidity 45-50% and the wind velocity 20-25 cm/sec was kept constant[7-9]. Normal animals kept at room temperature served as controls.

Stress Symptoms

The core body temperature, occurrence of salivation and behavioural prostration was recorded in each rat during the period of 4 h heat exposure. Microhaemorrhages in gastric mucosa at post mortem was examined as indices of heat stress[12].

Blood-Brain Barrier Permeability, Brain Edema and Cell Changes

The blood-brain barrier (BBB) permeability in vivo was measured using Evans blue albumin, ^{131}I-sodium tracers[9,12]. These tracers were administered into the right femoral vein under urethane anaesthesia immediately after heat exposure and allowed to circulate for 5 min. The intravascular tracer was washed out by a brief 0.9% saline rinse through heart followed by perfusion with a lanthanum containing fixative. Selected tissue pieces from one hemisphere of the brain were taken and embedded in epon and processed for transmission electron microscopy7,8. The passage of tracer across the cerebral endothelium was examined using lanthanum as electron dense tracer at ultrastructural level[8]. Some tissue pieces were embedded in paraffin and some pieces were cut on vibratome sections for immunostaining of heat shock protein using standard procedures[6,11].

Tissue pieces from remaining half of the brain were dissected out, weighed and counted in a gamma counter at energy window 500-800 KeV. The radioactivity in brain samples were expressed as percentage of whole blood radioactivity that was taken immediately before perfusion[12]. After counting the radioactivity, tissue pieces were dried in an oven in order to determine their dry weight for calculation of brain edema using water content[7].

Cerebral Blood Flow

In separate groups of rats cerebral blood flow (CBF) was determined using radiolabelled iodinated microspheres[9,12]. About one million microspheres were injected into the left atrium and the reference blood was collected through right femoral artery at the rate of 0.8 ml/min. About 90 sec after infusion, the animals were killed by decapitation and various brain regions were removed, weighed and counted. The CBF was calculated according to the formula: CBF ml/g/min = CPM.g^{-1}brain tissue x RBF (reference blood flow)÷ Total CPM in the reference blood sample[12].

Drug Treatment

In separate groups of animals, cyproheptadine (Merck, Sharp and Dohme, UK) or ketanserin (Janssen Pharmaceuticals, Belgium) was injected in a dose of 1 mg, 5 mg or 10 mg per kilogram intraperitoneally. One group of animals were exposed to heat stress 30 min after drug injection[5,13]. One group of animals were kept at room temperature to serve as drug treated controls. At the end of 4 h heat stress or 4.5 h after drug administration, the BBB permeability, brain edema, CBF and cell changes were examined.

Statistical Analysis

ANOVA followed by Dunnet's test for multiple group comparison was used to evaluate statistical significance of the data obtained. A p-value less than 0.05 was considered to be significant.

RESULTS

Stress Symptoms

Profound hyperthermia, behavioural symptoms and many microhaemorrhages in the gastric mucosa were noted in untreated animals subjected to 4 h heat stress (Table I). Pretreatment with cyproheptadine and ketanserin (10 mg/kg) attenuated the hyperthermic response without affecting the behavioural symptoms and microhaemorrhages in the gastric mucosa (Table I). The drug treatment alone did not influence these parameters in normal animals.

Blood-Brain Barrier Permeability

Subjection of untreated animals to 4 h heat stress resulted in marked extravasation of Evans blue and radioactive iodine (Fig 1a) in many brain regions. Pretreatment with cyproheptadine and ketanserin in doses of 10 mg per kilo gram significantly reduced the extravasation of these tracers in brain after heat stress. Lower doses of the drug (1 mg or 5 mg) were not effective in reducing the BBB permeability. The drug treatment alone did not influence the tracer distribution across the cerebral vessels in normal animals.

Table I. Influence of cyproheptadine and ketanserin on stress symptoms in heat stressed rats

Type of experiment	Stress symptoms			
	Rectal temperature °C	Salivation	Prostration	Haemorrhagic spots in stomach no.
Control (5)	37.4±0.21	nil	nil	nil
Heat stress (6)	41.38±0.32**	+++	+++	60±12
Cyproheptadine (5)[a]	39.44±0.41**[b]	+++	+++	40±18
Ketanserin (6)[a]	39.64±0.22**[b]	+++	+++	32±20

a = drugs were given in separate groups of rats (10 mg/kg, i.p.) 30 min prior to heat stress.
+++ = severe
b = significantly different from heat stressed controls (P < 0.05)
** = P < 0.01 ANOVA followed by Dunnet's test for multiple group comparison.

Brain Edema

Brain water content was significantly elevated in many brain regions after heat stress (Fig 1b). Pretreatment with ketanserin and cyproheptadine (10 mg/kg) thwarted the accumulation of brain water in these animals (Fig 1). Lower doses of the drugs were less effective in reducing brain edema caused by heat stress. The drug treatment alone did not influence brain water content in normal animals.

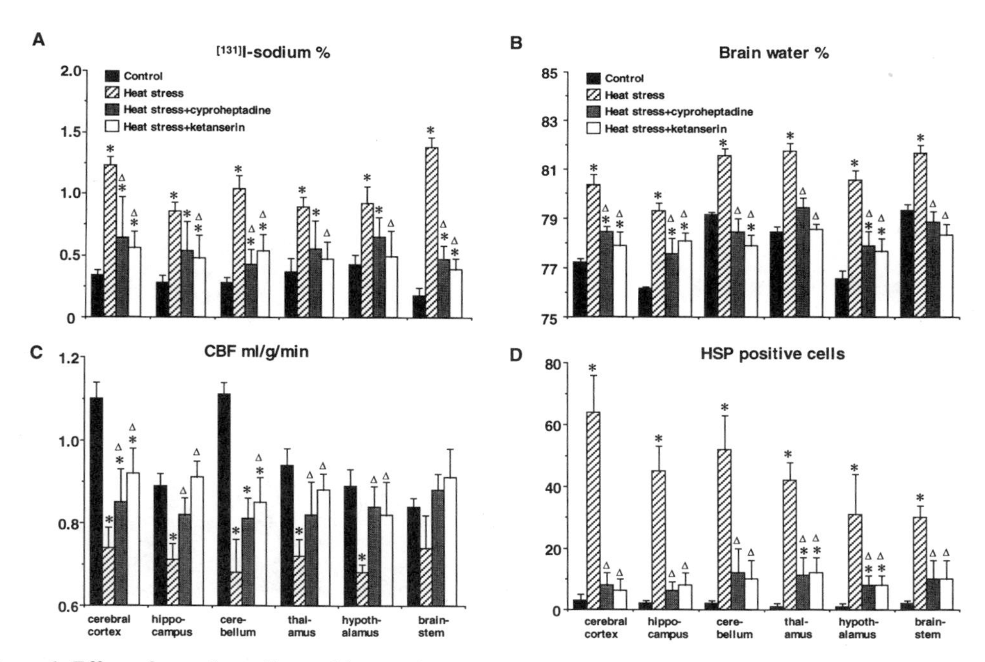

Figure 1. Effect of cyproheptadine and ketanserin on heat stress induced alterations in regional blood-brain barrier permeability (A), brain edema (B), cerebral blood flow (C) and heat shock protein response (D) in young rats. * = P < 0.01 compared from control. Δ = P < 0.05, compared from untreated heat stressed rats (ANOVA followed by Dunnet's test for multiple group comparison).

Cerebral Blood Flow

The CBF was decreased by 30-40% in many brain regions following heat stress (Fig. 1c). Pretreatment with ketanserin and cyproheptadine (10 mg/kg) significantly elevated the CBF near normal values in many brain regions of heat stressed animals (Fig 1). The drug treatment alone did not influence the CBF significantly.

Heat Shock Protein Immunostaining

Marked upregulation of heat shock protein (HSP, 72 kD) immunoactivity occurs in many parts of the brain of animals subjected to heat stress. A semiquantitative data on HSP positive cells in some brain regions is shown in Fig 1d. Prior treatment with ketanserin or cyproheptadine (10 mg/kg) significantly reduced the number of HSP positive cells in most of the brain regions of heat stressed animals (Fig 1d).

Cell Changes

Ultrastructural studies in heat stressed animals showed marked perivascular edema, membrane disruption, synaptic damage, neuronal, glial cell distortion and myelin vesiculation (Fig 2a). Lanthanum can be found in the basal lamina and within the cerebral endothelium, although the tight junctions were intact. Pretreatment with ketanserin or cyproheptadine minimised the cell changes in heat stressed rats (Fig 2b). In these animals lanthanum was mainly confined within the lumen of the microvessels and occurrence of perivascular edema is greatly reduced.

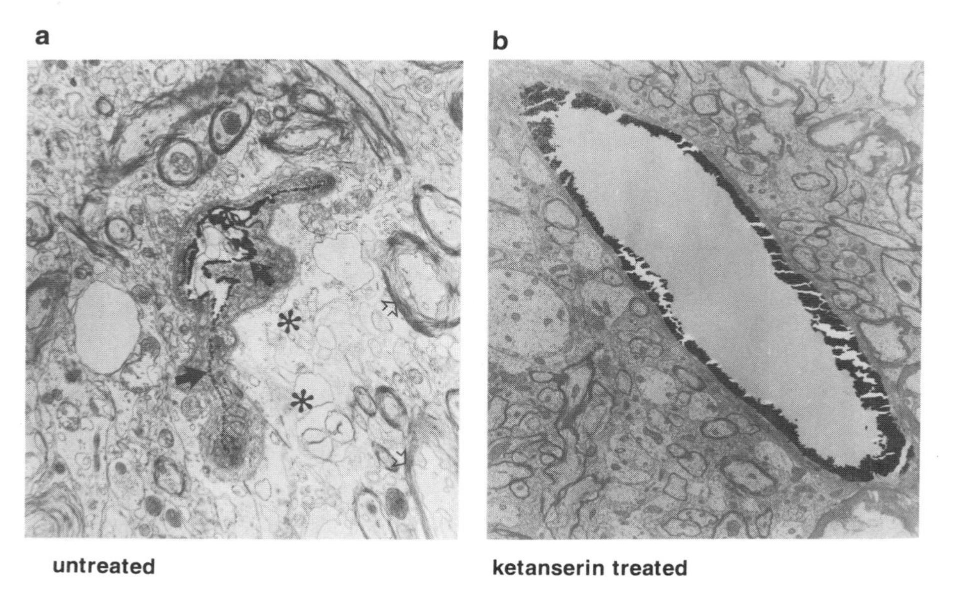

Figure 2. Low power electron micrograph from the cerebral cortex (layer III) of one untreated heat stressed (a) and one ketanserin pretreated heat stressed rat (b). Edema, membrane damage (*), collapse of microvessel (filled arrows) and myelin vesiculation (blank arrows) is very common in untreated stressed rat. These cellular and microvascular changes are markedly attenuated by ketanserin pretreatment. bar = 1 μm.

DISCUSSION

The salient new finding of the present study is that serotonin can influence the pathophysiology of the BBB permeability and consequently brain damage via $5\text{-}HT_2$ receptors. This is evident from the fact that pretreatment with two 5-HT receptor antagonists cyproheptadine and ketanserin reduced the BBB breakdown caused by heat stress in a dose dependent manner. Since drug treatment alone did not influence these parameters in normal rats, it appears that the drugs are capable of antagonising the action of serotonin in heat stress which is found to be elevated in this condition in earlier studies[9,12].

It has been suggested that both ketanserin and cyproheptadine possess a weak histamine H_1 receptor antagonistic activity[3,4,13]. Since histamine H_1 receptors do not seem to participate in reducing the increased BBB permeability and brain edema formation in heat stress[9], it appears that the protective effects of ketanserin and cyproheptadine in the present study is mainly due to their action of blocking $5\text{-}HT_2$ receptors in the CNS. However, it remains to be seen whether blockade of $5\text{-}HT_1$ receptors also can influence the BBB permeability and brain edema in heat stress, a feature which requires additional investigation.

Another intriguing finding of this study is a close relationship between breakdown of the BBB permeability and development of brain edema and cell changes in heat stress. Our results thus indicate that breakdown of the BBB could be instrumental in precipitating pathological changes in the brain. This is evident from the fact that in untreated rats heat stress resulted in profound brain edema and cell changes which are markedly reduced by cyproheptadine and ketanserin pretreatment.

The probable mechanisms by which alteration in the BBB function will induce brain pathology is not clear. However it appears that a breakdown of the BBB will result in leakage of various humoral, immunological, ionic, and the chemical factors into the cerebral compartment from the vascular compartment. This will induce a cascade of electrophysiological, biochemical and immunological reactions leading to the pathology of the CNS. Entry of serum proteins into the neuronal microenvironment will induce vasogenic edema. Neurochemicals, gaining access to the CNS compartment may cause profound vascular and neuronal reactions. All these factors alone or in combination can lead to the secondary brain damage[14].

5-HT is a mediator of BBB permeability and brain edema[14]. Infusion of 5-HT in normal animals induces breakdown of the BBB permeability and cause brain edema[10]. Increased levels of 5-HT is found in animals subjected to heat exposure[9,12]. Thus a possibility exists that 5-HT by itself when gaining access into the cerebral compartment will induce the cell reactions[15]. A marked reduction of HSP immunostaining following heat stress by pretreatment with cyproheptadine and ketanserin suggest that serotonin is somehow involved in stress protein response and the resulting cell reaction[6]. These findings further support the idea that the amine is involved in the pathological mechanisms of hyperthermia.

In summary, our results show that blockade of $5\text{-}HT_2$ receptors by cyproheptadine and ketanserin has a neuroprotective effect in heat stress. This effect of the drugs appears to be mediated by their ability to minimise vascular reaction. These observations suggest an important role of 5-HT in the pathophysiology of BBB and brain pathology in heat stress.

ACKNOWLEDGMENTS

This study is supported by grants from Swedish Medical Research Council project nos. 2710, 9459, 9710 and 10 523; Alexander von Humboldt Foundation, Bonn, Germany; The University Grants Commission, New Delhi, India and Astra Hässle, Mölndal, Sweden.

The skilful technical assistance of Kärstin Flink, Ingmarie Olsson, Madeleine Thörnwall, Gunilla Tibling, Elisabeth Scherer, Hana Plukhan and Franziska Drum is greatly appreciated.

REFERENCES

1. Austin M G, Berry J W (1956) Observation on one hundred cases of heatstroke. J Am Med Ass 161: 1525-1529.
2. Black K L (1995) Biochemical opening of the blood-brain barrier. Adv Drug Del Rev 15: 37-52.
3. Chaouloff F (1993) Physiopharmacological interactions between stress hormones and central serotonergic system. Brain Res Rev 18: 1-32.
4. Janssen P A J (1983) 5-HT_2 receptor blockade to study serotonin-induced pathology. Trends Pharmacol Sci 5: 198-206.
5. Johansson B B, Owman C, Widner H (1990) Pathophysiology of the blood-brain barrier. Fernström Foundation Series 14, Elsevier, Amsterdam.
6. Sharma H S, Olsson Y, Westman J (1995) A serotonin synthesis inhibitor, p-chlorophenylalanine reduces the heat shock protein response following trauma to the spinal cord. An immunohistochemical and ultrastructural study in the rat. Neurosci Res 21: 241-249.
7. Sharma H S, Westman, J, Nyberg F, Cervós-Navarro J, Dey P K (1994) Role of neurochemicals in the blood-brain barrier permeability, cerebral blood flow, vasogenic oedema and cell changes in heat stress. Experimental observations in the rat. J Physiol (Lond.) 480: 12 P.
8. Sharma H S, Westman J, Nyberg F, Cervós-Navarro J, Dey P K (1994) Role of serotonin and prostaglandins in brain edema induced by heat stress. An experimental study in the rat. Acta Neurochir (Suppl). 60: 65-70.
9. Sharma H S, Nyberg F, Cervós-Navarro J, Dey P K (1992) Histamine modulates heat stress induced changes in blood-brain barrier permeability, cerebral blood flow, brain oedema and serotonin levels: An experimental study in conscious young rats. Neuroscience 50: 445-454.
10. Sharma H S, Olsson Y, Dey P K (1990) Blood-brain barrier permeability and cerebral blood flow following elevation of circulating serotonin level in the anaesthetized rats. Brain Res 517: 215-223.
11. Sharma H S, Westman J, Cervós-Navarro J, Gosztonyi G (1992) Acute systemic heat exposure increases heat shock protein (HSP-70 kD) immunoreactivity in the brain and spinal cord of young rats. Clin Neuropathol 11: 174-175.
12. Sharma H S, Dey P K (1987) Influence of long-term acute heat exposure on regional blood-brain barrier permeability, cerebral blood flow and 5-HT level in conscious normotensive young rats. Brain Res 424:153-162.
13. Stone C A, Wenger H C, Ludden C T, Stavorski J M, Ross C A (1961) Antiserotonin-antihistaminic properties of cyproheptadine. J Pharmac Exp Ther 131: 73-84.
14. Wahl M, Unterberg A, Baethmann A, Schilling L (1988) Mediators of blood-brain barrier function and formation of vasogenic brain edema. J Cerebr Blood Flow Metab 8: 621-634.
15. Winkler T, Sharma H S, Stålberg E, Olsson Y, Dey P K (1995) Impairment of blood-brain barrier function by serotonin induces desynchronisation of spontaneous cerebral cortical activity: Experimental observations in the anaesthetised rat. Neuroscience 68: 1097-1104.

21

CORRELATION BETWEEN THE *IN VITRO* AND *IN VIVO* BLOOD-BRAIN BARRIER PERMEABILITIES OF H_2 RECEPTOR ANTAGONISTS

Evangeline Priya Eddy,[1] Roman A. Olearchyk,[1] Frederick M. Ryan,[1] Timothy K. Hart,[2] and Philip L. Smith[1]

[1] Department of Drug Delivery
[2] Department of Toxicology
SmithKline Beecham Pharmaceuticals
King of Prussia, Pennsylvania 19406

SUMMARY

In vitro models of the Blood-Brain Barrier (BBB) offer the possibility of a first level screen on new chemical entities for predicting potential brain permeability *in vivo*. An *in vitro* BBB model was established, utilizing cultured bovine brain microvessel endothelial cells (BBMECs). BBMEC cultures were assessed morphologically and biochemically for maintainance of *in vivo* cerebral microvessel endothelial characteristics. Barrier function was assessed by growing cultures on polycarbonate membrane filter inserts (Costar 12mm two piece Snapwell™ transwell plates; 0.4 um pore size) and placing these inserts between diffusion chambers and measuring permeability of C^{14}- sucrose, a paracellular marker with low permeability and H^3-propranolol, a transcellular permeability marker with high permeability. Between 10-15 days in culture, sucrose permeability was the lowest and ranged between 1.6-3.0 $x10^{-5}$ cm/sec (0.058-0.12 cm/hr) while propranolol permeability was unchanged from 6-12 days in culture. The validity and usefullness of the model in predicting *in vivo* brain penetration was assessed by measuring the *in vitro* permeability of seven H_2 receptor antagonists whose *in vivo* CNS permeability was assessed by Young et al.[4] A very good correlation was obtained between the *in vitro* and *in vivo* permeability for these molecules (R^2=0.898). The results suggest that the *in vitro* model of the BBB using BBMECs provides a useful screening system for selecting molecules which will possess appropriate characteristics to access the brain.

RÉSUMÉ

Les modèles cellulaires in vitro de la barrière hémato-encéphalique (BHE) permettent de prédire la pénétration cérébrale in vivo de nouvelles molécules et ainsi d'effectuer un

Biology and Physiology of the Blood–Brain Barrier, edited by Couraud and Scherman
Plenum Press, New York, 1996

premier criblage de ces composés thérapeutiques.Un tel modèle a été mis en place à partir de cultures de cellules endothéliales de microvaisseaux cérébraux bovins (BBMEC). Le maintien des caractéristiques de la BHE in vivo a été étudié dans ces cultures. La fonction de barrière physiologique de la BHE a été étudiée en cultivant ces BBMEC sur des membranes poreuses de polycarbonate dans des inserts de culture (Costar 12 mm, Snapwell, porosité 0.4 μm). Les membranes sont ensuite transférées dans une chambre de diffusion pour mesurer la perméabilité de la monocouche cellulaire pour le sucrose C14, un marqueur de la perméabilité paracellulaire, et pour le propanolol 3H, un marqueur de la perméabilité transcellulaire. La perméabilité des cultures au sucrose est la plus faible après 10-15 jours de culture, entre 1.6 et 3.0×10^{-5} cm/s(0.058-0.12 cm/h), tandis que le passage du propanolol est stable entre 6 et 12 jours de culture. la validation de ce modèle pour prédire la pénétration cérébrale in vivo a été étudiée en déterminant la perméabilité in vitro de 7 antagonistes du récepteur H2 , dont la perméabilité in vivo est connue (Young et al., J. Med. Chem.1988,31,656-671.). Une très bonne corrélation entre les valeurs de la perméabilité de ces antagonistes in vitro et in vivo a été obtenue (R2 = 0.898). Ces résultats démontrent que des cultures primaires de cellules endothéliales de microvaisseaux cérébraux bovins peuvent constituer un modèle in vitro de BHE permettant un criblage de médicaments à visée cérébrale.

1. INTRODUCTION

Although *in vivo* studies provide a physiological situation in studying the transport of drugs across the Blood-Brain Barrier (BBB), there are several limitations. The development and use of *in vitro* cell culture models for assessing BBB drug transport and metabolism has several advantages and applications especially in the pharmaceutical industry. These include: 1) requirement of small amounts of test compound; 2) rapid assessment of relative permeability of molecules and their analogs across the BBB; 3) elucidation of molecular mechanisms of transport (transporters, metabolism and regulation of barrier permeabily); 4) minimization of time consuming and expensive animal studies; 5) easier analytical method development due to the buffers employed and 6) the potential for use of human cells.

Bovine brain microvessel endothelial cell (BBMEC) monolayers that constitute the BBB have been grown in primary cultures and used as an *in vitro* model to study drug transport mechanisms, regulation of permeability and metabolism[1,2]. The P-glycoprotein efflux system has also been shown to be expressed by these cells in culture[3]. The anatomical basis of the BBB shows that it is not a static barrier but one that has special features subject to endogenous regulation and hence a challenge for drug delivery to the brain.

The present study was undertaken to validate the usefulness of the *in vitro* BBMEC system as a tool to identify potential brain penetrants. BBMEC cultures were assessed morphologically by light and transmission electron microscopy and biochemically by the identification of biochemical markers. BBMEC cultures were stained histochemically for the presence of alkaline phosphatase, immunohistochemically for the presence of the von Willebrand Factor and by the uptake of fluorescent labelled DiI-Ac-LDL. Barrier properties of the BBMEC monolayers were evaluated by testing the permeability of a low permeability paracellular marker, sucrose, and a high permeability transcellular marker, propranolol. *In vitro* permeability data obtained for seven H_2-receptor antagonists was correlated with *in vivo* CNS permeability assessed previously by Young et al.,[4]. This model can become a useful tool in the pharmaceutical industry as a possible first level screen of new chemical entities (NCE) for predicting potential brain endothelial cell permeability.

2. MATERIALS AND METHODS

[^{3}H]-Propranolol and [^{14}C]-Sucrose were obtained from New England Nuclear-DuPont (Boston, MA.). H_2 receptor antagonists, SKF 93319, SKF 93479, SKF 94117, SKF 94445, SKF 94674, SKF 95282 and SKF 34427 were obtained from Dr. Robert Mitchell, Analytical Sciences, SmithKline Beecham Pharmaceuticals, U.K. Alkaline phosphatase, kIT# 86-C, Anti-Human von Willebrand Factor, Percoll, dextran, heparin, Amphotericin B, gentamicin and fibronectin and endothelial growth supplements were purchased from Sigma Chemical Company (St.Louis, MO.), DiI-Ac-LDL was obtained from Biomedical Technologies Inc., (Stoughton, MA.). Ham's F12 medium and minimum essential medium (MEM) were purchased from Gibco (Grand Island, NY.), dispase and collagenase/dispase were obtained from Boehringer-Mannheim Biochemica (Indianapolis, IN.). Collagen strands were removed from rat tails and collagen solution at 3mg/ml was prepared in 0.1% acetic acid [7]. The collagen solution was centrifuged and the supernatent used as the coating solution.

2.1. Cell Culture

Primary cultures of bovine brain microvessel endothelial cells were prepared according to precedures described earlier [2,5,6,7] with only slight modifications. The incubation time for collagenase/dispase digestion was reduced from 4 hours to 3 hours with shaking in a water bath at 37°C. After the gradient centrifugation step in Percoll, the removal of the middle band containing the endothelial cells, was changed. Only cell clumps were removed rather than the whole middle layer of the 50% percoll gradient. Cells were washed three times to remove percoll with medium containing 10% horse serum (Hyclone Laboratories Inc.,Logan, Utah). Primary cultures were frozen in freezing medium containing 10% DMSO and stored in liquid nitrogen.

Frozen primary cultures were thawed and resuspended in culture medium[2], washed once and counted. Culture dishes (60mm) were prepared to establish a basement membrane according to previously described procedures [2]. Cells were seeded at a density of $1x10^6$ cells/dish and grown in culture medium (Ham's F-12/ MEM (1:1) with gentamicin, amphotericin B, heparin and endothelial growth supplements) at 37°C, 5% CO_2 and 95% relative humidity. Confluent monolayers were fixed in 2% glutaraldehyde and processed for transmission electron microscopy.

Confluent BBMEC monolayers were tested for the presence of alkaline phosphatase activity by using the Sigma kit # 86-C, the presence of the von Willebrand Factor by using Sigma's Immuno Chemical Anti-Human von Willebrand factor developed in rabbit. The uptake of DiI-AC-LDL was tested using DiI-Ac-LDL at a concentration of 10ug/ml for 4 hours at 37°C. Fluorescent uptake was observed with a fluorescent microscope using standard rhodamine excitation:emission filters.

2.1.1. Barrier Function Characterization. BBMECs were grown on 0.4 um pore size polycarbonate filter inserts of 12mm two-piece Snapwell™ transwells (Costar, Cambridge, MA.) whose surfaces were previously prepared as described above for culture dishes with collagen and fibronectin. Each well was seeded with 70,000-80,000 cells/cm^2 and grown in Ham's F-12/EMEM medium for 10-15 days. Monolayers were tested for barrier function from day 6 in culture onwards, by placing the snapwell inserts in diffusion chambers (The Diffusion Chamber Systems, Precision Instrument Design,) containing Dulbecco's PBS with 10mM glucose. [^{3}H]-propranolol and [^{14}C]-sucrose transport was measured by placing sufficient concentration of the radiolabelled drug along with cold drug (200uM) on the donor side of the chamber and sampling from both the donor and receiver sides every 15 minutes

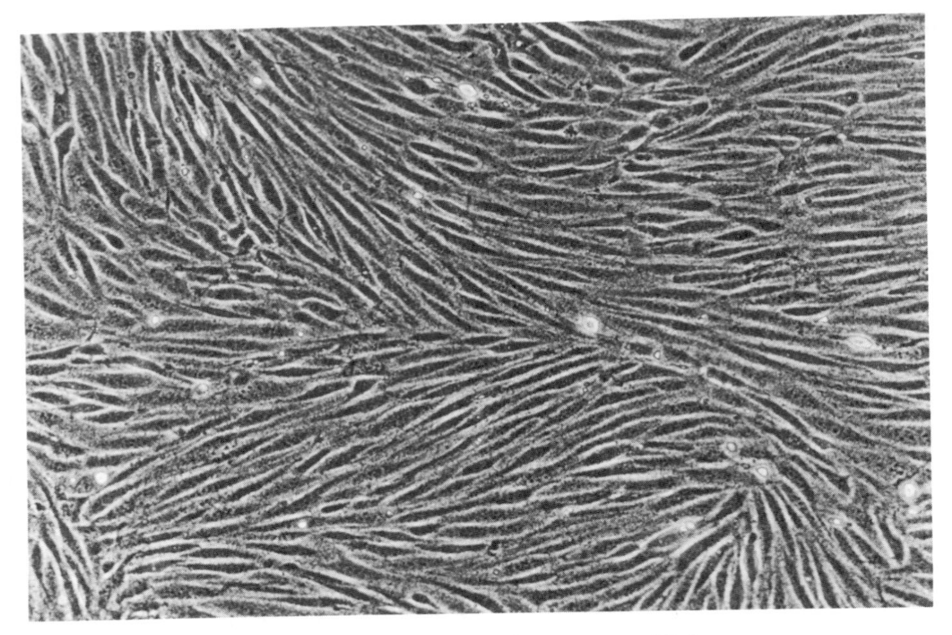

Figure 1. BBMEC momolayers *in vitro.* (20x)

for 1 hour. Radioactivity was measured in a Beckman Scintillation Counter using Ready safe liquid scintillation cocktail (Beckman Instruments Inc., Fullerton, CA U.S.A.). Drug permeability, P, was calculated using the following equation.

$$P = \frac{V_r}{A \bullet C_o} \bullet \frac{dC_r}{dt} \tag{1}$$

where V_r is the receiver volume, A is the surface area of the exposed tissue, C_o is the radioactive counts/drug concentration in the donor solution and dC_r/dt is the rate of change of counts/drug concentration on the receiver side [8]. All values are presented as mean and standard error of the mean of values from triplicate chambers.

2.1.1.1. In vitro/in vivo correlation studies. The permeability of H_2 receptor antagonists across BBMEC monolyers were measured between 10-15 days in culture when confluency was reached and tight barriers were formed. Concentration of drug used for testing was 200uM and placed in the donor chamber on the apical side of the monolayer and flux rates measured for 60min with sampling every 15min as described above for sucrose. H_2 receptor antagonists were analysed by reverse-phase HPLC on a C18 column. The *in vitro* permeability, P (cm/h) was calculated using Eq 1 described above.

Data for *in vivo* brain to blood ratio for H_2 receptor antagonists was taken from Young et al.[4]

3. RESULTS AND DISCUSSIONS

The *in vitro* BBMEC system has been shown to be useful in evaluating relative permeabilities of potential brain penetrants.The results reported here demonstrate a good

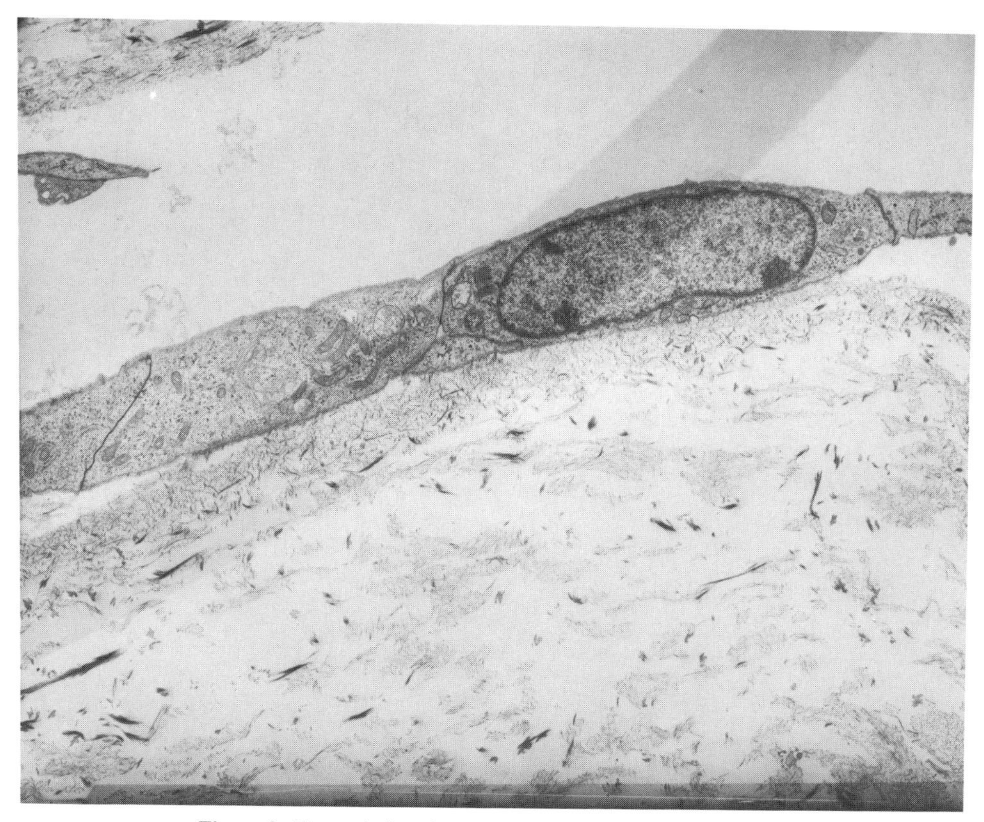

Figure 2. Transmission electron microscopy of BBMEC *in vitro*.

correlation between *in vitro* BBMEC permeability and the *in vivo* BBB penetration of a set of H_2 receptor antagonists. The BBMEC system has been shown to retain several morphological properties of the *in vivo* BBB, as demonstrated by the typical endothelial-like morphology obtained by light and transmission electron microscopy (Figures 1, 2 & 3). The figures indicate the presence of abundant mitochondria, few fenestra and presence of tight intercellular junctions. In addition, the presence of several biochemical endothelial cell markers have been retained in culture such as, the presence of alkaline phosphatase, von Willerbrand factor and γ-glutamyl transpeptidase (data not shown) and the specific uptake of the DiI-Ac-LDL.These results are in good agreement with results from prior studies[1,2,9]. The BBMEC system is a primary culture and our results indicate that several biochemical markers are maintained. However, on passaging the primary cell lines, these biochemical markers may be lost due to dedifferentiation and hence factors, such as the presence of astrocytes/astrocytic input may be necessary additions to the culture to maintain BBB characteristics[9,10,11,12,13].

Although the BBMEC system retains the specialized BBB function of possessing tight intercellular junctions, they are not comparable to the *in vivo* situation[14]. Sucrose has been used as a low permeability paracellular marker. The high permeability of C^{14}-sucrose (13.9 x 10^{-5} cm/sec or 0.5 cm/hr) at 6 days in culture indicates that the monolayers have not reached confluency (Figure 4), however, at days 10-15 in culture, C^{14}-sucrose permeabilities (Figures 4 & 5) decreased significantly (1.6-3.0 x 10^{-5} cm/sec or 0.058-0.12 cm/hr) indicative

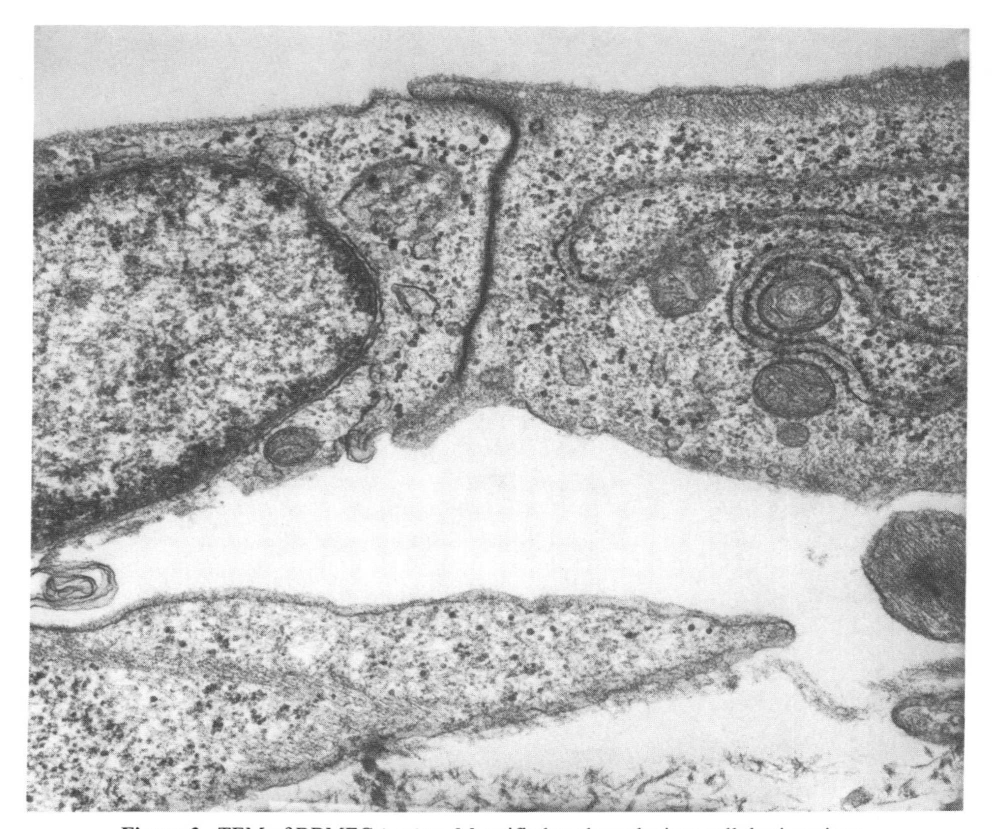

Figure 3. TEM of BBMEC *in vitro*. Magnified to show the intercellular junctions.

of the formation of a monolayer with tight junctions limiting paracellular transport. Sucrose permeability of monolayers above this range of 3.0 x 10^{-5} cm/sec or 0.12 cm/hr are considered not acceptable for use in transport studies. H^3-propranolol was used as a high permeability transcellular marker and did not change significantly over the 6-12 days in culture (Figure 5) indicative of its transcellular transport properties across the endothelial monolayer.

The expression of the P-glycoprotein efflux system on the luminal side of the cerebral endothelial cells is one of the possible reasons for the limitation of BBB.permeability *in vivo* of some lipophilic drugs including several anti-cancer agents into the brain[15]. Figure 6 demonstrates the expression and maintainance of the P-glycoprotein efflux system in the BBMEC at 10-15 days in culture. These results therefore, indicate that the BBMEC sytem has retained several morphological ,biochemical and functional properties of the *in vivo* BBB confirming other studies[1,2] and may be useful as an *in vitro* model to study BBB permeability mechanisms.

The BBMEC system was further studied to validate the usefulness of the *in vitro* model to predict BBB permeability by testing the *in vitro* permeability of a set of H_2 receptor antagonists with different molecular structures (Figure 7) whose *in vivo* BBB penetration was determined by Young et al.[4] (ref). Figure 8 demonstrates a good correlation (r2= 0.898) between the *in vitro* permeability of the H_2 receptor antagonists and the *in vivo* BBB penetration. These results indicate that drug transport *in vitro* correlates with drug transport

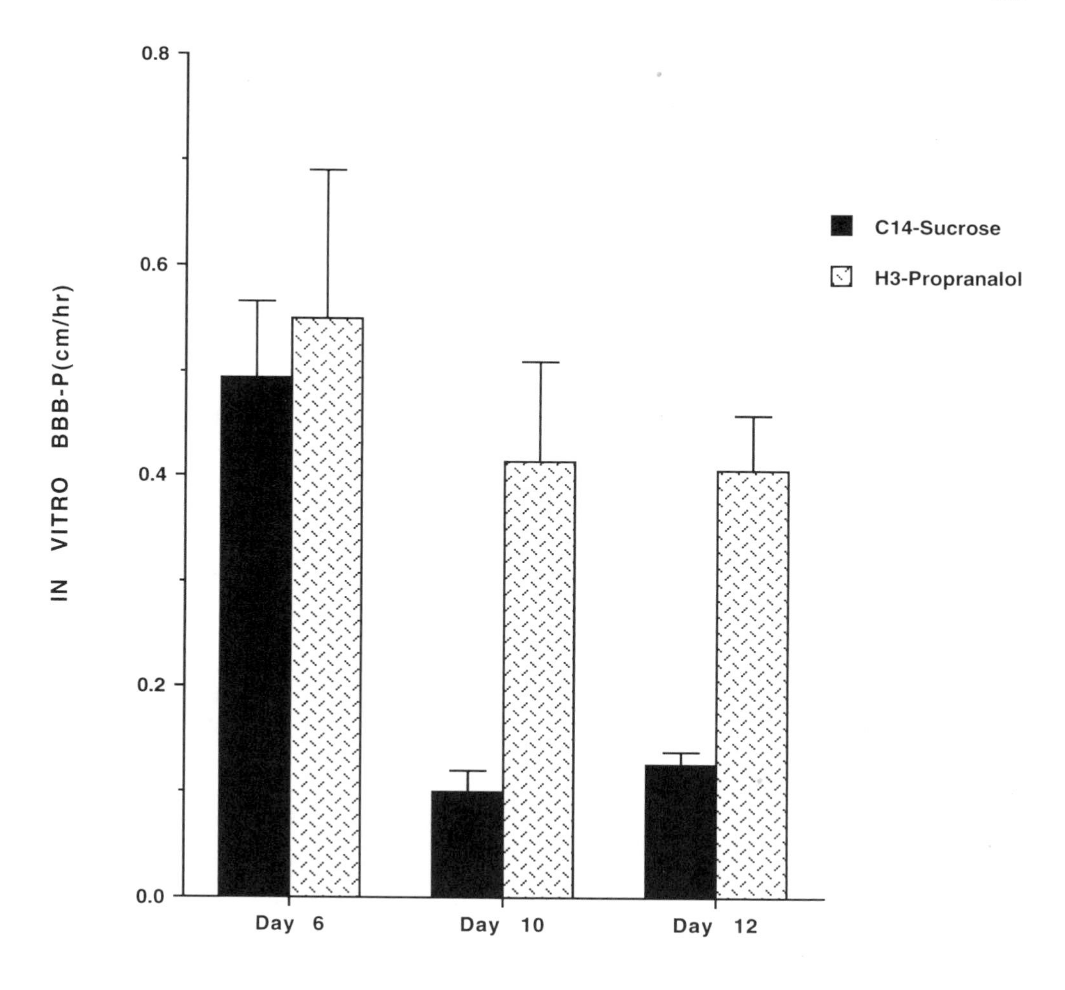

Figure 4. Permeability of sucrose and propranolol across BBMEC monolayers.

in vivo and suggests that the *in vitro* BBMEC system can be a useful tool in the selection of molecules with access into the brain.

This model can be used as a first level screen for pharmaceutical NCEs which can give us information for compounds that can penetrate or be excluded from the CNS. However, several issues need to be addressed regarding the reliability and efficiency of using this system to predict *in vivo* penetration.

1. Protein binding: Drug permeability can be limited by its capacity to bind proteins and hence render them unavailable for transport. Studies are ongoing to standardize the inclusion of serum proteins in the assay medium such as BSA to be able to address the issue of protein binding effects on the permeability of drugs.
2. Use of astrocyte co-cultures/astrocyte conditioned medium: It is known that astrocytes are in close proximity with the endothelial cells and may be involved

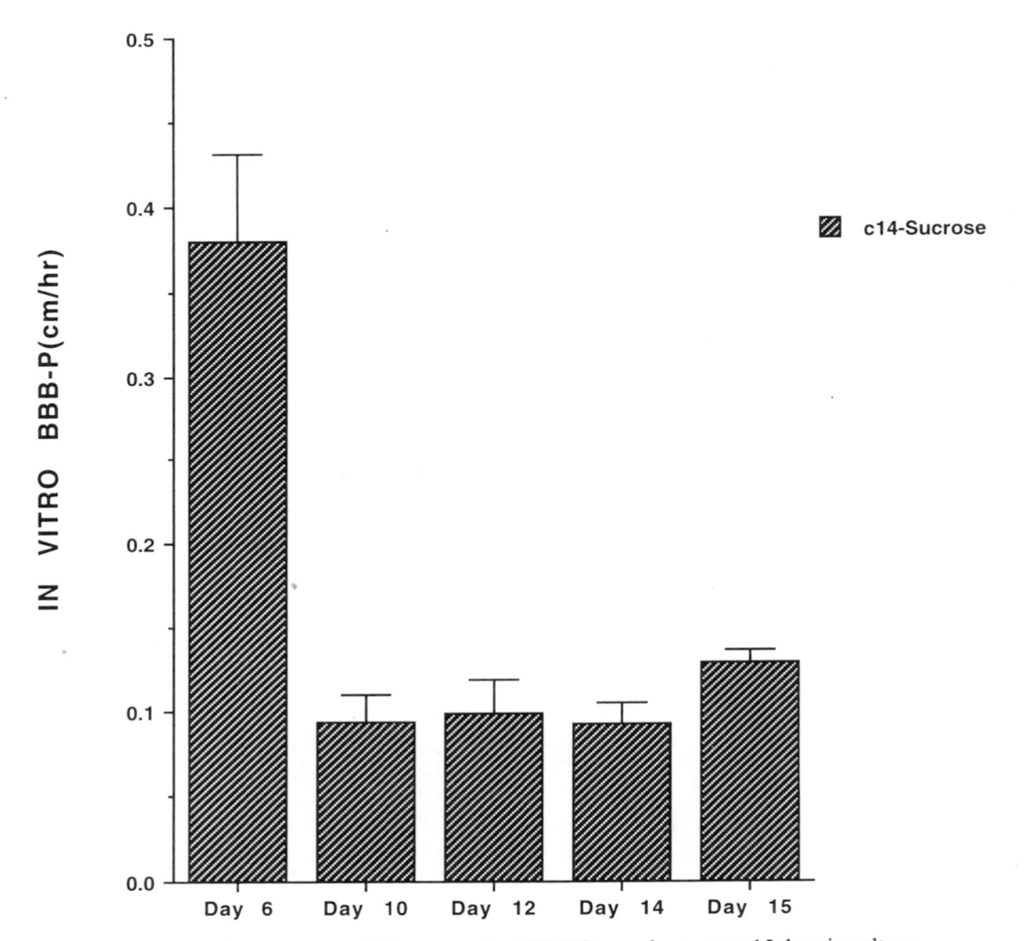

Figure 5. Sucrose permeability across the BBMEC monolayer up to 15 days in culture.

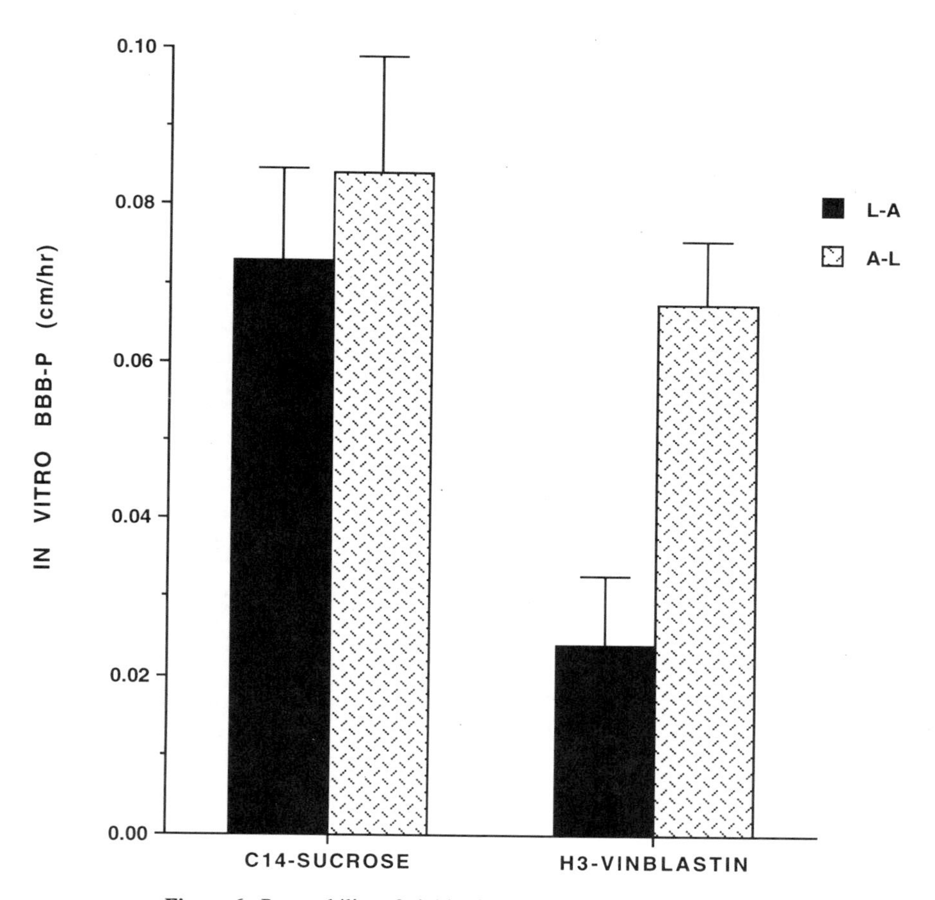

Figure 6. Permeability of vinblastine across BBMEC monolayers

* L= luminal side of the monolayer
A= abluminal side of the monolayer

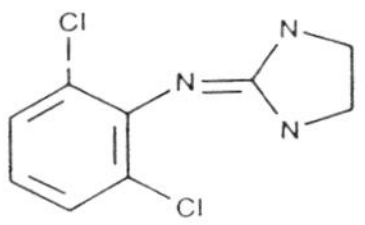

SK&F 34427

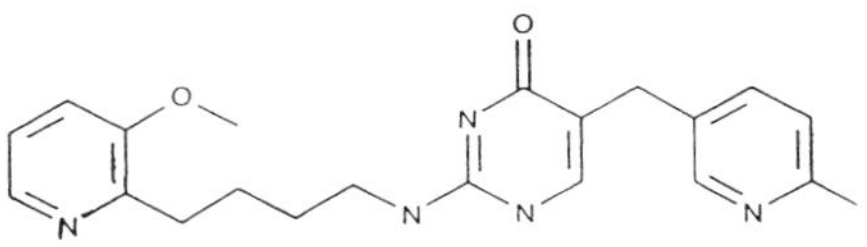

SK&F 93319

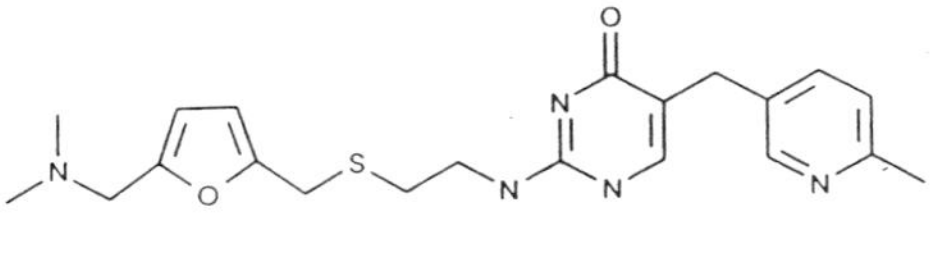

SK&F 93479

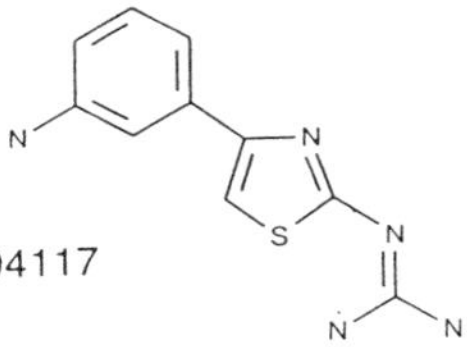

SK&F 94117

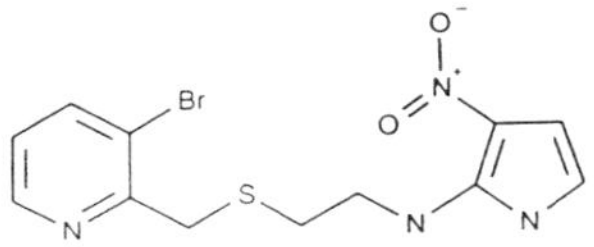

SK&F 94445

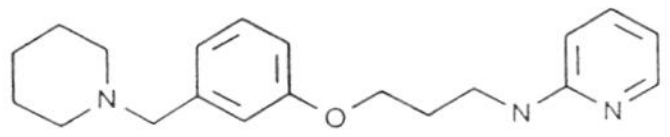

SK&F 94674

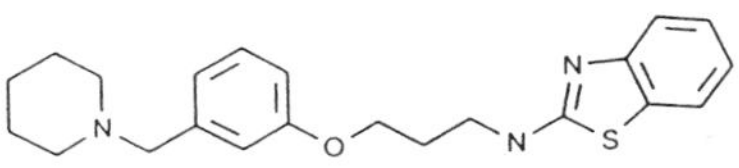

SK&F 95282

Figure 7. Structures of H_2 receptor antagonists

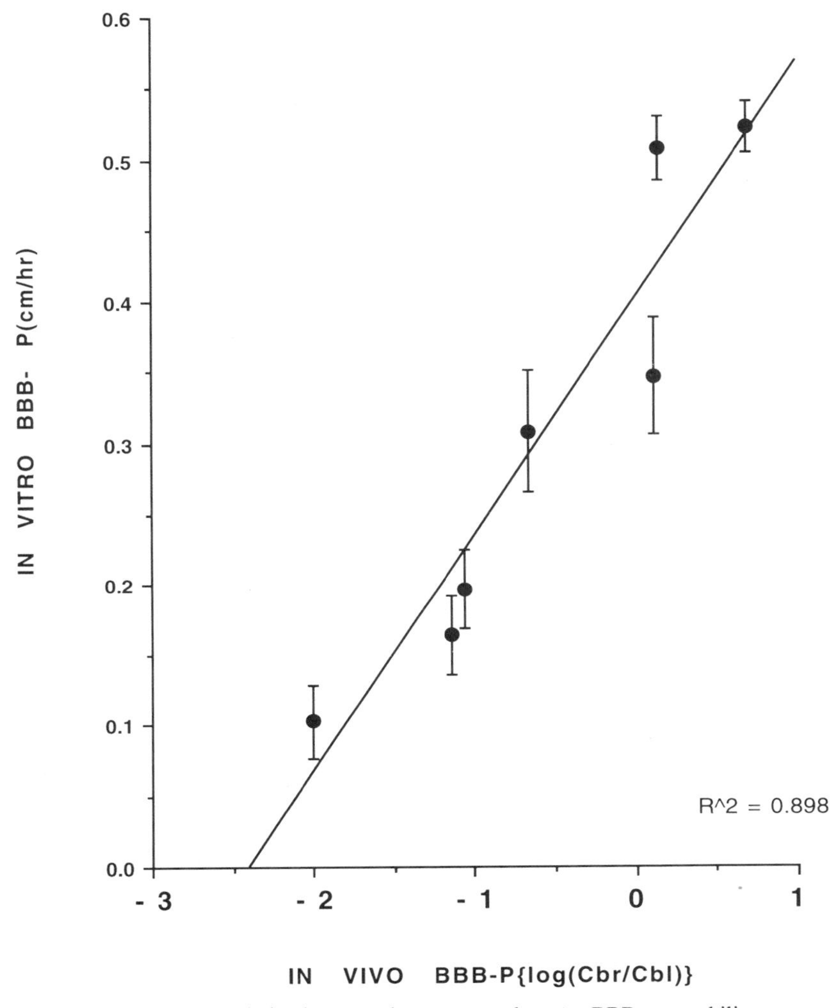

Figure 8. Correlation between the *in vitro* and *in vivo* BBB permeability.

in the regulation of barrier properties of the BBB[1,2,9,10,11,12,13]. Studies have been planned to address the significance of these factors on endothelial cell permeability. However, in this study the BBMEC without the astrocyteco-culture/astrocyte conditioned medium have demonstrated differences in drug permeability and provide a good correlation with *in vivo* results. Other investigators [9,10,13] have indicated that electrical resistance increases significantly with astrocyte conditioned medium/astrocyte co-cultures. However, it should be noted that the effects on sucrose permeability did not increase correspondingly (compareing our studies with others [9,10], who use astrocyte cocultures/astrocyte conditioned medium). Therefore, measurements of sucrose permeability have been the criteria in these studies for the evaluation of barrier properties rather than electrical resistances.

3. *In vitro/in vivo* database: The *in vitro/in vivo* studies reported here involve a small set of compounds. The database needs to be increased to enhance reliability of the

model. Work is ongoing towards extending the correlation to a large number of compounds.

In summary, the BBMEC sytem has been shown to have potential use in the rapid evaluation of relative permeability of drugs to predict BBB penetration. This model offers the opportunity to study transport mechanisms and manipulation of the BBB for drug delivery to the brain.

4. REFERENCES

1. K. L. Audus, R. L. Bartel, I. J. Hidalgo and R.T. Borchardt. *Pharm. Res.* **7**:(5), 435-451(1990).
2. K. L. Audus, L. Ng, W. Wang and R. T. Borchardt. In: *Model Systems for Biopharmaceutical Assessment of Drug Absorption and Metabolism* (R. T. Borchardt, P. L. Smith and G. Wilson, eds.) Plennum Publishers Corporation, N.Y.(**in press**, 1995).
3. A. Tsuji, T. Takabatake, Y. Tenda, I. Tamai, T. Yamashima, S. Moritani, T. Tsuruo and J Yamashita. *Life Sci.* **51**: 1427-1437 (1992)
4. R. C. Young, R. C. Mitchell, T. H. Brown, C. R. Ganellin, R. Griffiths, M. Jones, K. K. Rana, D. Saunders, I. R. Smith, N. E. Sore and T. J. Wilks. *J.Med.Chem.* **31**: 656-671(1988).
5. D. W. Miller, K. L. Audus and R. T. Borchardt. *J. Tiss. Cult. Meth.* **14**: 217-224 (1992).
6. K. L. Audus and R. T. Borchardt. *Pharm. Res.* **3**: 81-87 (1986).
7. K. L. Audus and R. T. Borchardt. *Ann. N. Y. Acad. Sci.* **507**: 9-18 (1987).
8. A. R. Hilgers, R. A. Conradi and P. S. Burton. *Pharm. Res.* **7**: 902-910 (1990).
9. W. M. Pardridge, D. Triguero, J. Yang and P. A. Cancilla. *J. Pharm. Exp. Ther.* **253**: 884-891 (1990).
10. M-P. Dehouck, P. Jolliet-Riant, F. Bree, J-C. Fruchart, R. Cecchelli and J-P. Tillement. *J. Neurochem.* **58**: 1790-1797 (1992).
11. F. Joo. *J. Neurochem.* **58**: 1-17 (1992)
12. F. Joo. *Neurochem. Int.* **23**: 499-521 (1993).
13. L. L. Rubin, et al., *J. Cell Biol.* **115**: 1725-1736 (1991).
14. M. W. Brightman and T. S. Reese. *J. Cell Biol.* **40**: 648-677 (1969).
15. C. Cordon-Cardo, J. P. O'Brien, D. Casals, et al., *Proc. Natl. Acad. Sci.* USA. **86**: 695-698 (1989).

22

THE BLOOD-BRAIN BARRIER AND BRAIN MORPHOLOGY FOLLOWING INSERTION OF A DIALYSIS PROBE

Barbro B. Johansson,[1] Irena Westergren,[1] and Claes Nordborg[2]

[1] Department of Neurology
Lund University Hospital
S-221 85 Lund, Sweden
[2] Laboratory of Neuropathology
Department of Pathology
Sahlgren University Hospital
S-413 45 Göteborg, Sweden

SUMMARY

Insertion of a dialysis probe in the cerebral cortex damages the BBB in the surrounding cortex. Extravasation of plasma constituents from the probe site leads to gliosis in the ipsilateral white matter, thalamus and the lateral geniculate body far from the site of the probe.

RÉSUMÉ

L'insertion d'une sonde de dialyse dans le cortex cérébral provoque une lésion de la BHE dans le tissu cortical environnant. L'extravasation de constituants plasmatiques à partir du site d'implantation entraîne une gliose dans la substance blanche ipsilatérale, le thalamus et le corps géniculé latéral, à distance du site d'implantation.

1. INTRODUCTION

Brain dialysis, when a semipermeable dialysis fibre is implanted in a selected brain area in living animals, is a popular tool in experimental brain research.[1] We have reported that the blood-brain barrier (BBB) remains open for at least 24 hours after probe insertion[2], an observation that is in agreement with early studies showing that a 50 mm thin needle or glass capillary alters the BBB for days[3,4]. Extravasated plasma constituents will alter the environment of neurons and glial cells and there is evidence from other experimental models

Biology and Physiology of the Blood–Brain Barrier, edited by Couraud and Scherman
Plenum Press, New York, 1996

that plasma leakage into the brain can induce permanent morphological changes[5-8]. The aim of the present study was to see if the BBB remains disturbed for a longer period and to evaluate morphological changes distant from the probe, i.e. changes that could not be due to the mechanical trauma of probe insertion.

2. MATERIAL AND METHODS

Male Sprague-Dawley rats were anesthetized with an i.p. injection of methohexital. The head was fixed in a stereotactic frame, the right side of the skull was exposed and a hole was drilled according to the coordinates for the right parietal cortex 2.8 mm posterior and 4.1 mm lateral to the bregma. A microdialysis probe (CMA/10, Carnegie Medicine, Sweden, Cat.no 830 9502, dialysis membrane diameter 0.5 mm, steel shaft diameter 0.6 mm) was mounted into the micromanipulator of the stereotactic instrument, lowered into the parietal cortex (3.5 mm beneath the skull) and fixed to the skull with screws and dental cement. Three (n 3), seven (n 5) or 21 (n 5) days later the rats were anesthetized and the brains were perfused in situ through the left heart ventricle with a 4 % formaldehyde solution in 0.1 mol/L phosphate butter at aH 7.4 and 37°C after an initial flush with physiological saline. 2% Evan's blue (3ml/kg) which in vivo binds to serum albumin, was injected i.v. 20 min before the animals were killed. The brains were sections in 2 mm thick coronal sections in a brain cutter, dehydrated and embedded in paraffin. For immunohistochemistry, 5 mm thick sections were placed on chrome gelatine coated slides. Extravasated plasma albumin was demonstrated with antirat albumin (1:8000) and the glial reaction with antibodies to glial fibrillary acidic protein, GFAP (1:1000, both from Dakopatts A/S, DK 2600 Glostrup, Denmark). Bound antibodies were visualized with a avidin-biotin-peroxidase complex (Cectastain ABC-kit, Vector Laboratories, Burlingame, CA) using diaminobenzidine as a chromogen. Alternative sections were stained with hematoxylin-eosin and acid fuchsin/celestine blue for routine light microscopy.

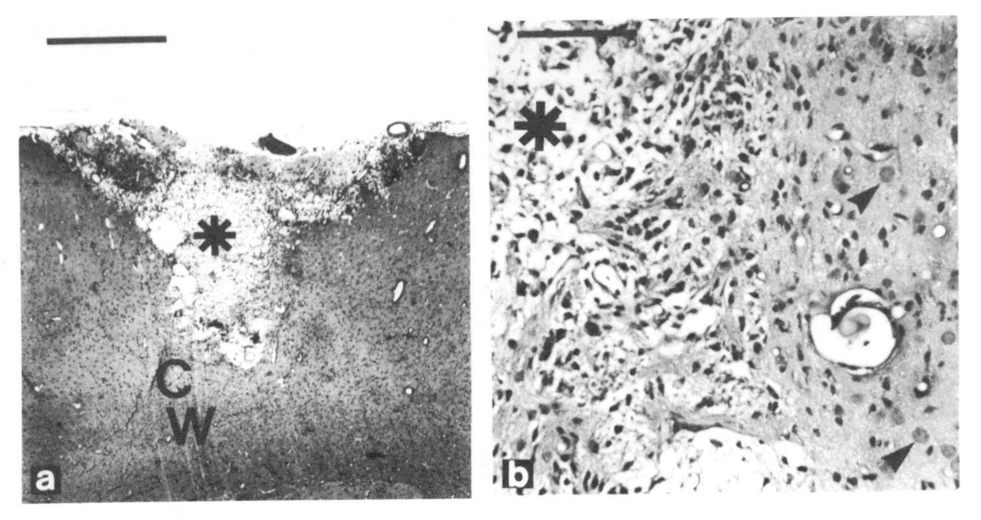

Figure 1. a. Cortical lesion (*) one week after the application of the probe. C lower cortex; W subcortical white matter. Bar: 780 mm. b. Detail showing the gliotic border between brain tissue (right) and cavity of the lesion. Arrowheads: nerve cells. (Haematoxylin and eosin. Bar:105 mm)

3. RESULTS

Evans blue/albumin was observed around the dialysis probe and in the surrounding cortex indicating an ongoing albumin leakage 3 and 7 days but not 21 days after probe insertion. Albumin immunoreactivity could be detected also after 21 days suggesting that the extravasated endogenous albumin or immunoactive degradation products remained in the brain after closure of the BBB. Alternatively, the Evans blue/albumin circulation time of 20 minutes might have been to short to detect a minor BBB leakage. In rats killed after 3 days the albumin reactivity was noted in the ipsilateral cortex and white matter with spread to lateral thalamus and lateral geniculate body. Immunoreactivity was more intense at 7 days and considerable albumin remained at 3 weeks as noted in figure 2 a. Albumin occasionally crossed the midline via corpus callosum and spread to contralateral parasagittal cortex. Most of the albumin reactivity was extracellular but albumin uptake was also seen in the neurons or as a marginal accentuation of albumin immunoreactivity around nerve cell bodies as seen in fig 2 b. Albumin positive neurons were particularly common around the probe and in the Purkinje cell layer in the cerebellum but were also seen in the lateral thalamus. The albumin spread to distal cortical regions apparently occurred both subpially and through uptake in the deeper cortical layers from the white matter. Albumin reached the Purkinje cells via the cerebrospinal fluid as indicated by its dendritic uptake in the superficial cerebellar cortex. With acid fuchin/celestine blue staining the diffusely albumin-positive cells were often eosiniphilic and shrunken.

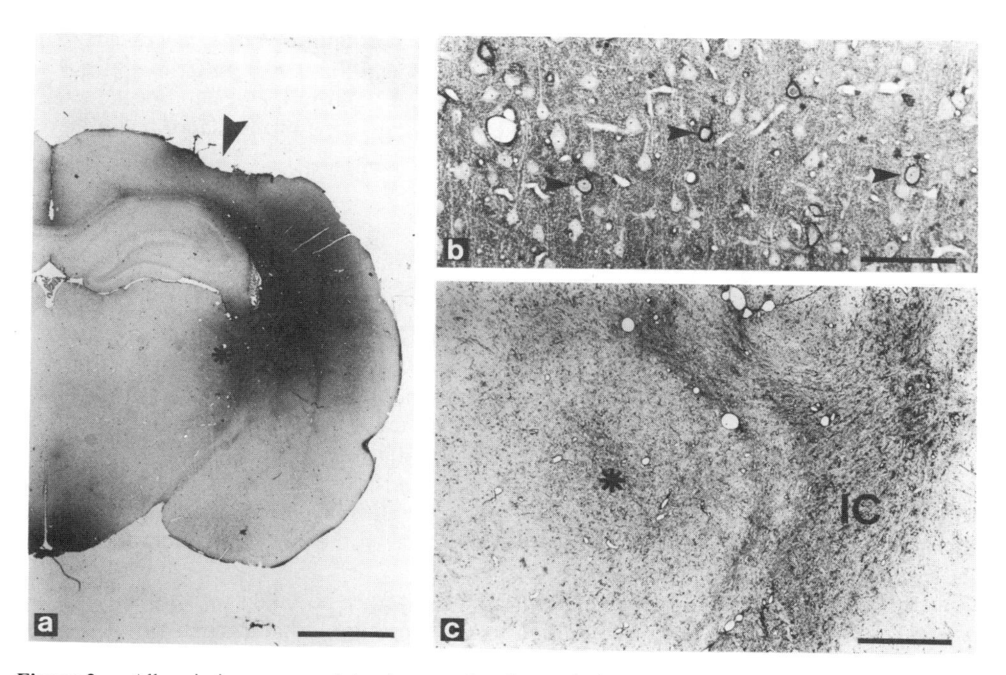

Figure 2. a. Albumin immunoreactivity three weeks after probe insertion. The dark area at the bottom is the median eminence, a region that does not have a BBB. Arrowhead: site of the probe b. Marginal accentuation of albumin immunoreactivity around nerve cell bodies in the cortex lateral to the probe lesion one week after probe insertion (Anti-albumin and haematoxylin. a. bar : 2050 mm; b bar: 105 mm). c. Gliosis in the internal capsule (IC) and lateral thalamus (*) three weeks after probe insertion.Anti-GFAP and haematoxylin.Bar: 515 mm.

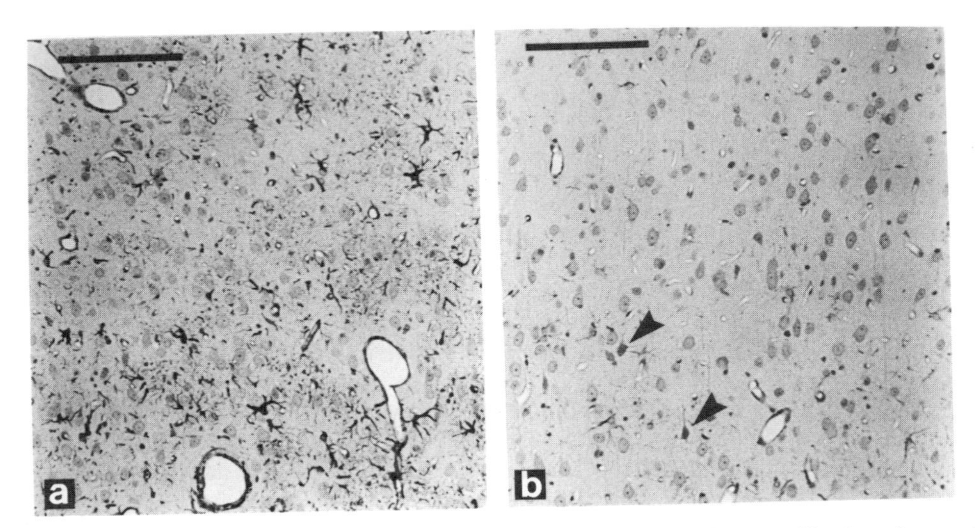

Figure 3. Gliosis in the ipsilateral cortex (a) compared to the contralateral cortex (b) where few small astrocytes (arrowheads) were observed. (Anti-GFAP and haematoxylin, bar: 155 mm).

In rats killed three days after probe insertion a moderate perifocal and parasagittal cortical gliosis was noted. Increased GFAP immunoreactivity was also noted below the probe in the white matter. Seven and 21 days after the probe insertion the gliosis was extensive in the white matter including the internal capsule(fig 2 c) and the ipsilateral cortex (fig 3 a). Gliosis was also evident in lateral thalamus (marked with a star in fig 2 c) and/or the lateral geniculate body in six of the ten rats killed seven or 21 days after probe insertion and in the substantia nigra in one rat.

4. DISCUSSION

The extensive spread of extravasated serum albumin from the area surrounding the probe is in agreement with our earlier study limited to 24 hours after the probe insertion. [2] Gliosis in the ipsilateral cortex and white matter remote from the probe is consistent with the hypothesis that the edema fluid induces gliosis, either by some serum component or by excitatory amino acids released as a result of the altered environment.[10] Glutamate and other excitatory amino acids have been shown to increase in dialysate concomitantly with edema formation after opening the BBB[10] and glutamate has been postulated to be related to cytolytic neuronal changes in relation to brain edema[10-12] . Since there was no evidence for cytolytic neuronal damage in the present model serum constituents such as albumin may be more likely inducers of the gliosis.

When extravasated albumin reaches the CSF it may be taken up subpially in the opposite hemisphere and in the cerebellum. It has been speculated that this may be one reason for Purkinje cell loss in patients with epileptic seizures.[9] A detailed morphological analysis would be needed to establish if the BBB opening induced with the present model leads to a Purkinje cell loss.

In spite of the trauma and the disturbed BBB caused by the insertion of a dialysis probe, many studies have shown that a transmitter release induced by various activations and presumably reflecting intracellular events, can be detected in the dialysate. However, when using this technique it is important to know that plasma constituents and drugs with little or no passage through an intact BBB may have increased access to the brain due to the BBB lesion in the surroundings of the probe. Whether or not the dialysis probe is fixed to the scull may be relevant for the degree of damage. A dialysis probe that is not fixed to the skull and can move with the brain may induce less damage.

6. ACKNOWLEDGMENTS

The study was supported by grants from the Swedish Medical Council (14X-4968), King Gustaf V and Queen Victoria Research Foundation, the Bank of Sweden Tercentenary Foundation, the Elsa Schmitz' Research Fund and the Rut and Erik Hardebo's Donation Fund.

7. REFERENCES

1. Benveniste H. (1989) Brain microdialysis. J. Neurochem. 52, 1667-1679.
2. Westergren I, Nyström B, Hamberger A, Johansson BB (1995) Intracerebral dialysis and the blood-brain barrier. J. Neurochem. 64:229-234.
3. Persson L. (1976) Cellular reactions to small cerebral stab wounds in the rat frontal lobe. An ultrastructural study. Virchows Arch. [B] 22, 21-37.
4. Persson L., Hansson H.-A., and Sourander P. (1976) Extravasation, spread and cellular uptake of Evans blue-labelled albumin around a reproducible small stab- wound in the rat brain. Acta Neuropathol. (Berl) 34:125-136.
5. Fredriksson C, Auer RN. Kalimo H, Nordborg C, Olsson Y Johansson BB (1985) Cerebrovascular lesions in stroke-prone spontaneously hypertensive rats. Acta Neuropathol 68:284-294.
6. Sokrab T.E.O., Johansson B.B., Kalimi H. Olsson Y. (1998). A transient opening of the blood-brain barrier can lead to brain damage. Extravasation of serum proteins and cellular changes in rats subjected to aortic compression. Acta Neuropathol (Berl) 75:557-565.
7. Salahuddin TS, Johansson BB, Kalimo H, Olsson Y (1988) Structural changes in the rat brain after carotid infusions of hyperosmolar solutions. An electron microscopic study. Acta Neuropathol. (Berl) 77:5-13.
8. Johansson BB Nordborg C, Westergren I (1990) Neuronal injury after a transient operning of the blood-brain barrier: modifying factors. in Pathophysiology of the blood-brain barrier. Eds Johansson BB, Owman Ch, Widner H. Elsevier Science Publishers.
9. Sokrab TEO, Kalimo H, Johansson BB (1990). Parenchymal changes related to plasma protein extravasa-tion in experimental seizures. Epilepsia 31:1-8.
10. Westergren I., Nyström B., Hamberger A., Nordborg C., and Johansson B. B. (1994). Concentrations of amino acids in extracellular fluid after opening of the blood-brain barrier by intracarotid infusion of protamine sulfate. J. Neurochem. 62, 159-165.
11. Nordborg C, Sokrab TEO, Johansson BB (1991) The relationship between plasma protein extravasation and remote tissue changes after experimental brain infarction. Acta Neuropath 82:118-126, 1991.
12. Nordborg C, Sokrab TEO, Johansson BB (1994) Oedema-related tissue damage after temporary and permanent occlusion of the middle cerebral artery. Neuropathol Appl Neurobiol 20:56-65.

23

BLOOD-BRAIN BARRIER *IN VITRO*

Rapid Evaluation of Strategies for Achieving Drug Targeting to the Central Nervous System

M. P. Dehouck,[1] B. Dehouck,[1] L. Fenart,[1] and R. Cecchelli [1,2]

[1] INSERM U325
Institut Pasteur
Lille, France
[2] Universite De Lille I
Villeneuve d'Ascq, France

SUMMARY

To provide an in vitro system for studying brain capillary functions, we have developed a process of coculture that closely mimics the in vivo situation by culturing brain capillary endothelial cells on one side of a filter and astrocytes on the other side. The strong correlation between the in vivo (Oldendorf method) and in vitro (coculture) drug transport, the relative ease with which such cocultures can be produced in large quantities and the reproducibility of the system provide an efficient system for the screening of centrally active drugs.

RÉSUMÉ

De façon à étudier les fonctions des capillaires cérébraux support de la barrière hémato-encéphalique (BHE), nous avons mis au point un modèle de BHE in vitro qui consiste en une coculture de cellules endothéliales de capillaires cérébraux et d'astrocytes de part et d'autre d'un filtre. La parfaite corrélation entre les résultats obtenus pour le transport des médicaments in vivo (méthode d'Oldendorf) et in vitro (coculture), la facilité avec laquelle les cocultures peuvent être produites en grandes quantités et la reproductibilité du modèle permettent l'utilisation de ce modèle pour le screening des médicaments à visée cérébrale.

1. INTRODUCTION

The passage of substances across the blood-brain barrier (BBB) is regulated by the endothelial cells (ECs) which in brain capillaries, in contrast to the other vascular beds, are

Biology and Physiology of the Blood–Brain Barrier, edited by Couraud and Scherman
Plenum Press, New York, 1996

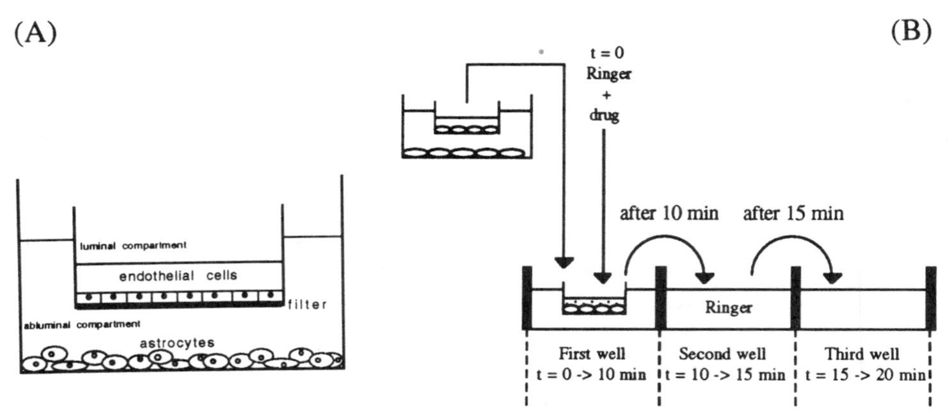

Figure 1. Coculture system: Description of the model (A); Transendothelial transport studies (B).

sealed together by continuous tight junctions and have little transcellular vesicular transport. In addition to these morphological barrier properties, the presence of monoamine oxidase and P-glycoprotein impeding the delivery of a certain number of molecules to the brain, constitute the so called "metabolic barrier." Investigations into the functional characteristics of brain capillaries have been facilitated by the use of cultured brain endothelial cells. To provide an in vitro system for studying brain capillary fonctions, we have developed a process of coculture that closely mimics the in vivo situation by culturing brain capillary endothelial cells on one side of a filter and astrocytes on the other side. Under these conditions, endothelial cells retained all the endothelial cell markers and the characteristics of the BBB including tight junctions, paucity of pinocytotic vesicles, monoamine oxidase and γ-glutamyl transpeptidase activity.

2. EXPERIMENTAL METHODS

2.1. Coculture of Brain Capillary ECs and Astrocytes

Cocultures (Fig.1A) were obtained using the method of Dehouck et al. (1990) with some modifications [1]. The astrocytes were plated on six-well plastic plates. Three weeks after seeding, the cultures of astrocytes became stabilized. Then culture plate inserts (Millicell-CM; pore size, 0.4μm; diameter 30mm; Millipore), coated on the upper side with rat tail collagen, were set in the plates containing astrocytes. Brain capillary ECs were seeded on the upper side of the inserts. Experiments were performed 5 days after confluency.

2.2. Transendothelial Transport Studies

One filter containing a confluent monolayer of ECs was transferred into the first well of a six-well plate containing Ringer (Fig.1B). A Ringer containing labeled or unlabeled drug was placed in the luminal compartment. At 10, 15, 20, 30, 45 minutes, the filter was transferred to another well. Abluminal concentrations were determined either by radioactivity or HPLC measurements. To calculate the permeability coefficient (Pe) the clearance principle was used [2].

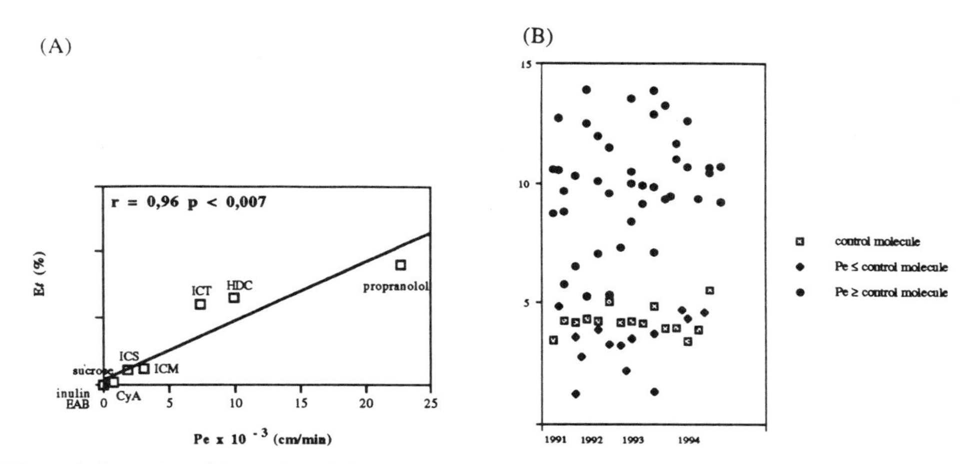

Figure 2. Screening of drugs: Correlation between in vivo Et values and in vitro Pe values (A); Reproducibility of the coculture model (B).

3. RESULTS AND DISCUSSION

The resistance of these monolayers averaged 700 to 800 Ω.cm^2.

Investigating the movement of [^{3}H] inuline across the monolayers, studies have shown that the permeability of endothelial cells is twice as high when they were not cocultured with astrocytes.

In order to assess drug transport across the BBB, we have compared the brain extraction (Et) using the in vivo Oldendorf method with the permeability (Pe) of our coculture for different compounds corresponding to a wide range of lipid solubility (Fig. 2A). A high correlation (r=0,96) was observed between Et values and in vitro Pe values, showing that the coculture model can be used for studying drug transport to the brain [3].

The screening of a series of 75 different drugs, in 15 experiments, using 7 different colonies of cells was carried out (Fig. 2B). The results demonstrate the reproducibility of the model for the control molecule and enable us to determine in a very short time, whether the chemical modification of the control molecule can increase its brain penetration.

4. CONCLUSION

The strong correlation between the in vivo and in vitro values, the reproducibility of the system and the relative ease with which such cocultures can be produced in large quantities provide an efficient system for the screening of centrally active drugs.

5. REFERENCES

1. Dehouck M.P., S. Méresse, P. Delorme, J.C. Fruchart and R. Cecchelli An easier, reproducible, and mass-production method to study the BBB in vitro. J. Neurochem., **54** 1798-1801 (1990)
2. Dehouck M.P., P. Jolliet-Riant, F. Brée, J.C. Fruchart, R. Cecchelli and J.P. Tillement Drug transfer across the BBB:correlation between in vitro and in vivo models. J. Neurochem., **58** 1790-1797 (1992)

3. Dehouck M.P., B. Dehouck, C. Schluep, M. Lemaire and R. Cecchelli Drug transport to the brain:comparison between in vitro and in vivo models of the BBB. European J. of Pharmaceutical Research, in press (1995)

24

NEW TECHNIQUES TO STUDY TRANSEPITHELIAL AND TRANSENDOTHELIAL RESISTANCES OF CULTURED CELLS

Joachim Wegener, Helmut Franke, Stephan Decker, Martin Erben, and Hans-Joachim Galla

Department of Biochemistry
University of Münster
D-48149 Münster, Germany

1. SUMMARY

Measurement of transepithelial electrical resistances (TER) is a very common tool for investigations concerning epithelial and endothelial barrier properties. We here introduce two new and powerful techniques for the determination of TERs that might open further applications. The first method takes advantage of the wide spread DC technique and allows to study individual colonies of a cell monolayer grown on waterpermeable filter supports. The second technique is based on AC impedance analysis and enables to determine TERs of cell monolayers grown on gold surfaces.

1. RÉSUMÉ

La mesure de la résistance transépithéliale (RTE) est un outil très courant pour la détermination des propriétés des barrières physiologiques épithéliales et endothéliales. Nous avons mis au point deux nouvelles techniques efficaces de mesure des RTE qui, de plus, pourraient permettre de nouvelles applications. La première méthode consiste à utiliser la technique bien connue en courant continu et permet la mesure de la RTE d'un petit nombre de cellules organisées en colonies distinctes sur des supports semi-imperméables. La seconde technique utilise l'analyse de l'impédance en courant alternatif et permet de déterminer les RTE de monocouches cellulaires ensemencées sur des supports recouverts d'or.

Biology and Physiology of the Blood–Brain Barrier, edited by Couraud and Scherman
Plenum Press, New York, 1996

2. INTRODUCTION

The predominant physiological function of epithelial and endothelial tissue is to build up and maintain a highly selective permeability barrier between two fluid compartments that need to be separated. The blood-brain-barrier, built up by brain capillary endothelial cells, and the blood-cerebrospinal-fluid-barrier, formed by epithelial cells of the choroid plexus, are two examples for the extreme significance of the epithelial and endothelial barrier properties. Both of these barriers separate the brain tissue from the circulating blood flow and therefore guarantee the homeostasis in brain tissue necessary for its correct function. One physical parameter to quantify the barrier function - as far as ion movement is concerned - is the transepithelial or transendothelial electrical resistance (TER). We here present two new methods to investigate TERs of cultured epithelial or endothelial cells, that might open further applications.

3. DIMINISHED AREA DEVICE TO STUDY SMALL SURFACE AREAS

This method is in principle based on the most popular DC-technique that employs waterpermeable filter inserts as culture surfaces. So far resistance measurements were carried out only for the whole cell covered filter as for example described by Hein et al. (1992) [1]. But whole filter analysis disregards inhomogenities within the cell layer or contaminations by other cell types [2], not forming tight junctions. These contaminations will result in a dramatical decrease of the electrical resistance. We developed a new method which allows TER measurements in defined small cell monolayer areas, sized about 0.8 mm^2 (Fig. 2). Our new setup enables us to scan the cell-covered filter and to determine local TERs in a range up to 500 Ωcm^2. Investigations of tight epithelia like MDCK-I cells (Madin-Darby Canine

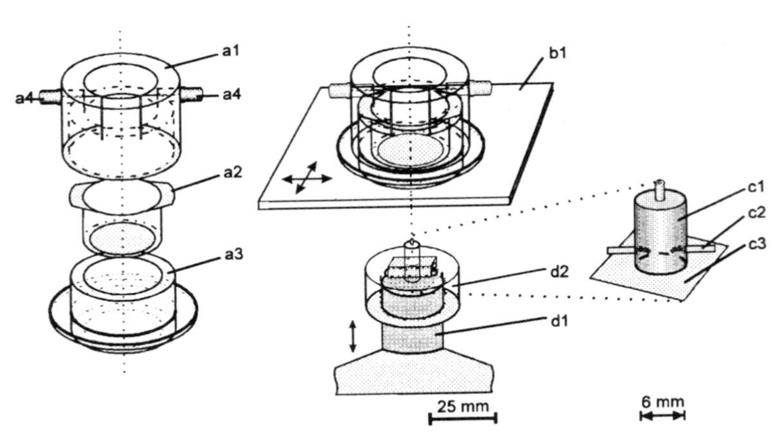

Figure 1. Diminished area device for electrical resistance measurements. The device includes an upper (apical) and a lower (basolateral) part. The upper electrode cap (a1) with one voltage and one current electrode (a4) is placed on an upper filter holder (a3) which carries the cell culture insert (a2). The whole upper (apical) chamber is placed on the microscope stage (b1) which allows to adjust the filter within the microscope's light beam for a simultaneous microscopic inspection of the cells. After selection of a given area under the microscope, the basolateral chamber (c1) with the lower voltage and current electrodes (c2) is placed on the microscope adapter (d1) and fixed with the lower compartment holder (d2).The narrow tube on the top of the lower compartment (dotted line) is moved with the microscope objective revolver to touch the filter avoiding strong forces.

Table 1. TERs of the epithelial and endothelial cell monolayers that were investigated in the present study in comparison to literature data

Source	TER / ($\Omega \cdot cm^2$)				
	MDCK-I	MDCK-II	PBEC	BBEC*	Plexus
Impedance analysis	400–3500	40–100	30–80		80–150 (pig)
Diminished area device		100–150	65 ± 17	62 ± 21	100–170 (pig)
Stevenson et al.[4]	2500–5000	50–70			
Richardson et al. [5]	4160	71			
Hein et al. [1]	1500–4500	50–100			
Erben [7]	2000	130–160			
Griepp et al. [6]		80–150			
Southwell et al. [10]					99 ± 15 (rat)
Rubin et al. [8]				61 ± 2	
Rutten et al.[9]				157–783	

*BBEC abbreviates bovine brain capillary endothelial cells.

Kidney) with TERs of more than 2000 Ωcm^2 [2] will require an enhanced electrical seal of the basolateral tube (compare fig. 1) due to the high absolute resistance within the small measuring area.

Typical scans across a cell-covered filter of porcine brain capillary endothelial cells using the diminished area device lead to resistances in a range from 50 to 105 Ωcm^2. High resistance values are obtained at spots where capillaries were visible in the microscopic picture. The average TER is $R = 65 \pm 17$ Ωcm^2 ($n = 25$) (compare table 1). Neglecting the high resistance values, assigned to areas covered by capillary fragments, the average TER is $R = 60 \pm 9$ Ωcm^2 ($n = 22$). Analogous scans with epithelial cells from porcine choroid plexus resulted in resistances of 13017 cm^2 ($n = 12$).

The new method has been applied to determine a reversible modulation of endothelial tightness by amino acids and Ca^{2+} in accordance with earlier measurements using epithelial cells [1]. Assuming the lipid model of tight junctions [12], substances that induce the hexagonal phase II of phospholipids (e.g. Ca^{2+}), should increase the electrical resistance of a cell monolayer, whereas molecules that stabilize the formation of the lamellar phase of lipids (e.g. basic amino acids) should lower the electrical resistance [4]. Addition of lysine (200 mM) and lysine methyl ester (200 mM) to confluent monolayers of porcine brain endothelial cells *in vitro* decreased resistances up to 90% from initial values. Reduction of extracellular Ca^{2+}-concentration revealed a clear drop in TER values of porcine brain endothelial cells while elevated Ca^{2+} concentrations were observed to increase transendothelial resistances (data not shown). Both results correspond to previous whole filter measurements carried out on MDCK cell line [5].

4. IMPEDANCE ANALYSIS OF EPITHELIAL AND ENDOTHELIAL CELLS GROWN ON GOLD SURFACES

A second method to gain information about the tightness and modulation of epithelial and endothelial barrier function is based on AC impedance analysis. In our experimental device the epithelial or endothelial cells under investigation are cultured on gold surfaces that serve as measuring electrodes. These gold electrodes are the core component of our self-developed measuring chambers that allow to culture epithelial and endothelial cells under normal culture conditions and to determine their TERs simultaneously without any

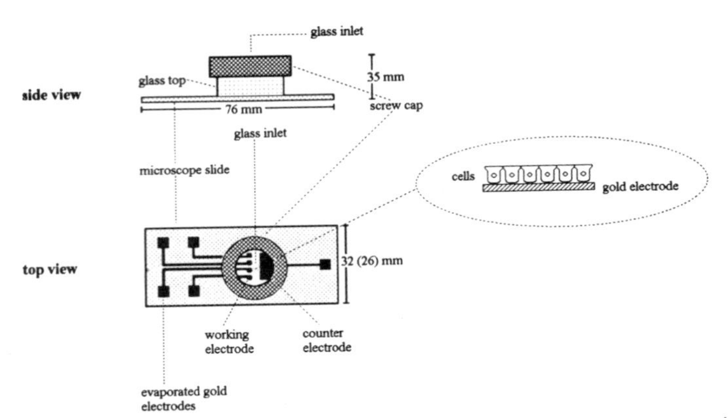

Figure 2. Schematic drawing of the self-developed measuring chambers. The chambers consist of ordinary microscope slides or comparable glass slides that are covered with specially designed, thin gold films (50 nm thickness) serving as measuring electrodes. The metal films were prepared thin enough to be transparent. Thus, growing of the cells on top of the measuring electrodes can be observed by common light microscopes.

disturbance of the culture. In contrast to the commonly used methods this technique allows to investigate TERs of cell monolayers grown on impermeable supports, thus conditions, cells are normally exposed to in ordinary culture dishes, are maintained. Figure 2 shows a schematic drawing of the measuring chambers.

Impedance spectra were acquired in the frequency range from 1 s^{-1} to 10^5 s^{-1} using a continuous wave impedance spectrometer, that was developed in our laboratory. Figure 3 shows a simplified equivalent circuit that illustrates the measuring principle. Instead of detecting the frequency dependent impedance of the electrochemical system by current-voltage measurements we determine the amplitude ratio of two voltages, called U_f and U_0, whose relation with the impedance of the electrochemical system M can be deduced from Kirchhoffs laws. The resistor R_0 - a component of the measuring principle - is adjustable to the experimental conditions. The measuring procedure results in a frequency spectrum of these amplitude ratios.

To evaluate and interpret the obtained impedance spectra, the parameters of an appropriate equivalent circuit, that represents the electrical properties of the electrochemical system, are fitted to the recorded data by means of a non-linear-least-square-fit (NLSQ) according to the Levenberg-Marquardt-algorithm. The equivalent circuit used for data analysis includes a resistor R_{cell} that was found to be analogue to the TER and it is depicted within the impedance spectrum given in fig. 4.

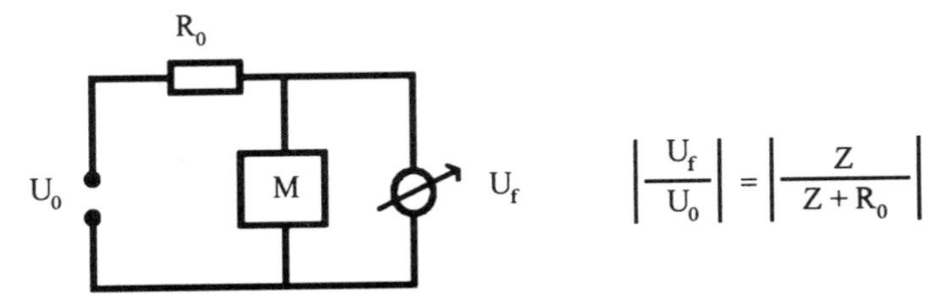

$$\left|\frac{U_f}{U_0}\right| = \left|\frac{Z}{Z+R_0}\right|$$

Figure 3. Simplified equivalent circuit to illustrate the principle of impedance determination. Z represents the impedance of the electrochemical system.

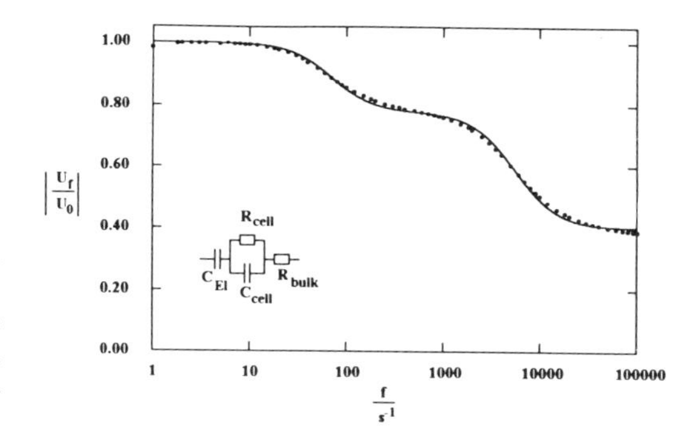

Figure 4. Typical impedance spectrum of a confluent monolayer of PBECs grown gold surfaces. Points represent experimental data, the solid line represents the transfer function of the given equivalent circuit fitted to the experimental data (non-linear-least-square-fit).

Impedance analysis was performed with strains I and II of the epithelial cell line MDCK, porcine brain capillary endothelial cells (PBEC) and epithelial cells from porcine choroid plexus. Figure 4 exemplarily shows a typical impedance spectrum of a confluent PBEC-monolayer and its analysis via the illustrated equivalent circuit.

Results of data fitting : R_{cell} = (66 7) cm^2, C_{cell} = (0.61 0.06) F/cm^2. The parameters C_{El} and R_{bulk} are not related to the cell layer.

Besides the determination of TERs for confluent monolayers in a steady state like situation we also determined the time development of impedance spectra from seeding a cell suspension into the measuring chambers to a finally confluent monolayer on top of the electrodes. The recorded data (not shown here) illustrate the slow coverage of the electrode surface by proliferating cells until the confluent monolayer is established.

5. CONCLUSION

As detailed in table 1 TER values of all cell types that were investigated in the present study are in good agreement with published data. Therefore both methods proved to be reliable and sensitive for the routine determination of TERs. Each technique is characterized with very special features that might open special applications. The diminished area device allows to select individual colonies of a cell covered filter insert by morphological inspection and therefore enables to study subconfluent monolayers at early days in culture. The main advantages of the AC impedance analysis technique are the ability (i) to use multi-electrode arrays that allow to determine TERs at different locations of a given cell monolayer, (ii) to carry out impedance analysis under normal culture conditions and (iii) to obtain TER values of cell monolayers grown on impermeable supports, thus conditions, cells are normally exposed to in ordinary culture dishes, are maintained.

6. REFERENCES

1. Hein M., Mädefessel C., Haag B., Teichmann K., Post A., Galla H.-J., Chem. Phys. Lipids, 63 (1992) 223-233
2. Risau W., Dingler A., Albrecht U., Dehouck M.P., Ceccelli R.J., Neuochem., 58 (1992) 667-672
3. Erben M., Decker S., Franke H., Galla H.-J., J. Biochem. Biophys. Methods, in press
4. Stevenson B.R., Anderson J.M., Goodenough D.A., Mooseker M.S., Cell Biol., 107 (1988) 2401-2408
5. Richardson J.C.W., Scalera V., Simmons N.L., Biochim. Biophys. Acta, 673 (1981) 26-36
6. Griepp E.B., Dolan W.J., Robbins E.S., Sabatini D.D., J. Cell Biol., 96 (1983) 693-702
7. Erben M., Dissertation, Universität Münster, Germany, 1993
8. Rubin L.L. et al., J. Cell Biol., 155:6 (1991) 1725-1735

9. Rutten M. J., Hoover R. J., Karnovsky M. J., Brain Res., 425 (1987) 301-310
10. Southwell B. et al., Endocrinology, 132 (1993) 2116-2126
11. Hein M., Post A., Galla H.-J., Chem. Phys. Lipids 63 (1992) 213-222
12. Kachar, Reese, Nature 296(1982)464-466

25

THE BLOOD-CSF BARRIER *IN VITRO*

Ursula Gath, Ansgar Hakvoort, and H.-J. Galla

Institut für Biochemie
D-48149 Münster, Germany

SUMMARY

We established an *in vitro*-model for the blood-CSF barrier by culturing epithelial cells from the porcine choroid plexus. The cells express their epithelial properties in culture and form confluent monolayers on permeable membranes, maintaining the cell polarity. Moreover the cells in culture produce cerebrospinal fluid.

RÉSUMÉ

Nous avons mis en place un modèle cellulaire in vitro de la barrière entre le sang et le liquide cérébrospinal à partir de cultures de cellules épithéliales du plexus choroïde de porc.Ces cultures conservent leurs propriétés spécifiques de cellules épithéliales et forment des monocouches cellulaires confluentes sur des supports perméables. De plus, ces cellules en culture sécrètent du liquide céphalo-rachidien.

INTRODUCTION

The choroid plexus is the major source of the cerebrospinal fluid (CSF) and acts together with the arachnoid membrane as a barrier between blood and CSF (Cserr, 1971). The component responsible for the CSF production and the blood-CSF barrier is the layer of epithelial cells, overlying a highly vascularized tissue stroma. These epithelial cells show the morphological characteristics of transporting cells, with microvillis at the apical pole and special infoldings of the basal membrane (Maxwell and Pease, 1956). Furthermore the cells exhibit a polar distribution of enzymes, which regulate the controlled pathway of substances between blood and CSF.

For a detailed investigation of the transport systems of the blood-CSF barrier, an *in vitro* model of the barrier is highly desirable. Some transport mechanism of the choroid plexus have been studied *in vitro* (Spector and Goldberg, 1982), but these investigations were carried out with tissue fragments in culture wich have the main disadvantage, that the localisation of the transport enzymes (apical or basolateral membrane) can not be determind.

Biology and Physiology of the Blood–Brain Barrier, edited by Couraud and Scherman
Plenum Press, New York, 1996

We now established an *in vitro* model of the barrier using epithelial cells isolated from porcine choroid plexus and cultivated on permeable membranes, separating the culture flask into a basal and apical medium containing chamber. An undispensible prerequisite was to work with cultures of pure epithelial cells. We characterized the cells by immunochemistry. Furthermore we checked for any changes due to the *in vitro* situation of the cells by comparing the antigenic character of cultured cells with that of thin frozen fractions of porcine choroid plexus.

MATERIAL AND METHODS

Preparation and Cultivation of Porcine Epithelial Cells

The epithelial cells were isolated by a modified method described by Crook et al. (1981).

The plexus were removed from the lateral ventricle of porcine brain. After a cold trypsination (trypsin solution 0,25%) for 2,5 hours, the solution was warmed up to 37°C and incubated at this temperature for 30min. The digestion was inhibited by addition of newborn calf serum and the enzymatically unreleased cells were separated. After a first centrifugation, the epithelial cells were separated from the erythrocytes by centrifugation in a percoll-solution (density: 1,08 g/cm^3). The suspension was seeded onto uncoated glass slides with MEM-Medium. To prevent the growth of foreign cells, we supplemented the medium with 20 μM cytosinearabinoside.

Cultivation onto Permeable Filter Membranes

The filter membranes (Transwell-Membranes from Costar) with a diameter of 12 mm were coated with laminin (Sigma) from solution of 20 mg/ml by incubation for about 20 h at room temperature. Afterwards the isolated cells were seeded onto the membranes with an area of 50 cm^2/g wet weight of the plexus choroideus. The tightness of the cell monolayer was controlled by light microscopy using carbolfuchsin (Sigma).

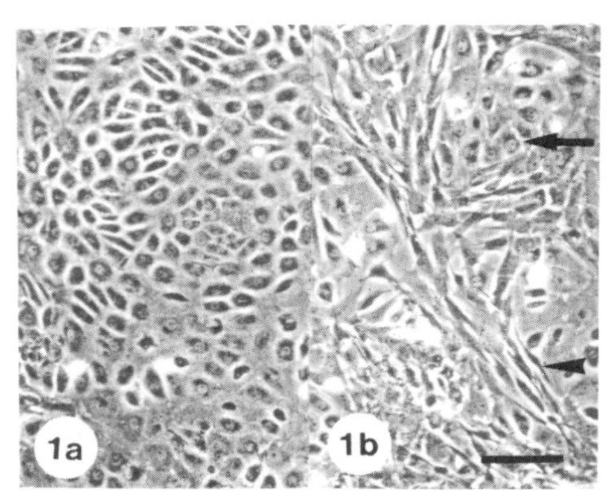

Figure 1. Cells isolated from the porcine choroid plexus were observed by phase contrast microscopy. a) Epithelial cell cultures supplemented by cytosinearabin-oside b) Epithelial cells (filled arrow) in vitro without cytosinearabinoside are over-grown by fibroblast-like cells (arrow head). Bar=500 mm.

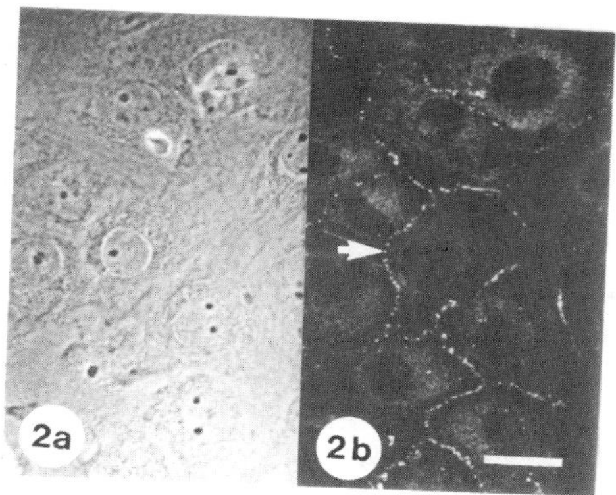

Figure 2. Pure epithelial cell cultures were stained with an anti-desmoplakin antibody. a) Phase contrast microscopy b) Fluorescence microscopy of the immunostaining demostrating the presence of small spots of desmoplakin at the cell boundaries (arrow). Bar=22μm

Thin Frozen Fractions of Choroid Plexus

The plexus were removed and rapidly frozen in liquid nitrogen. Thin frozen fractions were prepared by using a Cryostat Microtom (American Optical Corporation). These tissue fractions (7-8 μm) were placed onto glass microscope slides and stored at -75°C.

Immunocytochemistry

The fixation for the immunostaining of desmoplakin was carried out with aceton for 10 min at -20°C, for the staining of fibronectin we applied ethanol for 10 min at room temperature, the fixation for the staining of the Na^+K^+ATPase and ZO-1 was performed with methanol/aceticacid (97/3 v/v). Prior to the antibody incubation, the samples were treated with 3% bovine serum albumin for 10 min at 37°C. The commercial sources of the antibodies, the experimental concentrations and incubation conditions are as following:

- Monoclonal antibody (Mab) anti-desmoplakin from Boehringer (Mannhaim), 2 μg/ml, 3h at 37°C;

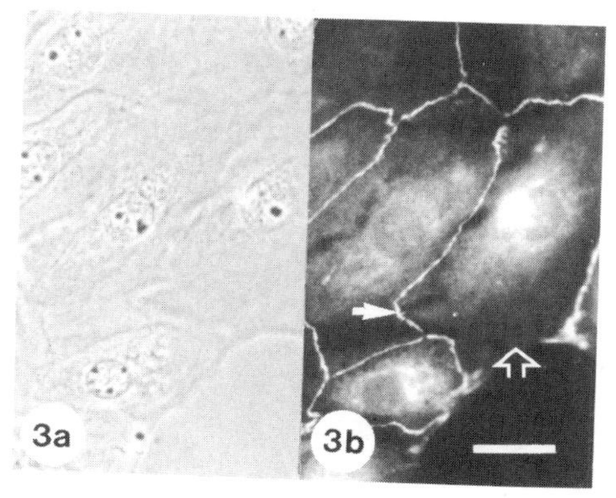

Figure 3. ZO-1 in cultured epithelial cells a) Phase contrast microscopy b) Identification of ZO-1 between adjacent cells (filled arrow) by immunofluorescence. The cell boundaries without contact remain unstained (unfilled arrow). Bar= 16 μm.

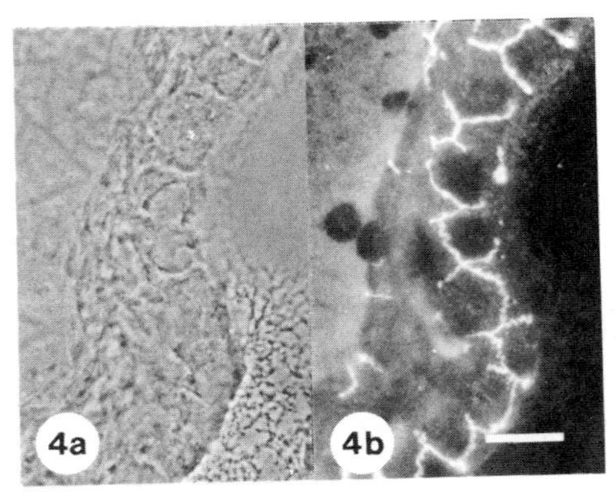

Figure 4. ZO-1 in thin frozen fractions of plexus choroideus a) Phase contrast microscopy b) Immunofluorescence with ZO-1 antibody. Bar= 16 μm

- Mab anti-ZO-1 from Chemicon, H. Biermann GmbH (Bad Nauheim), 10 μg/ml, 1h at 37 °C;
- polyclonal antibody (Pab) anti-Na^+K^+ATPase from Chemicon, H. Biermann GmbH (Bad Nauheim), 10 μg/ml, 1h at 37°C;
- Pab anti-fibronectin from Sigma (Deisendorf), 5 μg/ml, 1h at 37 °C;
- secondary Pab anti-mouse-Ig-rhodamine from Boehringer (Mannheim), 20 μg/ml, 1h at 37°C;
- secondary Pab anti-rabbit-Ig-rhodamine from Boehringer (Mannheim), 20 μg/ml, 1h at 37°C,
- secondary Pab anti-rat-Ig-rhodamine from Sigma (Deisendorf), 100 μg/ml, 1h at 37°C.

RESULTS

The porcine epithelial cells were isolated from the plexus tissue as sheets of approximatly 30-40 cells. After one day *in vitro* the cells became adherent and started to proliferate

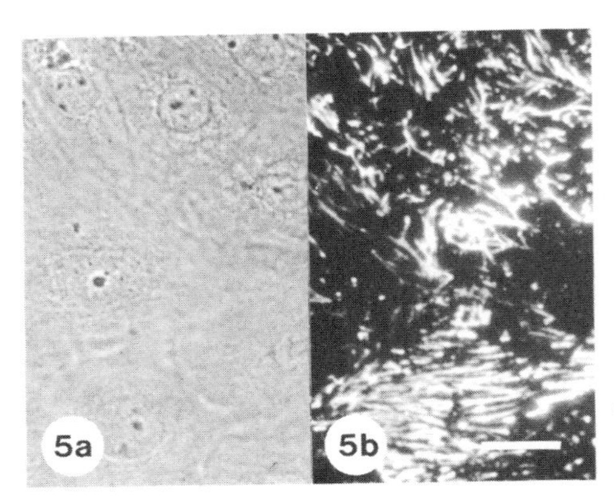

Figure 5. Identification of fibronectin in 7 days old cultures a) Phase contrast microscopy b) The immunostaining of fibronectin becomes visible as a network of small fibrills. Bar=16 μm

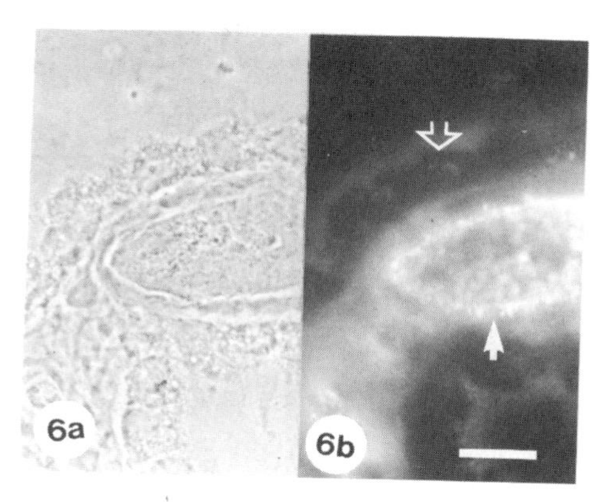

Figure 6. Identification of fibronectin in thin frozen fractions by immnunofluorescence a) Phase contrast b) The immunofluorescence obtained with the anti-fibronectin antibody showed an intense staining round a blood vessel at the surface of the epithelial cells (filled arrow). The epithelial cells themselves were not stained (unfilled arrow). Bar=16 μm.

developing in a cobblestone morphology. However, in addition to the epithelial cells, fibroblast-like cells from the stroma of the choroid plexus appeared in the culture (fig.1b). These cells started to overgrow the epithelial culture, due to of their high proliferation rate. We managed to supress the growth of these contaminating cells by the supplemention of the culture medium with cytsinearabinoside, with the result of pure cultures of epithelial cells (fig. 1a).

The epithelial character of the cultured cells was shown by the identification of desmoplakin (fig 2), which is clearly visible at the cell boundaries. The protein ZO-1, a component of tight-junctions, was identified in the cultured cells (fig. 3) and in the tissue of choroid plexus (fig 4) and gives a first hint of the preserved polarity of the cells in culture. One component of the extracellular matrix of the epithelial cells *in vivo* is fibronectin. In the plexus tissue fibronectin is found only in the blood vessels but not in the epithelial cells (fig. 6b). The network of distinct fibrils is shown in the culture of epithelial cells (fig. 5b). Similary we were able to demonstrate the expression of $Na^{+}K^{+}$ATPase *in vivo* and *in vitro* (data not shown).

For further transport measurements it is essential to grow confluent and impermeable monolayers on a filter. In a special set up (fig. 7) the cell monolayer separates two compartments filled with medium mimicking the apical and basolateral side. With this set

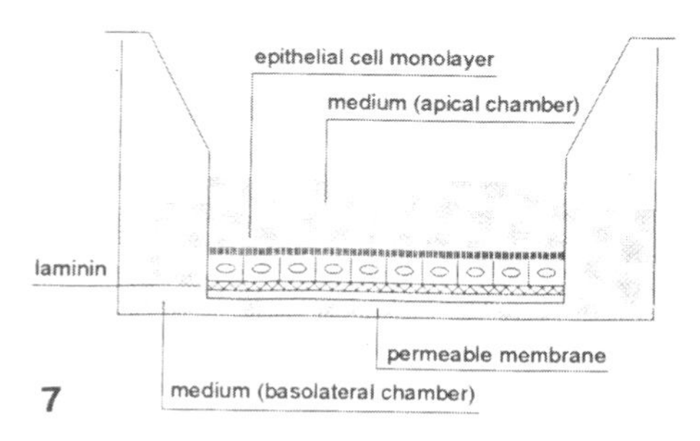

Figure 7. Schematic set up of the culture system of epithelial cells growing on permeable membranes and separating two fluid compartments.

Figure 8. Two chamber culture system a) The level of medium in the apical chamber became higher (2mm) (filled arrow), than the one in the basolateral chamber (unfilled arrow) after two days in culture due to the production of CSF *in vitro*. b) Control filter without cells, medium level (unfilled arrow).

up we observed the developement of an increasing medium level in the upper (apical) compartment which clearly shows that the cell monolayer is producing CSF or CSF compounds and that the cell monolayers are impermeable even under slight hydrostatic pressure.

DISCUSSION

Pure cultures of epithelial cells were obtained from porcine choroid plexus. Addition of cytosinearabinoside prevents the contamination with fibroblast-like stroma cells and confluent monolayers are obtained on laminin coated polycarbonate-filters.

By immunohistochemistry we demonstrated, that the cells maintain their epitheloid characteristics like *tight junctions* (ZO-1) and desmosomes (desmoplakin). Furthermore the identification of fibronectin as one important compound of the extracellular matrix and of Na^+K^+ATPase as prerequisite for liquor production, proved the differentiated state of the cells *in vitro*. The localization of fibronectin at the culture surface and the asymetric distribution of Na^+K^+ATPase (data not shown) clearly shows, that the epithelial cells are polarized *in vitro*.

Confluent monolayers on permeable membranes were obtained, if the culture surface is coated with laminin. On uncoated filters cells only form small cell islands, but do not spread across the filter. Cells of confluent monolayers are able to mimic the production of cerebrospinal fluid. This leads to an increase of the fluid amount in the apical chamber within 2-3 days (fig. 8).

Permeability studies with peroxidases as well as FITC labled dextranes are possible in this cellular system.

REFERENCES

Crook R.B., Kasagami H., Prusiner S.B., J. Neurochem., **34** (1981): 845-854
Cserr H.F., Physiol.Rev., **51** (1971): 273-311
Maxwell D.S., Pease D.C. J., Biophys. Biochem., Cytol., **2** (1956): 467-474
Spector R., Goldberg M.J., J. of Neurochem. **38** (1982): 594-596

26

PENETRATION OF ^{3}H TIAZOFURIN FROM BLOOD INTO GUINEA PIG BRAIN

Z. B. Redzic,[1] I. D. Markovic,[2] S. S. Jovanovic,[2] D. M. Mitrovic,[1] and Lj. M. Rakic[1]

[1] Institute of Biochemistry
School of Medicine
Belgrade, Yugoslavia
[2] ICN Galenika Institute
Biomedical Research Department
Belgrade, Yugoslavia

SUMMARY

Transport of tiazofurin (2-β-D-ribofuranosyl thiazole-4-carboxamide) across the blood-brain barrier (BBB) was studied using brain vascular perfusion method in the guinea pig. The obtained results demonstrate that brain clearance of ^{3}H tiazofurin significantly differs from zero, suggesting that this molecule penetrates from blood into the brain. The values of tiazofurin brain clearance are very close to the values obtained for neuropeptides and other so called "slow penetrating molecules" (regarding the blood brain barrier). Addition of increasing concentrations of unlabelled tiazofurin to the perfusing medium caused a significant decrease in the uptake of [^{3}H] labelled tiazofurin. Therefore, penetration of tiazofurin from blood into brain seems to be a saturable process. Presence of increasing concentrations of unlabelled adenosine has similar effect as the presence of unlabelled tiazofurin in the perfusing medium. However, it did not cause complete inhibition of tiazofurin brain uptake.

Kinetic parameters of transport, calculated from the experimental data points, suggest that: a) transport of tiazofurin from blood into guinea pig brain is mostly mediated by the nucleosides' transport system, but another mechanism is also involved in this transport ($K_d^a>0$); b) tiazofurin, compared to adenosine, has very small affinity for nucleoside carriers at luminal side of the guinea pig BBB ($Km>>K_i^a$); c) free diffusion of tiazofurin from blood to brain parenchyma does not exist ($K_d \approx 0$).

Perfusion with medium containing low concentration of Na^+ (<1mmol/l) did not cause significant decrease in tiazofurin brain uptake. Hence, Na^+- dependent cotransport is probably not involved in tiazofurin brain uptake.

Our results show that TZF transport from blood to the brain could be considered as a complex process which is still to be elucidated in order to improve its therapeutic use.

Biology and Physiology of the Blood–Brain Barrier, edited by Couraud and Scherman
Plenum Press, New York, 1996

RÉSUMÉ

Nous avons étudié chez le cobaye le transport de la thiazofurine (2-β-D-ribofuranosyl-thiazole-4-carboxamide) à travers la BHE par la méthode perfusion vasculaire cérébrale. Nos résultats montrent que la clairance de la 3H-thiazofurine est nettement différente de zéro, ce qui suggère que cette molécule passe du sang vers le cerveau. Les taux de clairance cérébrale de la thiazofurine sont très voisins de ceux obtenus pour des neuropeptides et d'autres molécules dites " à pénétration lente "(dans la BHE). L'addition de taux croissants de thiazofurine non marquée au milieu de perfusion provoque une chute significative de la capture de thiazofurine marquée. Cette pénétration du sang vers le cerveau semble donc relever d'un processus saturable; l'effet est le même par addition d'adénosine non marquée, mais sans toutefois aboutir à une totale inhibition de la capture de la thiazofurine marquée.

Les paramètres cinétiques de ce transport, calculés à partir des données expérimentales, suggèrent les conclusions suivantes: a/- chez le cobaye, le transport de la thiazofurine du sang vers le cerveau; est médié essentiellement par un système de transport de nucléosides, mais un autre mécanisme intervient également (K^i_d>0); b/- par comparaison avec l'adénosine, la thiazofurine a une très faible affinité pour les transporteurs de nucléosides au niveau de la face luminale de la BHE du cerveau de cobaye (Km >> K^i_d); c/- il n'y a pas de libre diffusion de la thiazofurine du sang vers le parenchyme du cerveau (Kd ≅0). La capture cérébrale de thiazofurine n'est pas réduite de façon significative par la perfusion d'un milieu à faible concentration en Na (< 1 mmol/l). Donc le co-transport Na-dépendant n'est sans doute pas impliqué dans la capture cérébrale de la thiazofurine.

1. INTRODUCTION

Tiazofurin (2-ß-D-ribofuranosyl thiazole-4-carboxamide NSC 286193) is a nucleoside analogue, which exhibits potent antitumor activity[1]. The mechanism of action of tiazofurin is due to its metabolic conversion to thiazole-4-carboxamide adenine dinucleotide (TAD)[2]. TAD is a selective inhibitor of the inosine monophosphate dehydrogenase (IMPDH) activity[2].

Clarifying the mechanism which enables the transfer of tiazofurin from blood to central nervous system (CNS) is necessary for two reasons. Obtaining these data seemed to be essential to redefine therapeutic scheme in order to minimize the observed neurotoxicity of tiazofurin. Secondly, the transport through the blood-brain barrier (BBB) appears to be the rate limiting step for achieving therapeutic concentration in brain tissue. Defining the rate of tiazofurin unidirectional influx from plasma into brain at various concentrations in plasma enables us to foresee the probabilities for possible addition of therapeutic indications targeted at CNS tumors. We also considered important investigating the probable competition of endogenous metabolites with tiazofurin related to the transfer through the luminal side of the BBB.

2. METHODOLOGY

The experiments were performed on 92 guinea pigs. Animals were devided into five groups:

a. Experimental group I - control - 15 animals were used to determine clearance of ^3H tiazofurin in guinea pig brain;

b. Experimental group II - 36 animals were used to determine parameters of Michaelis-Menten kinetics of TZF transport through the BBB (K_m, V_{max}, K_d);

c. Experimental group III - 24 animals were used to define the kinetic parameters of TZF transport inhibition by adenosine (K_i and K_d^a);

d. Experimental group IV - 8 animals were used to define brain clearance of ^{3}H tiazofurin in the presence of dipyridamole in perfusing medium;

e. Experimental group V - 9 animals were used in experiments performed to elucidate if TZF transport from blood into the brain depended on presence of Na^+ in plasma.

The technique of brain vascular perfusion in guinea pig has been previously reported in details[3]. Briefly, adult guinea pigs were anesthetized and right common carotid artery cannulated with polyethene tubing connected to the perfusion circuit. Perfusion fluid consisted of washed sheep erythrocytes suspended in a saline medium (hematocrit ~20%): Erythrocyte-free medium was used for experiments performed with experimental group II (to avoid strong accumulation of adenosine in the red cells). In experimental group III, Na^+ ions in perfusing medium were replaced with choline$^+$. The perfusion medium was pumped from a reservoir through the water bath. Isotopically labeled [^{3}H] tiazofurin (ICN Pharmaceuticals, Costa Mesa, CA, USA, specific activity 2 Ci/mMol) was introduced into a perfusion circuit by a slow-drive syringe "Harvard Apparatus 22" at a constant rate of 0.2 ml/min. Immediately after the start of the perfusion, the contralateral carotid artery was tied, and jugular veins severed to allow drainage of the perfusate. At the appropriate times the perfusion was terminated by decapitation of the animal and the samples of cerebral cortex and nucleus caudate prepared for scintillation counting.

2.1. Calculations

Brain clearance of [^{3}H] TZF after different perfusion times can be calculated as *(3)*:

$$\frac{C_{br}(T)}{C_{pl}(T)} = K_{in}T + V_i \tag{1}$$

where C_{br} is the concentration of [^{3}H] TZF in brain tissue (expressed as dpm/g) at appropriate time t, C_{pl} is its concentration in artificial plasma (dpm/ml), K_{in} is the unidirectional transport constant and V_i is the initial volume of distribution.

Michaelis Menten equation can be expressed as[5]:

$$K_{in} = \frac{V_{\max}}{\left(K_m + C_{pl}\right)} + K_d \tag{2}$$

where K_m is Michaelis-Menten constant expressed in mol/l, V_{max} is maximal transport rate expressed in mol/min/g and K_d is a constant of nonspecific diffusion expressed in ml/min/g.

In the presence of a competitive inhibitor of transport kinetic parameters of inhibition can be calculated from the equation[5]:

$$K_{in}^a = \frac{\left(V_{\max} / K_m\right)}{\left(1 + C_{pl}^a / K_i\right)} + K_d^a \tag{3}$$

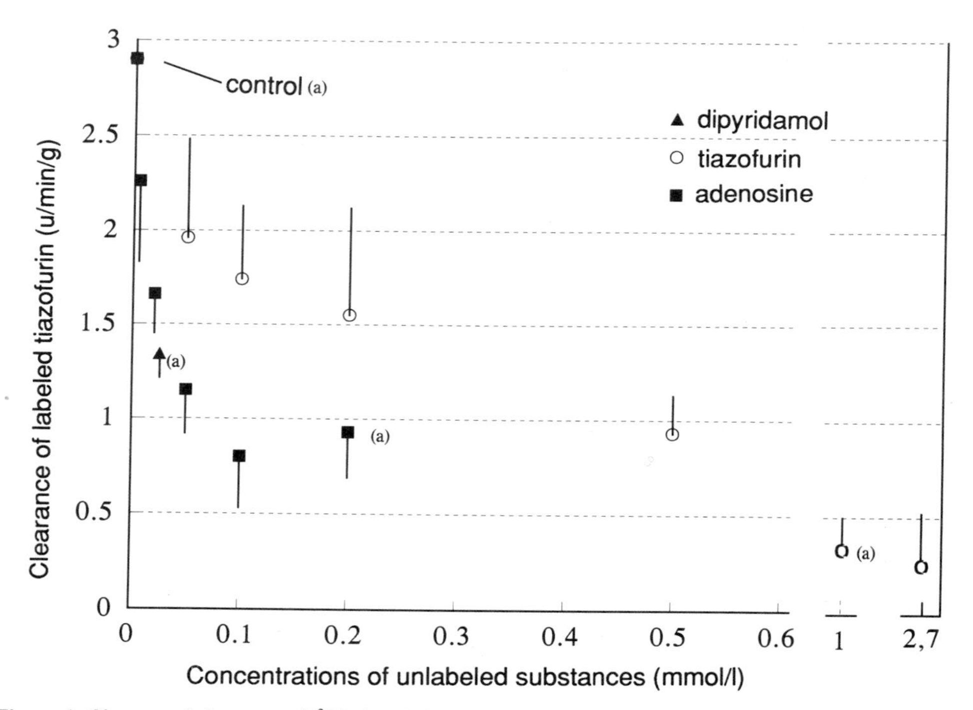

Figure 1. Changes of clearance of [^{3}H] tiazofurin in guinea pig cerebral cortex versus various concentrations of unlabelled tiazofurin (○), adenosine (■) and dipyridamole (▲) in the perfusing medium. Every point represents the mean value (± SE) from the results of experiments obtained on 4-6 animals.

where K_{in}^a is the clearance of [^{3}H] TZF in the presence of unlabelled adenosine, C_{pl}^a is the concentration of inhibitor (unlabelled adenosine) in perfusing medium, K_i is the inhibition constant of adenosine to [^{3}H] TZF transport and K_d^a is the diffusional constant of TZF transport in the presence of adenosine.

3. RESULTS

Our results show that tiazofurin penetrates very slowly from blood into the guinea pig brain. Its clearance in the cerebral cortex is similar to the clearances of so called "slow

Table 1. Kinetic parameters (maximal transport velocity, Michaelis-Menten constant K_m, constant of nonspecific diffusion K_d) for ^{3}H tiazofurin blood-to-brain transport and parameters of inhibition of [^{3}H] tiazofurin blood-to-brain transport by adenosine (inhibition constant K_i, and diffusion constant of transport inhibition by adenosine K_d^a)

Region	K_m (μM/l)	V_{max} (pmol/min/g)	K_d (μl/min/g)	K_d^a (μl/min/g)	K_i (μM/l)
nc. caudatus	119.57 ± 40.1	325.03 ± 113.93	0.220 ± 0.188	0.70 ± 0.18	6.36 ± 3.14
cortex	150.17 ± 51.6	417.50 ± 151.53	0.136 ± 0.206	0.71 ± 0.18	11.74 ± 4.85

Values were obtained by fitting data points from experimental groups II (36 separate experiments) and III (24 separate experiments) to eq. (2) and (3). Values are mean ± SE.

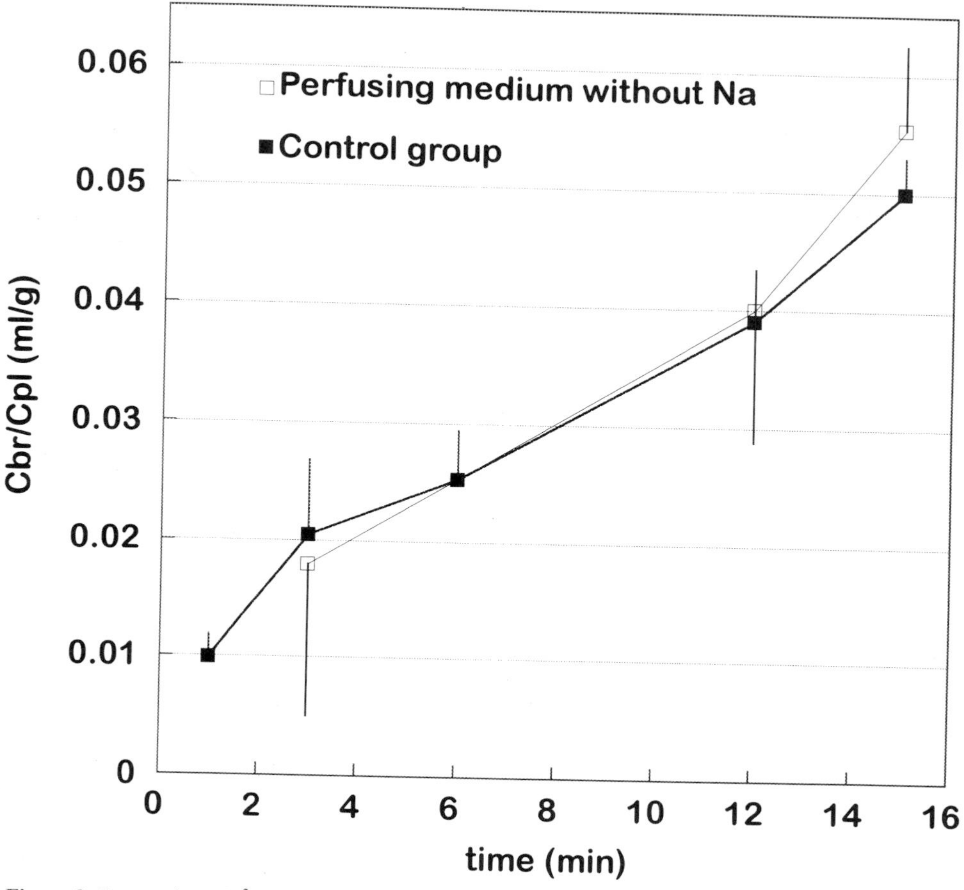

Figure 2. Penetration of [^{3}H] tiazofurin from blood into brain tissue in control group (■) and group where perfusion medium was depleted of Na^+ ions (□). Every point represents mean value (± SD) of the volume of distribution obtained from 2-4 separate experiments. Values for control group are from ref.

penetrating substances" (i.e. neuropeptides) (Fig. 1). Introducing of increasing concentrations of unlabelled tiazofurin in the perfusing medium caused decrease in clearance of ^{3}H tiazofurin. Therefore, tiazofurin transport through blood brain barrier is a **saturable** process. Both addition of increasing concentrations of unlabelled adenosine to the perfusing medium and presence of dipyridamole (nucleosides' transport inhibitor) had similar effect (Fig. 1). It seems that tiazofurin blood-to-brain transport is mediated mostly by the nucleoside transport system.

Free diffusion of tiazofurin from blood to brain parenchyma does not exist ($K_d \approx 0$ - Tab. 1). It is in accord with very low liposolubility of tiazofurin. Transport of tiazofurin from blood into guinea pig brain is mostly mediated by the **nucleosides'** transport system, but another mechanism is also involved in this transport ($K_d^a > 0$ - Tab. 1). Tiazofurin, compared to adenosine, has **very small affinity** for nucleoside carriers at luminal side of the guinea pig BBB ($Km >> K_i^a$).

Substitution of Na^+ ions with choline in perfusing medium did not change the clearance of ^3H tiazofurin in the cerebral cortex (Fig. 2). Consequently, transport of tiazofurin from blood into the guinea pig brain seems to be **sodium independent**.

4. CONCLUSIONS

Tiazofurin penetrates very slowly from blood into the guinea pig brain by a **saturable mechanism**. It seems that this process is **sodium independent**. Transport of tiazofurin from blood into guinea pig brain is mostly mediated by the **nucleosides'** transport system, but another mechanism is also involved in this transport. Tiazofurin, in relation to adenosine, has **very small affinity** for nucleoside carriers at luminal side of guinea pig BBB. The capacity of the transport system, if fully saturated, could provide the active concentration of the drug in brain tissue **after only 10-20 minutes** since the i.v. application of tiazofurin.

REFERENCES

1. Szekeres T, Fritzer M, Pillwein K, Felzmann T and Chiba P. Cell cycle dependent regulation of IMP dehydrogenase activity and effect of tiazofurin. Life Sci 1992; 51 (16):1309-15.
2. Yamada Y., Natsumeda Y., Yamaji Y. et al. IMP Dehydrogenase: Inhibition by the antileukemic drug, tiazofurin. Leuk Res (1989) 13 (2): 179-84.
3. Zlokovic BV, Begley D, Durièiæ B and Mitroviæ D. M. Measurement of solute transport across the blood brain barrier in the perfused guinea pig brain; method and application to N-methyl-α-amino isobutyric acid. J Neurochem (1986) 46:1444-1451.
4. Redzic Z. B., Markovic I.D., Jovanovic S. S., Mitrovic D., M., Zlokovic B. V., Rakic Lj. M. Slow penetration of [3H] tiazofurin into guinea pig brain by a saturable mechanism.(1995). 17(6) (in press).
5. Pardridge W. M., Yoshikawa T., Kang Y. S., Miller L. P. Blood-brain transport and brain metabolism of adenosine and adenosine analogs. J Pharmacol Exp Ther. (1994) 268 (1): 14-8.

27

THE PERMEABILITY OF BORONOPHENYLALANINE

A Compound Used for the Treatment of Cerebral Gliomas

H. Patel[1] and J. W. Hopewell[2]

[1] Clinical Neurological Sciences
Southampton General Hospital
Southampton, United Kingdom
[2] Research Institute
Churchill Hospital
Oxford, United Kingdom

SUMMARY

The integrity of the blood-brain barrier may severely restrict the passage of drugs intended for cells within the central nervous system. These experiments were designed to show whether one particular drug used for the treatment of gliomas, boronophenylalanine (BPA), is affected by the state of the BBB.

The results show that the degree of BBB disruption does not influence the amount of BPA accumulating in normal brain tissue.

RÉSUMÉ

L'absence de perméabilité membranaire de la barrière hémato-encéphalique (BHE) peut restreindre gravement le passage de dérivés thérapeutiques destinés à traiter les cellules contenues dans le SNC. Cette étude a pour but de vérifier si l'action d'un dérivé particulier, utilisé pour le traitement des gliomes, la boronophénylalanine (BPA), est affectée par l'état de la BHE. Les résultats montrent que le degré de rupture de la BHE ne modifie pas la quantité de BPA accumulée dans les tissus normaux du cerveau.

INTRODUCTION

Boronophenylalanine (BPA) is a boronated form of the amino acid phenylalanine and is currently undergoing clinical trials as part of a new treatment for cerebral gliomas. This novel

Biology and Physiology of the Blood–Brain Barrier, edited by Couraud and Scherman
Plenum Press, New York, 1996

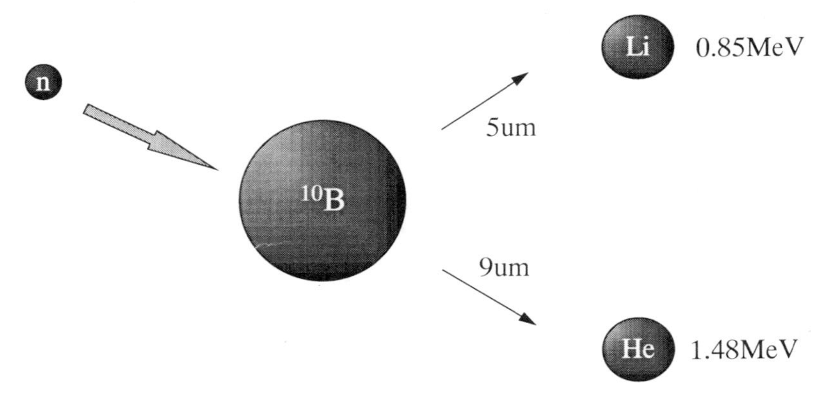

Figure 1. This shows the energies and path lengths of the breakdown products from an interaction of a Boron atom and a neutron.

therapy termed Boron Neutron Capture Therapy (BNCT), is a two step process which culminates in the delivery of almost cell specific, lethal radiation to cancer cells. The first step involves injecting the patient with a compound containing boron-10, a naturally occurring, non-radioactive, non-toxic isotope of boron. This compound should concentrate in tumour cells with no appreciable accumulation in normal cells. The next step is to irradiate the tumour with low energy neutrons. When a neutron hits a boron atom it disintegrates, releasing an alpha particle and a lithium atom, both of which are highly lethal to cells (Fig.1). Fortunately these fragments only travel about 10 microns, which is about the diameter of one cell. Thus neutrons cause indirect destruction of those cells containing boron but pass through other cells.

The ability of agents to effectively accumulate in tumours is a function of a number of different factors, firstly the integrity of the blood-brain barrier (BBB) for vessels both within the tumour and the oedematous area surrounding the brain and secondly, the ability of the tumour cells to take up the agent. The present experiments are part of a series designed to determine how these factors may interact and culminate in effective tumour concentrations of boronated drugs. Since the amount of drug measured within a tumour is the sum of these factors, preliminary data are presented on the permeability of cerebral vessels to BPA after a cryoinjury. This type of injury was used to simulate an oedematous area in normal brain as may be found around a tumour. This allowed an investigation of BPA accumulation without the complications produced by tumour bulk. Both albumin leakage as a marker for general BBB disruption, and the accumulation of BPA have been measured.

METHOD

A cryolesion injury was produced in male Fischer 344 rats by placing a hollow metal probe 5mm diameter, cooled in and filled with liquid nitrogen, on the surface of the skull for 90 seconds. Animals were used either 24 hours or 7 days post injury. On the day of the experiment animals were split into two groups. One group received an intravenous injection of 50mg of BPA and were killed one hour later. Blood and relevant areas of brain tissue were analysed for boron content by Inductively Coupled Plasma Mass Spectrometry (ICPMS). The second group received saline injections and the brains processed using standard histological methods to assess albumin permeability. Computer assisted image analysis was performed to assess the degree of albumin leakage.

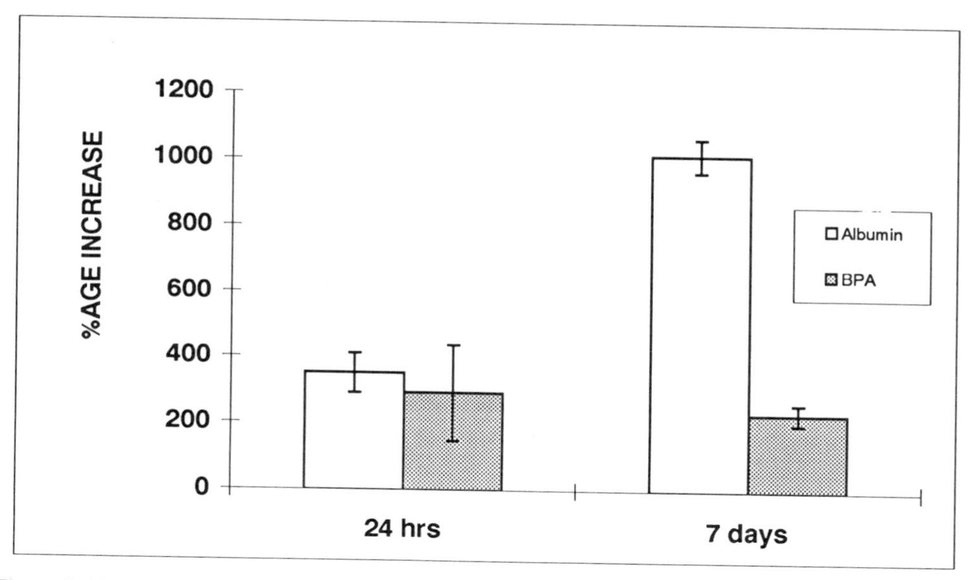

Figure 2. The increase in permeability of either Albumin or BPA in the lesioned hemisphere when compared to the contralateral hemisphere. The mean of each group, ± s.e.m. is indicated. n=4-7.

RESULTS

The results (Fig 2) show that 24 hours after injury, albumin staining in the lesioned hemisphere when compared to the contralateral side, is increased. There is however a much greater increase in albumin permeability 7 days post injury. In contrast, the permeability of BPA does not change significantly between the two time points investigated.

CONCLUSION

These results indicate that a proportion of BPA does cross into the brain parenchyma. However, this is probably a non-specific effect, not dependant on the degree of BBB damage. This is shown by the fact that, though at day 7 the BBB was largely disrupted, there was little change in the BPA permeability. Thus the area around a tumour, where the BBB is damaged will still have appreciable amounts of BPA. This may then in turn be taken up by any so called guerilla cells present.

Similar experiments to calculate BPA permeability within an experimental tumour model are being conducted. In addition, it is hoped to conduct experiments to calculate the vascular perfusion and density within both the cryolesion and in tumours.

28

DIFFERENCES BETWEEN THE CALCULATED FRACTION AVAILABLE FOR BBB TRANSPORT AND THE *IN VITRO* MEASURED FREE FRACTION COULD BE EXPLAINED BY CAPILLARY HETEROGENEITY

C. Videbæk,[1] P. Ott,[2] O. B. Paulson,[2] and G. M. Knudsen[2]

[1] Department of Hepatology
[2] Department of Neurology
State University Hospital Rigshospitalet
Blegdamsvej 9, DK-2100
Copenhagen, Denmark

SUMMARY

It has earlier been shown that presence of albumin in the brain capillaries apparently enhances uptake of protein bound substances (Pardridge and Fierer, *J. Neurochem.* 54: 971-976,1990; Pardridge and Landaw, *Am. J. Physiol.* 249: E5534-E542, 1985; Westphal Receptors and hormone action. Acad. Press 443-472, 1978; Jones et al. *J. Pharmacol. Exp. Ther.*, 245: 816-822, 1988; Cornford et al. *J. Neurochem.*, 44: 1541-1550, 1985). This study evaluated the protein-ligand interaction for two benzodiazepine antagonists, iomazenil and flumazenil, in rats by comparaison of the calculated fraction available for BBB transport (f_{avail}) and the *in vitro* measured free fraction (f_{eq}).

RÉSUMÉ

Il a été montré précédemment que lorsque des substances sont fixées dans le sang à l'albumine, leur captation au niveau des capillaires cérébraux est augmentée (Pardridge and Fisher, *J. Neurochem.*,54:974-976, 1990; Pardridge and Landaw, *Amer. J. Physiol.*,249: 534-E542,1985; Westphal *Receptors and hormone action.* Acad. Press 443-472, 1978; Jones et al., *J. Pharmacol. Exp. Ther.*, 245:816-822, 1988; Cornford et al., *J. Neurochem.*, 44:1541-1550, 1985). Nous avons évalué l'interaction de deux antagonistes de la benzodiazépine, l'iomazenil et leflumazenil à l'albumine et calculé la fraction libre disponible

Biology and Physiology of the Blood–Brain Barrier, edited by Couraud and Scherman
Plenum Press, New York, 1996

pour le transport à travers la BHE (f_{libre}). Nous avons aussi mesuré cette fraction libre *in vitro* (f_{eq}).

METHODS

Repeated measurements of blood-brain barrier permeability for iomazenil and flumazenil were performed in 44 rats by the double-indicator technique. During anesthesia femoral vein and artery were cannulated, the external carotid artery and all major extra cerebral branches from the carotid artery were ligated (Hertz and Bolwig, *Brain Res.*, 107:333-343, 1976). A polyethylene catheter was placed in the external carotid artery with the tip at the carotid bifurcation and a sawn off needle was placed in the confluence sinus for cerebral veinous blood sampling. A 20 μl bolus, containing a BBB impermeable reference substance, either [^{36}Cl] (Amersham, U.K.) or [^{24}Na] (Riso, Roskilde, Denmark) and a test substance ([^{125}I]iomazenil (Paul Scherrer Institute, Villigen, Switzeland) or [^{3}H]flumazenil (Dupont, NEN Research Products, Boston, MA., USA), was rapidly injected into the carotid artery and blood samples were passively collected from the confluence sinus. For evaluation of bolus mixing with blood during the capillary passage the size of the saline bolus was altered between 20 μl and 120 μl. In BBB experiments where CBF changes were induced, 5% albumin boluses were used.

Cerebral blood flow (CBF) was measured by the intracarotid ^{133}Xe-injection technique. CBF was increased by adding approximately 6% CO_2 to the inhaled gas mixture, and decreased by hyperventilation.

The apparent permeability-surface product (PS_{app} = -CBF x ln (1 - Extraction)) was measured under different conditions: Changing bolus injection volume, CBF, or bolus injectate composition. f_{avail} was obtained by comparaison of PS_{app} obtained in absence and presence of 5% albumin in the injectate, respectively. f_{eq} was measured in eight rats by equilibrium dialysis.

Several phenomena may play a role in the evaluation of brain uptake of protein bound substances, i.e., non-equilibrium conditions, unstirred water layer effects, and capillary heterogeneity. For evaluation of the potential influence of these phenomena, simulation models were applied and the results were compared to the observed data.

RESULTS AND DISCUSSION

f_{avail} for iomazenil and flumazenil was 62% and 82%, respectively, whereas f_{eq} was significantly lower, 42% and 61%. PS_{app} for iomazenil and flumazenil increased significantly by 89% and by 161% after relative CBF increases of 259% and 201%, respectively.

Non-equilibrium protein binding could only explain about 25% of the observed increase in PS_{app} in hypercapnia. In hypocapnia binding desequilibrium could explain almost half of the observed PS_{app} reduction. Inclusion of an unstirred water layer did not influence the data at all.

Different distributions of capillary lengths were applied to the data for iomazenil. When capillary lengths were considered to be exponentially distributed a good accordance between modelled and observed results was obtained. $f_{avail} > f_{eq}$ and PS_{app} increased with CBF, and when the same exponential distribution was applied to flumazenil, a very nice accordance between the calculated and the observed data was found.

THE STRONGEST EXPRESSION OF P-GLYCOPROTEIN IS IN THE BLOOD-BRAIN BARRIER

Edith Beaulieu, Michel Demeule, Diana A. Averill-Bates, and Richard Béliveau

Laboratoire d'oncologie Moléculaire
Département de Chimie-Biochimie
Université du Québec à Montréal
C.P. 8888, Succursale A
Montréal, Québec, Canada, H3C 3P8

SUMMARY

P-glycoprotein (P-gp), an active efflux pump of antitumor drugs, is strongly expressed in endothelial cells of the blood brain barrier (BBB). Two proteins (155 and 190 kDa) were detected by Western blot analysis of beef and rat capillaries with the monoclonal antibody (MAb) C219. In order to characterize the nature of these proteins, biochemical studies were performed in comparison with P-gp from the $CH^{R}C5$ tumoral cell line. Results suggest that only p155 is the P-gp in BBB and that MAb C219 cross-reacts with a 190 kDa MDR-unrelated glycosylated protein. Western blot analysis shows presence of P-gp in several tissues and presence of p190 in some tissues. P-glycoprotein was predominantly expressed in brain capillaries compared to other tissues.

RÉSUMÉ

La P-glycoprotéine (P-gp), pompe membranaire à efflux sortant ATP-dépendante, est fortement exprimée dans les cellules endothéliales de la barrière hémato-encéphalique (BHE). L'utilisation de l'anticorps monoclonal (Mab) C219 a permis de mettre en évidence par immunotransfert l'expression de deux protéines (155 et 190 kD) dans les capillaires de rat et de boeuf. Afin de caractériser la nature de ces protéines, des études biochimiques ont été réalisées en comparant l'expression de la P-gp dans les capillaires cérébraux et dans la lignée tumorale CHRC5. Les résultats suggèrent que seule la révélation par l'anticorps C219 de la protéine de 155 kD est spécifique de la P-glycoprotéine. L'anticorps C219 semble avoir une réaction croisée avec une glycoprotéine de 190 kD n'appartenant pas à la famille des protéines de résistance multidrogue. Des expériences d'immunotranfert montrent que la

Biology and Physiology of the Blood–Brain Barrier, edited by Couraud and Scherman
Plenum Press, New York, 1996

P-glycoprotéine est présente dans de nombreux tissus et que la protéine P190 est présente seulement dans quelques tissus. L'expression de la P-gp est la plus forte dans les capillaires cérébraux en comparaison des autres tissus étudiés.

INTRODUCTION

Multidrug resistance (MDR) constitutes a major obstacle in the clinical treatment of cancer[1]. The overexpression of a membrane glycoprotein of 150-170 kDa[2-4], P-glycoprotein (P-gp), is considered to be the cause of this resistance to multiple chemotherapeutic drugs[5]. The presence of P-gp has been demonstrated in normal tissues such as adrenal medulla, liver, kidney, colon[6,7] and endothelial cells of brain capillaries[7-9] where it is strongly expressed. Most of the tissues studies have been made by Northern blot, Southern blot or immunohistochemistry. The protein expression could be a better clue for P-gp role than DNA or mRNA presence. Immunohistochemistry detects protein expression but it can lead to false positive because of the cross reactivity.

Among the various antibodies available to detect P-gp in human tumors and in animal studies, MAb C219 is by far the most widely used. The extensive use of this antibody is justified by its high affinity for the three classes of P-gp isoforms identified so far in most species where it was studied, including human, rat, mouse, beef, hamster and cell lines of various origins[2,10]. However, cross-reactivity of MAb C219 have already been reported with a protein of 65 kDa by Western blot and with a muscular protein by immunohistochemistry[11,12]. In this paper, we present a biochemical characterization of P-gp in purified BBB endothelial cells. A tumoral cell line from chinese hamster ovaries (CH^RC5) which expresses the MDR phenotype was used as a reference[13-15]. Western blot analysis with MAb C219 is also performed in various tissues showing presence of P-gp and p190 in several tissues.

RESULTS AND DISCUSSION

Two proteins bands of comparable intensity with apparent molecular weight of 155 and 190 kDa from beef and rat brain capillaries were immunodetected with the MAb C219 as described previously[16]. The detection of p155 was significantly reduced by heating the samples from both species (Fig. 1). It has been shown that P-gp from CH^RC5 cells was also sensitive to heat treatment[17].

The physicochemical characteristics of these two proteins (190 and 155 kDa) were strikingly different. Extraction of brain capillary proteins with a variety of detergents indicated that the two proteins detected by MAb C219 showed different patterns of solubi-

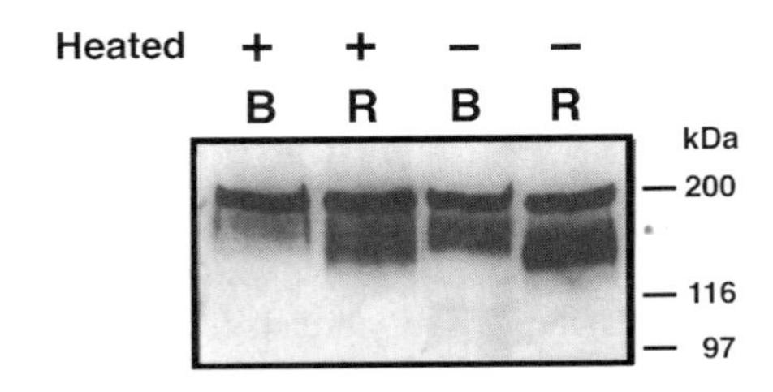

Figure 1. Detection of P-gp from brain capillaries. Proteins (10 μg) from beef (B) and rat (R) were heated 3 min or solubilized 30 min at room temperature before SDS-PAGE. Proteins were transferred onto PVDF membranes and analyzed by Western blot with MAb C219 as described previously [21].

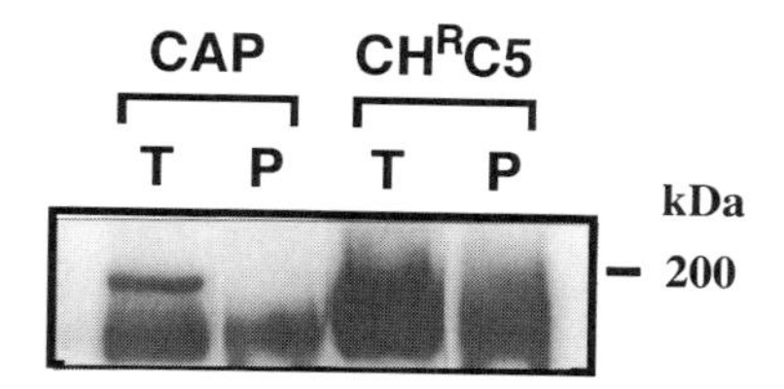

Figure 2. *Immunoprecipitation of brain capillary proteins and P-gp from CH^RC5 cells with MAb C219.* Solubilized proteins from rat brain capillaries (CAP) and CH^RC5 cells (CH^RC5) were immunoprecipitated with MAb C219 as described in the Materials and Methods section. Total (T), and the immunoprecipitated proteins (P) were separated by SDS-PAGE and Western blot analysis performed as described previously [21].

lization[16]. The detergents that solubilized efficiently p155 from brain capillaries also solubilized P-gp from CH^RC5 cells. P190 behaved very differently since SDS was the only detergent that solubilized it. Both brain capillary proteins detected with MAb C219 are glycoproteins but p155 and p190 reacted very differently to the N-deglycosylation treatment[16]. P155 was more strongly affected in its migration pattern than p190 and the shift in the apparent molecular weight of p155 from capillaries was similar to that obtained for the P-gp of CH^RC5 cells after the same deglycosylation treatment. Finally, among brain capillary proteins, only p155 was immunoprecipitated by MAb C219 (fig. 2). These results suggest that only p155 is P-gp.

Previous studies of P-gp expression in various tissues have been performed by immunohistochemistry[18]. In this paper, Western blot analysis are performed with MAb C219 on total membrane proteins from rat tissues (fig. 3). Presence of P-gp is shown in all tissues except pancreas where it is not detected. The highest level of P-gp is detected in brain capillaries. Presence of P-gp in almost all tissues suggest a fundamental physiological role

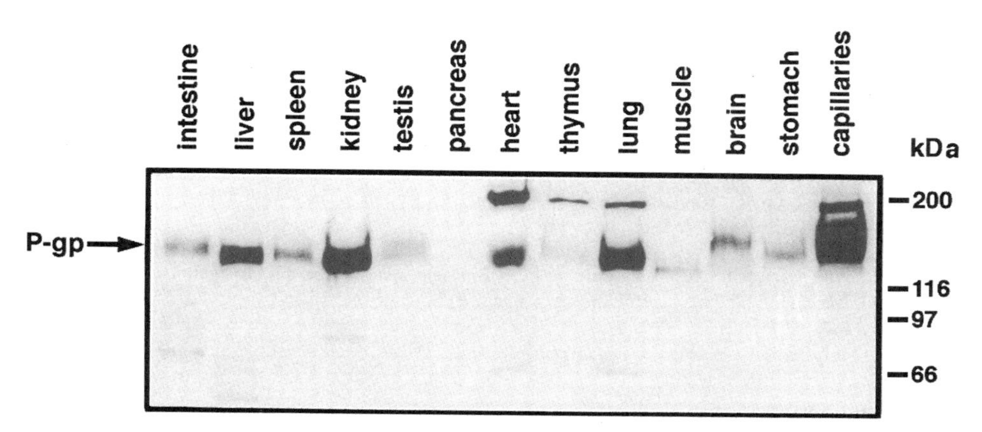

Figure 3. *Detection of P-gp in rat tissues by MAb C219.* Proteins (30 μg) from total membranes of rat tissues or proteins (15 μg) from rat brain capillaries were loaded on 1.5 mm gels. Electrophoresis, transfer and immunodetection were performed as described prviously [21].

for this protein, possibly in secretion or transport[4]. Strong P-gp expression in the brain capillaries suggest its implication in the formation of the pharmacological barrier limiting the access of drugs to the brain[19]. P-gp was found to limit the amount of drug accumulating in the brain[20]. Drug interaction with P-gp in brain capillaries was recently studied[21]. Cyclosprin A and its non-immunosuppressive analog, PSC 833 were found to be the most potent inhibitor of drug binding to P-gp. Chronic treatment of rats with CsA caused a strong induction of P-gp expression in the kidney and the gut, but not in brain capillaries, suggesting that CsA or PSC 833 could help to enhance the efficiency of chemotherapeutic agents in patients with brain tumor.

ACKNOWLEDGMENTS

This work was supported by grants from the Natural Sciences and Engineering Research Council of Canada and by Sandoz Canada.

REFERENCES

1. Pastan, I. and Gottesman, M.M. (1987) N. Engl. J. Med. 316, 1388-1393.
2. Kartner, N., Evernden-Porelle, D., Bradley, G. and Ling, V. (1985) Nature 316, 820-823.
3. Juliano, R.L. and Ling, V. (1976) Biochim. Biophys. Acta 455, 152-162.
4. Endicott, J.A. and Ling, V. (1989) Annu. Rev. Biochem. 58, 137-171.
5. Georges, E., Sharom, F.J. and Ling, V. (1990) Adv. Pharmacol. 21, 185-215.
6. Fojo, A.T., Ueda, K., Slamon, D.J., Poplack, D.G., Gottesman, M.M. and Pastan, I. (1987) Proc. Natl. Acad. Sci. USA 84, 265-269.
7. Cordon-Cardo, C., O'Brien, J.P., Casals, D., Rittman-Grauer, L., Biedler, J.L., Melamed, M.R. and Bertino, J.R. (1989) Proc. Natl. Acad. Sci. USA 86, 695-698.
8. Tsuji, A., Terasaki, T., Takabatake, Y., Tenda, Y., Tamai, I., Yamashima, T., Moritani, S., Tsuruo, T. and Yamashita, J. (1992) Life Sci. 51, 1427-1437.
9. Jetté, L., Têtu, B. and Béliveau, R. (1993) Biochim. Biophys. Acta 1150, 147-154.
10. Georges, E., Bradley, G., Gariepy, J. and Ling, V. (1190) Proc. Natl. Acad. Sci. USA 87, 152-156.
11. Kaway, K., Kusano, I., Ido, M., Sakurai, M., Shiraishi, T. and Yatani, R. (1994) Bioche. Biophy. Res. Commu. 2, 804-810.
12. Thiebaut, F., Tsuruo, T., Hamada, H., Gottesman, M.M., Pastan, I. and Willingham, M.C. (1989) Histochem. Cytochem. 37, 159-164.
13. Marie, J.P. (1990) Médecine/Science 6, 443-448.
14. Kartner, N. and Ling, V. (1989) Sci. Am., March 1989, 44-52.
15. Kartner, N., Riodan, J.R. and Ling, V. (1983) Science 222, 1285-1288.
16. Beaulieu, E., Demeule, M., Pouliot, J.-F., Averill-Bates, D.A., Murphy, G.F. and Béliveau, R. 1995. Biochim. Biophys. Acta 1233, 27-32.
17. Greenberger, L.M., Williams, S.S., Georges, E., Ling, V. and Horwitz, S.B. (1988) J. Natl. Cancer Inst. 80, 506-510.
18. Thiebaut, F., Tsuruo, T., Hamada, H., Gottesman, M.M., Pastan, I. and Willingham, M.C. 1987. Proc. Natl. Acad. Sci. USA. 84, 7735-7738.
19. Tatsuta, T., Naito, M., Oh-hara, T., Sugawara, I. and Tsuruo, T. (1992) J. Biol. Chem. 267, 20383-20391.
20. Schinkel, A.H., Smit, J.J.M., van Tellingen, O., Beijnen, J.H., Wagenaar, E., van Deemter, L., Mol, C.A.A.M., van der Valk, M.A., Robanus-Maandag, E.C., te Riele, H.P.J., Berns, A.J.M. and Borst, P. (1994) Cell 77, 491-502.
21. Jetté, L., Murphy, G.F., Leclerc, J.-M. and Béliveau, R. (1995). Biochem. Pharmacol. (In press)

30

FUNCTIONAL EXPRESSION OF MULTIDRUG RESISTANCE P-GLYCOPROTEIN IN CELLULAR MODELS OF PHYSIOLOGICAL BARRIERS

Delphine Lechardeur, Pierre Wils, Bertrand Schwartz, and Daniel Scherman

UMR 133 CNRS/Rhône-Poulenc Rorer
Centre de Recherche de Vitry-Alfortville
13, quai Jules Guesde, BP 14 94403 Vitry/Seine, France

SUMMARY

The P-glycoprotein mdr (P-gp) is expressed not only in tumoral cells, but also in several non-transformed cells, specially in the apical plasma membrane of the intestinal, kidney or hepatic epithelium or of the blood-brain barrier endothelial cells. This apical localization is thought to be responsible for the net flux of hydrophobic compounds toward the intestinal lumen or from the brain to the blood, reflecting the protective role of P-gp at the epithelial and endothelial barriers. Many efforts are presently spent by several groups for the obtention of cellular models of intestinal epithelium or of the blood-brain barrier, in order to determine drug absorption by a predictive *in vitro* assay.

We have observed the functional expression of the mdr P-glycoprotein in the human intestinal epithelial cell line, Caco-2. On confluent monolayers in dual culture chambers, the expression of P-gp was apical, leading to a polarized basal-to-apical transport of P-gp substrates such as vinblastine, taxotere and pristinamycine I-A. Moreover, we have studied the P-gp expression in confluent monolayers of primary cultures of rat and bovine brain capillary endothelial cells. On the contrary of other specific cerebral endothelial cell markers such as gamma-glutamyl transpeptidase, the P-gp espression is maintained and functional in primary cultures of bovine cerebral endothelial cells. P-gp expression could also be induced in rat brain capillary endothelial cell line immortalized by chromosomal insertion of the SV 40 T gene controlled by the human vimentin promoter.

RÉSUMÉ

La P-glycoprotéine de résistance multidrogue (P-gp) n'est pas seulement exprimée dans certaines cellules tumorales mais aussi au niveau de nombreuses cellules normales telles

Biology and Physiology of the Blood–Brain Barrier, edited by Couraud and Scherman
Plenum Press, New York, 1996

que la membrane apicale des cellules épithéliales de l'intestin, du rein ou du foie ou des cellules endothéliales de la barrière hémato-encéphalique. L'expression polarisée de la P-gp serait responsable du flux net de nombreuses molécules lipophiles de l'épithélium intestinal vers la lumière intestinale ou des cellules endothéliales de la BHE vers le sang ce qui reflèterait un rôle physiologique de protection pour la P-gp contre les xénobiotiques au niveau des barrières cellulaires formées par les épithéliums et les endothéliums. Le développement de modèles cellulaire permettant la prédiction in vitro de lapénétration intestinale ou cérébrale de molécules à visée thérapeutique est en pleine expansion. Le rôle protecteur de la P-gp au niveau des barrières physiologiques montre que de tels modèles ne peuvent être validés que si la P-gp y est exprimée de façon fonctionnelle et polarisée. Nous avons étudié l'expression fonctionnelle de la P-gp dans la lignée cellulaire d'épithélium intestinal humain. la lignée Caco-2. Cultivées sur des supports poreux dans dans des inserts de culture, les cellules Caco-2 à confluence expriment la P-gp au niveau apical ainsi que le montre le transport polarisé de la membrane basale vers la membrane apicale de substrats de la P-gp tels que la vinblastine, le taxotère ou la pristinamycine 1A. De plus, nous avons aussi étudié l'expression de la P-gp dans des cultures primaires de cellules endothéliales de capillaires cérébraux de rat er de boeuf. A la différence d'autres marqueurs spécifiques de la BHE tels que la gamma-glutamyl transpeptidase, l'expression de la P-gp est maintenue et fonctionnelle dans les cultures d'origine bovine. L'expression de la P-gp a pu aussi être induite dans une lignée de cellules endothéliales cérébrales de rat immortalisée par insertion chromosomique du gene T de SV 40 sous contrôle du promoteur humain de la vimentine.

INTRODUCTION

P-Glycoproteins are 130 to 180 kDa ATP-dependent plasma membrane efflux pump encoded by the multidrug resistance MDR genes. P-Glycoprotein is responsible for the MDR phenotype of tumor cell lines selected *in vitro* for their resistance to a variety of cytotoxic molecules such as vinca alkaloids, actinomycin, doxorubicin, epidophyllotoxin or cyclosporin A (Beck, 1987). P-Glycoprotein (P-gp) pumps these drugs out of the multidrug resistant cells and thus prevents drug intracellular accumulation and toxicity (Shimabuku et al., 1992). The P-glycoprotein has been detected in a variety of normal human and rodent tissues and cell types including adrenal cortex, epithelial cells from small intestine, colon, biliary caniculi or kidney proximal tubules, and endothelial cells from the brain, derma, testis and retina (Hsing et al., 1992; Fojo et al., 1987; Cordon-Cardo et al.,1989; Cordon-Cardo et al., 1990; Pileri et al., 1991; Greenwood, 1992). The function and endogenous substrate(s) of P-gp in normal cells remain unknown, but the P-gp localization on the apical face of epithelial cells forming physiological barriers and of the blood-brain-barrier endothelium suggests that P-gp may limit intestinal absorption or brain penetration of hydrophobic xenobiotics. Disruption of the mouse mdr1a P-glycoprotein has been shown to lead to an increased brain sensitivity to neurotoxic drugs (Schinkel et al., 1994). This indicates that the blood-brain barrier may function as an active barrier against the brain penetration of many lipophilic compounds.

In order to develop convenient and versatile methods to predict the oral absorption or brain penetration of drug candidates, major efforts are presently spent by several groups for the obtention of *in vitro* models of intestinal epithelium and blood-brain-barrier (BBB) (Wils et al., 1993; 1994a, 1994b; Dehouck et al., 1992; Rubin et al., 1991). These models should display relevant *in vivo* characteristics: i) establishment of intercellular tight junctions leading to a high transmonolayer electric resistance; ii) expression of specific enzymes or transporters . In particular, the polarized expression of P-gp on the apical side of the epithelial or endothelial cell monolayer should result in a vectorial basal-to-apical flux of P-gp substrates. For BBB models for instance, this should reflect the *in vivo* situation, in which the apical P-gp expression in brain capillary endothelial cells is commonly thought to be responsible for the very low brain penetration of drugs such as vinca alkaloids or cyclosporin (Greig et al., 1990; Begley et al., 1990).

I. P-GLYCOPROTEIN EXPRESSION IN *IN VITRO* MODELS OF INTESTINAL EPITHELIAL CELLS

We have studied the transepithelial transport of drugs through intestinal cell monolayers in the dual chamber system presented in Figure 1 (for methods, see Wils at al., 1994a, 1994b and 1995).

The human colon carcinoma cell lines HT29-18-C1 and Caco-2 express many characteristics of differentiated cells of the normal small intestine (Pinto et al., 1983; Huet at al., 1987) and are widely used as an *in vitro* model for the study of drug transport across the intestinal epithelium (Hidalgo et al., 1989; Artursson, 1990; Wils et al., 1993, 1994a, 1994b). No functional expression of P-glycoprotein was found in the differentiated HT29-18-C1 cell line. The expression of P-glycoprotein in the Caco-2 cell line was very high, as judged by immunoblotting and by active efflux of vinblastine (Hunter et al., 1993; Wils at al., 1994b). The transport of vinblastine was polarized in the basolateral to apical direction (Figure 2). It was temperature and energy dependent, and was reduced by P-glycoprotein inhibitors such as verapamil, chlorpromazine and reserpine. This added further evidence that the polarized transport of vinblastine across Caco-2 monolayers was mediated by P-glycoprotein.

The anticancer drug docetaxel (Taxotere®) was also transported in a polarized manner by Caco2 monolayers in the dual chamber system: the basolateral to apical permeability was 20-fold higher than in the reverse direction. This polarized transport was inhibited by verapamil, chlorpromazine and reserpine, thus demonstrating that docetaxel is a substrate of P-glycoprotein (Wils et al., 1994b).

Using this same technique, we also demonstrated that Pristinamycin IA, a cyclo-peptidic macrolactone antibiotic belonging to the streptogramin family, also interacted with multidrug transporter P-glycoprotein, and was indeed a P-gp substrate, a finding which might have important consequences for the pharmacokinetics of this drug (Wils et al., 1995). It has

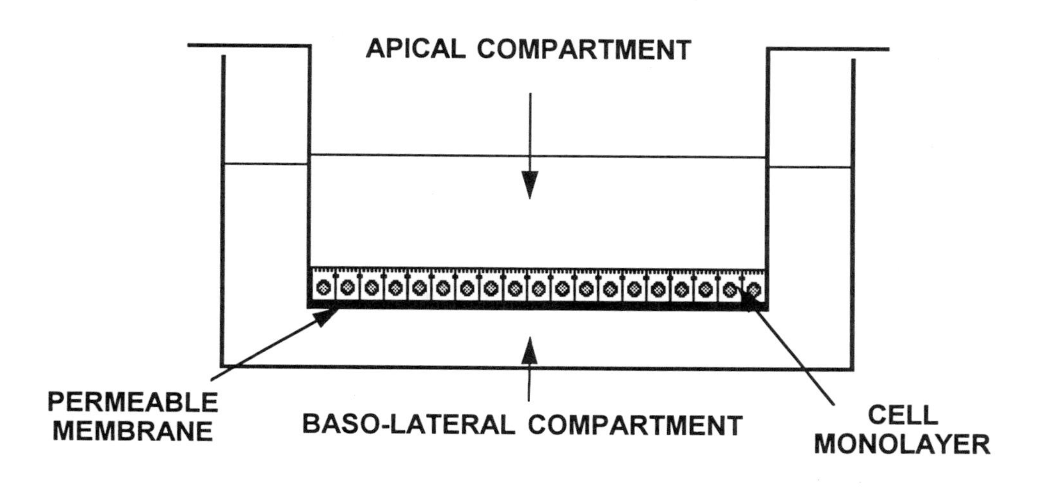

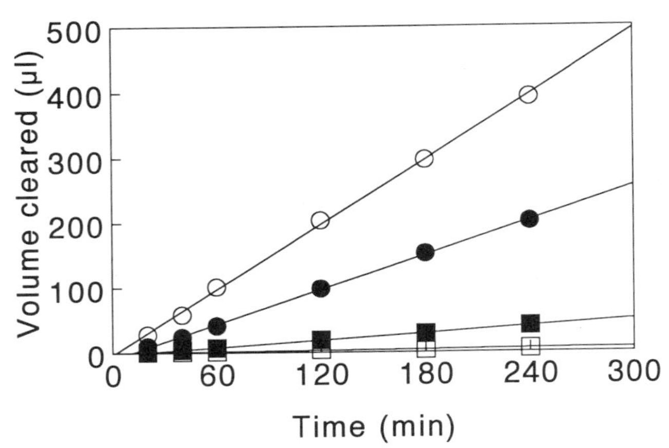

Figure 2. Transepithelial passage of [3H]vinblastine (50 nM) across Caco-2 monolayers. The volume of vinblastine cleared in the AP to BL (squares), or BL to AP (circles) direction was plotted against time of incubation (open symbols). Passages in the presence of 25 µM verapamil (closed symbols) are also illustrated. Data are mean ± S.D. (n = 3 wells). Error bars are smaller than the symbols.

to be noted that Pristinamycin IA represents with cyclosporin the only example of a cyclic peptide P-gp substrate.

These works illustrate first that P-glycoprotein is expressed in a polarized fashion on the apical side of epithelial Caco-2 monolayers, which reflects the in vivo situation. We also observed that the level of P-gp expression in the Caco-2 in vitro system was of the same order of magnitude than that immunodetected in isolated intestinal mucosa (Lechardeur et al., in press). Second, our results show that this property of the Caco-2 system can be used for a sensitive assay of the recognition of a compound by *mdr* P-glycoprotein. This could be relevant for screening anticancer drugs which are not substrate of the multidrug resistance proton pump (Wils et al., unpublished data).

II. FUNCTIONAL EXPRESSION OF P-GLYCOPROTEIN IN BLOOD-BRAIN-BARRIER CELLULAR MODELS

The blood-brain barrier (BBB) is constituted by the sheet of cerebral capillary endothelial cells. These cells differ from most other peripheral endothelial beds by the presence of intercellular tight junctions and by the paucity of pinocytotic and transcytotic traffic. Because of these unique properties, the passage across the BBB of most water-soluble molecules including nutrients and hydrophilic drugs occurs only through specific carrier systems or membrane receptors. On the contrary, lipophilic compounds enter the brain by passive diffusion across the lipidic membrane of the cerebral endothelial cells. The permeability through the BBB of lipophilic compounds is thus directly correlated to their lipophilicity, which is reflected by their water/octanol partition coefficient (Levin, 1980; Van Bree et al., 1988). However, several compounds such as vinblastine, vincristine or cyclosporine, which are known to be substrates of the P-glycoprotein, display a much lower brain penetration than expected from their octanol/water partition coefficient (Begley et al., 1990; Greig et al., 1990; Pardridge et al., 1990; Safa and Tamai, 1990). Since Cordon-Cardo et al. (1989) have presented immunohistochemical evidence of the presence of the P-gp at the luminal face of brain capillary endothelial cells, the role of that pump in restricting the blood to brain penetration of a variety of lipophilic compounds has to be considered.

Interestingly, P-glycoprotein expression in endothelial cells is restricted to the endothelium forming physiological barriers, such as those of brain, retina, derma or testis capillaries. In the case of brain, the possibility has been suggested that the specific P-gp expression may be induced by factors originating from neighbouring cerebral cells, as suggested by Janzer and Raff (1987). Since all previously identified blood-brain barrier markers are rapidly lost when cerebral capillary endothelial cells are maintained in primary culture, we have studied the influence of the loss of cerebral environment on P-gp expression in brain capillary cells.

P-glycoprotein was detected by immunochemistry in freshly isolated purified bovine brain cerebral capillaries (Figure 3; Lechardeur et al., in press). P-glycoprotein was also detected in 5-7 days primary cultures of bovine brain capillary cells, at a level comparable to that of freshly isolated capillaries. Such cultures have been proposed as an *in vitro* blood-brain-barrier model (Audus and Borchardt, 1986) The P-glycoprotein was, however, immunodetected at a lower molecular weight than that found in freshly isolated capillaries (Lechardeur and Scherman, 1995). Enzymatic deglycosylation leads to the same 130 kDa protein for both fresh and cultured samples, suggesting that P-gp post-translational modifications were altered in primary cultures. However, studies on the uptake and efflux of the Pgp substrate [^{3}H]vinblastine, and on the effect of various *mdr* reversing agents on these uptake and efflux clearly indicated that the efflux pump function of the P-glycoprotein was maintained in primary cultures of bovine cerebral capillary endothelial cells (Lechardeur and Scherman, 1995). P-glycoprotein might thus represent the first blood-brain barrier marker which is maintained in cerebral endothelial cells cultured in the absence of factors originating from the brain parenchyma.

An immortalized brain capillary endothelial cell line displaying blood-brain barrier characteristics might represent a useful tool for studying the blood-brain barrier endothelial cell differentiation and for the *in vitro* prediction of drug brain penetration. We have established a rat cerebral capillary endothelial cell line (CR3) by genomic introduction of the immortalizing SV40 large T gene under the control of the human vimentin promoter, a method previously used for the obtention of various differentiated immortalized cell lines (Schwartz et al., 1991). The CR3 cell line displayed endothelial morphological and biochemical characteristics for up to 30 passages (Figure 4). However, the CR3 cell line did not spontaneously express the specific blood-brain barrier markers gamma-glutamyl transpeptidase and *mdr* P-glycoprotein. However, when the cells were treated with the cell differentiating agent all-*trans*-retinoic acid, the blood-brain barrier markers were induced. Retinoic acid treated CR3 cells may thus represent a useful tool for biological and pharmacological research related to the blood-brain barrier (Lechardeur et al., 1995).

Finally, in the immortalized rat brain endothelial cell line RBE4 (Roux et al., 1994), P-glycoprotein expression was also detected both biochemically and functionnally (D. J. Begley, D. Lechardeur, Z-D. Chen et al., submitted to publication).

CONCLUSION

Strong evidence has established that *mdr* P-gp is expressed apically in various epithelium and endothelium, and plays a role as an active enzymatic barrier against the tissue penetration of xenobiotics. Relevant *in vitro* models of cellular barriers should thus express the P-gp pump in a restricted apical localization and at a level comparable to *in vivo* values. When such a property is established, the cellular models in dual culture chamber systems could provide useful tools for the early prediction of drug bioavailability and for the

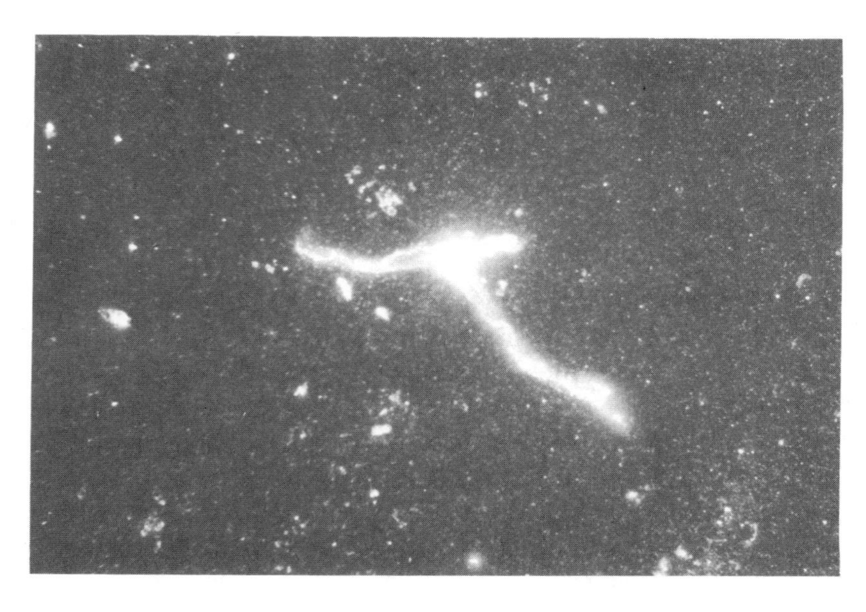

A

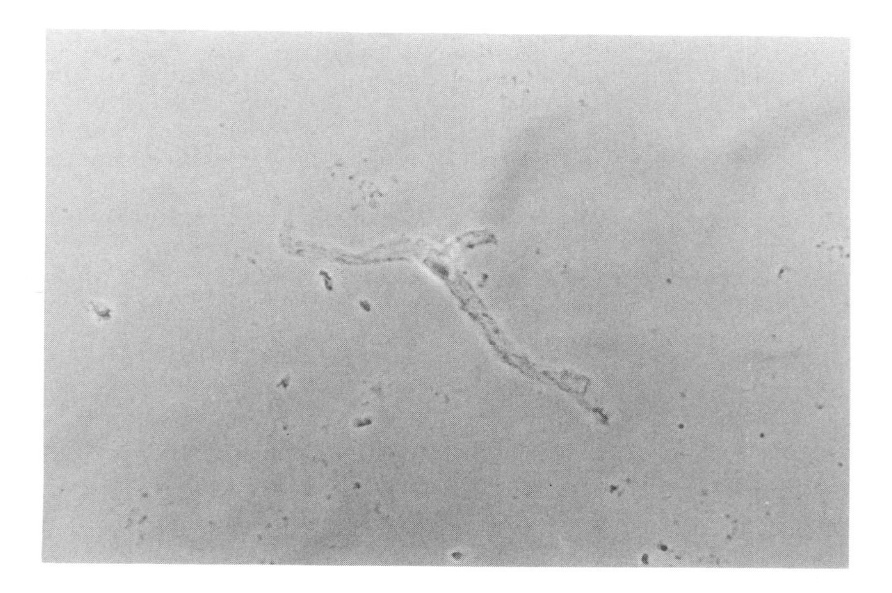

B

Figure 3. Immunohistochemical detection of P-glycoprotein in isolated rat brain capillaries. (A) a strong staining was observed on freshly isolated rat brain capillaries in the presence of the MRK16 anti-P-gp monoclonal antibody and fluorescein-conjugated anti-mouse secondary antibody. (B) Phase contrast microscopy (x400). No fluorescence was observed in controls where only the secondary antibody had been added (data not shown).

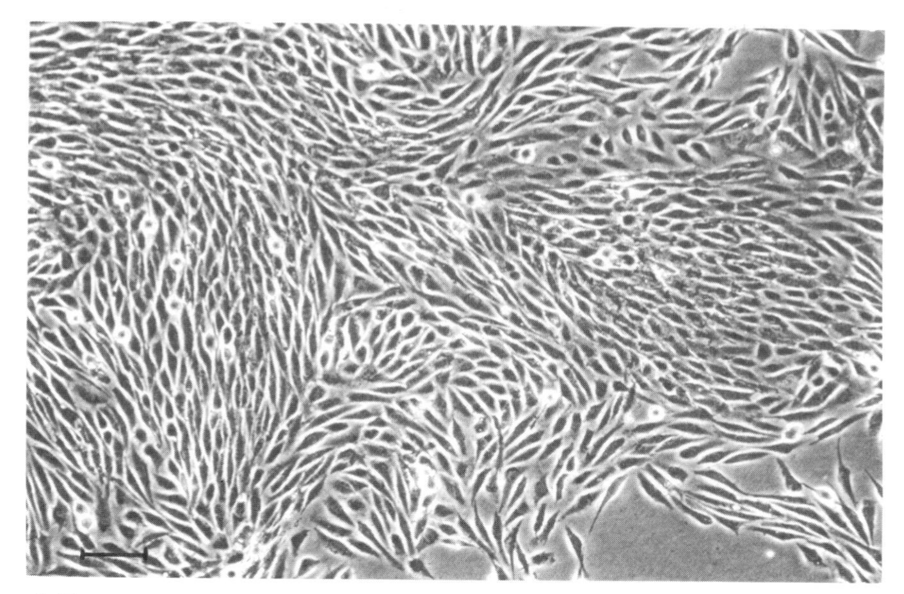

Figure 4. Phase contrast microscopy of the immortalized rat brain capillary endothelial CR3 cell line (see Lechardeur et al., 1995, for experimental details).

screening of new antitumor compounds. Finally, P-glycoprotein might represent an interesting marker for the study of epithelial or endothelial terminal differentiation.

ACKNOWLEDGMENTS

This work has been done as part of the Bio Avenir programme supported by Rhône-Poulenc with the participation of the French Ministry of Industry and the French Ministry of Research". D. Lechardeur was supported by a grant from the the French Ministry of Research". The authors thank Drs. C. Huet and D. Louvard for the gift of the HT29-18-C1 cell line and Dr. A. Zweibaum for the Caco-2 cell line, and D. Paulin for the construction comprising the SV40 T antigen (t deleted) under the control of the human vimentin promoter.

REFERENCES

Atursson P (1990) Epithelial transport of drugs in cell culture. I: a model for studying the passive diffusion of drugs over absorbtive (Caco-2) cells. *J Pharm Sci*, *79*: 476-482.

Audus KL (1986) Borchardt RT. Bovine brain microvessel endothelial cell monolayers as a model system for the blood-brain barrier. *Ann N Y Acad Sci*, *17*: 9-18.

Beck WT (1987) The cell biology of the multidrug resistance. *Biochem Pharmacol*, *36*: 2879-2887.

Begley DJ, Squires LK, Zlokovic BV, Mitrovic DM, Hughes CCW, Revest PA, Greenwood J (1990) Permeability of the blood-brain barrier to the immunosuppressive cyclic peptide cyclosporin A. *J Neurochem*, *55*: 1222-1230.

Cordon-Cardo C, O'Brien JP, Casals D, Rittman-Grauer L, Biedler JL, Melamed MR, Bertino JR (1989) Multidrug-resistance gene (P-glycoprotein) is expressed by endothelial cells at blood-brain barrier sites. *Proc Natl Acad Sci USA*, *86*: 695-698.

Cordon-Cardo C, O'Brien JP, Boccia J, Cassals D, Bertino JR, Melamed MR (1990) Expression of the multidrug resistance gene product (P-glycoprotein) in human normal and tumor tissues. *J. Histochem. Cytochem.*, *38*: 1277-1287.

Dehouck MP, Meresse S, Dehouck B, Fruchart JC, Cecchelli R (1992) In vitro reconstituted blood-brain barrier *J. Controlled Release*, *21*: 81-92.

Fojo AT, Ueda K, Slamon D, Poplack DG, Gottesman MM, Pastan I (1987) Expression of multidrug-resistance gene in human tumors and tissues. *Proc Natl Acad Sci USA.*, *84*: 265-269.

Greenwood J (1992) Characterization of a rat retinal endothelial cell culture and the expression of P-glycoprotein in brain and retinal endothelium in vitro. *J. Neuroimmunol.*, *39*: 123-132.

Greig NH, Soncrant TT, Shetty U, Momma S, Smith QR, Rappoport, SI. (1990) Brain uptake and anticancer activities of vincristine and vinblastine are restricted by their low cerebrovascular permeability and binding to plasma constituents in rat. *Cancer Chemother Pharmacol*, *26*: 263-268.

Hidalgo IJ, Raub TJ and Borchardt RT (1989) Characterization of the human colon carcinoma cell line (Caco-2) as a model system for intestinal epithelial permeability. *Gastroenterology* , *96*: 736-749.

Hsing S, Gatmaitan Z, Arias IM. (1992) The function of Gp170 the multidrug-resistance gene product, in the brush border of rat intestinal mucosa. *Gastroenterology*, *102*: 879-885.

Huet C, Sahuquillo-Merino C, Coudrier E and Louvard D (1987) Absorptive and mucus-secreting subclones isolated from a multipotent intestinal cell line (HT-29) provide new models for cell polarity and terminal differentiation. *J Cell Biol*, *105*: 345-357.

Hunter J, Jepson MA, Tsuruo T, Simmons NL and Hirst BH (1993) Functional expression of P-glycoprotein in apical membranes of human intestinal Caco-2 cells. *J Biol Chem 268*: 14991-14497.

Janzer R C and Raff MC (1987) Astrocytes induce blood-brain barrier properties in endothelial cells. *Nature*, *325*: 253-257.

Lechardeur D, Schwartz B, Paulin D and Scherman D (1995) Induction of blood-brain barrier differentiation in a rat brain derived endothelial cell line. *Exp. Cell Res*, *220:* 161-170.

Lechardeur D and Scherman D (1995) Functional expression of the P-glycoprotein *mdr* in primary cultures of bovine capillary endothelial cells. *Cell Biol Tox*, *11*: 219-230.

Lechardeur D and Scherman D Detection of the multidrug resistance P-glycoprotein in healthy tissues: the example of the blood-brain barrier. *Annales de Biologie Cliniques* (in press).

Levin UA (1980) Relationships of octanol/water partition coefficient and molecular weight to rat brain capillary permeability. *J Med Chem*, *23*: 682-684.

Pardridge WM (1990) Peptide drug delivery to the brain. *New-York: Raven Press*: 280-302.

Pileri SA, Sabattini E, Falini B, Tazzari PL, Gherlinzoni F, Michieli MG, Damiani D, Zucchini L, Gobbi M, Tsuruo T and Baccarani M (1991) Immunohistochemical detection of the multidrug transport protein P170 in human normal tissues and malignant lymphomas. *Histopathology*, *19*: 131-140.

Pinto M, Robine-Léon S, Appay MD, Kedinger M, Triadou N, Dussaulx E, Lacroix B, Simon-Assmann P, Haffen K, Fogh J and Zweibaum A (1983) Enterocyte-like differentiation and polarization of the human colon carcinoma cell line Caco-2 in culture. *Biol Cell*, *47*, 323-330.

Roux F., Durieu-Trautmann O, Chaverot N, Claire M, Mailly P, Bourre J.M, Strosberg AD and Couraud PO (1994) Regulation of gamma-glutamyl transpeptidase and alkaline phosphatase activities in immortalized rat brain microvessel endothelial cells. *J Cell Physiol*, *159*: 101-113.

Rubin LL, Hall DE, Porter S, Barbu K, Cannon C, Horner HC, Janatpour M, Liaw CW, Manning K, Morales J, Tanner LI, Tomaselli KJ and Bard F (1991) A cell culture model of the blood-brain barrier. *J Cell Biol*, 115: 1725-1735.

Safa AR and Tamai I (1990) Competitive interaction of cyclosporins with the vinca alcaloid-binding site of P-glycoprotein in multidrug resistant cells. *J Biol Chem*, *265*: 16509-16513.

Schinkel AH, Smit JJL, Van Tellingen O, Beijnen JH, Wagenaar E, Van Deemter L, Mol CAAM, Van der Valk MA, Robanus-Maandag EC, Te Riele HPJ, Berns AJM and Borst P (1994) Disruption of the mouse mdr1a P-glycoprotein gene leads to a defieciency in the blood-brain barrier and to increased sensitivity to drugs. *Cell*, *77*: 491-502.

Schwartz B, Vicart P, Delouis C. and Paulin D (1991). Mammalian cell lines can be efficiently established in vitro upon expression of the SV40 large T antigen driven by a promoter sequence derived from the human vimentin gene. *Biol Cell*, *73*::7-14.

Shimabuku AM, Nishimoto T, Ueda K, Komano T (1992) P-glycoprotein. *J Biol Chem*, *267*: 4308-4311.

Van Bree JB, De Boer AG, Danhof M, Ginsel LA, Breimer DD (1988) Characterization of an in vitro blood-brain barrier: effects of molecular size and lipophilicity on cerebrovascular endothelial transport rates of drugs. *J Pharmacol Exp Ther*, *247*: 1233-1239.

Wils P, Legrain S, Frenois E and Scherman D (1993) HT29-18-C1 intestinal cells: a new model for studying the epithelial transport of drugs. *Biochim Biophys Acta*, *1117*: 134-138.

Wils P, Warnery A, Phung-Ba V, Legrain S and Scherman D (1994a) High lipophilicity decreases drug transport across intestinal epithelial cells. *J Pharmacol Exp Therr*: *269*, 654-658.

Wils P, Phung-Ba V, Warnery A, Lechardeur D, Raeissi S, Hidalgo IJ, Scherman D 1994b) Polarized transport of docetaxel and vinblastine mediated by P-glycoprotein in human intestinal epithelial cell monolayer. *Biochem Pharmacol.*, *48*: 1528-1530.

Wils P., Phung-BA V., Warnery A., Raeissi S., Hidalgo I., et Scherman D. (1995) Interaction of pristinamycin IA with P-glycoprotein in human intestinal epithelial cells.*Eur. J Pharmacol*, *288*: 187-192.

A METHOD TO ASSESS FUNCTIONAL ACTIVITY OF P-GLYCOPROTEIN *IN VITRO* BASED ON THE ENERGY REQUIREMENTS OF THE TRANSPORTER

A. Reichel, Z.-D. Reeve-Chen, D. J. Begley, and N. J. Abbott

Physiology Group
Biomedical Sciences Division
King's College London
Strand, London WC2R 2LS
United Kingdom

SUMMARY

P-glycoprotein (Pgp) is constitutively expressed in the luminal membrane of endothelial cells forming the blood-brain barrier. Pgp acts as an active drug efflux pump and hence frustrates treatment of brain diseases by preventing a large variety of chemically unrelated drugs from entering the central nervous system. The screening for non-toxic, specific inhibitors requires a large number of compounds to be tested for their possible interaction with Pgp. As brain endothelial cells retain expression of Pgp in culture, *in vitro* assays are the obvious choice. The quality of such a screening model relys on a maintained expression of Pgp which needs to be characterised beforehand. Thus, we designed a simple protocol based on energy requirements of Pgp mediated drug efflux to assess the functional activity of the transporter quantitatively.

RÉSUMÉ

La P-glycoprotéine (P-gp) est exprimée constitutionnellement au niveau de la membrane luminale des cellules endothéliales qui forment la barrière hémato-encéphalique. La P-gp agit comme une pompe active pour expulser les médicaments et donc empêche le traitement des maladies cérébrales, en empêchant un certain nombre de médicaments, de structures chimiques diverses, de pénétrer dans le système nerveux central. Il faut donc tester un grand nombre de composés par rapport à leur interaction possible avec la P-gp, pour définir des inhibiteurs spécifiques non toxiques.Comme les cellules endothéliales cérébrales en culture expriment encore la P-gp, il est judicieux de faire des tests *in vitro*. La qualité

Biology and Physiology of the Blood–Brain Barrier, edited by Couraud and Scherman
Plenum Press, New York, 1996

d'un tel critère de choix repose sur cette expression maintenue de la P-gp, qui doit être vérifiée. Nous avons donc défini un protocole simple basé sur les besoins énergétiques de la P-gp pour l'efflux des drogues pour contrôler quantitativement l'activité fonctionnelle du transporteur.

I. INTRODUCTION

Brain endothelial cells forming the blood-brain barrier (BBB) have been shown to constitutively express P-glycoprotein (Pgp) in their luminal cell membranes[1]. Pgp is a member of the large family of ATP binding cassette (ABC) transporter molecules. Within this family Pgp consists of a group of closely related membrane proteins which are involved in multidrug resistance. The transporter acts as an active drug efflux pump with affinity for a wide range of chemically unrelated compounds[2]. As many cytotoxic and other drugs are substrates for Pgp their use in the chemotherapeutic treatment of brain diseases can be frustrated. A better understanding of the cell biology of the drug efflux pump is required to design (1) non-toxic agents which reverse the activity of Pgp, and thus allow a greater rate of entry into the brain for drugs during treatment of brain diseases, but also (2) drugs which retain their interaction with Pgp to preserve a peripheral rather than a central action.

So far, there is no satisfactory model available to test which compounds are substrates or specific inhibitors for Pgp at BBB sites. Therefore, an *in vitro* screening assay for examining new and existing drugs will be a very helpful tool to discover agents which reverse Pgp-associated multidrug resistance at blood levels not producing side effects. Predictions derived from such an *in vitro* system rely on a reproducible expression of Pgp. We have used a culture of immortalised rat brain endothelial cells (RBE4)[3] which has been shown previously to retain expression of Pgp[5] in order to design a simple protocol which would allow routine characterisation of the functional activity of Pgp.

The accumulation of colchicine in confluent cultured cell monolayers has been used to study the functional activity of Pgp as it is known to be a substrate, sufficiently lipophilic (log PC_{oct} = 1.28) to readily cross cell membranes, and available as radio-labelled isotope. The cellular accumulation of [^{3}H]-colchicine is the result of three processes: (1) lipid-mediated influx of colchicine into the cells, (2) active efflux by Pgp and (3) apparent accumulation as result of non-specific adsorption of tracer to the cell and plastic surface and/or entry into spaces between cells and also carryover during washing procedures (Fig. 1). The latter can be corrected for by adding [^{14}C]-sucrose, which is of similar molecular weight as colchicine to an extracellular marker, to the incubation medium.

To distinguish between passive influx and active efflux of [^{3}H]-colchicine we inhibited the formation of intracellular ATP, as under these conditions Pgp activity ceased and the net movement of [^{3}H]-colchicine is due only to passive influx according to the diffusion gradient. Consequently, the resulting difference in the [^{3}H]-colchicine uptake is a measure of the functional activity of Pgp. The ATP depletion method used is to incubate the cells in a medium containing 2 mM 2,4-dinitrophenol (DNP), an uncoupling agent of oxidative phosphorylation, and 10 mM 2-deoxyglucose (2-DG), a glucose antimetabolite, which in combination completely inhibit oxidative catabolism and have been shown to reduce cellular ATP levels by 98% after 30 min[4].

II. METHOD

Confluent monolayers of RBE4 cells growing in 24 well plates were pre-treated with 2 mM DNP and 10 mM 2-DG in culture medium for 30 minutes at 37^0C. In the control wells

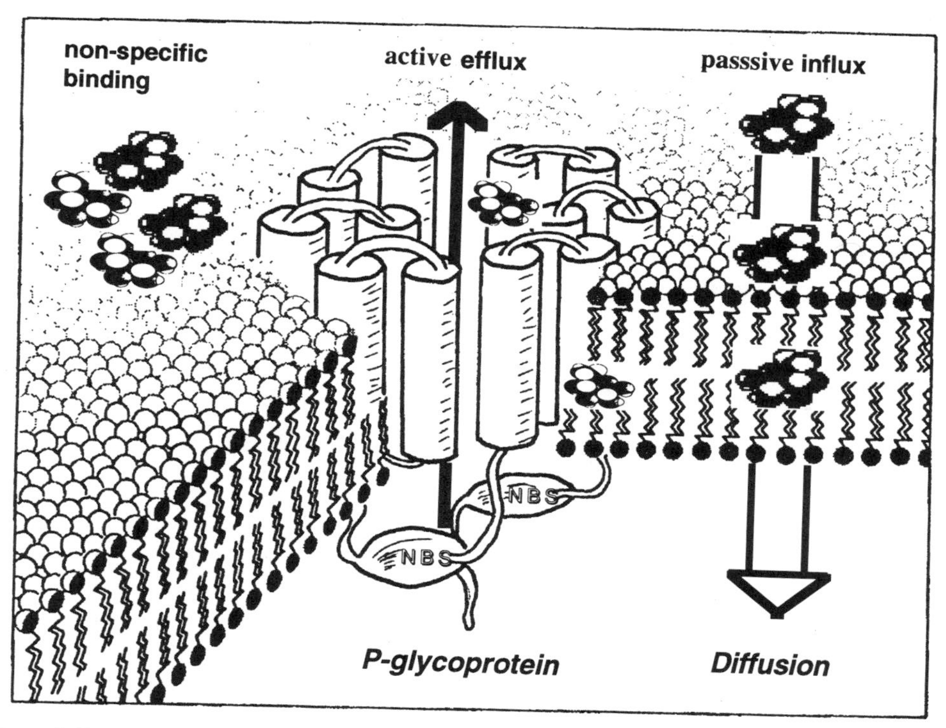

Figure 1. Factors contributing to the cellular accumulatin of [^{3}H]-colchicine including (i) passive influx of the drug into cells via diffusion, (ii) active efflux mediated by Pgp and (iii) non-specific binding to the cell surface. (NBS: nucleotide binding site).

culture medium has been changed at the same time. Cells were then incubated with 18 nM [^{3}H]-colchicine (0.3 μCi/well) in Hank's balanced salt solution containing 5% fetal calf serum (control), and additionally 2 mM DNP and 10 mM 2-DG (experiment). The uptake of [^{3}H]-colchicine by RBE4 cells was studied over a 60 minute period. [^{14}C]-sucrose (0.07 μCi/well) was added to the incubation medium to correct for non-specific accumulation. HEPES (10 mM) has been used to achieve a constant pH 7.4. The [^{3}H]-colchicine uptake was calculated as ratio of cellular vs. incubation medium radioactivity and was expressed as a distribution volume (V_d) in μl per mg protein.

III. RESULTS

Pre-treatment of RBE4 cells with 2 mM DNP and 10 mM 2-DG for 30 min significantly ($p<0.0001$) increased the cellular accumulation of [^{3}H]-colchicine at all time points studied (Fig. 2). Pretreatment with 2 mM DNP alone did not significantly affect Pgp activity (data not shown).

Non-specific accumulation has been determined using [^{14}C]-sucrose in the incubation medium. A distribution volume of about 3 μl/mg of protein is due to this factor (Fig. 2). In ATP deprived cells the accumulation of [^{3}H]-colchicine reflects mainly the passive influx of the tracer produced by the diffusion gradient. The accumulation of [^{3}H]-colchicine in control cells, however, is a result of two fluxes, i.e. passive influx but also active efflux of the tracer mediated by Pgp. Consequently, the difference in the uptake of [^{3}H]-colchicine is due to the

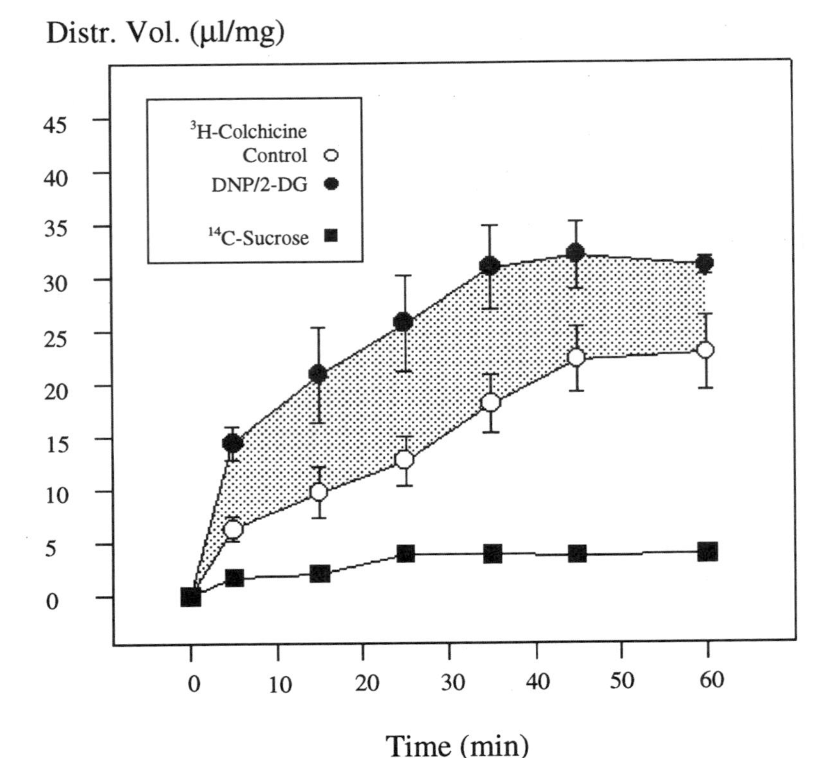

Figure 2. Effect of 2 mM DNP and 10 mM 2-DG on the uptake of [^{3}H]-colchicine by RBE4 cells (Distribution volume, μl/mg protein, mean ± S.D., n=4 wells). Also shown is the distribution volume (V_d) of [^{14}C]sucrose. The grey area reflects the functional activity of Pgp.

functional activity of Pgp. Thus the percentage reduction in [^{3}H]-colchicine uptake caused by Pgp as a result of metabolic inhibition reflects the active efflux mediated by the transporter. This percentage reduction provides a good measure of the functional activity of Pgp and may allow quantitative comparisons between different culture models.

IV. DISCUSSION

It was the aim of this study to design a simple experimental protocol to characterise the functional activity of Pgp *in vitro,* using ATP-depriving agents and conditions which are crucial for a reliable assay.

Our results show that ATP-deprivation using a combination of DNP and 2-DG is an effective way to inhibit active efflux of drugs mediated by Pgp. The failure of DNP alone to inhibit Pgp activity suggests that the small amounts of ATP and GTP generated during earlier stages of oxidative catabolism may be sufficient for continued operation of the Pgp-mediated efflux pump. A steady pH throughout the experiment is also crutial as Pgp activity appears to be pH sensitive. Furthermore, for studies over longer time periods the addition of glucose (10 mM) for control is recommended to allow replenishing ATP supply.

Using colchicine as a Pgp substrate it should be kept in mind that the cellular accumulation of [^{3}H]-colchicine contains a non-reversible component which is due to tight intracellular binding of the alkaloid to free tubulin molecules[5]. Thus even a very high Pgp activity will not reduce the cellular accumulation of [^{3}H]-colchicine down to the V_d value of the extracellular marker.

We conclude that the measurement of the cellular accumulation of [^{3}H]-colchicine, in combination with [^{14}C]-sucrose to correct for non-specifically adsorbed molecules, and ATP deprivation to distinguish between passive diffusion and active Pgp-mediated drug efflux, can be used as a simple routine test to characterise the activity of Pgp in cultured brain endothelial cells. The percentage reduction of the drug influx caused by Pgp-mediated efflux is a result of the activity as well as the number of Pgp molecules in the cell membrane. Therefore, it can be used as a quantitative measure to compare the Pgp activity in different cell lines. In contrast, Pgp blockers reduce transport activity to an unknown extent and thus are a less rigorous way to assess the activity of the pump. The protocol also overcomes the problem of the quantitative interpretation of Pgp expression derived from gel-electrophoresis studies in terms of the actual functional activity of Pgp[6].

ACKNOWLEDGMENT

We are grateful to P.-O. Couraud and F. Roux for the supply of RBE4 cells. The work was supported by MRC, Merck, Sharp & Dohme and Yamanouchi Pharmaceutical Co. and SmithKline Beecham. A. Reichel held a fellowship from the German Academic Exchange Service.

V. REFERENCES

1. Cordon-Cardo C., O'Brien J.P., Casals D., Ritman-Grauer L., Beidler J.L., Melamed M.R., and Bertino J.R. (1989) Multidrug resistance gene (P-glycoprotein) is expressed by endothelial cells at blood-brain barrier sites. *Proc. Natl. Acad. Sci. U.S.A.* 86, 695-698
2. Ruetz S. and Gros P. (1994) A mechanism for P-glycoprotein action in multidrug resistance: are we there yet? TiPS 15, 260-263
3. Roux F.S., Durieu-Trautmann O., Chaverot N., Claire M., Mailly P., Bourre J-M., Strosberg A.D. and Couraud P-O. (1994) Regulation of gamma-glutamyl transpeptidase and alkaline phosphatase activities in immortalised rat brain microvessel endothelial cells. *J. Cell. Physiol.* 159, 101-113
4. Sanger J.W., Sanger J.M. and Jockusch B.M. (1983) Different responses of three types of actin filament bundles to depletion of cellular ATP levels. *Eur. J. Cell Biol.* 31, 197-204
5. Salmon E.D., McKeel M. and Hays T. (1984) Rapid rate of tubulin dissociation from microtubules in the mitotic spindle in vitro measured by blocking polymerization with colchicine. *J. Cell Biol.* 99, 1066-1075
6. Begley D.J., Lechardeur D., Chen Z-D., Rollinson C., Bardoul M., Roux F.S., Scherman and N.J. Abbott (1995) Functional expression of P-glycoprotein in an immortalised cell line of rat brain endothelial cells, RBE4. (submitted)

32

P-GLYCOPROTEIN ACTIVITY IN AN *IN VITRO* BLOOD-BRAIN BARRIER MODEL

B. Joly,[1,2] O. Fardel,[1] V. Lécureur,[1] C. Chesné,[3] C. Puozzo,[2] and A. Guillouzo[1]

[1] INSERM U 49
Hopital Pontchaillou
35033 Rennes cedex, France
[2] Institut de Recherche Pierre Fabre
81100 Castres, France
[3] Biopredic
Technopole Atalante Villejean
35000 Rennes, France

SUMMARY

Drug transport across the blood-brain barrier (BBB) is a key step in the treatment of cerebral diseases; it is largely dependent on membrane permeability of brain capillary endothelial cells. In order to investigate the functional features of the BBB *in vitro*, a coculture model of bovine brain capillary endothelial cells and new-born rat astrocytes has been recently established (Dehouck *et al*, J. Neurochemistry, 54: 1798-1801, 1990). Using this system, we have analysed brain capillary endothelial cell (BCEC) permeability for vinblastine, an anticancer drug substrate for the multidrug transporter P-glycoprotein (P-gp). Vinblastine endothelial permeability coefficient was found to be low, as for various compounds (i.e. inulin, sucrose, sulpiride), known to poorly accumulate in the brain. In the presence of verapamil, a known inhibitor of P-gp function, vinblastine transfer across BCECs was strongly enhanced. These data therefore suggest that BCECs display P-gp activity that is directly involved in low endothelial permeability of anticancer drugs such as vinblastine.

RÉSUMÉ

Le passage de médicaments à travers la barrière hémato-encéphalique (BHE) est une étape importante dans le traitement de maladies cérébrales; il dépend de la perméabilité membranaire des cellules endothéliales des capillaires cérébraux. Un modèle de coculture de cellules endothéliales de capillaires cérébraux et d'astrocytes de rat nouveaux-nés, a été récemment développé (Dehouck *et al*, J. Neurochemistry, 54: 1798-1801, 1990), permettant

Biology and Physiology of the Blood–Brain Barrier, edited by Couraud and Scherman
Plenum Press, New York, 1996

d'étudier les caractéristiques de la BHE *in vitro*. Nous avons utilisé ce modèle pour apprécier la perméabilité des cellules endothéliales à la vinblastine, un médicament anticancéreux, substrat du transporteur "multidrogue", la P-glycoproteine (P-gp). Nos résultats montrent que le coefficient de perméabilité endothéliale à la vinblastine est faible, tout comme il l'est, pour de nombreux autres composés (inuline, sucrose, sulpiride), connus pour s'accumuler faiblement au niveau cérébral. En présence de vérapamil, un inhibiteur de l'activité de la P-gp, le passage de vinblastine à travers la monocouche de cellules endothéliales est fortement augmenté, ce qui indique que les cellules endothéliales cérébrales, possèdent une activité P-gp pouvant jouer un rôle direct dans la faible perméabilité endothéliale aux médicaments anticancéreux, tels que la vinblastine.

INTRODUCTION

Compared with other organs of the body, the central nervous system shows a unique behaviour in the transport of chemicals. Many drugs are not freely transferred from, and to, the blood. This phenomenon has led to the concept of the blood-brain barrier (BBB). The anatomical features of the BBB, particularly the presence of very tight interendothelial junctions and the lack of transendothelial vesicular transport, means that, unlike noncerebral vessels, the cerebral endothelium has the permeability properties of a continuous plasma membrane, with the virtual elimination of passive diffusion across the endothelium (14). In addition, a number of proteins are specifically expressed by brain capillary endothelial cells (BCECs), which are required for metabolic protection and transport activities at the BBB interface (2). This specialization of BCECs is thought to be induced by the brain microenvironment including astrocytes.

Another mechanism that can prevent passage of various lipophilic substances has been demonstrated recently at the cerebral endothelial cell level (3). This mechanism is based on the presence of P-glycoprotein (P-gp), a 170 kD plasma transmembrane phosphoglycoprotein, thought to act as an ATP-dependent pump. This glycoprotein actively excludes a broad spectrum of structurally and functionally unrelated cytotoxic drugs, and confers a multidrug resistance phenotype (17). A high degree of lipophilicity, does not therefore automatically ensure passage of a solute across the BBB.

A large number of *in vivo* models, that have been used to study transport of drugs across the BBB, mimics the physiological transfer conditions; however, these models, do not provide direct assess to transport phenomena (4) and estimated transport parameters are inevitably influenced by physiological factors that cannot be kept under control during the experiment (e.g. cerebral blood flow, hormone levels and stress). To avoid the complexity of the *in vivo* BBB and to standardize transport parameters, *in vitro* models have been proposed. Dehouck *et al* (5), has established a model of BBB, that mimics the *in vivo* situation by coculturing bovine BCECs and new-born rat astrocytes. Using this system, we have analysed endothelial permeability for vinblastine, an anticancer drug substrate for P-gp and provided evidence that BCECs display P-gp activity.

MATERIALS AND METHODS

Chemicals

[^{14}C]sucrose (sp. act. 350 mCi/mmol), [^{3}H]inulin (sp. act. 3.2 Ci/mmol), [^{3}H]sulpiride (sp. act. 68.7 Ci/mmol), [^{3}H]imipramine (sp. act. 50.3 Ci/mmol) and [^{3}H]vinblastine (sp. act. 11.4 Ci/mmole) were purchased from Amersham (France); verapamil was obtained from Sigma.

Cell Cultures

Bovine brain endothelial cells were isolated and characterized as described by Méresse *et al* (13).

Newborn rat astrocytes were isolated according to Booher and Sensenbrenner (1), by forcing the brain tissue gently through a nylon mesh (48 μm pore) into a small nutrient medium reservoir.

Cocultures of BCECs and astrocytes were obtained using the method of Dehouck *et al* (6). BCECs were plated at a concentration of 4. 10^5 cells on the upper side of a filter (Millicell-CM; pore size 0.4 μm; diameter 30 mm; Millipore) coated with rat tail collagen. The filter was put in the well in which the astrocytes were seeded. The medium used for the coculture was DMEM supplemented with 15% calf serum, 2 mM glutamine, 50 μM gentamicin and bFGF (1 ng/ml added every other day). Under these conditions, BCECs formed a confluent monolayer within 8 days. The coculture integrity was checked using a Millicell -ERS apparatus (Millipore) and its electrical resistance was consistently found to be more than 500 ohms.cm2.

Transendothelial Transport Studies

For the experiments, the filter was transferred to the first well of a six-well plate and 2 ml Ringer HEPES medium containing the drug were placed in the upper compartment. Triplicate cocultures were assayed for each drug. After 5, 10, 15, 20, 30, 40 and 60 min exposure to the compounds, the insert was transferred to another well of a six-well plate in order to minimize possible passage of compounds, from the lower to the upper compartment. Incubations were performed in a rocking platform at 37°C in a humidified atmosphere of 95% air and 5% CO_2. Drug accumulation in the lower chamber was evaluated by scintillation counting. The endothelial permeability coefficient (Pe) was then calculated using the clearance principle, (10).

RESULTS

As previously described, (10) confluent BCECs were absolutely free of pericyte contamination, and formed a monolayer of small, tightly packed, non overlapping, and contact inhibiting cells on the upper side of the filter and astrocytes grew in the lower chamber.

By using this coculture model, transfer of vinblastine across BCECs was determined and compared to that of four other compounds (i.e. sucrose, inulin, imipramine, sulpiride), used as reference agents. As shown in Table 1, a low endothelial permeability coefficient (0.64 10^{-3} cm/min) was found for vinblastine. Inulin, sucrose and sulpiride also displayed similar weak endothelial permeability while a much higher permeability coefficient was observed for imipramine. Transfer of vinblastine across BCECs was then analysed in the presence of verapamil, a known inhibitor of the P-gp. As shown in Figure 1, addition of verapamil resulted in a strong increase of vinblastine endothelial permeability.

Table 1. Permeability coefficient values of different substances through the *in vitro* blood-brain barrier model. Calculations have been made as described in the Materials and Methods section. [^{3}H]Inulin (0.31 µCi/ml), [^{14}C]sucrose (3.3 µM), [^{3}H]imipramine (80 ng/ml), [^{3}H]sulpiride (2µg/ml) and [^{3}H]vinblastine (200nM) were tested

Compounds	Permeability (10-3 cm/min)
Inuline	0.45
Sucrose	0.79
Sulpiride	0.43
Imipramine	2.90
Vinblastine	0.64

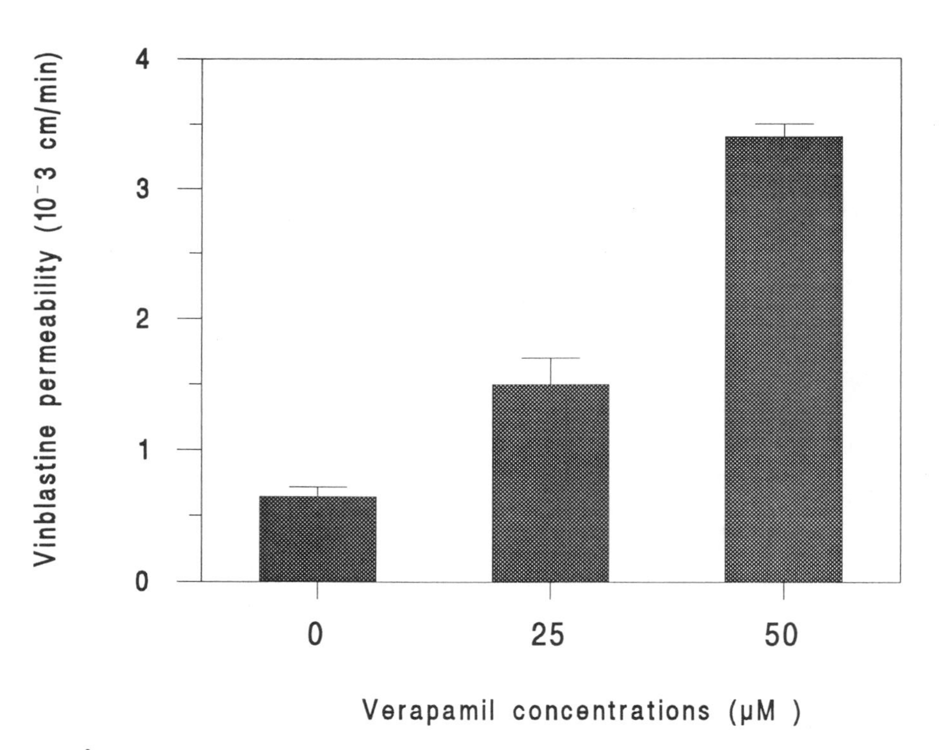

Figure 1. [^{3}H]vinblastine (200nM) permeability in the presence of different concentrations of verapamil. The values are the mean ± SEM of three independent experiments in triplicate.

DISCUSSION

Although vinca alcaloid anticancer drugs are classified as highly lipophilic compounds, the passage of these drugs across the BBB is unexpectedly low (11). Vinblastine belongs to this substance category. The mechanism that prevents its passage is not well understood. Using an *in vitro* BBB model, we demonstrated in the present study that vinblastine transfer across functional BCECs is low, thus suggesting that low capillary endothelial permeability is a key event in the restricted transport of the anticancer drug into the brain. In addition, vinblastine passage across BCECs was found to be strongly enhanced by verapamil. This effect occurred for concentrations known to markedly inhibit P-gp function in various cell types (8). Moreover, vinblastine has been demonstrated to be a substrate for P-gp (9). Taken together, all these results support the idea that cocultured BCECs displayed P-gp activity, that is directly responsible for the low vinblastine permeability. P-gp has been shown to be present in the luminal capillary endothelium of the brain (16). It can therefore be hypothetized that this P-gp expression is functional *in vivo* as in our BBB coculture model and thus is involved in the restricted passage into the brain of anticancer drug effluxed by P-gp. A recent study using generated mice with a homozygous disruption of the mdr1 a P-gp gene also supports this conclusion (15). Similarly to vinblastine, sucrose, inulin and sulpiride, displayed low endothelial permeability coefficient. All these compounds have been shown to poorly accumulate into the brain (12; 5). By contrast, imipramine, which readily enters the brain (7), showed high permeability across cocultured BCECs. These results suggest that brain endothelial permeabilities determined using the BBB *in vitro* model paralleled data obtained from *in vivo* experiments. The coculture model of BCECs and rat astrocytes thus appears to represent a promising tool for investigating *in vitro* the role of BBB in the delivery to the brain of drugs, including anticancer drugs.

ACKNOWLEDGMENTS

We are grateful to the Institut de Recherche Pierre Fabre for supporting the research described in this paper.

REFERENCES

1. Booher, J., and Sensenbrenner, M., Growth and cultivation of dissociated neurons and glial cells from embryonic chick, rat, and human brain in flask cultures, *Neurobiology.*, 2: 97-105, 1972.
2. Broadwell, R.D., and Salcman, M., Expanding the definition of the blood-brain barrier to proteins. *Proc. Natl. Acad. Sci. USA*, 78: 7820-7824, 1981.
3. Cordon-Cardo, C., O'Brien, J.P., Casals, D., Rittman-Grauer, L., Biedler, J.L., Melamed, M.R., and Bertino, J. R., Multidrug resistance genes (P-glycoprotein) is expressed by endothelial cells at BBB sites. *Proc. Natl. Acad. Sci. USA*, 86: 695-698, 1989.
4. Crone, C., The permeability of capillaries in various organs as determined by use of indicator method, *Acta Physiol. Scand.*, 58: 292-305, 1963.
5. Dehouck, M.P., Jolliet-Riant, P., Brée, F., Fruchart, J.C., Cecchelli, R., and Tillement, J.P., Drug transfer across the blood-brain barrier: correlation between *in vitro* and *in vivo* models, *J. Neurochemistry.*, 58: 1790-1797, 1992.
6. Dehouck, B., Dehouck, M.P., Fruchart, J.C., and Cecchelli, R., Upregulation of the low density lipoprotein receptor at the blood-brain barrier: intercommunications between brain capillary endothelial cells and astrocytes, *J. Cell Biol.*, 126: 465-473, 1994.
7. De Montis, M.G., Gambarana, C., Meloni, D., Taddei, I., and Tagliamonte, A., Long-term imipramine effects are prevented by NMDA receptor blockade, *Brain Res.*, 606: 63-67, 1993.

8. Ford, J.M., and Hait, W.N., Pharmacology of drugs that alter multidrug resistance in cancer, *Pharmacol Rev.* 58: 137-171, 1990.
9. Gottesman, M.M., and Pastan, I., The multidrug transporter, a double-edged sword, *J. Biol. Chem.*, 262: 12163-12166, 1988.
10. Joly, B., Fardel, O., Cecchelli, R., Chesné, C., Puozzo, C., and Guillouzo, A., Selective drug transport and P-glycoprotein activity in an *in vitro* blood-brain barrier model, *Toxic. in Vitro*, 9: 357-364, 1995.
11. Levin, V.A., Relationship of octanol/water partition coefficient and molecular weight to rat brain capillary permeability, *J. Med. Chem.*, 23: 682-684, 1980.
12. McClelland, G.R., Cooper, S.M., and Pilgrim, A.J., A comparison of the central nervous system effects of halloperidol, chlorpromazine and sulpiride in normal volunteers, *Brit. J. Clin. Pharmacol.*, 30: 795-803, 1990.
13.Méresse, S., Delbart, C., Fruchart, J.C., and Cecchelli, R., Low density lipoprotein receptor on endothelium of brain capillaries, *J. Neurochemistry.*, 53: 340-345, 1989.
14. Reese, T.S., and Karnovski, M.J., Fine structure localization of a blood-brain barrier to exogenous peroxidase, *J. Cell. Biol.,* 34: 207-217, 1967.
15. Schinkel, A.H., Smit, J.J.M., Van Telligen, O., Beijnen, J.H., Wagenaar, E., Van Deemter, L., Mol, C.C.A.M., Van der Valk, M.A., Robanus-Maandag, E.C., Te Riele, H.P.J., Berns, A.J.M., and Borst, P., Disruption of the mouse mdr1a P-glycoprotein gene leads to deficiency in the blood-brain barrier and increased sensitivity to drugs, *Cell*, 77: 491-502, 1994.
16. Thiebaut, F., Tsuoro, T., Hamada, H., Gottesman, M.M., Pastan, I., and Willingham, M., Cellular localization of the multidrug-resistance gene product P-glycoprotein in normal human tissues, *Proc. Natl. Acad. Sci. USA*, 84: 7735-7738, 1987.
17. Ueda, K., Cornwell, M.M., Gottesman, M.M., Pastan, I., Roninson, I.,B., Ling, V., and Riordan, J.R., The mdr1 gene responsible for multidrug resistance codes for P-glycoprotein, *Biochem. Biophys. Res. Com.*, 141: 956-962, 1986.

BRAIN PENETRATION OF SDZ PSC 833 IN RATS

A Comparison with Cyclosporin A

Sandrine Desrayaud and Michel Lemaire

Drug Metabolism and Pharmacokinetics
Sandoz Pharma Ldt
Basle, Switzerland

SUMMARY

The brain penetration of a novel MDR-reversing cyclosporin, SDZ PSC 833, was studied by measuring brain and blood concentrations after intravenous administration either as a bolus or as a constant-rate infusion. At lower blood concentrations of SDZ PSC 833, the brain penetration defined as the brain-to-blood concentration ratio (K_p), was very low in spite of the high lipophilicity of this compound. At higher blood concentrations, however, the brain penetration of SDZ PSC 833 was markedly increased. Since the blood pharmacokinetic of SDZ PSC 833 was found to be linear in the dosage range studied, these results demonstrated a non-linear brain penetration of SDZ PSC 833. The brain passage of cyclosporin A was also found to obey a non-linear kinetic. However the potency of SDZ PSC 833 to inhibit the efflux mechanism at the blood-brain barrier (BBB) was higher than that of the cyclosporin A since 10 times higher doses of cyclosporin A were required to obtain the same K_p values as SDZ PSC 833. Moreover, the coadministration of SDZ PSC 833 increased the brain penetration of cyclosporin A whereas the latter did not modify that of SDZ PSC 833. The increase in K_p values for SDZ PSC 833 observed at high blood levels of SDZ PSC 833 are consistent with the hypothesis that SDZ PSC 833 governs its own brain passage possibly by inhibition of the P-glycoprotein pump present in the brain microcapillary endothelial cells.

RÉSUMÉ

Le SDZ PSC 833, un nouvel analogue de la cyclosporine A dépourvu de propriétés immunosuppressives, semble avoir la propriété d'annuler la résistance des cellules tumorales vis-à-vis de nombreux agents anticancereux. La pénétration cérébrale de ce composé a été étudiée en mesurant les concentrations sanguines et cérébrales obtenues soit après un bolus intraveineux, soit au cours d'une perfusion intraveineuse à vitesse constante. Pour de faibles

Biology and Physiology of the Blood–Brain Barrier, edited by Couraud and Scherman
Plenum Press, New York, 1996

concentrations sanguines de SDZ PSC 833, sa pénétration cérébrale, définie par le rapport des concentrations cérébrales et sanguines K_p, est très faible malgré la forte liposolubilité de ce composé. Par contre, pour des concentrations sanguines élevées, la pénétration cérébrale du SDZ PSC 833 est considérablement augmentée. Etant donné que la pharmacocinétique sanguine du SDZ PSC 833 est linéaire pour la gamme de doses étudiées, ces résultats démontrent que la pénétration cérébrale du SDZ PSC 833 obéit à une pharmacocinétique non linéaire. Cette non-linéarité a également été démontrée pour la cyclosporine A. Cependant, la capacité du SDZ PSC 833 à inhiber le mécanisme d'efflux au niveau de la barrière hémato-encéphalique est plus importante que celle de la cyclosporine A. En effet, des doses de cyclosporine A 10 fois supérieures à celles du PSC sont nécessaires pour obtenir le même K_p. De plus, la coadministration de SDZ PSC 833 entraîne une augmentation de la pénétration cérébrale de la cyclosporine A, alors que l'inverse ne peut être démontré. Cette augmentation des valeurs de K_p pour le SDZ PSC 833 observée pour des concentrations sanguines de SDZ PSC 833 élevées confirme l'hypothèse selon laquelle le SDZ PSC 833 contrôle son propre passage cérébral probablement en inhibant la pompe P-glycoprotéine présente dans les cellules endothéliales des microcapillaires cérébraux.

1. INTRODUCTION

SDZ PSC 833, a new non-immunosuppressive cyclosporin analog has been reported to be much more potent than cyclosporin A for reversing multidrug resistance[1,2,3]. The entry of cyclosporin A into the brain, in other words its ability to become active on the brain tumor cells, was shown to be restricted by the blood-brain barrier, despite the high lipophilic nature of this compound[4,5]. *In vivo* and *in vitro* studies[6,7,8] suggested that the low brain penetration of cyclosporin A was due to the combined effects of its extensive blood binding and the active efflux from the blood-brain barrier by the P-glycoprotein. Therefore, the objective of this study was to assess the brain penetration of SDZ PSC 833 and cyclosporin A in order to compare their potency to inhibit P-glycoprotein-mediated efflux at the blood-brain barrier. To examine the brain passage of SDZ PSC 833 and cylcosporin A, the rat brain and blood concentrations were measured after iv bolus and during iv constant-rate infusion.

2. MATERIALS AND METHODS

I. Materials

[^{14}C]SDZ PSC 833 (44 μCi/mg) and [^{3}H]cyclosporin A (12.2 mCi/mg) were supplied by Sandoz Pharmaceuticals (Basel, Switzerland). [^{14}C]SDZ PSC 833 was labeled in the position 1 of L-valine2 whereas [^{3}H]cyclosporin A was labeled in the Abu-β position. The radiochemical purity of isotopes was assessed by HPLC and was greater than 95%.

II. Animals

Male Wistar rats weighing 250±10 g (KFM) were used for all experiments. The animals were fasted overnight prior to intravenous administration of drugs.

III. Intravenous Bolus Injection Studies

Brain distribution of SDZ PSC 833 and cyclosporin A was examined after iv administration of the labeled compounds into the femoral vein. The solvent, used as a vehicle for SDZ PSC 833 and cyclosporin A, was a mixture polyethylene glycol 200 - ethanol (40:10 v/v).

Each group of rats (n=3) received the following dosages: [^{14}C]SDZ PSC 833 (0.1, 0.3, 1, 3, 10 and 30 mg/kg); [^{3}H]cyclosporin A (0.1, 0.3, 1, 3, 10 and 30 mg/kg). Further, the role of SDZ PSC 833 on the brain penetration of cyclosporin A was studied using the following drug combinations: [^{3}H]cyclosporin A (10 mg/kg)-[^{14}C]SDZ PSC 833 (10 mg/kg) and [^{3}H]cyclosporin A (0.1 mg/kg)-SDZ PSC 833 (10 mg/kg). Two hours after drug injection, the animals were sacrificed by exsanguination under light anesthesia, thereafter brain was removed and radioactivity was counted in blood and brain samples. At this time, SDZ PSC 833 and cyclosporin A are present in both blood and brain essentially as parent drug[9]; therefore the blood and brain concentrations of radioactivity were considered as representative of parent drug concentrations.

IV. Constant–Rate Infusion Study

The experiments were performed under general anesthesia using a first ip dose of 0.7 g/kg urethane given as a 20% w/v solution in 0.9% w/v saline, which was followed, 20 minutes later, by an identical dose. [^{14}C]SDZ PSC 833 was dissolved in ethanol-polyethylene glycol 200-isotonic glucose (2:50:16 v/v/v) and infused into the femoral vein for 2, 3, 4, 6 or 8 hours at the rate of 5 μg/min (5 μl/min) by means of a CMA/100 microinjection pump. [^{14}C]SDZ PSC 833 was diluted with SDZ PSC 833 in order to achieve a radioactive dose of 15 μCi/rat. After each infusion time, 3 animals were sacrificed, blood was sampled by cardiac punction whereas whole brain was removed and homogenized. Radioactivity was counted in both blood and brain samples. During the 8 h iv infusion, blood and brain concentrations of radioactivity were considered as representative of unchanged SDZ PSC 833.

3. RESULTS

I. Dose Dependency of SDZ PSC 833 and Cyclosporin a Brain Distribution

In the wide dose range examined, i.e. 0.1 to 30 mg/kg, the SDZ PSC 833 and cyclosporin A blood levels of radioactivity measured 2 hr after injection showed a good dose proportionality (Fig. 1).

Due to this good dose-blood levels relationship, the brain penetration of both compounds was characterized by the brain/blood distribution ratio K_p (Fig. 2). The values corresponding to SDZ PSC 833 indicated a low K_p of about 0.2 in the dosage range 0.1 to 1 mg/kg followed by a rapid increase up to 1.4 after a 10 mg/kg dosage; thereafter the K_p value remains constant after a 30 mg/kg dose. A similar dose-dependent brain penetration was observed with cyclosporin A at higher doses: very low K_p values of about 0.1 were found in the dose range 0.1 to 10 mg/kg followed by an increase up to 1.1 after the 30 mg/kg dosage.

II. Influence of SDZ PSC 833 on the Brain Penetration of Cyclosporin A

Figure 3 shows that the coadministration of [^{14}C]SDZ PSC 833 (10 mg/kg) and [^{3}H]cyclosporin A (10 mg/kg) resulted in a 5 times increase of K_p values for cyclosporin A; a similar effect was even observed after the coadministration of 0.1 mg/kg dosage of cyclosporin A. On the contrary, the K_p value of 1.2 observed for [^{14}C]SDZ PSC 833 after

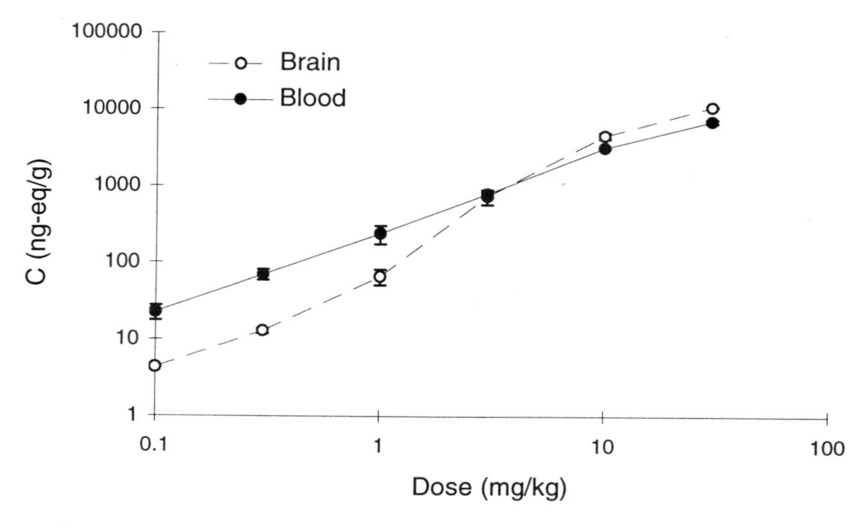

Figure 1. Brain and blood concentrations of radioactivity 2hr after iv injection of [14C]SDZ PSC 833. Data are means ± SD, n=3.

the coadministration of 10 mg/kg cyclosporin A was identical to the K_p value observed after single administration of [^{14}C]SDZ PSC 833.

III. SDZ PSC 833 Brain Penetration after IV Infusion

The non-linearity of the brain penetration of SDZ PSC 833 is clearly demonstrated in Figure 4 by plotting the brain/blood distribution ratios (K_p) versus blood concentrations. At blood levels below 0.8 μg/ml, a very low, if any, brain penetration of SDZ PSC 833 was observed ($K_p < 0.3$). However, when the blood concentration increased from 0.8 to 1.1 μg/ml, the brain penetration (K_p) markedly increased from 0.3 to 1.8. Thereafter, for blood levels higher than 1.1 μg/ml, similar K_p values are observed, indicating a linear brain distribution.

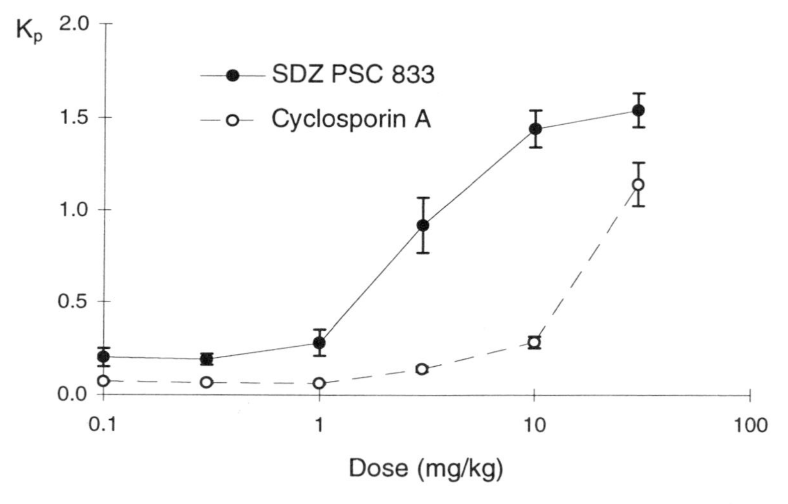

Figure 2. Brain/blood distribution ratios of radioactivity 2 hr after injection of [14C]SDZ PSC 833 or [3H]cyclosporine A. Data are means ± SD, n=3.

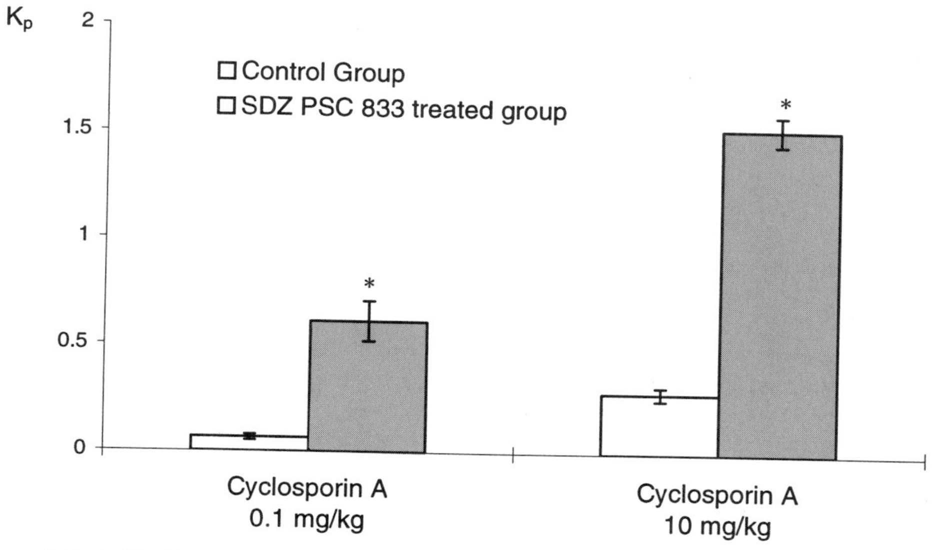

Figure 3. Brain/blood distribution ratios of [3H]cyclosporin A with/without coadministration of SDZ PSC 833. The concentrations of radioactivity were measured 2 hr after iv injection of the labeled drug dose alone (control group) or in combination with SDZ PSC 833 (10 mg/kg). Data are means ± SD, n=3. *$p<0.05$, significantly different from the control.

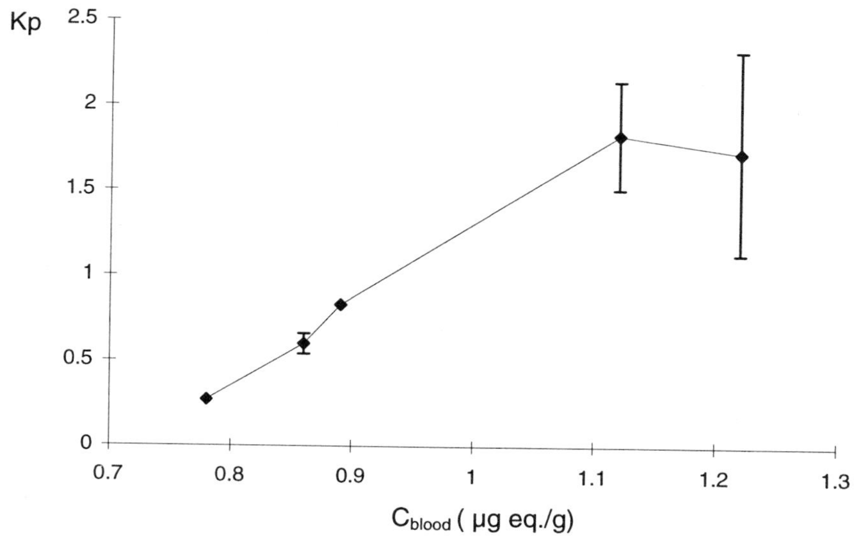

Figure 4. Relationship between blood concentrations of [14C]SDZ PSC 833 and its brain/blood concentration ratios (Kp) during iv infusion (5 µg/min). Data are means ± SD, n=3.

4. DISCUSSION

This study was designed to assess the brain penetration of a potent P-glycoprotein inhibitor, SDZ PSC 833, and to compare it with that of the cyclosporin A.

Our *in vivo* approaches clearly demonstrated the non-linearity of the brain penetration of SDZ PSC 833. Indeed, after administration of various iv doses (0.1 to 30 mg/kg) (Fig. 1), the blood concentrations increased in a dose proportional way whereas the brain concentrations of SDZ PSC 833 increased in a non-linear fashion. These results were confirmed by those obtained in the iv constant-rate infusion study (Fig. 4); the latter showed that the transport of SDZ PSC 833 into the brain was restricted by the blood-brain barrier at low blood levels whereas at higher blood levels (0.8 to 1.1 µg/ml), the brain penetration of SDZ PSC 833 markedly increased suggesting that the amount of SDZ PSC 833 present in the blood should be sufficient to inhibit the efflux from the brain. Moreover, similar K_p values for SDZ PSC 833 were observed after 10 and 30 mg/kg iv doses (Fig. 2) or for blood concentrations higher than 1.1 µg/ml (Fig. 4); this indicate that the brain passage of SDZ PSC 833 obeys a passive diffusion process in this high blood concentration range. These findings support the hypothesis that the non-linearity of the brain penetration of SDZ PSC 833 is due to a blood concentration-dependent modulation of the blood-brain barrier permeability.

The brain penetration of cyclosporin A was also found to be dose-dependent (Fig. 2), indicating that the penetration of cyclosporin A into the brain increased at higher doses. The low brain penetration of cyclosporin A, at low blood levels, has been explained by its extensive blood binding[10] and its high molecular weight[11]. However recent *in vitro* studies[7] performed with bovine brain capillary endothelial cells suggest that P-glycoprotein, detected in the membrane of these cells by means of anti-P-glycoprotein antibodies, participated in the blood-brain barrier transport of cyclosporin A. Indeed, the uptake of cyclosporin A by these cells was significantly increased in the presence of several multidrug resistant reversing agents (verapamil, quinidine, vincristine, vinblastine) which are also competitive P-glycoprotein substrates. *In vivo* studies[6] demonstrated that the cyclosporin A transport is an ATP energy-dependent mechanism which is in accordance with the P-glycoprotein characteristics. In addition, in an *in vitro* study, Shirai *et al.*[8] report that P-glycoprotein in the brain capillary endothelial cells could transport cyclosporin A across the endothelium from the basal to the apical side. These results indicate that the P-glycoprotein localized at the luminal side of the capillary endothelial cells[12,13,14,15] prevents the penetration of cyclosporin A into the brain in an ATP-dependent manner. The similarity of the brain penetration profiles of SDZ PSC 833 and cyclosporin A (Fig. 2) and the fact that the coadministration of SDZ PSC 833 increased the brain penetration of cyclosporin A (Fig. 3) suggest that both cyclosporins were pump out from the brain into the bloodstream by P-glycoprotein at the blood-brain barrier under physiological conditions and inhibited the efflux pump at high blood concentrations. However, in order to obtain a similar brain penetration, the iv dose of cyclosporin A has to be roughly 10 times higher than the SDZ PSC 833 dose (Fig. 2). This different brain distribution of both cyclosporins may be explained by the different unbound fractions observed in rat blood, i.e. 3% for cyclosporin A[16] and 50% for SDZ PSC 833[17]. The role of lipophilicity on a saturable efflux system in the BBB was suggested recently with a series of model peptides[18]; thus the higher lipophilicity of SDZ PSC 833 compared to cyclosporin A could also explain its higher BBB passage. Finally, this difference in brain penetration may result from the higher affinity of SDZ PSC 833 for the P-glycoprotein. SDZ PSC 833 may have a higher *in vivo* potency than cyclosporin A in blocking the P-glycoprotein pump at the BBB, as judged from the observations that lower doses of SDZ PSC 833 were required to attain the same K_p value and from the fact that the coadministration of SDZ PSC 833 increased the brain distribution of cyclosporin A while cyclosporin A did not influence that of SDZ PSC 833.

In conclusion, the present *in vivo* study demonstrated that the non-linearity of the brain penetration of SDZ PSC 833 is due to a blood concentration-dependent modulation of the blood brain barrier permeability. Furthermore, it is reasonable to think that SDZ PSC 833 like cyclosporin A governs its own brain passage by inhibition of the P-glycoprotein pump present in the brain microcapillary endothelial cells. The potency of SDZ PSC 833 to inhibit the P-glycoprotein-mediated efflux at the blood brain barrier is higher than that of cyclosporin A. The inhibition of P-glycoprotein by SDZ PSC 833 and cyclosporin A could be of great significance in the treatment of brain cancers.

5. REFERENCES

1. Boesch, D., Gaveriaux, C., Jachez, B., Pourtier-Manzanedo, A., Bollinger, P., and Loor, F.: In vivo circumvention of P-glycoprotein-mediated multidrug resistance of cells with SDZ PSC 833. Cancer Res. 51: 4226-4233, 1991.
2. Gaveriaux, C., Boesch, D., Jachez, B., Bollinger, P., Payne, T., and Loor, F.: SDZ PSC 833, a non-immunosuppressive cyclosporin analog, is a very potent multidrug- resistance modifier. J. Cell. Pharmacol. 2: 225-234, 1991.
3. Boekhorst, P.a.w., Kapel J.v., Schoester, M., and Sonneveld, P.: Reversal of typical multidrug resistance by cyclosporin and its non-immunosuppressive SDZ PSC 833 in Chinese hamster ovary cell exppressing the mdr1 phenotype. Cancer Chemother. Pharmacol. 30; 238-142, 1992.
4. Niederberger, W., Lemaire, M., Maurer, G., Nussbaumer, K., and Wagner, O.: Distribution and binding of cyclosporin in blood and tissues. Transplant. Proc. 15: 2419- 2421, 1983.
5. Cefalu, W.t., and Pardridge, W.m.: Restrictive transport of a lipid soluble peptide (cyclosporin) through the blood-brain barrier. J. Neurochem. 44: 1954-1956,1985.
6. Sakata, A., Tamai, I., Kawazu, K., Deguchi, Y., Ohnishi, T., Saheki, A., and Tsuji, A.: *In vivo* evidence for ATP-dependent and P-glycoprotein mediated transport of cyclosporin A at the blood-brain barrier. Biochem. Pharmacol. 48: 1989-1992, 1994.
7. Tsuji, A., Tamai, I., Sakata, A., Tenda, Y., and Terasaki, T.: Restricted transport of cyclosporin A across the blood-brain barrier by a multidrug transporter, P-glycoprotein. Biochem. Pharmacol. 46: 1096-1099, 1993.
8. Shirai, A., Naito, M., Tatsuta, T., Dong, J., Hanaoka, K., Mikami, K., Oh- Hara, T., and Tsuruo, T.: Transport of cyclosporin A across the brain capillary endothelial cell monolayer by P-glycoprotein. Biochem. Biophys. Acta 1222: 400-404, 1994.
9. Wagner, O., Schreier, E., Heitz, F., and Maurer, G.: Tissue distribution, disposition, and metabolism of cyclosporine in rats. Drug Metab. Dispos. 15: 377-383, 1987.
10. Lemaire, M., Pardridge, W.m., and Chaudhuri, G.: Influence of blood components on the tissue uptake indices of cyclosporin in rats. J. Pharmacol. Exp. Ther. 244: 740-743, 1988.
11. Pardridge, W.M.: Peptide lipidization and liposomes. *In* Peptide drug delivery to the brain, ed. by W.M. Pardridge, pp. 123-148, Raven Press, New-York, 1991.
12. Tatsuta, T., Naito, M., Oh-hara, T., Sugawara, I., and Tsuruo, T.: Functional involvement of P-glycoprotein in blood-brain barrier. J. Biol. Chem. 267: 20383-20391, 1992.
13. Tsuji, A., Terasaki, T., Takabatake, Y., Tenda, Y., Tamai, I., Yamashima, T., Moritani, S., Tsuruo, T., and Yamashita, J.: P-glycoprotein as the drug efflux pump in primary cultured bovine brain cappillary endothelial cells. Life Sci. 51: 1427-1437, 1992.
14. Thiebaut, F., Tsuruo, T., Hamada, H., Gottesman, M.M., Pastan, I., and Willingham, M.C.: Immunohistochemical localization in normal tissues of different epitopes in the multidrug transport protein P170: evidence for localization in brain capillaries and crossreactivity of one antibody with a muscle protein. J. Histochem. Cytochem. 37: 159- 164, 1989.
15. Cordon-cardo, L., O'brien, J.P., Casals, D., Rittman-grauer, L., Biedler, J.L., Melamed, M.R., and Bertino, J.r.: Multi-resistance gene (P-glycoprotein) is expressed by endothelial cells at blood-brain barrier sites. Proc. Natl. Acad. Sci. USA 86: 695- 698, 1989.
16. Lemaire, M., and Tillement, J.P.: Role of lipoproteins and erythrocytes in the in vitro binding and distribution of cyclosporin A in the blood. J. Pharm. Pharmacol. 34: 715-718, 1982.

17. Lemaire, M.: SDZ PSC 833: Distribution and binding to blood components. Document, Sandoz Pharma Ltd., 1992.
18. Chikhale, E.G., Burton, P.S., and Borchardt, R.T.: The effect of verapamil on the transport of peptides across the blood-brain barrier in rats: kinetic evidence for an apically polarized efflux mechanism. J. Pharmacol. Exp. Ther. 273: 298-303, 1995.

34

PEPTIDE-LIKE DRUGS MAY BE EXCLUDED FROM THE BRAIN BY P-GLYCOPROTEIN AT THE BLOOD–BRAIN BARRIER

M. A. Barrand,[1] K. Robertson,[1] S. F. von Weikersthal,[1] and D. Horwell[2]

[1] Department of Pharmacology
University of Cambridge
Tennis Court Road
Cambridge, CB2 1QJ United Kingdom
[2] Department of Chemistry
Parke-Davis Research
Addenbrooke's Hospital Site
Hills Road, Cambridge
United Kingdom

SUMMARY

Peptide-like compounds, particularly a NK_1 antagonist and to a small extent CCK_A and CCK_B antagonists, appear to interact with P-glycoprotein since they show effects on intracellular drug accumulation and can displace photoaffinity labelling of the protein. They may thus be P-glycoprotein substrates, liable to expulsion from the blood-brain-barrier.

RÉSUMÉ

Quelques composés de type peptidique, en particulier un antagoniste du récepteur NK1, et à un moindre degré des antagonistes des récepteurs CCK_A et CCK_B, semblent interagir avec la P-gp, puisqu'ils agissent sur l'accumulation intracellulaire de médicaments et peuvent déplacer le photomarquage de la protéine. Ils peuvent donc être considérés comme des substrats de la P-gp, susceptibles d'être expulsés de la BHE.

1. INTRODUCTION

P-glycoprotein expressed on endothelial cells forming part of the blood-brain-barrier plays a critical role in limiting entry of toxic agents to the brain[1] but it may also restrict entry of therapeutic drugs. The potential of peptides as neurotherapeutic agents is limited because

Biology and Physiology of the Blood–Brain Barrier, edited by Couraud and Scherman
Plenum Press, New York, 1996

of poor access across the blood-brain-barrier. Although specific peptide transport systems exist[2], it is known that peptides may also be substrates for P-glycoprotein[3] and so one reason for poor access to the brain may be the presence of this drug efflux transporter at the blood-brain-barrier. A number of alternative non-peptide structures have been synthesised that show high affinity as antagonists for CCK_A, CCK_B and NK_1 receptors in the brain[4] and have potent anxiolytic properties. The design of these peptoid structures has been based around the minimum fragment in each endogenous peptide found to be essential for function[5]. Some of these peptoid compounds show relatively poor brain penetrability[6]. In the present study, we look at the ability of some of these compounds to interact with P-glycoprotein since a compound showing such an interaction might well be a substrate for the drug transporter and so be hindered from entering the brain by being expelled at the blood-brain barrier. Thus we have examined the effects of these compounds on intracellular accumulation of the P-glycoprotein substrates, vincristine, daunorubicin and cyclosporin A and on photoaffinity labelling of P-glycoprotein and have compared these results with those obtained with the cyclic peptide, cyclosporin A and with two CNS active drugs, tacrine and gabapentin. Cyclosporin A is a P-glycoprotein substrate showing limited access to the brain[7] whereas gabapentin is actively taken up and enters the brain readily[8].

2. METHODS

Effects of the dipeptoids on drug efflux activity of P-glycoprotein were investigated in two separate cell types; primary cultures of rat brain endothelial cells known to contain moderate levels of mdr1a and mdr1b isoforms of P-glycoprotein[9] and multidrug resistant variant cells (EMT6/AR1.0) of the mouse mammary tumour cell line (EMT6/P) that show high level expression of mdr1a P-glycoprotein[10]. The predominant isoform expressed in rat and mouse brain capillaries in vivo is mdr1a[1, 9, 11].

Cells grown to near confluence in 24 well plates were pretreated for 30 min with or without the test compounds over a concentration range of 1-100μM before incubation for 90 min in the presence of $[^3H]$-vincristine, $[^3H]$-daunorubicin or $[^3H]$-cyclosporin A each at 30nM. The amount of drug accumulated inside the cells was measured by scintillation counting following cell lysis.

To assess the ability of the dipeptoids to displace photoaffinity labelling of P-glycoprotein, membranes containing mdr1a P-glycoprotein were prepared from EMT6/AR1.0 cells and covalently labelled by light activation with 0.5μM $[^3H]$-azidopine[12] in the absence or presence of 5μM or 50μM concentrations of the peptide receptor antagonists. Membrane proteins were then separated by electrophoresis, the gels fixed, dried and exposed to film for 2-3 weeks.

3. RESULTS

The non-peptide antagonist (Cam-4261) of NK_1 receptors caused dose dependent increases (2-8 fold with 1-50μM) in drug accumulation in both cell types expressing P-glycoprotein. Antagonists of CCK_A (Cam-1481) and CCK_B (Cam-1028) receptors had much weaker effects (<3 fold at 50μM) as did the acridine-like compound, tacrine. The order of potency for the antagonists was NK_1>>CCK_A>CCK_B=Tacrine. Gabapentin which is known to become concentrated in the brain had no effect on drug accumulation.

The cyclic peptide, cyclosporin A, at 5 and 50μM and the peptoid NK_1 antagonist at 50μM were able to displace $[^3H]$-azidopine photoaffinity labelling of P-glycoprotein in membranes prepared from multidrug resistant mouse tumour cells indicating that both drugs are able to interact directly with P-glycoprotein. The other drugs were without effect at 50μM.

4. CONCLUSION

Thus some of the dipeptoids synthesised for use as centrally-acting therapeutic agents can interact with P-glycoprotein. As possible substrates for the drug efflux transporter, they may therefore be hindered in their access to the brain.

The work described and the salaries of KR and SFW were supported by a grant from the Wellcome Trust

REFERENCES

1. Schinkel AH, Smit JJM, van Tellingen O, Beljnen JH, Wagenaar E, van Deemter L, Mol CAAM, van der Valk MA, Robanus-Maandag EC, te Riele HPJ, Berns AJM and Borst P (1994) Cell, 77, 491-502.
2. Banks WA & Kastin AJ (1990) Am J Physiol., 259, E1-E10.
3. Sharma RC, Inoue S, Roitelman J, Schimke RT and Simon RD (1992) J.Biol.Chem., 267, 5731-5734.
4. Horwell DC, Hughes J, Hunter JC, Pritchard MC, Richardson RS, Roberts E and Woodruff GN (1991) J Med Chem., 34, 404-411.
5. Horwell DC (1995) TIBTECH, 13, 132-134.
6. Patel S, Chapman KL, Heald A, Smith AJ and Freedman SB (1994) Eur J Pharmacol., 253, 237-244.
7. Begley DJ, Squires LK, Zlokovic BV, Mitrovic DM, Hughes CCW, Revest PA and Greenwood J (1990) J Neurochem., 55, 1222-1230.
8. Thurlow RJ, Brown JP, Gee NS, Hill DR and Woodruff GN (1993) Eur J Pharmacol., 247, 341-345.
9. Barrand MA, Robertson K & von Weikersthal SF (1995) FEBS Lett., in press.
10. Barrand MA & Twentyman PR (1992) Br J Cancer, 65, 239-245.
11. Jetté L, Pouliot J-F, Murphy GF and Béliveau R (1995) Biochem J., 305, 761-766.
12. Friche E, Demant EJF, Sehested M & Nissen NI (1993) Br J Cancer, 67, 226-231.

SIGNIFICANCE OF THE SECOND MESSENGER MOLECULES IN THE REGULATION OF CEREBRAL ENDOTHELIAL CELL FUNCTIONS

An Overview

Ferenc Joó

Laboratory of Molecular Neurobiology
Institute of Biophysics
Biological Research Center
H-6701 Szeged, Hungary

Several receptors for different vasoactive substances have been shown to be coupled to the microvascular adenylate cyclase and to increase the intraendothelial amount of cyclic AMP (Joó, 1986, 1992, 1993). As a consequence of elevated intraendothelial cyclic AMP, changes in permeability (Joó, 1972; Rubin et al. 1991), in endothelial growth and differentiation (Kempski et al. 1987), in nitric oxide and endothelin secretions (Durieu-Trautmann et al. 1993) and in a hyperpolarization-activated current carried by both Na^+ and K^+ ions (Janigro et al. 1994) have been reported. Elevations of cyclic AMP concentrations may be important in regulating the endothelial barrier in two-ways: (i) by *narrowing the paracellular pathways* for small molecular weight substances (Rubin et al. 1991), and (ii) activating the *transendothelial transport* for macromolecules (Joó, 1972). Recently, Grammas et al. (1994) found significantly elevated cAMP levels in microvessels from Alzheimer's disease compared to nondemented elderly controls.

A unique feature of intact microvessels in the brain is the apparent absence of macromolecular transport (Ehrlich, 1885). Histamine has been shown in previous *in vivo* studies (Dux and Joó, 1982) to increase the permeation of circulating albumin through the blood-brain barrier, but the exact mechanism of action remained unclear. Results of recent studies (Deli et al. 1995) showed that, in contrast to peripheral endothelial cells, the permeability of tight junctions connecting the neigbouring endothelial cells does not change on the effect of histamine. On the other hand, histamine is capable of inducing a selective transendothelial albumin penetration through the cerebral endothelium. The effects of histamine to the cerebral endothelium are mediated either through histamine H_2-receptors coupled to the microvascular adenylate cyclase (Karnushina et al. 1980) or via histamine H_1-receptors changing the level of intracellular Ca^{2+}, as revealed by Revest et al. (1991).

Biology and Physiology of the Blood–Brain Barrier, edited by Couraud and Scherman
Plenum Press, New York, 1996

Uptake by and release of histamine from the cerebral endothelium has also been documented (Huszti et al. 1995). The uptake is saturable, Na^+-dependent with a yield of K_m 0.3 ± 0.02 μM and a V_{max} 4.6 ± 0.04 pmol/mg protein per min. Interestingly, the release of histamine occurred mainly luminally, although the cultured cerebral endothelial cells were able of taking up histamine from both the luminal and abluminal sides. Histamine receptors seem to be involved in the genesis of brain oedema (Joó and Klatzo, 1989) and specific blockers were shown (Dux et al. 1984, 1987) to be able of preventing brain oedema formation. Recently, it has been documented (Tósaki et al. 1994) that ranitidine could interfere with the process of brain oedema formation even when given 1 h after the onset of cerebral ischemia. This means that the therapeutic window is wide enough to interfere effectively by means of ranitidine treatment with the histamine-mediated molecular mechanisms which lead eventually to the enhanced transport of water and electrolytes from blood to brain.

Guanylate cyclase, the synthesizing enzyme of cyclic GMP was found to be activated in brain microvessels by Ca^{2+}, but not influenced with calmodulin or carbamyl choline (Palmer, 1981). The enzyme was seen (Homayoun et al. 1989) to respond to atrial natriuretic peptide (ANP) in a dose-dependent manner. ANP specifically inhibits amiloride-sensitive sodium uptake into isolated cerebral capillaries in vitro (Ibaragi et al. 1989) and ameliorates brain oedema formation (Dóczi et al. 1987). Large increases in microvascular cGMP level was detected by Marsault and Frelin (1992) on the effect of exogenous nitric oxyde (NO) donor molecules (sin-1 and sodium nitroprusside). The finding indicates that brain capillary endothelial cells, in contrast to aortic ones, may act as recipient cells for NO produced by neurons.

Changes in intraendothelial Ca^{2+} concentrations can be trasmitted to calcium/calmodulin-dependent protein kinase II (Deli et al. 1993) and protein kinase C (PK C) (Markovac and Goldstein, 1988), or both. Substance P can stimulate PK C translocation (Catalán et al. 1989), phorbol esters increase hexose uptake (Drewes et al. 1988), and fluid-phase endocytosis (Guillot et al. 1990). The role of PK C in the activation of ICAM-1 expression on endothelial cells was also evidenced (Lane et al. 1989). In a recent study (Krizbai et al. 1995), six isoforms of PK C were seen in freshly purified microvessels, PK C-α, -ß, -δ, -ε, -η, and-ζ; primary cultures of endothelial cells expressed PK C-α, -β, -δ and -ε isoenzymes, whereas the immortalized cell line expressed only PK C-α, -δ, and η. The rat aortic endothelium contained PK C-α, and -δ isoforms, only.

Recently, (Vigne et al. 1994) presented evidence for the operation of cross talk among cyclic AMP, cyclic GMP and Ca^{2+}-dependent intracellular signalling mechanisms.

ACKNOWLEDGMENT

The research was supported in part by the Hungarian Research Fund (OTKA T-14645, F-5207, F-12722, F-013104); Ministry of Public Welfare (ETT T-04 029/93) and U.S.-Hungarian Joint Fund (JFNo.392).

REFERENCES

Catalán, R.E., A.M. Martínez, M.D. Aragonés, and I. Fernández (1989) Substance P stimulates translocation of protein kinase C in brain microvessels. Biochem.Biophys.Res.Commun. *164*:595-600.

Deli, M., M.-P. Dehouck, R. Cecchelli, C.S. Abraham, and F. Joó (1995) Histamine induces a selective albumin permeation through the blood-brain barrier in vitro. Inflammation Res. *44*:S56-S57.

Deli, M., F. Joó, I. Krizbai, I. Lengyel, M.G. Nunzi, and J.R. Wolff (1993) Calcium/calmodulin-stimulated protein kinase II is present in primary cultures of cerebral endothelial cells. J.Neurochem. *60*:1960-1963.

Dóczi, T., F. Joó, P. Szerdahelyi, and M. Bodosi (1987) Regulation of brain water and electrolyte contents: the possible involvement of central atrial natriuretic factor. Neurosurgery *21*:454-458.

Drewes, L.R., M.A. Broderius, and D.Z. Gerhart (1988) Phorbol ester stimulates hexose uptake by brain microvessel endothelial cells. Brain Res.Bull. *21*:771-776.

Durieu-Trautmann, O., C. Fédérici, C. Créminon, N. Foignant-Chaverot, F. Roux, M. Claire, A.D. Strosberg, and P.O. Couraud (1993) Nitric oxide and endothelin secretion by brain microvessel endothelial cells: Regulation by cyclic nucleotides. J. Cellular Physiol. *155*:104-111.

Dux, E. and F. Joó (1982) Effects of histamine on brain capillaries: fine structural and immunohistochemical studies after intracarotid infusion. Exptl.Brain Res. *47*:252-258.

Dux, E., P. Temesvári, F. Joó, G. Ádám, F. Clementi, L. Dux, J. Hideg, and K.A. Hossmann (1984) The blood-brain barrier in hypoxia: ultrastructural aspects and adenylate cyclase activity of brain capillaries. Neuroscience *12*:951-958.

Dux, E., P. Temesvári, P. Szerdahelyi, Å. Nagy, J. Kovács, and F. Joó (1987) Protective effect of antihistamines on cerebral oedema induced by experimental pneumothorax in newborn piglets. Neuroscience *22*:317-321.

Ehrlich, P. (1885) Das Sauerstoff-bedürfnis des Organismus. Eine farbenanalytische Studie. Berlin: A. Hirschwald.

Grammas, P., A.E. Roher, and M.J. Ball (1994) Increased accumulation of cAMPin cerebral microvessels in Alzheimer disease. Neurobiol. of Aging *15*:113-116.

Guillot, F.L., K.L. Audus, and T.J. Raub (1990) Fluid-phase endocytosis by primary cultures of bovine brain microvessel endothelial cell monolayers. Microvasc.Res. *39*:1-14.

Homayoun, P., W.D. Lust, and S.I. Harik (1989) Effect of several vasoactive agents on guanylate cyclase activity in isolated rat brain microvessels. Neurosci.Lett. *107*:273-278.

Huszti, Z., M.A. Deli, and F. Joó (1995) Carrier-mediated uptake and release of histamine by cultured rat cerebral endothelial cells. Neurosci. Letters *184*:185-188.

Ibaragi, M., M. Niwa, and M. Ozaki (1989) Atrial natriuretic peptide modulates amiloride-sensitive Na+ transport across the blood-brain barrier. J.Neurochem. *53*:1802-1806.

Janigro, D., G.A. West, T.-S. Nguyen, and H.R. Winn (1994) Regulation of blood-brain barrier endothelial cells by nitric oxyde. Circ. Res. *75*:528-538.

Joó, F. (1972) Effect of N^6,O^6-dibutyril cyclic 3′,5′-adenosine monophosphate on the pinocytosis of brain capillaries in mice. Experientia *28*:1470-1471.

Joó, F. (1986) New aspects to the function of cerebral endothelium. Nature *321*:197-198.

Joó, F. (1992) The cerebral microvessels in tissue culture, an update. J.Neurochem. *58*:1-17.

Joó, F. (1993) The blood-brain barrier in vitro: The second decade. Neurochem. Int. *23*:499-521.

Joó, F. and I. Klatzo (1989) Role of cerebral endothelium in brain oedema. Neurol.Res. *11*:67-75.

Karnushina, I.L., J.M. Palacios, G. Barbin, E. Dux, F. Joó, and J.C. Schwartz (1980) Studies on a capillary-rich fraction isolated from brain: histaminic components and characterization of the histamine receptors linked to adenylate cyclase. J.Neurochem. *34*:1201-1208.

Kempski, O., B. Wroblewska, and M. Spatz (1987) Effects of forskolin on growth and morphology of cultured glial and cerebrovascular endothelial and smooth muscle cells. Int.J.Dev.Neurosci. *5*:435-445.

Krizbai, I., G. Szabó, M. Deli, K. Maderspach, C. Lehel, Z. Oláh, J.R. Wolff, and F. Joó (1995) Expression of protein kinase C family members in the cerebral endothelial cells. J. Neurochem. *65*:495-462.

Lane, T.A., G.E. Lamkin, and E. Wancewicz (1989) Modulation of endothelial cell expression of intracellular adhesion molecule 1 by protein kinase C activation. Biochem.Biophys.Res.Commun. *161*:945-952.

Markovac, J. and G.W. Goldstein (1988) Transforming growth factor beta activiates protein kinase C in microvessels isolated from immature rat brain. Biochem.Biophys.Res.Commun. *150*:575-582.

Marsault, R. and C. Frelin (1992) Activation by nitric oxyde of guanylate cyclase in endothelial cells from brain capillaies. J.Neurochem. *59*:942-945.

Palmer, G.C. (1981) Distribution of guanylate cyclase in rat cerebral cortex: neuronal, glial, capillary, pia-arachnoid and synaptosomal fractions plus choroid plexus. Neuroscience *6*:2547-2553.

Revest, P.A., N.J. Abbott, and J.I. Gillespie (1991) Receptor mediated changes in intracellular Ca^{2+} in cultured rat brain capillary endothelial cells. Brain Res. *549*:159-161.

Rubin, L.L., D.E. Hall, S. Porter, K. Barbu, C. Cannon, H.C. Horner, M. Janatpour, C.W. Liaw, K. Manning, J. Morales, L.I. Tanner, J. Tomaselli, and F. Bard (1991) A cell culture model of the blood-brain barrier. J.Cell.Biol. *115*:1725-1735.

Tósaki, A., P. Szerdahelyi, and F. Joó (1994) Treatment with ranitidine of ischemic brain edema. Eur.J.Pharm. *264*:455-458.

Vigne, P., L. Lund, and C. Frelin (1994) Cross talk among cyclic AMP, cyclic GMP, and Ca2+-dependent intracellular signalling mechanisms in brain capillary endothelial cells. J. Neurochem. *62:2269-2274.*

THE ROLE OF PROTEIN KINASE C AND MARCKS PROTEIN PHOSPHORYLATION IN RAT CEREBROMICROVASCULAR ENDOTHELIAL CELL PROLIFERATION INDUCED BY ASTROCYTE-DERIVED FACTORS

Danica B. Stanimirovic, Rita Ball, Josée Wong, and Jon P. Durkin

Cellular Neurobiology Group
Institute for Biological Sciences
National Research Council of Canada
Ottawa, ONT K1A 0R6

SUMMARY

Serum-free medium conditioned by rat cortical astrocytes was found to prevent apoptosis induced by growth factor-deprivation, accelerate DNA synthesis, induce transient activation of protein kinase C (PKC), and increase the endogenous phosphorylation of the PKC-specific substrate, the 85 kD MARCKS protein, in rat cerebromicrovascular endothelial cells (RCEC). The trophic and stimulatory factor(s) in astrocyte conditioned media (ACM) were heat- and trypsin-sensitive and found to have an apparent molecular weight greater than 10 kD. The potent PKC activator, 12-O-tetradecanoyl phorbol 13-acetate (TPA), also stimulated RCEC proliferation, whereas the inhibition of PKC by staurosporine caused a concomitant loss in ACM-induced PKC translocation, MARCKS protein phosphorylation and DNA synthesis. These findings implicate PKC activation as a critical early event in cerebral endothelial cell proliferation triggered by astrocyte-derived mitogen(s).

RÉSUMÉ

Il a été trouvé qu'un milieu sans serum traité par des astrocytes du cortex de rat prévient l'apoptose induite par la privation de facteurs de croissance, accélère la synthèse d'ADN, induit une activation transitoire de la proteine kinase C (PKC), et augmente la phosphorylation endogène du substrat spécifique de la PKC, soit de la protéine MARCKS de 85 KD, chez les cellules endothéliales cérébromicrovasculaires du rat (RCEC). Le(s)

Biology and Physiology of the Blood–Brain Barrier, edited by Couraud and Scherman
Plenum Press, New York, 1996

facteur(s) trophique(s) et stimulant(s) du milieu traité par les astrocytes (ACM) est (sont) sensible(s) à la chaleur et à la trypsine et a (ont) un poids moléculaire apparent supérieure à 10 kD. L'activateur puissant de la PKC, le 12-O-tétradécanoyle phorbole 13-acétate (TPA) stimule la prolifération de RCEC, lorsqu'une qu'une inhibition de la PKC par la staurosporine rend l'ACM inefficace à induire la translocation de la PKC, la phosphorylation de la protéine MARCKS et la synthèse de l'ADN. Ces résultats indiquent que l'activation de la PKC est un événement critique arrivant tôt lorsque la prolifération des cellules endothéliales du cerveau est induite par un (des) mitogène(s) provenant des astrocytes.

I. INTRODUCTION

It has been demonstrated that complex interactions between the central nervous system (CNS) tissue and cerebral endothelial cells (CEC) *in vivo* influence the expression of properties associated with the functional blood-brain barrier (BBB) [6]. A growing body of evidence obtained from *in vitro* studies indicates that astrocytes and/or astrocyte-derived factors increase the number and complexity of tight junctions and enhance the transendothelial electrical resistance of cerebral endothelial cell monolayers [2] and induce the expression and activities of the BBB-associated enzymes γ-glutamyl transpeptidase (γ-GTP) and alkaline phosphatase (ALP) [3, 13]. In addition, interactions between cerebral endothelial cells and astrocytes may play a pivital role in the phenotypic changes which take place in CEC during neural capillary angiogenesis [11, 13]. It has been demonstrated that the formation of capillary-like tubular structures and incipient basement membrane by CEC in culture, an *in vitro* model of the *in vivo* angiogenic process, can be induced by the combined actions of extracellular matrix and, still unidentified, astrocyte-derived factor(s) [11, 13].

In this study we provide evidence that secretable astrocyte-derived factor(s) prevent apoptosis and stimulate DNA synthesis in neonatal rat cerebromicrovascular endothelial cells (RCEC) by a mechanism that likely involves the transient activation of PKC and the PKC-mediated phosphorylation of specific endothelial cell proteins.

II. MATERIAL AND METHODS

II.I. Cell Cultures and Astrocyte–Conditioned Media (ACM)

Cerebral capillaries and microvessels, isolated from the brains of 2-4 days old Sprague-Dawley rats by separation on 20% dextran and sequential filtration through 350, 112, and 20 μm Nitex meshes [8] were cultured in rat-tail collagen- coated tissue culture dishes in complete medium (M199 [Gibco BRL] supplemented with 1% basal media Eagle's (BME) amino acids, 1% BME vitamins, 1% glucose, 0.05% peptone) containing 20% FBS. Cultures of highly purified RCEC were obtained by cloning endothelial cell colonies migrating from the adhered microvessels (using cloning rings) and propagating them in complete medium. RCEC cultures derived by these procedures were positively identified as such by immunocytochemical staining for Factor VIII-related antigen and angiotensin-converting enzyme.

Primary rat cortical type 1 astrocytes were isolated from the same groups of animals as were RCEC using the modified differential adhesion method of McCarthy and DeVellis (1980) [14]. Astrocytes were grown in poly-L-lysine coated tissue culture flasks in Dulbecco's minimal essential medium containing 10% FBS and 2 mM L-leucine methylester to eliminate microglia. Confluent astrocyte cultures between the 2nd and 5th passage were used 10-15 days after plating to condition serum-free, unsupplemented M199 medium for three consecutive days. The resultant astrocyte conditioned medium (ACM) was subjected to either heat-exposure (95^{0}C for 30 min), trypsinization (2 mg/ml trypsin for 30 min followed by inactivation of trypsin activity by 2 mg/ml soybean trypsin inhibitor for 30 min), or ultrafiltration through Centriprep-10 (Amicon Inc., Beverly, MA) membranes retaining material of apparent molecular weight greater than 10 kD. In some experiments, astrocytes were pretreated with 1 μM cycloheximide or 1 μM dexamethasone for 24 hours prior to the conditioning of medium M199 by these cultures.

II.II. Viability Assays

The effect of ACM on the ability of RCEC to survive in serum- and growth factor-free media was determined in experiments in which RCEC were initially subjected to serum-free M199 or serum-free ACM for 48-72 hours, and cellular nuclei then stained for 10 min by either propidium iodide (1 μg/ml) to mark the dead cells, or 33342 Hoechst (4 μg/ml) to assess the extent of nuclear fragmentation.

II.III. DNA Synthesis

The influence of ACM on DNA synthesis in RCEC was determined by the measurement of [^{3}H]thymidine incorporation into cellular DNA. Subconfluent RCEC (2 days after plating) were rendered proliferatively quiescent by incubation in serum-free M199 medium for 24 hours. The cells were then stimulated for 15 min or 24 hours with either ACM, 10% FBS, or 12-O-tetradecanoyl phorbol 13-acetate (TPA), and simultaneously pulsed with [^{3}H]-thymidine (25 Ci/mmol; 2.5 μCi/ml) over 24 hours. In some experiments, the PKC inhibitors staurosporine (20 nM), bisindolylmaleimide GF 109203X (BIS; 5 μM), or calphostin C (2 μM) (all obtained from Calbiochem, San Diego, CA) were added to the RCEC cultures 30 min prior to the stimulation. After 24 hours of incubation, the cells were washed in PBS, lysed in 0.1% Triton X-100, and extracted with 0.3% trichloroacetic acid. The precipitates were filtered through Whatman GF/C glass microfibre filters and counted for TCA-precipitable radioactivity. Protein contents of the cell lysates were determined by the method of Lowry et al (1951) [12].

II.IV. Protein Kinase C Activity and Marcks Protein Phosphorylation

To measure protein kinase C (PKC) activity, serum-deprived RCEC were stimulated as described above, then washed and lysed in ice-cold hypotonic medium [6]. PKC activity was determined in both membrane extracts and cytosolic fractions prepared as previously described [6] using a modified conventional PKC activity assay in which the incorporation of ^{32}P into a PKC-selective peptide substrate, Ac-FKKSFKL-NH_2 was determined. This peptide corresponds to residues 160-166 of the phosphorylation site domain (155-175) of the MARCKS (myristoilated adenine-rich kinase-C substrate) protein, a highly specific endogenous PKC substrate [1]. In parallel experiments, the effect of ACM on the extent of endogenous phosphorylation of 85-kD MARCKS protein was determined in intact RBEC previously exposed to [^{32}P]-orthophosphate for 2 hours in serum-free M199 to label the intracellular ATP pool. The 85- kDa MARCKS protein was extracted using a modification[7] of the method of Robinson et al. (1993) [16] after the cells were stimulated with ACM for 15 min. The extent of MARCKS protein phosphorylation was determined by separation on 10% sodium dodecyl sulphate-polyacrylamide gel (SDS-PAGE) and visualized by autoradiography using Kodak XAR film.

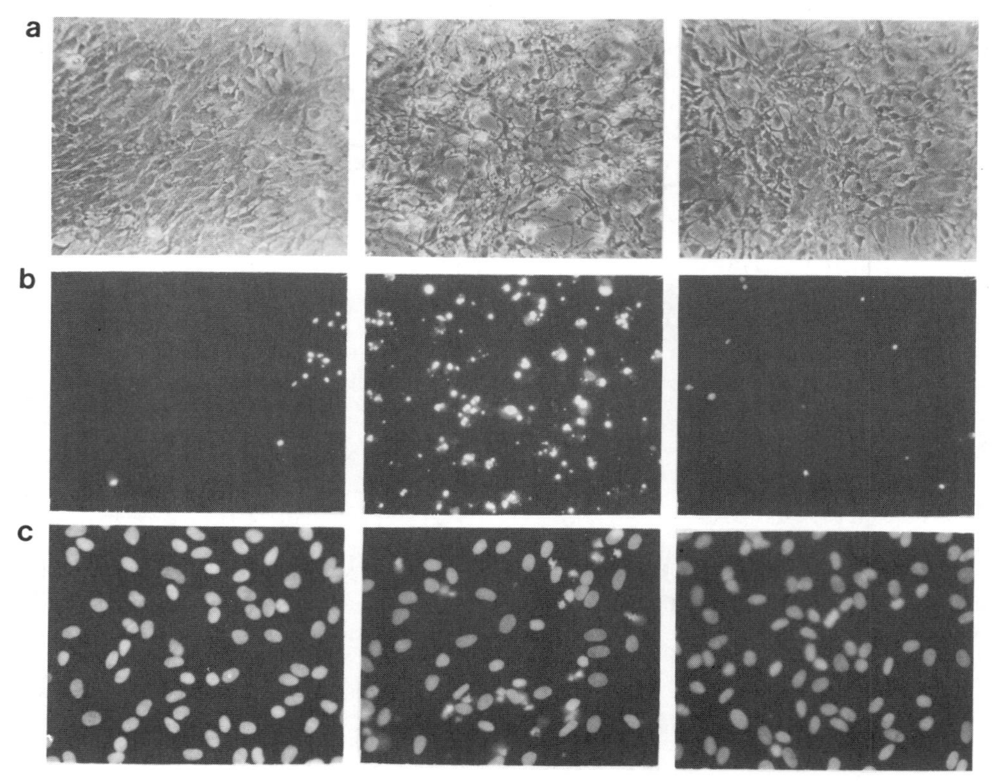

Figure 1. Effects of ACM on the survival of RCEC exposed to growth factor-deprived conditions. Cells were meintained in either complete growth media (first column), serum- and growth factor-free media M199 (middle column), or ACM (third column) for 48 hours. A) Phase contrast micrographs of cells subjected to the treatments described above. ACM prevented the shrinkage and contraction of RCEC seen in growth-factor deprived media. B) Propidium iodide staining of cell nuclei indicated reduction in the number of dead cells in ACM-treated cultures compared to M199-treated cultures. C) 33342 Hoechst staining indicated extensive fragmentation of RCEC nuclei characteristic of apoptotic cell death in growth-factor deprived media. ACM significantly reduced the number of fragmented nuclei relative to unconditioned medium.

III. RESULTS AND DISCUSSION

III.I. ACM Prevents Apoptosis of RCEC

In order to determine the effects of ACM on RCEC survival in growth factor-deprived conditions, we subjected RCEC to serum- and growth factor-free media M199 or ACM for 48 hours. Both the morphological appearance of the cells and propidium iodide staining indicated that while a significant (40-60%) number of cells died during the 48 hour exposure to serum- and growth factor-free M199, they did not when maintained ACM (Figure 1). The majority of RCEC subjected to growth-factor-deprived conditions appeared to die from apoptosis, since extensive nuclear fragmentation was observed in cultures stained with 33342 Hoechst (Figure 1). By contrast, cultures maintained in ACM showed no sign of significant apoptosis over a 48 hour period (Figure 1). These findings indicate that astrocytes secrete factors which prolong RCEC survival and prevent RCEC apoptosis induced by the

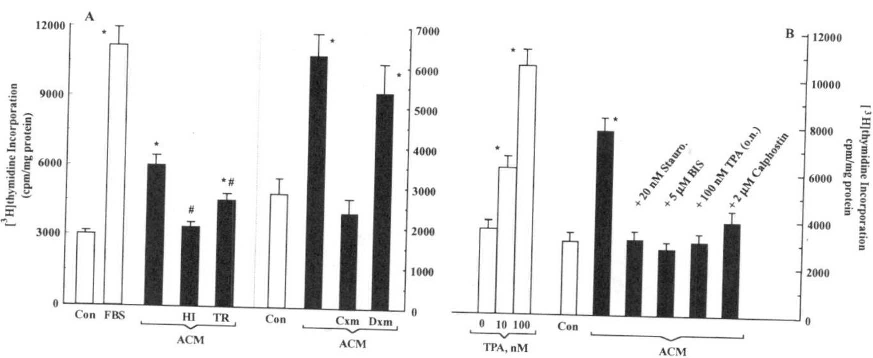

Figure 2. The effects of astrocyte conditioned media (ACM) and 12-O-tetradecanoyl phorbol 13-acetate (TPA) on [^{3}H]thymidine incorporation in rat cerebromicrovascular endothelial cells (RCEC). RCEC were rendered quiescent by a 24 hours incubation in serum-free medium (Con) before being stimulated with A) 10% FBS, untreated ACM, or ACM previously heated (HI) or subjected to brief trypsinization (TR), or ACM collected from astrocyte cultures pretreated with 1 μM cycloheximide (Cxm) or 1 μM dexamethasone (Dxm) for 24 hours, or B) 10-100 nM TPA alone, or ACM in the presence of the PKC inhibitors. The PKC inhibitors were added to the cells 30 min before the addition of ACM. The cells were exposed to [^{3}H]-thymidine as described in the text and the amount of radiolabel incorporated into cellular DNA between 0-24 hours was determined. The values represent the means ± S.E.M.of six replicate dishes and are representative of the results of 3 separate experiments. *-indicates significant difference (P<0.01; ANOVA) as compared to control (Con). #-indicates significant difference (P<0.01; ANOVA) as compared to untreated ACM.

removal of growth-factors from the culture media. This "trophic" activity of ACM on RCEC was concentrated in the >10 kD fraction of ACM obtained by ultrafiltration, but was not found in fraction of <10 kD (data not shown).

III.II. ACM Induces DNA Synthesis in RCEC

In order to examine whether ACM is mitogenic for RCEC, we investigated the influence of ACM on DNA synthesis in RCEC. The exposure of subconfluent and proliferatively quiescent RCEC to ACM enhanced the incorporation of [^{3}H]thymidine into DNA by 2-2.5 fold during the ensuing 24 hours (Fig. 1A&B) relative to cultures maintained in serum-free medium (Figure 2A). By comparison, the stimulation of DNA synthesis effected by 10% serum was 4-fold above control values (Figure 2A). The mitogenic activity associated with ACM was lost by heat treatment or brief trypsinization (Figure 2A). Furthermore, ACM collected from astrocytes previously treated by the protein synthesis inhibitor, cycloheximide, failed to induce DNA synthesis in RCEC (Figure 2A), suggesting that *de novo* protein synthesis by astrocytes was necessary for the expression of mitogenic activity in ACM. A 24 hour pretreatment of astrocytes by the glucocorticoid, dexamethasone, known to modulate the expression of various genes and to suppress the synthesis of eicosanoids including the mitogenic PGE_2, did not have an appreciable effect on DNA synthesis induced in RCEC by ACM (Figure 2A). These data suggest that some of the initial steps of neural angiogenesis, such as CEC proliferation, may be activated essentially by protein-like factors synthesized and secreted by astrocytes.

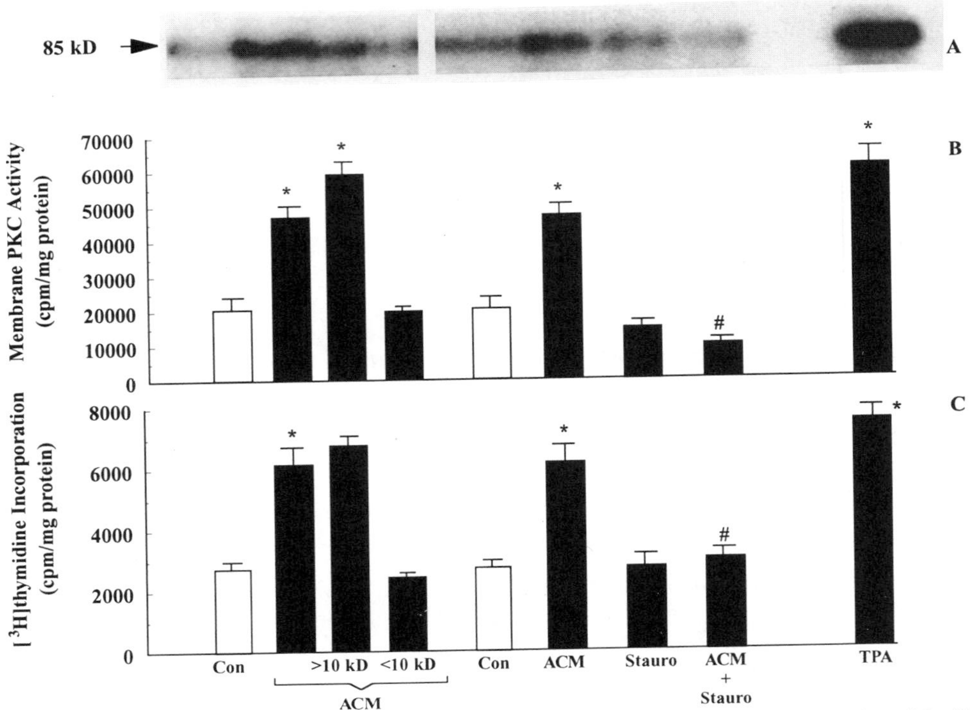

Figure 3. The effect of astrocyte-conditioned media (ACM) on A) the endogenous phosphorylation of the 85 kD MARCKS protein, B) membrane protein kinase C (PKC) activity, and C) DNA synthesis in rat cerebromicrovascular endothelial cells (RCEC). All three parameters were determined in parallel experiments. Cells were left untreated (Con) or subjected to either complete ACM, >10 kD or <10 kD molecular weight fraction of the same ACM, 20 nM staurosporine, 20 nM staurosporine plus ACM, or 1 μM TPA for 15 min. Staurosporine was added to the cells 30 min prior to ACM stimulation. MARCKS protein phosphorylation, was determined and analysed as described in the Materials and Methods. PKC activity in isolated RCEC membranes, subjected to the same treatments as above, was determined by the incorporation of ^{32}P into a PKC-selective peptide substrate, Ac-FKKSFKL-NH_2 as described in the Materials and Methods. DNA synthesis was determined as described in the text. The values in B) and C) represent the means ± S.E.M.of six replicate dishes and are representative of the results of 3 separate experiments. *-indicates significant difference ($P<0.01$; ANOVA) as compared to control (Con). #-indicates significant difference ($P<0.01$; ANOVA) as compared to untreated ACM.

III.III. ACM-Induced Mitogenesis Is Mediated by PKC

In order to better understand the mechanisms by which astrocyte-derived factors effect RCEC proliferation, the role of the key signaling enzyme, PKC, known to be involved in the proliferative responses of various cell types [4, 9], was determined.

The potent activator of PKC, TPA, byitself induced DNA synthesis in RCEC when present in the medium for either 24 hours (Figure 2B) or 15 min (Figure 3C). Moreover, ACM-induced DNA synthesis in RCEC was blocked by the PKC inhibitors, staurosporine, BIS, and calphostin C (Figure 2B), as well as by down-regulating PKC (100 nM TPA for 24 hours) prior to stimulation with ACM (Figure 2B) or 1 μM TPA (not shown). These results indicated that transient PKC activation is in itself sufficient to trigger a mitogenic response in RCEC, and that the mitogenic activity of ACM is dependent on PKC activation in RCEC.

We next investigated the effect of ACM on PKC activity and the *in situ* phosphorylation of the PKC-specific substrate, the 85 kD MARCKS protein. The exposure of RCEC to ACM stimulated the translocation of cytosolic PKC to membranes (Figure 3B) with a concomitant increase in endogenous phosphorylation of the MARCKS protein (Figure 3A). The transient (15 min) exposure of RCEC to ACM was also sufficent to cause a 2.2-fold increase in DNA synthesis (Figure 3C). Both the PKC stimulating- and mitogenic activities of ACM were retained in the >10 kD fraction obtained by ultrafiltration and lost in <10 kD molecular weight fraction of ACM (Figure 3 A,B,C). Most importantly, the potent kinase inhibitor staurosporine, inhibited not only PKC translocation and MARCKS protein phosphorylation, but also completely blocked the ACM-induced increase in [^{3}H]thymidine incorporation (Figure 3A,B,C). Collectively, these results strongly suggest that protein-like astrocyte-secreted mitogens utilize a PKC-signaling pathway(s) to initiate a proliferative response in neonatal RCEC *in vitro*.

The exact mechanism by which the transient activation of PKC in RCEC translates into DNA synthesis and eventually cell proliferation is not known. One possibility is that ACM effects a direct nuclear translocation of PKC and subsequent phosphorylation of laminin B [4], a step necessary for the entry of cells into S-phase of cell cycle [4]. Alternatively, PKC is known to activate the downstream mitogen-activated (MAP) kinase cascade [4, 9], or may phosphorylate proteins that are subsequently translocated to the nucleus [4], resulting in stimulation of DNA synthesis.

The nature of the astrocyte-derived mitogen(s) which stimulate PKC in RCEC remains to be determined. It is known that astrocytes secrete a wide range of growth factor-like substances some of which (i.e, fibroblast growth factor, insulin-like growth factor(s), vascular endothelial growth factor) [10, 15], are known angiogenic factors. However, these growth factors have been shown to primarily activate tyrosine-kinase receptors and none of them were able to stimulate PKC in RCEC in our hands (data not shown).

In summary, the transient activation of PKC appears to be an essential early event required for DNA replication in RCEC in response to astrocyte-derived mitogen(s). The phosphorylation of cellular PKC substrates such as the MARCKS protein, implicated in a multitude of cellular processes linked to cytoskeletal rearrangements, cell motility, cell cycle regulation and transformation [1], may be involved in the reorganizations of the microtubular network(s) by which endothelial cells dislodge and migrate away from the parent microvessel and proliferate.

IV. REFERENCES

1. Aderem, A.(1992) The MARCKS brothers: a family of protein kinase C substrates. Cell, 71: 713-716.
2. Arthur, F.E., Shivers, R.R., and Bowman, P.D. (1987) Astrocyte-mediated induction of tight junctions in brain capillary endothelium: an efficient in vitro model. Brain Res., 36: 155-159.
3. Beck, D.W., Roberts, R.L., and Olsom, J.J. (1986) Glial cells influence membrane-associated enzyme activity at the blood brain barrier. Brain Res., 381: 131-137.
4. Buchner, K. (1995) Protein kinase C in the transduction of signals toward and within the cell nucleus. Eur. J. Biochem., 228:211-221.
5. Cancilla, P.A., Bready, J., and Berliner, J. (1993) Brain endothelal-astrocyte interactions. In. The Blood Brain Barrier Cellular and Molecular Biology (W. M. Pardridge, Ed.), pp. 25-46, Raven Press, New York.
6. Chakravarthy, B.R., Bussey, A., Whitfield, J.F., Sikorska, M., Williams, R.E., and Durkin J.P. (1991) The direct measurement of protein kinase C in isolated membranes using a selective peptide substrate. Anal. Biochem., 196:144-150.
7. Chakravarthy, B.R., Isaacs, R.J., Morley, P., Durkin, J.P., and Whitfield, J.F.(1994) Stimulation of protein kinase C during Ca^{2+}-induced keratinocyte differentiation: selective blockade of MARCKS phosphorylation by calmodulin. J. Biol. Chem., 270: 1-7.

8. Diglio, C.A., Grammas, P., Giacomelli, F., and Wiener, J. (1982) Primary culture of rat cerebral microvascular endothelial cells. Isolation, growth, and characterization. Lab. Invest., 46: 554-563.
9. Hug, H., and Sarre, T.F. (1993) Protein kinase C isoenzymes: divergence in signal transduction? Biochem. J. 291:329-343.
10. Ijichi, A., Sakuma, S., and Tofilion, P.J. (1995) Hypoxia-induced vascular endothelial growth factor expression in normal rat astrocyte cultures. Glia 14: 87-93.
11. Lattera, J., and Goldstein, G.W. (1993) Brain microvessels and microvascular cells in vitro. In: The Blood Brain Barrier (W. M. Pardridge, Ed.), pp. 1-24, Raven Press, New York.
12. Lowry, O.H., Rosebrough, N.U., Farr, A.L., and Randall, R.J. (1951) Protein measurement with the folin phenol reagent. J. Biol. Chem., 193: 265-275.
13. Maxwell, K., Berliner, J.A., and Cancilla, P.A. (1987) Induction of γ-glutamyl transpeptidase in cultured cerebral endothelial cells by a product released by astrocytes. Brain Res. 410: 309-314.
14. McCarthy, K.D. and DeVellis, J. (1980) Preparation of separate astroglial and oligodendroglial cell cultures from rat cerebral tissue. J. Cell. Biol. 85: 890-902.
15. Patel, A.J. and Gray, C.W. (1993) Neurotrophic factors produced by astrocytes involved in the regulation of cholinergic neurons in the central nervous system. In: Biology and Pathology of Astrocyte-Neuron Interactions (Fedoroff et al., eds), pp.103-115. Plenum Press, New York.
16. Robinson, P.J., Liu, J.-P., Chen, W., and Wenzel, T.(1993) Activation of protein kinase C *in vitro* and in intact cells or synaptosomes determined by acetic acid extraction of MARCKS. Anal. Biochem., 210: 172-178.

CALCIUM AND PROTEIN KINASE C SIGNALING IN RESPONSE TO VASOACTIVE PEPTIDES IN HUMAN CEREBROMICROVASCULAR ENDOTHELIAL CELLS

Danica B. Stanimirovic,[1] Paul Morley,[1] Edith Hamel,[2] Rita Ball,[1] Geoff Mealing,[1] and Jon P. Durkin[1]

[1] Cellular Neurobiology Group
Institute for Biological Sciences
National Research Council of Canada
Ottawa, Ontario K1A 0R6
[2] Montreal Neurological Institute
Montreal, Quebec, H3A 2B4

SUMMARY

Vasoactive peptides endothelin-1 (ET-1) and bradykinin (BK) were shown to induce immediate increases in intracellular calcium concentrations, $[Ca^{2+}]_i$, and endogenous phosphorylation of the protein kinase C (PKC)-specific substrate, 85 kD MARCKS protein, in human cerebromicrovascular endothelial cells (HBEC). The peptides-induced $[Ca^{2+}]_i$ surges were not affected by incubating the cells in Ca^{2+}-free medium or by pretreating them with Ca^{2+}-channel blocker D600 (50 μM). BK-stimulated $[Ca^{2+}]_i$ increases were completely inhibited, whereas ET-1-induced $[Ca^{2+}]_i$ increases were insensitive to ADP-ribosylation of G-proteins by pertussis toxin. Both peptides-triggered $[Ca^{2+}]_i$ surges ánd MARCKS protein phosphorylation were abolished by the inhibitor of inositol phospholipid hydrolysis, U-73122 (2.5-5 μM).

RÉSUMÉ

Il est démontré que les peptides vasoactifs, l'endotheline (ET-1) et la bradykinine (BK) induisent une augmentation rapide du calcium intracellulaire, $[Ca^{2+}]_i$, et une phosphorylation endogène du substrat spécifique de la protéine kinase C (PKC), soit la protéine MARCKS de 85 KD, chez les cellules endothéliales cérébromicrovasculaires humaines

Biology and Physiology of the Blood–Brain Barrier, edited by Couraud and Scherman
Plenum Press, New York, 1996

(HBEC). L'augmentation transitoire du $[Ca^{2+}]_i$ induite par ces peptides n'est pas dépendante du calcium extracellulaire et n'est pas affectée par le bloqueur de canaux calciques, le D-600 (500 μM). L'ADP-ribosylation de la protéine G induite par la toxine de pertussis, inhibe complètement l'augmentation du $[Ca^{2+}]_i$ stimulée par la BK et n'a aucun effet sur celle provoquée par l'ET-1. La variation du $[Ca^{2+}]_i$ produite par ces peptides, de même que la phosphorylation de la protéine MARCKS sont abolies par le U-73122 (2.5-5 μM), un inhibiteur de l'hydrolyse des phosphatidylinositols.

I. INTRODUCTION

Cerebral endothelial cells are joined together by an elaborate network of interendothelial tight junctions which are crucial in maintaining a strict blood-brain permeability barrier (BBB). The opening of tight junctions and the consequent increase in BBB permeability appears to be dependent on microtubular rearrangements induced by changes in intracellular calcium concentrations, $[Ca^{2+}]_i$ [8, 10]. Cerebral endothelial cells express a variety of Ca^{2+}-binding microtubule-associated proteins including F-actin, vimentin and caldesmon[10]. F-actin filaments have been shown to undergo a complex reorganization during cerebral endothelial cell proliferation, differentiation, and in response to hypoxic conditions[2].

Cerebral endothelial cells have also been shown to express receptors for various naturally occurring vasoactive and neuromodulatory peptides, such as endothelins [15], arginine-vasopressin[13], angiotensin II [13], and bradykinin [14]. Among these peptides, endothelin-1 (ET-1), a potent vasoconstrictor [9], has been implicated in the pathogenesis of cerebrovascular disorders such as hypertension and stroke [9], whereas bradykinin (BK) has been shown to induce relaxation of cerebral vessels and receptor-mediated permeabilization of the BBB [12].

In order to elucidate the potential role that early signals elicited by ET-1 and BK may play in functional responses of the cerebral vasculature, we investigated the impact these peptides have on $[Ca^{2+}]_i$ and endogenous PKC activity in intact human cerebromicrovascular endothelial cells (HBEC).

II. MATERIAL AND METHODS

II.I. Cell Culture

HBEC were isolated using a modification of the procedure of Gerhart et al (1988) [5] from dissociated microvessels and capillaries obtained from small samples of human temporal lobe excised surgically from patients being treated for idiopathic epilepsy. Cells were grown in 0.5% gelatin-coated tissue culture dishes in growth medium containing 65% M199, 10% fetal calf serum, 5% human serum, 20% murine melanoma cell-conditioned media, 5 μg/ml of insulin, 5 μg/ml of transferrin, 5 ng/ml selenium, and 10 μg/ml of endothelial cell growth supplement (Collaborative Biomedical Products, Bedford, MA) at 37°C in an atmosphere of 5% CO_2 in air. Endothelial cells derived by these procedures were positively identified by immunocytochemical staining for Factor VIII-related antigen and incorporation of acetylated low density lipoprotein.

II.II. $[Ca^{2+}]_I$ Measurement

To determine $[Ca^{2+}]_i$, cells were plated on 24-mm glass coverslips (Canadawide Scientific, Ottawa, Ontario, Canada) and incubated in 1 ml of growth medium at 37°C in an atmosphere of 5% CO_2 in air. Cells were used for the determination of $[Ca^{2+}]_i$ 2 to 4 days after plating. $[Ca^{2+}]_i$ was determined by measuring the fluorescence signal from the Ca^{2+}-sensitive indicator fura-2 [6]. Briefly, the cells were loaded with fura-2 by incubation for 30 min at 37°C in a normal buffer solution (NBS; 140 mM NaCl, 5 mM KCl, 2.5 mM $CaCl_2$, 1.1 mM $MgCl_2$, 2.6 mM dextrose and 10 mM HEPES) containing 2.5 mM fura-2/AM. Experiments were conducted at room temperature on single cells or small groups of four to eight cells in a custom-made coverslip holder fitted to the stage of a Zeiss IM inverted microscope (Carl Zeiss Canada, Don Mills, Ontario, Canada) coupled to a CM3 cation measurement spectrofluorimeter (Spex Inc., Newark, NJ). Measurements were performed using 350 and 380 nm excitation wavelengths alternating at a frequency of 1 Hz. The concentration of $[Ca^{2+}]_i$ was reflected in the ratio of the fluorescence intensities of fura-2 emission at 505 nm induced by the alternating excitation wavelengths (350 nm and 380 nm) according to the equation described by Grynkiewicz et al., 1985 [6].

II.III. Endogenous Phosphorylation of the 85-kD Marcks Protein

The effects of ET-1 and BK on protein kinase C (PKC) activation in intact cells were determined by the extent of endogenous phosphorylation of the specific PKC substrate, 85-kD MARCKS protein. MARCKS protein phosphorylation was measured in intact HBEC exposed to $[^{32}P]$-orthophosphate (9000 Ci/mmol; 100 μCi/ml) for 2 hours in serum-free M199 medium to label the intracellular ATP pool. Cells were stimulated with peptides for 15 min and the 85-kDa MARCKS protein was subsequently extracted with 40% acetic acid using a modification [3] of the method of Robinson et al. (1993) [11]. Equal amounts of extracted cellular proteins were loaded and electrophoretically separated on 10% sodium dodecyl sulphate-polyacrylamide gel (SDS-PAGE). MARCKS protein phosphorylation was visualized by autoradiography using Kodak XAR film.

III. RESULTS AND DISCUSSION

Both ET-1 and BK induced a rapid and transient increase in $[Ca^{2+}]_i$ in HBEC (Figure 1A). These $[Ca^{2+}]_i$ surges occurred predominantly through the release of Ca^{2+} from intracellular stores, since the initial $[Ca^{2+}]_i$ increases induced by both peptides were not reduced in cells incubated in Ca^{2+}-free medium containing 1 mM EGTA (Figure 1B), or in cells preincubated with the Ca^{2+} channel blockers D600 (Figure 1C), lanthanum (1 mM), or nickel (1 mM) (data not shown). On the other hand, the relatively slow rate at which $[Ca^{2+}]_i$ returned to resting levels following peptide stimulation was significantly accelerated in cells incubated in Ca^{2+}-free medium, but not in the presence of the voltage-dependent Ca^{2+} channel (VDCC) blocker, D600 (Figure 1A,B,C). These results suggest that this secondary phase of the $[Ca^{2+}]_i$ response to both peptides was likely due to capacitative extracellular Ca^{2+} entry through non-VDCC routes. We have recently suggested [14] that rat cerebromicrovascular EC may lack dihydropyridine-sensitive VDCC, since the specific activator of these channels, Bay K-8644, as well as depolarization by KCl had no effect on ^{45}Ca uptake, and ET-1-induced ^{45}Ca entry in these cells could not be inhibited by L-type VDCC blockers.

Both peptide-induced $[Ca^{2+}]_i$ transients and inositol tri-phosphate (IP_3) increases (data not shown) were completely inhibited by the phospholipase C (PLC) inhibitor, U-73122 (Figure 2A), indicating that ET-1 and BK mobilize Ca^{2+} from intracellular stores in HBEC by activating a phosphoinositide-specific PLC. The ET-1- and BK-induced formation of IP_3 in rat and human cerebromicrovascular endothelial cells has been shown to occur through the activation of ET_A and B_2 receptors, respectively [4, 14, 15]. ADP

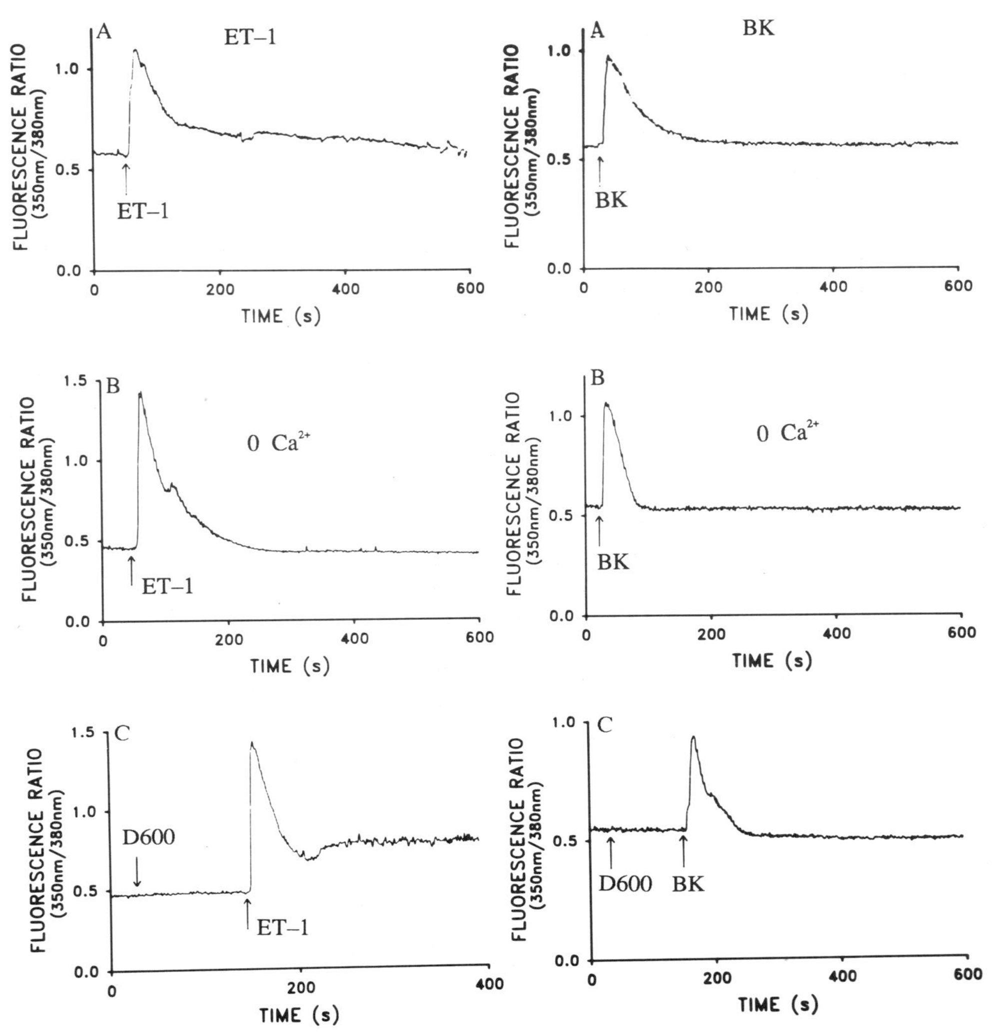

Figure 1. Changes in $[Ca^{2+}]_i$ induced by ET-1 and BK in HBEC. Cells were preloaded with fura-2 for 30 min and then A) stimulated with 20 nM ET-1 or 20 μM BK, B) exposed to 20 nM ET-1 or 20 μM BK in Ca^{2+}-free medium containing 1 mM EGTA, or C) pretreated with 50 μM of the voltage-dependent calcium channel blocker, D600, for 2 min before being stimulated with 20 nM ET-1 or 20 μM BK. Changes in $[Ca^{2+}]_i$ were followed as described in the Material and Methods.

ribosylation of G proteins by overnight pretreatment of HBEC with pertussis toxin (Ptx) did not influence ET-1-induced $[Ca^{2+}]_i$ surges (Figure 2B), whereas it completely abolished the $[Ca^{2+}]_i$ increase induced by BK in HBEC. Insensitivity of ET_A-receptor-mediated IP_3 formation to Ptx has previously been demonstrated in HBEC [15].

The data suggest that the respective ET-1 and BK receptors are coupled to different classes of G-proteins.

Both peptides also stimulated endogenous phosphorylation of 85-kD MARCKS protein (Figure 3). Since MARCKS protein is a specific PKC substrate not phosphorylated by other cellular kinases (cAMP- or calmodulin-dependent) [1,3], the results indicate that both

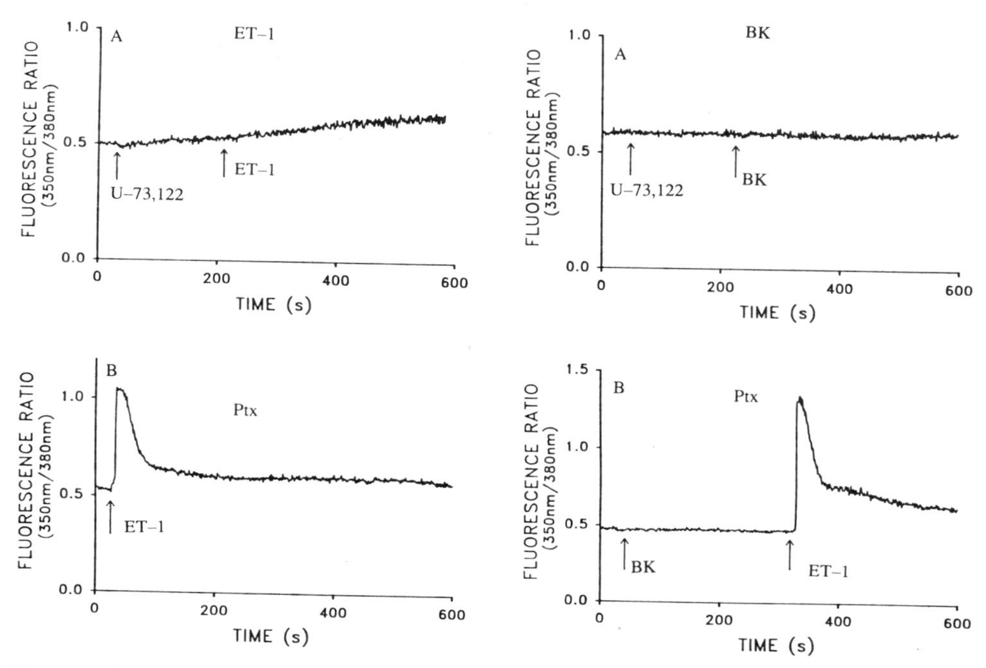

Figure 2. Effects of phospholipase C (PLC) inhibition and ADP- ribosylation on ET-1 and BK-induced $[Ca^{2+}]_i$ responses in HBEC. A) Fura-2 loaded cells were pretreated with 2.5 μM of the selective PLC inhibitor, U-73122, for 2.5 min before being stimulated by 20 nM ET-1 or 20 μM BK. B). Cells were pretreated with 500 ng/ml pertussis toxin (Ptx) for 16 hours before stimulation with 20 nM ET-1 or 20 μM BK at t=30 sec.

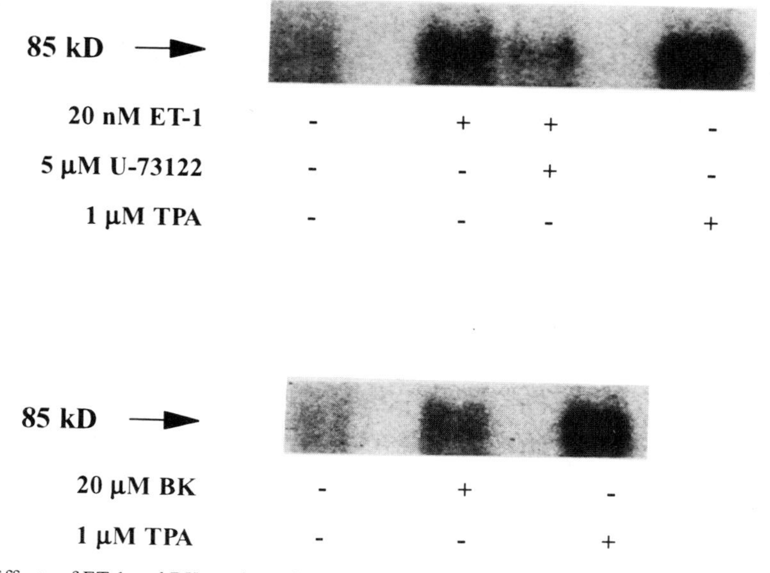

Figure 3. Effects of ET-1 and BK on the endogenous phosphorylation of 85-kD MARCKS protein in HBEC. Cells were exposed to 20 nM ET-1 or 20 μM BK in the absence or presence of 5 μM U-73122, or 1 μM of 12-O-tetradecanoyl phorbol 13-acetate (TPA), for 15 min. Endogenous phosphorylation of the MARCKS protein was determined by SDS-PAGE/autoradiography and quantified by densitometry as described in the Material and Methods.

ET-1 and BK activate PKC and stimulate the *in situ* phosphorylation of PKC-specific substrates in HBEC. The extent of MARCKS phosphorylation elicited by ET-1 was more pronounced than that induced by BK (Figure 3). ET-1-induced MARCKS protein phosphorylation was inhibited by U-73122 (Figure 3), suggesting that PKC is activated by diacylglycerol released by PLC.

Previous studies have demonstrated that PKC activation in cerebral endothelial cells facilitates fluid-phase endocytosis [7], stimulates secretion of vasoactive prostaglandins and endothelins [13], and causes rearrangement of endothelial cytoskeletal filaments [8]. However, the specific and functionally relevant PKC substrates in these cells have yet to be determined. In this study we have demonstrated that at least one specific PKC substrate, the 85-kD MARCKS protein, is present in HBEC and undergoes *in situ* phosphorylation in response to the naturally occurring vasoactive peptides ET-1 and BK. MARCKS is an acidic protein which binds to actin filaments, cross-linking them into rigid, membrane-attached entanglements [1]. MARCKS also binds calmodulin in the presence of Ca^{2+} [1]. It has been reported that the actin filament cross-linking activity of MARCKS is disrupted by phosphorylation and Ca^{2+}-calmodulin binding, resulting in the translocation of MARCKS from the membrane into the cytosol with a concomitant increase in actin plasticity [1]. Thus, it is possible that the observed ET-1- and BK-induced increases in $[Ca^{2+}]_i$ and MARCKS phosphorylation in HBEC may initiate changes in actin plasticity which are necessary to cause endothelial contraction/relaxation or permeabilization of tight junctions. However, it remains unclear whether the observed differences in Ca^{2+}- and PKC-signaling in HBEC by ET-1 and BK (i.e, different groups of G-proteins involved, more pronounced MARCKS protein phosphorylation by ET-1) could account for the contrasting vasomotor/functional effects of the two peptides observed *in vivo*. The functional responses of HBEC to these peptides are probably more specifically determined by signaling pathways activated downstream of PKC.

IV. REFERENCES

1. Aderem, A. (1992) The MARCKS brothers: a family of protein kinase C substrates. Cell 71: 713-716.
2. Augustin, H.G., Kozian, D.H., and Johnson, R.C. (1994) Differentiation of endothelial cells: Analysis of the constitutive and activated endothelial cell phenotypes. BioEssays 16: 901-906.
3. Chakravarthy, B.R., Isaacs, R.J., Morley, P., Durkin, J.P., and Whitfield, J.F. (1994) Stimulation of protein kinase C during Ca^{2+}-induced keratinocyte differentiation: selective blockade of MARCKS phosphorylation by calmodulin. J. Biol. Chem. 270: 1-7.
4. Doctrow, S.R., Abelleira, S.M., Curry, L.A. et al. (1994) The bradykinin analog RMP-7 increases intracellular free calcium levels in rat brain microvascular endothelial cells. J. Phramacol. Exp. Therap. 271:229-237.
5. Gerhart, D.Z., Broderius, M.A., and Drewse, L.R. (1988) Cultured human and canine endothelial cells from brain microvessels. Brain Res. Bull. 21:785-793.
6. Grynkiewicz, G., Poenie, M., Tsien, R.Y. (1985) A new generation of Ca^{2+} indicators with greatly improved fluorescence properties. J. Biol. Chem. 260: 3440-3450.
7. Guillot, F.L., and Audus, K.L. (1990) Angiotensin peptide regulation of fluid-phase endocytosis by primary cultures of bovine brain microvessel endothelial cell monolayers. J. Cerebral Blood Flow Metab. 10: 827-834.
8. Himmel, H.M., Whorton, A.R., and Straus, H.C. (1993) Intracellular calcium, currents, and stimulus-response coupling in endothelial cells. Hypertension 21:112-145.
9. Kobayashi, H., Hayashi, M., Kobayashi, M., Kabuto, M., Handa, Y., Kawano, H., and Ide, H. (1991) Cerebral vasospasm and vasoconstriction induced by endothelin. Neurosurgery 28: 673-679.
10. Lum, H., and Malik, A.B. (1994) Regulation of vascular endothelial barrier function. Am. J. Physiol. 267: L223-L241.
11. Robinson, P.J., Liu, J.-P., Chen, W., and Wenzel, T.(1993) Activation of protein kinase C *in vitro* and in intact cells or synaptosomes determined by acetic acid extraction of MARCKS. Anal. Biochem. 210: 172-178.

12. Schurer, L., Temesvari, P., Wahl, M., Unterberg, A., and Baethman, A. (1989) Blood-brain barrier permeability and vascular reactivity to bradykinin after pretreatment with dexamethasone. Acta Neuropathol. 77:576-581.
13. Spatz, M., Stanimirovic, D.B., Bacic, F., Uematsu, S., and McCarron, R.M. (1993) Relationship between endothelin-1 and prostanoids induced by vasoconstrictive peptides in human cerebromicrovascular endothelium. Am. J. Physiol. 266: C654-C660.
14. Stanimirovic, D.B., Nikodijevic, B., Nikodijevic, D., McCarron, R.M., and Spatz, M. (1994) Signal transduction and $^{45}Ca^{2+}$ uptake activated by endothelins in rat brain endothelial cells. Eur. J. Pharmacol. 288: 1-8.
15. Stanimirovic, D.B., Yamamoto, T., Uematsu, S., and Spatz, M. (1994) Endothelin-1 receptor binding and cellular signal transduction in cultured human brain endothelial cells. J. Neurochem. 62: 592-601.

THE EFFECT OF CONDITIONED MEDIUM OF PORCINE ASTROCYTES AND DIFFERENT CELL LINES ON THE ACTIVITY OF THE MARKER ENZYMES ALP AND γ-GT IN ENDOTHELIAL CELLS OF PORCINE BRAIN MICROVESSELS

Barbara Ahlemeyer, Sabine Matys, and Peter Brust

Research Center Rossendorf
Institute of Bioinorganic and Radiopharmaceutic Chemistry
PF 51 01 19, 01314 Dresden, Germany

SUMMARY

The conditioned media of subconfluent and confluent monolayers of porcine astrocytes and different cell lines were used to reinduce in primary cultured endothelial cells of porcine brain microvessels in vivo properties like high activities of the marker enzymes alkaline phosphatase (ALP) and γ-glutamyltranspeptidase (γ-GT). After 7 days in culture, the activity of ALP increased only by incubation of cerebral endothelial cells with conditioned media from confluent adult porcine astrocytes and CHO-cells (fibroblasts), whereas the activity of γ-GT was significantly increased by conditioned media of astrocytes, CHO-cells, chondrocytes and adenocarcinoma cells. However, the increase in γ-GT seems to be the result of an increase in the number of cells on the top of the monolayer with high γ-GT activities as revealed by histochemistry. The mechanism of the inductive effect on γ-GT seems to be not astrocyte specific and has to be clarified.

RÉSUMÉ

Des milieux de culture conditionnés par des monocouches confluentes et subconfluentes d'astrocytes de porc et de differentes lignées cellulaires ont été utilisées pour réinduire, dans des cultures primaires de microvaisseaux de porc, des propriétés de la barrière hémato-encéphalique in vivo telles les fortes activités enzymatiques de la phosphatase alcaline (PA) ou de la γ-glutamytranspeptidase (γ-GT). Après sept jours de culture, l'activité de la PA a pu être réinduite seulement par les milieux conditionnés par les cultures

Biology and Physiology of the Blood–Brain Barrier, edited by Couraud and Scherman
Plenum Press, New York, 1996

confluentes d'astrocytes de porc adulte et d'une lignée de fibroblastes (CHO). Une augmentation de l'activité de la γ-GT a pu, dans les mêmes conditions, être observée avec des milieux conditionnés d'astrocytes, de cellules CHO, de chondrocytes et de cellules d'adénocarcinome. Toutefois, l'augmentation de l'activité de la γ-GT semble être le resultat de la multiplication au dessus de la monocouche de cellules possédant une forte activité γ-GT, comme le montrent des expériences d'histochimie. L'effet inducteur des milieux de cultures sur l'activité de la γ-GT ne semble pas spécifique des astrocytes et reste à élucider.

1. INTRODUCTION

Endothelial cells of brain capillaries are the main constituents to form the blood-brain barrier (BBB) which restricts the transport between the blood and the extracellular fluid of the brain. It has been shown that cultured endothelial cells possess some features of the BBB in vivo such as different transporters, ion channels, receptors and enzymes[1]. However, in contrast to in vivo, the cellular permeability increased and the activity of the marker enzymes alkaline phosphatase (ALP) and γ-glutamyltranspeptidase (γ-GT) decreased as a function of culture time [2,3]. Therefore, several attempts have been made to improve this in vitro model e.g. by coculturing endothelial cells with astrocytes or by using astrocyte-conditioned medium. While some authors discussed that only a direct cell-cell contact is necessary to induce BBB features [4-9], some others have found that conditioned medium with and without astrocyte-derived matrix is already sufficient [10-16]. In the case of using conditioned media, a heterologous model was used, e.g. the astrocytes of neonatal rats or C6 rat glioma cells have been shown to increase electrical resistance or the activity of ALP and γ-GT in adult endothelial cells of capillaries or arteries from different species or in endothelial cell lines [10-16]. In addition, the results are difficult to interpret because some authors measured the increase in ALP, whereas others measured the increase in γ-GT or the electrical resistance.

Therefore, we tried to find out whether conditioned media of astrocytes from adult porcine brain (autologous model) and of some other proliferating cells are able to induce ALP as well as γ-GT in endothelial cells of adult porcine brain microvessels.

2. METHODS

Primary cultures of porcine brain microvessel endothelial cells were obtained according to Mischek et al. [17]. The cells were seeded on collagen type I/III coated petri dishes with a density of 1-2 x 10^4 cells/cm2 and were cultured for 7 days in medium 199 with 10% fetal calf serum, 100 U/ml penicillin, 100 µg/ ml streptomycin and 2.5 µg/ml amphotericin at 37°C with 5% CO_2 and 95% air. These primary cultures were confirmed to be endothelial cells > 95% by immunostaining method using factor VIII related antigen after 6 days in culture.

Astrocytes were derived from the white matter of adult porcine brain and were cultured in the same medium as the endothelial cells to avoid effects of medium components by the use of astrocyte-conditioned medium. The cells were identified to be 99% astrocytes by immunostaining using GFAP antigen. The cell lines CHO (chinese hamster ovary fibroblasts), Sp.LOE (mouse osteochondroma) and KTCTL-2 (human adenocarcinoma) were from the Deutsches Krebsforschungszentrum Heidelberg, Germany (provided by Dr. Gudrun Kampf). All cell lines were cultured in medium 199, supplemented with 10% fetal calf serum and antibiotics. The conditioned media of all these cells were collected 72 h after adding the fresh media to subconfluent and confluent cultures. Endothelial cells were incubated with a mixture of conditioned medium and fresh medium (1:2) 24 h after seeding until day 7.

In addition, the endothelial cells were cultured on the matrices of astrocytes, chondrocytes and CHO-cells from day 0 until day 7. ALP and γ–GT activities were determined as described by Meyer et al. [18].

3. RESULTS AND DISCUSSION

The activity of ALP increased 1.8 and 2.3-fold by incubation of cerebral endothelial cells after 7 days with conditioned medium from confluent adult porcine astrocytes and CHO-cells, respectively. Forskolin, which stimulates of cell differentiation, as well as the conditioned media of adenocarcinoma cells and chondrocytes had no effect. In contrast, the activity of γ-GT was significantly increased by conditioned medium of all the cells, e.g. from astrocytes, CHO-cells, chondrocytes and adenocarcinoma cells. The matrices of CHO-cells and chondrocytes inhibited cell growth and decreased the activity of γ-GT (all results Tab.1).

The inductive effects of porcine astrocyte-conditioned media on ALP and γ-GT is lower than those found by other authors in bovine endothelial cells, with 4-fold increases in ALP [15] and γ-GT [16]. However, the authors used conditioned media of rat C6-glioma cells and from neonatal rat astrocytes in combination with astrocyte-derived matrix, respectively to induce enzyme activities in cultured artery endothelial cells. It may be suggested that these differences are the reasons for the higher induction found by other authors [15,16].

Using conditioned media of confluent astrocytes and different cell lines, we observed that the number of cells on the top of the monolayer was increased after reaching confluence compared with control cultures. The activity of γ-GT in the cells on the top of the monolayer is found to be as high or higher than in cells on the bottom of the petri dish. These cells are in part endothelial cells (factor VIII-positive) and also pericytes/smooth muscle cells (actin-positive). Similar observations of cells on the top of the monolayer with high γ-GT

Table 1. The effect of conditioned media from confluent and subconfluent cultures of porcine astrocytes (ACM-C and ACM-SC), CHO-fibroblasts (CHOCM-C and CHOCM-SC), chondrocytes (SCM-C and SCM-SC) and adenocarcinoma cells (NCM-C and NCM-SC) and of different matrices of astrocytes (Ax), chondrocytes (Sx) and CHO-cells (CHOx) on the activities of ALP and γ-GT in cultured cerebral endothelial cells after 7 days in culture

	ALP [U/mg protein]	γ-GT [U/mg protein]	n
control	0.28 ± 0.06	0.98 ± 0.52	24
ACM-SC	0.43 ± 0.17	1.91 ± 0.65***	16
ACM-C	0.51 ± 0.29*	1.94 ± 1.06**	16
forskolin (50μM)	0.20 ± 0.02	0.63 ± 0.17	8
CHOCM-SC	0.48 ± 0.33	1.09 ± 0.81	12
CHOCM-C	0.65 ± 0.37*	1.64 ± 0.43*	12
SCM-SC	0.33 ± 0.18	1.53 ± 0.55	12
SCM-C	0.32 ± 0.20	2.02 ± 0.65***	12
NCM-SC	0.27 ± 0.03	2.02 ± 0.65***	8
NCM-C	0.30 ± 0.05	1.97 ± 0.43***	8
AX	0.21 ± 0.04	1.51 ± 0.40	8
CHO-X	0.23 ± 0.09	0.36 ± 0.18***	8
S-X	0.26 ± 0.08	0.27 ± 0.34**	8

All values are given as means ± S.D. of n experiments. Statistical significance of differences between control and treated groups * $p< 0.05$, ** $p< 0.01$, *** $p<0.001$

activities were already made by other authors [6,19,20]. We used conditioned media of subconfluent and confluent cultures to find out whether the induction of marker enzymes depend on the concentration of soluble factors (in subconfluent cultures a lower enrichment of these factors is suspected). As shown in Tab.1 conditioned media of confluent cultures stronger increased γ-GT staining than that of subconfluent cultures.

Therefore it may be suggested that soluble factors released by proliferating cells a) increase the γ-GT in all the endothelial cells or b) increase the number of endothelial cells on the top of the monolayer with high γ-GT activities or c) increase the number of contaminating cells which then increase the γ-GT in endothelial cells by direct cell-cell contact. Therefore, by using conditioned medium to improve BBB features in endothelial cells, attention has to be made whether the morphology and composition of the cell culture has been changed. The mechanism of this phenomenon has to be clarified.

4. CONCLUSIONS

1. The conditioned media from confluent porcine astrocytes and other cell lines increased the activities of the marker enzymes ALP and γ-GT in cerebral porcine endothelial cells after 7 days in culture.
2. The increase in γ-GT by conditioned media from confluent astrocytes and other cell lines seems to be the result of an increase in the number of cells on the top of the monolayer with high γ-GT activities as revealed by histochemistry.
3. The mechanism of this inductive effect seems to be not astrocyte-specific and has to be clarified.

5. ACKNOWLEDGMENT

This work was supported by a research grant from the Saxon State Ministry of Science and Arts (Grant 7541.82-FZR/309).

6. REFERENCES

1. Joó, F., The cerebral microvessels in culture, an update, *J. Neurochem.* 58:1, 1992.
2. Fukushima, H., Fujimoto, M., and Ide, M., Quantitative detection of blood-brain barrier-associated enzymes in cultured endothelial cells of porcine brain microvessels, *In Vitro Cell. Dev. Biol.* 26:612, 1990.
3. Meyer, J., Mischek, U., Veyhl, M., Henzel, K., and Galla, H.J., Blood-brain barrier characteristics enzymatic properties in cultured brain capillary endothelial cells, *Brain Res.* 514:305, 1990.
4. DeBault, L.E., and Cancilla P.A., γ-Glutamyltranspeptidase in isolated brain endothelial cells: induction by glial cells in vitro, *Science* 207:653, 1980.
5. Tao-Cheng, J.H., Nagy, Z., and Brightman, M.W., Tight juctions of brain endothelium in vitro are enhanced by astroglia, *J. Neurosci.* 7: 3293, 1987.
6. Laterra, J., Guerin, C., and Goldstein, G.W., Astrocytes induce neural microvascular endothelial cells to form capillary-like structures in vitro, *J. Cell. Physiol.* 144:204, 1990.
7. Meyer, J., Rauh, J., and Galla, H.J., The susceptibility of cerebral endothelial cells to astroglial induction of blood-brain barrier enzymes depends on their proliferative state, *J. Neurochem.* 57:1971, 1991.
8. Tontsch, U., and Bauer, H.C., Glial cells and neurons induce blood-brain barrier related enzymes in cultured cerebral endothelial cells, *Brain Res.* 539:247, 1991.
9. Hurwitz, A.A., Berman, J.W., Rashbaum, W.K., and Lyman, W.D., Human fetal astrocytes induce the expression of blood brain barrier specific proteins by autologous endothelial cells, *Brain Res.* 625:238, 1993.

10. Arthur, F.E., Shivers, R.R., and Bowman P.D., Astrocyte-mediated induction of tight junctions in brain capillary endothelium: an efficient in vitro model, *Dev. Brain Res.* 36:155, 1987.
11. Maxwell, K., Berliner, J.A., and Cancilla, P.A., Induction of γ-glutamyltranspeptidase in cultured cerebral endothelial cells by a product released by astrocytes, *Brain Res.* 410: 309, 1987.
12. Rubin, L.L., Hall, D.E., Porter, S., Barbu, K., Cannon, C., Horner, H.C., Janatpour, M., Liaw, C.W., Manning, K., Morales, J., Tanner, L.I., Tomaselli, K.J., and Bard, F., A culture model of the blood-brain barrier, *J. Cell Biol.* 115:1725, 1991.
13. Raub, T.J., Kuentzel, S.L., and Sawada, G.A., Permeability of bovine brain microvessel endothelial cells in vitro: barrier tightening by a factor released from astroglioma cells, *Exp. Cell. Res.* 199:330, 1992.
14. Wolburg, H., Neuhaus, J., Kniesel, U., Krauß, B., Schmid, E.M., Öcalan, M., Farrell, C., and Risau, W., Modulation of tight junction structure in blood-brain barrier endothelial cells, *J. Cell Sci.* 107:1347, 1994.
15. Takemoto, H., Kaneda, K., Hosokawa, M., Ide, M., and Fukushima, H., Conditioned media of glial cell lines induce alkaline phosphatase activity in cultured artery endothelial cells, *FEBS Lett.* 350:99, 1994.
16. Mizuguchi, H., Hashioka, Y., Fujii,A., Utoguchi, N., Kubo, K., Nakagawa, S., Baba, A., and Mayumi, T., Glial extracellular matrix modulates γ-glutamyltranspeptidase activity in cultured bovine brain capillary and bovine aortic endothelial cells, *Brain Res.* 651:155, 1994.
17. Mischek, U., Meyer, J., and Galla, H.J., Characterization of γ-glutamyl transpeptidase activity of cultured endothelial cells from porcine brain capillaries, *Cell. Tissue Res.* 256:221, 1989.
18. Meyer J. , Mischek, U., Vehyl, M., Henzel, K., and Galla, H.J., Blood-brain barrier characteristic enzymatic properties in cultured brain capillary endothelial cells. *Cell Tissue Res.* 256: 221, 1989.
19. Roux, F., Durieu-Trautmann O., Chaverot, N., Claire, M., Mailly, P., Bourre, J.M., Strosberg, A.D., and Couraud, P.O., Regulation of gamma-glutamyltranspeptidase and alkaline phosphatase activities in immortalized rat brain microvessel endothelial cells, *J. Cell. Physiol.* 159:101, 1994.
20. Wang, B.L., Grammas, P., and DeBault, L.E., Characterization of a γ-glutamyltranspeptidase positive subpopulation of endothelial cells in a spontaneous tube-forming clone of rat cerebral resistance-vessel endothelium, *J. Cell. Physiol.* 156:531, 1993.

39

HYPOXIA AUGMENTS NA^+-K^+-CL^- COTRANSPORT ACTIVITY IN CULTURED BRAIN CAPILLARY ENDOTHELIAL CELLS OF THE RAT

Nobutoshi Kawai, Richard M. McCarron, and Maria Spatz

Stroke Branch, NINDS, NIH
Bethesda, Maryland 20892

I. SUMMARY

The effect of hypoxia on K^+ uptake activity in cultured rat brain capillary endothelial cells (RBEC) was investigated by using $^{86}Rb^+$ as a tracer for K^+. Exposure of RBEC to hypoxia (95% N_2/5% CO_2, 24 hr) reduced Na^+,K^+-ATPase activity by 39%, whereas it significantly increased Na^+-K^+-Cl^- cotransport activity by 48%. Exposure of RBEC to oligomycin, a metabolic inhibitor, led to a complete inhibition of Na^+,K^+-ATPase and a coordinated increase of Na^+-K^+-Cl^- cotransport activity up to 2-fold. Oligomycin also increased the rate of K^+ efflux. The reduction of oligomycin-augmented Na^+-K^+-Cl^- cotransport activity by protein-tyrosine kinase inhibitors (genistein, 50 μM; herbimycin A, 10 μM) and the ineffectiveness of inhibitors of either protein kinase C (bisindolylmaleimide, 500 nM) or protein kinase A (H8, 20 μM) indicate the involvement of protein-tyrosine phosphorylation in this event. The data suggest that under hypoxic conditions when Na^+,K^+-ATPase activity is reduced, RBEC have the ability to increase K^+ uptake through activation of Na^+-K^+-Cl^- cotransport.

I. RÉSUMÉ

Nous avons étudié l'effet de l'hypoxie sur la capture du potassium par des cultures de cellules endothéliales de capillaires cérébraux de rat en utilisant le radioélement 86 Rb+ comme traceur du potassium. L'incubation des cultures en milieu hypoxique (95% N2/ 5% CO2) pendant 24 heures réduit l'activité du transporteur Na K ATP- dépendant de 39% tandis qu'elle augmente le cotransport Na,K,Cl de 48%. Le traitement des cellules par l'oligomycine, un inhibiteur métabolique, conduit à une complète inhibition du transporteur Na,K et une augmentation simultanée de 100% du cotransport Na, Cl, K. L'oligomycine augmente aussi l'efflux de potassium. L'induction de l'activité du cotransport Na, K, Cl peut être

Biology and Physiology of the Blood–Brain Barrier, edited by Couraud and Scherman
Plenum Press, New York, 1996

inhibée par des inhibiteurs de protéine tyrosine kinase (génistéine, 50μM; herbimycine A, 10 μM) mais pas par des inhibiteurs de protéine kinase C (bisindolylmaléimide, 500 nM) ou de protéine kinase A (H8, 20 μM). Ceci semble montrer que la phosphorylation de protéines par des protéine kinases intervient dans l'induction de ce cotransporteur. ces résultats suggèrent que, soumises à des conditions d'hypoxie, les cellules endothéliales cérébrales ont la possibité de compenser la diminution de l'activité du transporteur Na K par l'induction de l'activité du cotransporteur Na K Cl.

II. INTRODUCTION

It has been generally accepted that the movement of ions and water across the blood-brain barrier (BBB) is regulated by specific ion-transport systems localized in brain capillary endothelial cells, the main constituent of the BBB[1]. Na^+-K^+-ATPase is one of these ion-transport systems responsible for maintaining ionic homeostasis in the brain[1-3]. Another ion-transport system expressed in cerebral capillary endothelium is Na^+-K^+-Cl^- cotransport[4-6] which regulates cell volume, maintains endothelial cell integrity, and may participate in active passage of salt and water across the BBB. Although disturbances in the capillary Na^+-K^+-ATPase activity have been proposed as a mechanism for the formation of ischemic brain edema[7-9], other ion-transport systems at the BBB might also be involved in altering the water-electrolyte homeostasis under pathologic condition. The aim of this study was to investigate the effect of hypoxia on the ion-transport systems in cultured rat brain capillary endothelial cells (RBEC). The data suggest that under hypoxia, which leads to depletion of cellular ATP and reduction in Na^+,K^+-ATPase activity, RBEC retain the capacity to increase K+ uptake through Na^+-K^+-Cl^- cotransport.

III. EXPERIMENTAL METHODS

III.I. Cell Culture

A modified technique of Spatz et al.[10] was used for the isolation and cultivation of brain capillary endothelium derived from adult rat. At least two different cell lines were used in all experiments.

III.II. Hypoxic Conditions Were Made by Two Techniques. 1, oxygen was depleted by incubating RBEC in an atmosphere of 95% N_2/5% CO_2 for 24 hr; and 2, RBEC were treated with oligomycin (an inhibitor of mitochondrial ATP production, 1 μg/ml) for indicated periods of time.

III.III. Endothelial Cell ATP Content. was determined enzymatically as described by Lowry and Passonneau[11].

III.IV. $^{86}Rb^+$ Uptake. (as a marker for K^+ uptake) was measured as previously described[6]. In brief, RBEC grown in 96-well microtiter plates were incubated with $^{86}Rb^+$ (0.2 μCi/well) in 120 μl of Medium-199 (M199) for 10 min. After washing with M199 and solubilizing in 1% Triton X-100, the radioactivity was counted. K^+ uptake (expressed as nmol/mg protein/min) was calculated from K^+ content in the incubation buffer and the ratio of $^{86}Rb^+$ uptake into RBEC. Ouabain-sensitive and bumetanide-sensitive K^+ uptake was defined as Na^+-K^+-ATPase and Na^+-K^+-Cl^- cotransport activity, respectively.

Table 1. Effect of hypoxia on ATP content and K+ uptake in RBEC

Treatment	ATP content (nmol/mg pro protein)	K^+ uptake (nmol/mg protein/min)		
		Total	Na^+,K^+-ATPase	Na^+-K^+-Cl^- cotransport
Normoxia	13.7 ± 0.6	12.3 ± 0.7	4.6 ± 0.6	5.5 ± 0.5
Hypoxia	7.9 ± 1.4 **	14.9 ± 0.8*	2.8 ± 0.5*	8.1 ± 0.7**

Data are means ± SEM of six independent determinations performed in quadruplicate for ATP assays and of eight independent determinations performed in triplicate for K+ uptake studies. *, ** Values differ significantly from normoxia at $p < 0.05$ and $p < 0.01$, respectively, by Student's t-test.

III.V. $^{86}Rb^+$ Efflux. was determined in RBEC grown in 96-well microtiter plates preloaded with $^{86}Rb^+$ (0.1 μCi/well) in 80 μl M199 and equilibrated for 2 hr. Oligomycin (1 μg/ml, 30 min) or vehicle treated cells were rapidly washed and incubated in 100 μl M199 containing the same concentration of oligomycin or vehicle for indicated periods of time. The radioactivity in the supernatant and the cells was determined and the K^+ efflux rate was expressed as % of $^{86}Rb^+$ released into the supernatant.

IV. RESULTS AND DISCUSSION

The K^+ uptake into RBEC of control group (12.3 ± 0.7 nmol/mg protein/min) was sensitive to both ouabain (4.6 ± 0.6 nmol/mg protein/min) and bumetanide (5.5 ± 0.5 nmol/mg protein/min), respectively, indicating the existence of both Na^+-K^+-ATPase and Na^+-K^+-Cl^- cotransport systems in RBEC (Table). After 24 hr exposure of RBEC to hypoxia, ATP content was reduced by approximately 40% and Na^+,K^+-ATPase activity was significantly decreased (Table). The total K^+ uptake into RBEC, however, was slightly increased due to a significant increases of Na^+-K^+-Cl^- cotransport activity.

To further clarify the response of Na^+-K^+-Cl^- cotransport system to hypoxia, oligomycin was used to simulate ATP levels observed in hypoxic condition. Treatment of the RBEC with oligomycin decreased cellular ATP content (approximately by 65% after 30 min) in a time-dependent manner. No further reduction of ATP was observed after additional 30 min incubation (data not shown). The preserved cellular ATP content under these conditions was likely due to anoxic glucose metabolism since further reduction in cellular ATP level (by approximately 90%) was observed by treatment with oligomycin and 2-deoxy-D-glucose (20 mM) or with oligomycin alone in absence of glucose (data not shown). Oligomycin treatment almost completely inhibited Na^+,K^+-ATPase activity after 30 min preincubation (Figure A). This effect of oligomycin may be due to an ATP-depletion as well as a direct inhibition of Na^+,K^+-ATPase[12]. Despite a significant decrease of cellular ATP and a complete inhibition of Na^+,K^+-ATPase activity, the total cellular K^+ uptake rate into RBEC was significantly greater in oligomycin-treated cells than in control cells (Figure A). This increase in K^+ uptake rate was due to a marked increase of Na^+-K^+-Cl^- cotransport activity, consistent with the results obtained from RBEC exposed to 24 hr hypoxia.

After the equilibration, the amount of $^{86}Rb^+$ in the control and oligomycin-treated (1 μg/ml, 30 min) RBEC were not significantly different (4.1 ± 0.3 and 4.0 ± 0.4, x 10^2 mean dpm/μg protein, respectively; n = 5). The rate of K^+ efflux from oligomycin-treated RBEC was significantly greater than that of control (Figure B). Interestingly, preliminary studies showed that the increased K^+ efflux by oligomycin-treatment was abolished by 20 μM bumetanide, suggesting that oligomycin also increases efflux of K^+ possibly through Na^+-K^+-Cl^- cotransport. (data not shown).

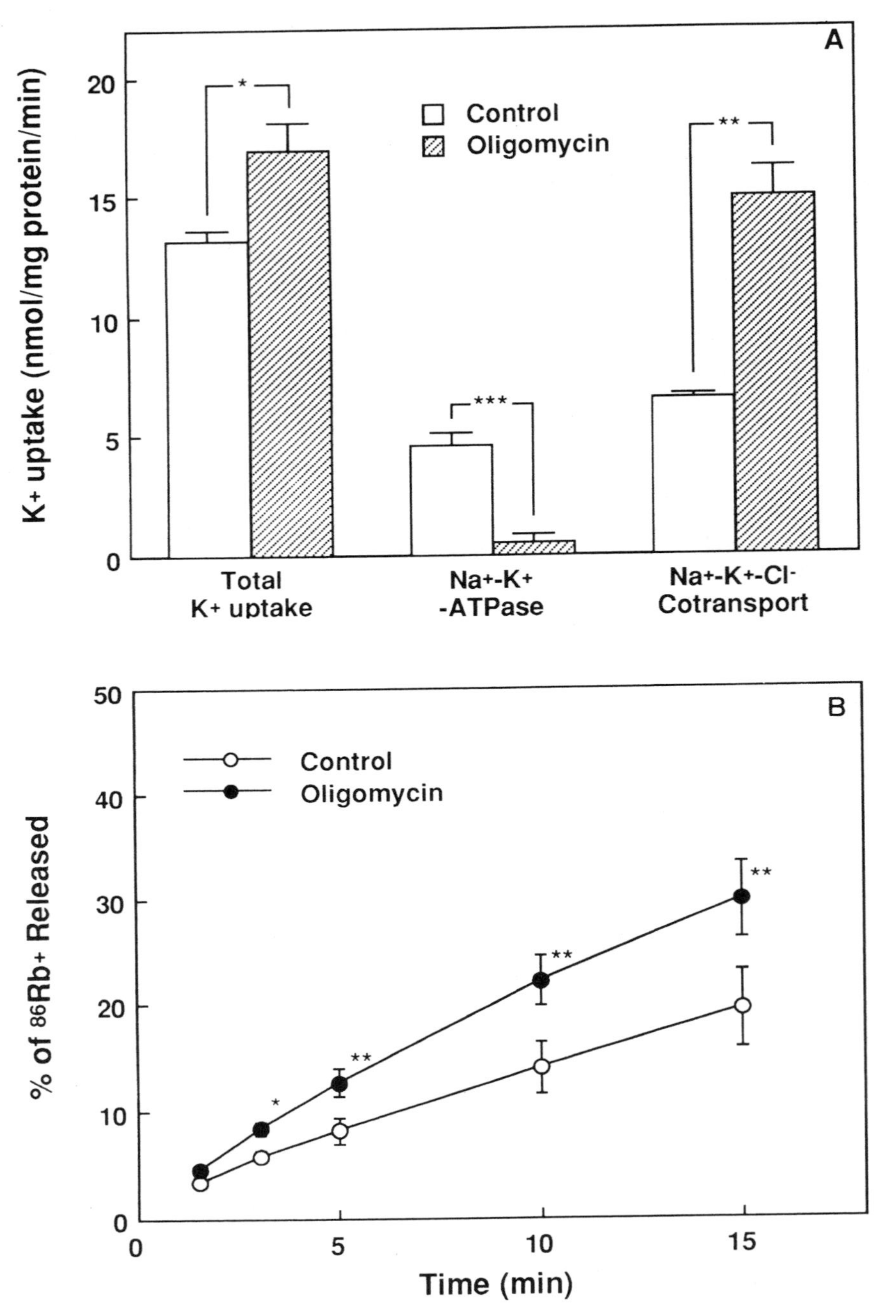

Figure 1. Effect of oligomycin on K^+ transport in RBEC. A Effect of oligomycin on K^+ uptake in RBEC. Data are means ± SEM of four independent determinations performed in triplicate. *, **, *** Values differ significantly from control at $p < 0.05$, $p < 0.01$, $p < 0.001$, respectively, by Student's t-test. B Effect of oligomycin on K+ efflux in RBEC. K^+ efflux rates are expressed as % of $^{86}Rb^+$ released into supernatant as described in method section. Data represent means ± SEM of four independent determinations performed in triplicate. *,** Values differ significantly from control at $p < 0.05$ and $p < 0.01$, respectively, by Student's t-test.

The precise mechanism by which hypoxia and oligomycin-treatment augment Na^+-K^+-Cl^- cotransport activity is presently unknown. Regulation of the cotransporter protein by direct phosphorylation, which may change the number of active cotransporters or any of the rate-limiting steps in the turnover cycle of the cotransporter[13,14], has been reported. In the present study, oligomycin-augmentation of Na^+-K^+-Cl^- cotransport activity was significantly reduced by protein-tyrosine kinase inhibitors, 50 μM genistein (by 72%) and 10 μM herbimycin A (by 48%), but was not affected by inhibitors of protein kinase C (bisindolyl-maleimide 500 nM) or protein kinase A (H8, 20 μM). Treatment with protein-tyrosine kinase inhibitors also completely inhibits endothelin-1-induced stimulation of Na^+-K^+-Cl^- cotransport activity in RBEC[5]. These data suggest that increased Na^+-K^+-Cl^- cotransport activity under hypoxia may involve protein-tyrosine phosphorylation that changes the number or the turnover cycle of cotransporter protein.

Many studies have shown that accumulation of tissue water (edema) accompanied by a net increase in Na+ and decrease in K+ content in the brain occurs during the early phase of cerebral ischemia prior to passive change in BBB permeability. It has been proposed that increased Na^+,K^+-ATPase activity in the brain capillary may participate in this disturbance of water-electrolyte homeostasis in the brain[5-7]. The present results demonstrate that RBEC exposed to hypoxia showed a reduced Na^+,K^+-ATPase activity but retained the capacity to increase K+ uptake through Na^+-K^+-Cl^- cotransport. Recently, Vigne et al. has suggested that both Na^+,K^+-ATPase and Na^+-K^+-Cl^- cotransport colocalized at the antiluminal side of the brain capillary may play a role in brain-to-blood K^+ transport[5]. Thus, augmentation of Na^+-K^+-Cl^- cotransport activity under hypoxic condition may provide a mechanism for the specific removal of K^+ from the ischemic brain. These results indicate the need for further studies to elucidate the role of Na^+-K^+-Cl^- cotransport system in pathophysiological responses of BBB.

V. REFERENCES

1. Betz A.L. and Goldstein G.W. Specialized properties and solute transport in brain capillaries. *Ann. Rev. Physiol.* 48: 241-250,1986.
2. Eisenberg H.M. and Suddith R.L. Cerebral vessels have the capacity to transport sodium and potassium. *Science* 206: 1083-1085,1979.
3. Goldstein G.W. Relation of potassium transport to oxidative metabolism in isolated brain capillaries. *J. Physiol.* 286: 185-195,1979.
4. Lin J.D. Effect of osmolarity on potassium transport in isolated cerebral microvesseles. *Life Sciences* 43: 325-333,1988.
5. Vigne P., Farre A.L., and Frelin C. Na^+-K^+-Cl^- cotransporter of brain capillary endothelial cells. *J. Biol. Chem.* 269: 19925-19930,1994.
6. Kawai N., Yamamoto T. Yamamoto H., McCarron R.M., and Spatz M. Endothelin 1 stimulates Na^+,K^+-ATPase and Na^+-K^+-Cl^- cotransport through ETA receptors and protein kinase C-dependent pathway in cerebral capillary endothelium. *J. Neurochem.* (In press).
7. Shigeno T., Asano T., Mima T., and Takakura K. Effect of enhanced capillary activity on the blood-brain barrier during focal cerebral ischemia in cats. *Stroke* 20: 1260-1266,1989.
8. Schielke G.P., Moises H.C., and Betz A.L. Blood to brain sodium transport and interstitial fluid potassium concentration during early focal ischemia of the rat. *J. Cereb. blood flow Metab.* 11: 466-471,1991.
9. Betz A.L., Keep R.F., Beer M.E., and Ren X. Blood-brain barrier permeability and brain concentration of sodium, potassium, and chloride during focal ischemia. *J.Cereb. Blood Flow Metab.* 14: 29-37,1994.
10. Spatz M., Bembry J., Dodson R.F., Hervonen H., and Murray M.R. Endothelial cell culture derived from isolated cerebral microvessels. *Brain Res.* 191: 577-582,1980.
11. Lowry O.H. and Passonneau J.V. A flexible system of enzymatic analysis. Academic Press, New York, 1972.
12. Esmann M. Oligomycin interaction with Na,K-ATPase: oligomycin binding and dissociation are slow processes. *Biochim. Biophys. Acta.* 1064: 31-36,1991.

13. Lytle C. and Forbush III B. The Na-K-Cl cotransport protein of shark rectal gland, II. Regulation by direct phosphorylation. *J. Biol. Chem.* 267: 25438-25443,1992.
14. Whisenant N., Khademazad M., and Muallem S. Regulatory interaction of ATP Na^+ and Cl^- in the turnover cycle of the NaK2Cl cotransporter. *J. Gen. Physiol.* 101, 889-908, 1993.

EFFECTS OF HISTAMINE ON THE ACID PHOSPHATASE ACTIVITY OF CULTURED CEREBRAL ENDOTHELIAL CELLS

Csilla A. Szabó,[1] István Krizbai,[1] Mária A. Deli,[1] Csongor S. Ábrahám,[2] and Ferenc Joó[1]

[1] Laboratory of Molecular Neurobiology
Institute of Biophysics
Biological Research Center
H-6701 Szeged, Hungary
[2] Department of Paediatrics
Albert Szent-Györgyi Medical University
H-6701 Szeged, Hungary

INTRODUCTION

Acid phosphatase (orthophosphoric monoester hydrolase, EC 3.1.3.2) is widely distributed in nature and has been studied thoroughly in both plants and animals (Hollander, 1971). The enzyme is activated in acidic environment, which is provided by the presence of H^+-ATPase, in different subcellular organelles of the exo- and endocytotic pathway (Mellman, Fuchs and Helenius, 1986). Consequently, the low pH provides favourable conditions for enzymatic hydrolyses in lysosomes; the proton gradient is used as an energy source for the coupled transport of biogenic amines, whereas the difference in pH between the endosome and the extracellular environment is used by the cell in receptor mediated endocytosis to provide symmetry to the recycling circuit between the two compartments (Mellman, Fuchs and Helenius, 1986).

Biochemical studies have revealed the presence of multiple molecular forms of acid phosphatase which differ in their molecular size, subcellular localization, sensitivity to inhibitors and substrate requirement (Rehkop and van Etten, 1975; Taga and van Etten, 1982). The high molecular weight form of enzyme (HMW, m.w.>100,000Da) can be completely inhibited by tartrate or fluoride and it is present mainly in the lysosomal fraction hydrolysing various phosphomonoesters, nonspecifically. The low molecular weight form (LMW, m.w.<20,000Da; tartate-resistent form) is present predominatly in the cytosolic fraction and the third form of APase activity can be detected in the presence of Zn2+ ions with a molecular weight of about 62,000Da (Shimohama and Fujimoto, 1993).

Acid hydrolase activity was first demonstrated in primary cultures of cerebral endothelial cells (CECs) by Baranczyk-Kuzma et al. (1989). However, it remained to be seen

Biology and Physiology of the Blood–Brain Barrier, edited by Couraud and Scherman
Plenum Press, New York, 1996

if the enzyme activity could be modified in or released from the CECs by vasoactive substances.

The present study was designed to check the possible effects of histamine on the activity of different molecular forms of acid phosphatase.

MATERIALS AND METHODS

Immortalized rat brain endothelial cells (RBE4 cell line; Roux et al. 1994) were cultured in 6 cm petri dishes and stimulated in 1 ml serum free DMEM (Dulbecco's modified Eagle's medium) for 1 hour in 37ºC with 10^{-5} M histamine with or without 10^{-6} M mepyramine (H_1-receptor blocker), 10^{-6} M ranitidine (H_2-receptor blocker) or in the presence of the antagonist only. The used concentations were selected on the basis of the affinity of agents to the receptors. After the incubation, the cells were washed with PBS (phosphate buffered saline), collected and homogenized in 500 µl PBS.

Acid phosphatase activity was measured by the rate of hydrolysis of p-nitrophenyl-phosphate (p-NPP) (Shimohama and Fujimoto, 1993). Different samples of the endothelial cells were incubated in solution containing 1 ml 0.1 M acetate buffer (pH 5.5) and 2.5 mM p-NPP at 37ºC for 1 hour with or without L-(+)tartrate (10 mM). After incubation, 0.3 ml of 1 M sodium hydroxyde was added to stop the reaction and the absorbance was read at 410 nm in a Shimadzu spectrophotometer. Enzyme activity expressed in mU/mg protein was determined using increasing concentrations of purified acid phosphatase with known activity (AcP Lin-trol, Sigma) and calibration curve. Protein concentration was determined according to Bradford using bovine albumin as a standard (Bradford, 1976).

Statistic analysis was performed using the Kruskal-Wallis one way analysis of variance on vanks followed by the Student-Newman-Kuels test, $p<0.05$ was considered significant difference.

RESULTS AND DISCUSSION

Histamine increased significantly the total APase activity (72.02±2.04 mU/mg protein vs. 60.24±4.79 mU/mg protein, mean±S.E.M., n=6, $p<0.05$) and tartrate-sensitive form (33.33±1.39 mU/mg protein vs. 26.41±5.24 mU/mg protein, mean±S.E.M., n=6, $p<0.05$) as well. The effect of histamine on total but not tartrate-sensitive activity could be inhibited by ranitidine (63.55±9.98 mU/mg protein, mean±S.E.M., n=6, $p<0.05$). The stimulatory effect of histamine on total acid phosphatase activity seemed, therefore, to be mediated by H_2-receptors. On the other hand mepyramine blocked the effect of histamine on tartrate-sensitive form (27.77±4.37 mU/mg protein, mean±S.E.M., n=6, $p<0.05$), only. Therefore, H_1-receptors are apparently involved in the regulation of activity of tartrate-sensitive isoforms. Mepyramine and ranitidine alone also increased the activity of enzyme which may be explained by the partial agonist effect of the drugs.

The precise role of the APase isoforms in cellular functions is still poorly understood. Changes in APase activity under different conditions have been demonstrated in several tissue types such as in gastric mucosa in response to indomethacine treatment (Nosalova and Navarova, 1994), in hepatocytes in response to insuline and cAMP (Stvolinskaya et al. 1992) or in brain in Alzheimer disease (Shimohama and Fujimoto, 1993). In cerebral endothelial cells approximately 52% of APase activity was associated with the microsome fraction and acid hydrolases were regarded potential factors in the endocytic pathway for transport of proteins though the blood-brain barrier (BBB) and as contributors to the BBB's enzymatic barrier function. Thus, histamine activation of acid phosphatase activity in the cerebral endothelial cells could possibly play a role during inflammatory reactions of the central nervous system.

ACKNOWLEDGMENT

The authors are grateful to Dr. P.O. Couraud (Laboratoire d'Immuno-Pharmacologie Moléculaire, CNRS UPR 0415, Université Paris VII ICGM, Paris, France) for providing the RBE4 cell line. The research was supported in part by the Hungarian Research Fund (OTKA T-14645, F-5207, F-12722, F-013104); Ministry of Public Welfare (ETT T-04 029/93) and U.S.-Hungarian Joint Fund (JFNo.392).

REFERENCES

Baranczyk-Kuzma, A., T.J. Raub, and K.L. Audus (1989) Demonstration of acid hydrolase activity in primary cultures of bovine brain microvessel endothelium. J.Cereb.Blood Flow Metab. *9*:280-289.

Bradford, M.M. (1976) A rapid and sensitive method for the quantitation of microgram quantities of protein utilizing the principle of protein dye binding. Anal. Biochem. *72*:248-254.

Hollander, V.P. (1971) P.D. Boyer (ed): The enzymes. New York: Academic, pp. 449-498.

Mellman, I., R.Fuchs and A.Helenius (1986) Acidification of the endocytotic and exocytotic pathways. Ann. Rev. Biochem. *55*:663-700.

Nosalova, V. and J. Navarova (1994) Indomethacine induced changes in mucosal lysosomal enzyme activity: effect of H2 antagonists. Agents Actions *41*:C95-C96.

Rehkop, D.M. and R.L. van Etten (1975) Human Liver Acid Phosphatases. Hoppe-Seyler's Z. Physiol. Chem. *356*:1775-1782.

Roux, F., O. Durieu-Trautmann, N. Chaverot, M. Claire, P. Mailly, J.M. Bourre, A.D. Strosberg, and P.-O. Couraud (1994) Regulation of gamma-glutamyl transpeptidase and alkaline phosphatase activities in immortalized rat brain microvessel endothelial cells. J. Cell Physiol. *159*:101-113.

Shimohama, S., S.Fujimoto (1993) Reduction of low-molecular-weight acid phosphatase activity in Alzheimer brains. Ann. Neurol. *33*:616-621.

Stvolinskaya, N., E. Poljakova, S. Nikulina, and B. Korovkin (1992) Effect of insulin on permeability of lysosome membrane in primary monolayer hepatocyte culture of newborn rats under anoxia conditions. Scand.J.Clin.Lab.Invest. *52*:791-796.

Taga, E.M. and R.L. van Etten (1982) Human Liver Acid Phosphatases: Purification and Properties of a Low-Molecular-Weight Isoenzyme. Arch. Biochem. Biophys. *214*:505-515.

41

A COMPARISON OF LYMPHOCYTE MIGRATION ACROSS THE ANTERIOR AND POSTERIOR BLOOD-RETINAL BARRIER *IN VITRO*

L. Devine, S. Lightman, and J. Greenwood

Department of Clinical Ophthalmology
Institute of Ophthalmology
University College London
Bath Street
London EC1V 9EL

1. SUMMARY

The traffic of lymphocytes into the neuroretina of the eye is an important aspect of the development of immune mediated diseases of this part of the central nervous system (CNS). At the interface between the systemic immune system and the parenchyma lies a blood-tissue barrier which in the eye is referred to as the blood-retinal barrier. This barrier is situated at two separate sites, one being composed of the vascular endothelium and the other of retinal pigment epithelium (RPE). The differential migration of lymphocytes across these two cellular barriers and the factors controlling this process are just beginning to be understood. We have shown that T cell line lymphocytes are capable of migrating through monolayers of both vascular endothelial and RPE monolayers in vitro. The principle adhesion molecules involved in this process being the LFA-1/ICAM-1 receptor-ligand pairing with VLA-4/VCAM-1 playing a role following cytokine activation of the monolayers.

1. RÉSUMÉ

Le trafic des lymphocytes dans la neurorétine de l'oeil est un aspect important du développement des maladies du système nerveux central (CNS) impliquant le système immunitaire. A l'interface entre le système immunitaire (circulation sanguine systémique) et le parenchyme se trouve une barrière sang-tissu qui au niveau de l'oeil porte le nom de barrière hémato-rétinienne. Cette barrière sépare l'endothélium vasculaire, d'une part et l'epithélium pigmenté de la rétine, d'autre part. Les différences de migration des lympho-

Biology and Physiology of the Blood–Brain Barrier, edited by Couraud and Scherman
Plenum Press, New York, 1996

cytes à travers ces deux barrières cellulaires, ainsi que des facteurs contrôlant ce processus commencent tout juste à être étudiés. In vitro, nous avons montré que des lignées lymphocytaires T sont capables de migrer à travers les monocouches de l'endothélium vasculaire et de l'epithélium pigmenté de la rétine. Les principales molécules d'adhésion impliquées dans ce processus au niveau de ces monocouches sont les complexes récepteur-ligand LFA-1/ICAM-1 et VLA-4/VCAM-1 après activation cytokines dépendante.

2. INTRODUCTION

The blood-retinal barrier (BRB) formed by the retinal vascular endothelium and the retinal pigment epithelium (RPE) plays an important role in the pathogenesis of ocular inflammatory disorders, such as posterior uveitis. The term uveitis, encompasses many clinical conditions but which refers to an inflammation of the uveal tract. It is a relatively common disorder occurring at a yearly rate of 20/100,000 in the UK (Forrester et al., 1990) and is now generally accepted that T cells play a dominant role in mediating posterior uveitis (Lightman and Chan, 1990; Lightman and Towler, 1992). Evidence for this comes from histological and immunohistochemical examination of eyes from patients with sympathetic ophthalmia, Vogt-Koyanagi (VKH) disease and birdshot chorioretinitis, where significant numbers of T cells were observed (Chan et al., 1988; Jakobiec et al., 1983) and the beneficial effects of cyclosporin in the clinical management of patients with uveitis. Moreover, as with experimental autoimmune encephalomyelitis (EAE) the experimental model for MS, the experimental model for uveitis, experimental autoimmune uveoretinitis (EAU), can be induced in naive rats by injection of $CD4^+$ T cells specific for the retinal antigen S-antigen (Caspi 1989). EAU has a number of other similarities with EAE (Calder and Lightman, 1992) and is characterised by destruction of the photoreceptors and a loss in integrity of the retinal layers.

One of the main problems associated with ocular inflammation is a dramatic increase in the permeability of the BRB which, although not being the central cause, does lead to the development of oedema and subsequent functional impairment of vision. There are a number of possible causes of this increase in permeability. Breakdown of the BRB in EAU occurs in association with lymphocytic infiltration in the retina and does not occur prior to this event (Lightman and Greenwood, 1992). This suggests that it is the migration of leucocytes that causes physical disruption of the BRB, although it is more likely that increased permeability is a result of locally released vasoactive inflammatory mediators from either infiltrated or tissue resident cells. The increase in leucocyte recruitment into the retina in EAU is associated with breakdown of the blood-retinal barrier (BRB) (Lightman and Greenwood, 1992) and oedema formation (de Kozak et al., 1981), both of which are involved in the pathogenesis of the disease.

The route of leucocyte entry into the retina remains uncertain. Due to the endothelia of the retinal vasculature being in close contact with circulating cells, leucocytes are more likely to cross the BRB at this site. Inflammatory cells crossing the BRB at the RPE must first be captured by the choroidal endothelia from where they must extravasate and penetrate a collagenous membrane called Bruch's membrane before crossing the RPE. Despite this being a more tortuous route there is ultrastructural evidence that leucocytes can cross the posterior barrier, but mainly during the latter stages of the disease (Greenwood et al., 1994; Lin et al., 1991; Dua et al., 1991). Moreover, in the guinea pig model of EAU the inflammatory cells are likely to enter the retina via the RPE as in this animal the retina is avascular (Caspi, 1989). In all models of EAU there is a significant recruitment of leucocytes by the choroidal vasculature leading to the choroid becoming packed with inflammatory cells. This may result in the separation of Bruch's membrane into its outer collagenous layer at the choriocapillaris and the inner cuticular layer at the RPE (Dua et al., 1991). From Bruch's membrane leucocytes can pass through the inner cuticular layer, migrate beneath and between the RPE cells eventually reaching the photoreceptors (Greenwood et al., 1994; Dua et al., 1991). There is also evidence that inflammatory cells have been seen apparently within RPE cells (Greenwood et al., 1994) but the direction of leucocyte migration in this case could not be determined.

It has been demonstrated that activated T lymphocytes can enter the CNS in a random manner, irrespective of antigen-specificity or MHC compatibility of the lymphocytes with the host nervous system (Hickey et al., 1991). The control of immune cell migration across the vasculature of the CNS has been attributed to the highly specialised nature of endothelium (Male and Pryce, 1989; Male et al., 1992; Calder and Greenwood, 1995; Wang et al., 1993). In the eye however, as already indicated, lymphocyte migration is controlled not only by the vascular endothelium but also by the retinal pigment epithelium which overlies the relatively leaky choroidal endothelium. In contrast to migration across the vascular endothelium, relatively little research has been carried out on migration across these cells.

3. COMPARISON OF LYMPHOCYTE MIGRATION ACROSS RPE AND RETINAL ENDOTHELIAL MONOLAYERS

Rat retinal endothelia and RPE were isolated and cultured as previously described (Greenwood, 1992; Chang et al., 1991). Lymphocyte migration across the cells of the BRB *in vitro*, and the factors controlling this migration were investigated using time-lapse videomicroscopy (Calder and Greenwood, 1993). Lymphocytes were added to confluent monolayers of either PVG derived RPE (Figure 1) or retinal endothelial cells (EC), or Lewis derived retinal EC and their interactions recorded over a 4 hour time course. The migration of untreated or mitogen activated peripheral lymph node (PLN) lymphocytes, and antigen specific T cell lines across both RPE and retinal endothelial cell monolayers were compared. The migration of three different $CD4^{+}$ T cell lines were investigated, one was specific for the retinal antigen S-antigen and the other two were raised against the irrelevant antigens purified protein derivative (PPD) and ovalbumin (OA).

The overall migration of the PPD and OA T cell line lymphocyte across both aspects of the BRB *in vitro* were not significantly different after four hours of co-culture. The S-antigen specific T cell line, however, exhibited significantly higher migration on through RPE monolayers (Table 1). The factors controlling this migration were similar, in that the state and mode of lymphocyte activation was crucial in determining the level of migration observed. Migration of antigen-activated long term cell lines was much greater across both retinal endothelium and RPE than PLN cells activated with mitogen. The rate at which lymphocyte migration occurred, however, was different. Lymphocyte migration across RPE monolayers occurred quickly, with lymphocytes migrating minutes after addition to the RPE, with the majority of migration occurring within the first 30 minutes and reaching a plateau after approximately two hours (Devine et al., 1996a). Migration across the retinal endothelium, on the other hand, was a much slower process, with only a few cells migrating in the first 90 minutes of co-culture, increasing more rapidly during the latter period of the co-culture. A possible explanation for this may be that the endothelial cells require activation (eg by cytokines produced by the lymphocytes) which may upregulate adhesion molecule expression (eg ICAM-1 expression) or function. The RPE, however, has been shown to express higher levels of ICAM-1 (Liversidge et al., 1990) than retinal endothelium, which may be sufficient to support the initial migration. Results from lymphocyte migration across retinal EC confirmed that, as had been shown previously, antigen specificity is irrelevant in determining the ability

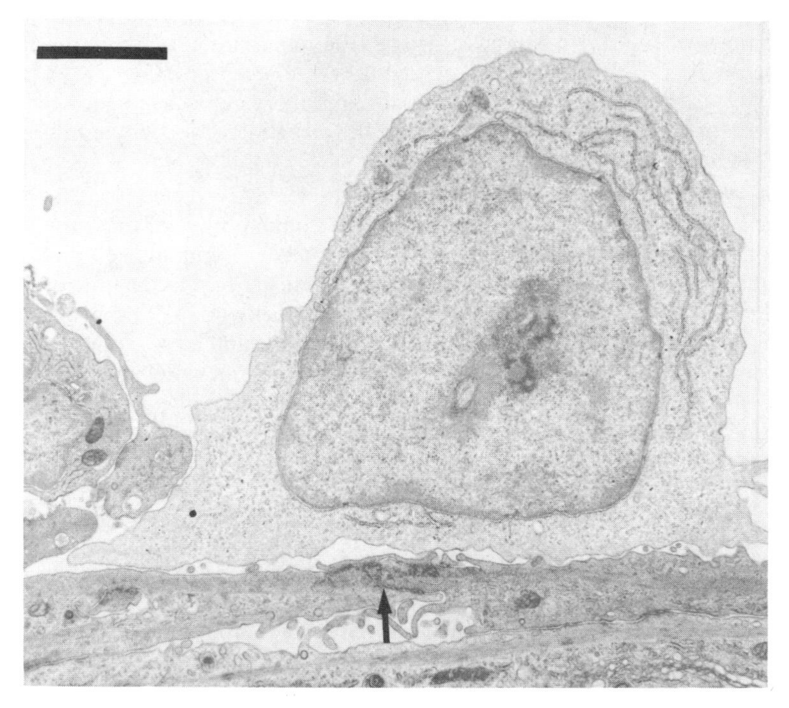

Figure 1. T cell line lymphocyte adhering to RPE monolayer. The lymphocyte can be seen spanning a junction between two RPE cells (arrow). Bar = 2μm.

of lymphocytes to cross endothelium (Hickey et al. 1991, Greenwood & Calder, 1993). The higher levels of migration observed by the S-antigen T cell line across RPE monolayers compared to that of either the PPD or OA T cell line suggests that this migration is antigen dependent. We believe, however, that this is more likely to be due to the greater antigenicity of S-antigen, resulting in lymphocyte lines which are more activated and hence more migratory. The reason why a differential ability to migrate is not observed with the endothelial monolayers is that the rate limiting factor may be due to functional aspects of the endothelium.

A difference in the level of migration across the retinal endothelium from different strains of rats was found (Table 1.). Genetic factors have been shown to be an important factor in the susceptibility of an individual to autoimmune conditions (Theofilopoulos, 1995) and of particular importance are differences in MHC class II expression. However, this study is the

Table 1. The percentage of lymphocyte migration over 4 hours across PVG derived RPE and retinal EC monolayers and Lewis derived retinal EC monolayers. The migration of untreated and concanavalin A treated PLN cells and three T cell lines are compared. Data taken from Greenwood and Calder, 1993 and Devine et al., 1996a. (mean ± SEM)

	PVG RPE	PVG retinal EC	Lewis retinal EC
Untreated PLN	0.8 ± 0.3	0	0.8 ± 0.4
Con A PLN	4.2 ± 1.6	0.6 ± 0.4	3.2 ± 1.4
PPD T cell line	32.9 ± 4.2	22.6 ± 4.9	44 ± 2.8
OA T cell line	30.9 ± 3.1	29.1 ± 2.8	38.6 ± 5.6
S-antigen T cell line	57 ± 4.9	22.5 ± 3.1	40.2 ± 4.4

only one we know of in which migration across endothelial cells isolated from different strains of rats differed. The strains of rats from which the retinal EC were isolated have different susceptibilities to EAU. Although EAU can be induced in both Lewis and PVG rats (de Kozak et al.,1981) the disease induced in PVG rats is less severe and more chronic in nature than that observed in Lewis. To determine if the difference in migration could be attributed to a difference in the level of adhesion molecules expressed on the endothelium, the expression of the adhesion molecule ICAM-1 was assessed on the endothelium from both strains of rat using flow cytometry. It was found that lower levels of ICAM-1 were expressed on endothelium from the less susceptible strain (ie PVG). Therefore this could be responsible for the lower level of lymphocyte migration observed with these endothelial cells.

4. ADHESION MOLECULES CONTROLLING LYMPHOCYTE MIGRATION AT THE BRB

The activation of RPE monolayers with IFN-γ, but not IL-1ß, caused a significant increase in migration which was not observed with IFN-γ activated retinal EC (Greenwood & Calder, 1993). IL-1ß activation did increase the migration of T cell line lymphocytes across retinal EC, however, which may indicate a difference in response of endothelial and epithelial cells. Cytokine activation results in an upregulation in adhesion molecule expression leading to enhanced lymphocyte migration during inflammation. Antibody blockade studies were used to determine which adhesion molecules were involved in controlling the migration across RPE and retinal EC monolayers. ICAM-1/LFA-1 interactions are the main ligands involved in lymphocyte migration across normal retinal vascular endothelium although VCAM-1/VLA-4 mediated migration can occur when the endothelium is activated (Greenwood et al., 1995). *In-vitro* studies have demonstrated that RPE also constitutively express ICAM-1, which was shown to be functional in the binding of neutrophils (Elner et al., 1992) and lymphocytes to monolayers of RPE (Liversidge et al., 1990; Elner et al., 1992). While ICAM-1 appeared to be the only adhesion molecule involved in neutrophil adhesion, an additional adhesion molecule was thought to be involved in the adhesion of lymphocytes. The adhesion molecules involved in lymphocyte migration across both aspects of the BRB are shown in table 2.

Unlike the retinal EC, significant VCAM-1 expression was not found on RPE monolayers following 24 hours of activation. After 72 hours activation with IFN-γ, however, VCAM-1 is expressed and the VCAM-1/VLA-4 interaction was found to support a significant degree of lymphocyte migration across the RPE monolayers (Devine et al., 1996b.)

Table 2. The percentage inhibition of lymphocyte migration across untreated and activated RPE and retinal EC monolayers. RPE were activated with IFN-γ and the retinal EC were activated with IL-1ß for 24 hours. Data taken from Greenwood et al., 1995 and Devine et al., 1996b. (means ± SEM)

	RPE		Retinal EC	
Antibody to:	Normal	Activated	Normal	Activated
ICAM-1	48.6 ± 3.5	68.2 ± 3.7	82.7 ± 4.4	47.3 ± 4.5
LFA-1α	61.0 ± 5.2	69.1 ± 3.5	89.2 ± 2.4	89.7 ± 2.3
LFA-1ß	63.2 ± 4.7	63.8 ± 6.5	71.3 ± 8.7	57.2 ± 3.2
VLA-4	No inhibition	No inhibition	No inhibition	34.4 ± 4.9
VCAM-1	No inhibition	No inhibition	No inhibition	42.0 ± 5.4

5. CONCLUSIONS

These studies demonstrate that the state of lymphocyte activation is paramount in determining their ability to migrate across retinal EC and RPE monolayers in vitro. The observation that the LFA-1/ICAM-1 receptor pairing is the major interaction mediating lymphocyte diapedesis through retinal EC monolayers is consistent with other studies on non-CNS endothelia. It was interesting to discover that lymphocyte migration across the barrier formed by the RPE was also dependent on the LFA-1/ICAM-1 despite the fact that this barrier lies within the ocular tissue and is not involved in recruiting cells from the circulation. The induction of VCAM-1 and its participation in migration across cytokine activated monolayers would be consistent with the increase in leucocyte traffic observed in retinal inflammatory disease.

6. REFERENCES

Calder, V.L. and Lightman, S.L. (1992). Experimental autoimmune uveoretinitis (EAU) versus experimental allergic encephalomyelitis (EAE): a comparison of T cell-mediated mechanisms. Clin. Exp. Immunol. 89:165-169.

Calder, V.L. and Greenwood, J. (1995). Role of the vascular endothelium in immunologically mediated neurological disease. In: Immunological aspects of the vascular endothelium. Savage, O.S. and Pearson, J.D. (eds) Cambridge University Press, Cambridge. pp. 96-123.

Caspi, R.R. (1989). Basic mechanisms in immune mediated uveitic disease. In: Immunology of Eye Diseases. Lightman, S. (ed.), Kluwer Academic Publishers, Dordrecht. pp. 61-87.

Chan, C., Palestine, A., Kuwabara, T. and Nussenblatt, R.B. (1988). Immunopathologic study of Vogt-Koyangai-Harada Syndrome. Am. J. Ophthalmol. 105:607-611.

Chang, C.W., Roque, R.S., Defoe, D.M. and Caldwell, R.B. (1991). An improved method for isolation and culture of pigmented epithelial cells from rat retina. Curr. Eye Res. 10:1081-1086.

de Kozak, Y., Sakai, J., Thillaye, B. and Faure, J.P. (1981). S-antigen-induced experimental autoimmune uveoretinitis in rats. Curr. Eye Res. 1:327-337.

Devine, L., Lightman, S.L. and Greenwood, J. (1996). Role of LFA-1, ICAM-1, VLA-4 and VCAM-1 in lymphocyte migration across retinal pigment epithelial monolayers in vitro. Immunology (submitted).

Devine, L., Lightman, S.L. and Greenwood, J. (1996). Lymphocyte migration across the anterior and posterior blood-retinal barrier in vitro. Cell. Immunol. (submitted).

Dua, H.S., McKinnon, A., McMenamin, P.G. and Forrester, J.V. (1991). Ultrastructural pathology of the "barrier sites" in experimental autoimmune uveitis and experimental autoimmune pinealitis. Br. J. Ophthalmol. 75:391-397.

Elner, S.G., Elner, V.M., Pavilack, M.A., Todd, R.F., Mayo-Bond, L., Franklin, W.A., Strieter, R.M., Kunkel, S.L. and Huber, A.R. (1992). Modulation and function of intercellular adhesion molecule-1 (CD54) on human retinal pigment epithelial cells. Lab. Invest. 66:200-211.

Forrester, J.V., Liversidge, J., Towler, H. and McMenamin, P.G. (1990). Comparison of clinical and experimental uveitis. Curr. Eye Res. 9:(Supp), 75-84.

Greenwood, J. (1992). Characterization of a rat retinal endothelial cell culture and the expression of P-glycoprotein in brain and retinal endothelium in vitro. J. Neuroimmunol. 39:123-132.

Greenwood, J. and Calder, V.L. (1993). Lymphocyte migration through cultured endothelial cell monolayers derived from the blood retinal barrier. Immunology 80:401-406.

Greenwood, J., Howes, R. and Lightman, S. (1994). The blood-retinal barrier in experimental autoimmune uveoretinitis. Lab. Invest. 70:39-52.

Greenwood, J., Wang, Y.F. and Calder, V.L. (1995). Lymphocyte adhesion and transendothelial migration in the CNS: the role of LFA-1, ICAM-1, VLA-4 and VCAM-1. Immunology 86:408-415.

Hickey, W.F., Hsu, B.L. and Kimura, H. (1991). T-lymphocyte entry into the central nervous system. J. Neurosci. Res. 28:254-260.

Jakobiec, F.A., Marboe, C.C., Knowles, D.M., Iwamoto, T., Harrison, W., Cleary, S. and Coleman, D.J. (1983). Human sympathetic ophthalmia. An analysis of the inflammatory infiltrate by hybridoma-monoclonal antibodies, immunohistochemistry, and correlative electron microscopy. Ophthalmology 90:76-95.

Lightman, S. and Chan, C.C. (1990). Immune mechanisms in choroido-retinal inflammation in man. Eye 4:345-353.

Lightman, S. and Greenwood, J. (1992). Effect of lymphocyte infiltration on the blood-retinal barrier in experimental autoimmune uveoretinitis. Clin. Exp. Immunol. 88:473-477.

Lightman, S. and Towler, H. (1992). Immunopathology and altered immunity in posterior uveitis in man: a review. Curr. Eye Res. 11:(Supp), 11-15.

Lin, W., Essner, E. and Shichi, H. (1991). Breakdown of the blood-retinal barrier in S-antigen-induced uveoretinitis in rats. Graefe's Arch. Clin. Exp. Ophthalmol. 229:457-463.

Liversidge, J., Sewell, H.F., and Forrester, J.V. (1990). Interactions between lymphocytes and cells of the blood-retina barrier: mechanisms of T lymphocyte adhesion to human capillary endothelial cells and retinal pigment epithelial cells *in vitro*. Immunology 71:390-396.

Male, D., Pryce, G., Linke, A. and Rahman, J. (1992). Lymphocyte migration into the CNS modelled *in vitro*.. J. Neuroimmunol. 40:167-172.

Male, D. and Pryce, G. (1989). Induction of Ia molecules on brain endothelium is related to susceptibility to experimental allergic encephalomyelitis. J. Neuroimmunol. *21*, 87-90.

Mirshahi, M., de Kozak, Y., Gregerson, D.S., Stiemer, R., Boucheix, C. and Faure, J.P. (1990). Genetic susceptibility of rat strains to induction of EAU in relation with the immune response to a particular epitope of S-antigen. Ocular Immunology Today, Proceedings of the Fifth International Symposium on the Immunology and Immunotherapy of the eye. pp. 195-198.

Theofilopoulos, A.N. (1995). The basis of autoimmunity: Part II Genetic predisposition. Immunol. Today 16:150-159.

Wang, Y.F., Calder, V.L., Greenwood, J. and Lightman, S.L. (1993). Lymphocyte adhesion to cultured endothelial cells of the blood-retinal barrier. J. Neuroimmunol. 48:161-168.

42

DISRUPTION OF MEMBRANOUS MONOLAYERS OF CULTURED PIG AND RAT BRAIN ENDOTHELIAL CELLS INDUCED BY ACTIVATED HUMAN POLYMORPHONUCLEAR LEUKOCYTES

Minoru Tomita,[1] Yasuo Fukuuchi,[1] Norio Tanahashi,[1] Masahiro Kobari,[1] Hidetaka Takeda,[1] Masako Yokoyama,[1] and Helena Haapaniemi[2]

[1] Department of Neurology
School of Medicine
Keio University
Tokyo, Japan
[2] Department of Neurology
University of Helsinki
Finland

SUMMARY

Employing VEC-DIC microscopy, we observed in an *in vitro* system that polymorphonuclear leukocytes (PMNLs) interacted with cultured endothelial cells (ECs) through the following steps:

1. the PMNL flowed in a round shape in the superfusion fluid;
2. the PMNL rolled on the EC for a couple of minutes;
3. the PMNL remained at a nonspecific location on the plasmic membrane of the EC for a few minutes;
4. the PMNL became firmly stuck through connections which were single-stemmed, multichanneled, or *en face*;
5. the PMNL apparently interacted with the EC, causing its retraction and detachment from the floor; and
6. finally, the PMNL swelled up in some but not all cases, being hugged by the "dead" EC, which had fallen into coagulation necrosis.

Biology and Physiology of the Blood–Brain Barrier, edited by Couraud and Scherman
Plenum Press, New York, 1996

Based on these observations, it is speculated that during the above process involving PMNLs, EC damage and shrinkage would cause cracks in the EC layer in the *in situ* microvessels, not only resulting in fluid leakage from the blood to the tissue, but also providing a gap for the diapedesis of newly arriving white cells, since PMNLs are always available in the blood stream even in ischemic tissue (Fig. 5).

RÉSUMÉ

Nous avons observé, en microscopie VEC-DIC, dans un système *in vitro*,que les leucocytes polymorphonucléaires (PMNL) interagissent sur les cellules endotheliales (EC) selon les processus suivants:

1. les PMNL prennent une forme sphérique dans le liquide de perfusion;
2. les PMNL roulent sur les EC pendant 2 minutes;
3. les PMNL demeurent sur la membrane plasmique des EC, dans un site non spécifique pendant quelques minutes;
4. les PMNL s'accolent fermement par des liaisons comportant un seul brin, plusieurs canaux, ou "en face";
5. les PMNL semblent agir sur les Ec , qui se rétractent et se décollent du support;
6. et, finalement, les PMNL sont souvent gonflés par les EC "morts" qui les enserrent et qui sont entrés en nécrose et coagulation.

Sur la base de ces observations, on suppose que pendant le processus décrit ci-dessus avec les PMNL, la rétraction des EC provoquerait des éclatements in situ dans les couches de EC des microvaisseaux, ce qui induirait non seulement une fuite de fluide du sang vers les tissus, mais ouvrirait aussi une brèche pour la diapédèse de nouvelles cellules blanches, puisque les PMNL sont toujours présents dans le flux sanguin, même dans les tissus ischémiés (Fig.5).

1. INTRODUCTION

White blood cells have been implicated in the pathogenesis of ischemic tissue damage to the brain via inflammatory changes of the microvasculature[1,2]. Several hours after the onset of ischemia, polymorphonuclear leukocytes (PMNLs) have been demonstrated to accumulate gradually in the ischemic microvasculature of the brain both in experimental animals and in human stroke patients[3,4]. Subsequently, PMNLs are said to infiltrate into the brain parenchyma[5,6]. With a delay of several hours after ischemia, the blood-brain barrier (BBB) is also known to become opened, as evidenced by leakage of macromolecules from the vascular wall accompanying water: this state is usually termed vasogenic edema. Such similarity in the time required for the above phenomena leads us to speculate that PMNLs may somehow participate in damaging the endothelial layers of the cerebral microvasculature, resulting in disruption of the BBB. The fundamental question is whether or not PMNLs could interact with such an entirely different cell-types, viz. brain endothelial cells, and damage its barrier function.

The aim of the present study was to evaluate how activated PMNLs adhered to, and damaged monolayers of cultured microvascular endothelial cells, employing video enhanced contrast and differential interference contrast (VEC-DIC) microscopy.

2. MATERIALS AND METHODS

The VEC-DIC equipment consisted of an inverted Nomarski microscope and a CCD (charge-coupled device) camera coupled with an image processing computer[7]. Briefly, cultured endothelial cells (see below) were observed under the inverted Nomarski microscope equipped with a x100 DIC (differential interference contrast) objective lens and a x2.5 insertion lens. Optical images were obtained using the CCD camera (TI-23P, NEC, Tokyo). The digital signals in the small detection area (7 x 9 mm) of this camera were fed into a computer which displayed the video image with enhanced contrast (VEC) on the monitor at a magnification of x12,000 (Fig. 1).

Thirteen experiments were carried out employing primary cultures of pig brain microvascular endothelial cells (pECs; supplied by courtesy of Prof. S. Murota, Department of Physiological Chemistry, Tokyo Medical and Dental University, Tokyo), primary cultures of rat fetal brain microvascular endothelial cells (rECs), and cultured human umbilical cord vein endothelial cells (HUVEC; Kurabo, Japan) for reference. The endothelial nature of the latter cells had been confirmed by immunostaining for von Willebrand factor and from the uptake of acetylated low density lipoproteins (LDL). Processing with ethylenediaminetetraacetic acid (EDTA)-trypsin (0.05%) (GIBCO Laboratories) was carried out, and the cells were seeded onto an uncoated 175-μm thick cover glass contained in a polystyrene dish (Corning Petri Dish, 35 x 10 mm; Iwaki Glass, Japan). Cells were grown in modified MCDB131 medium supplemented with 2% FBS, 10 mg/ml EGF, 1 mg/ml hydrocortisone, 50 mg/ml gentamicin, 0.25 mg/ml amphotericin B and 0.4% v/v bovine brain extract (E-GM UV; Kurabo, Japan), and incubated in a humidified atmosphere of 5% CO_2/95% at 37°C. The medium was changed 3 times per week. Cells were used at 2 weeks after seeding, when they had spread and adhered well, and become confluent.

Human leukocytes were obtained from heparinized venous blood of healthy human volunteers. Neutrophils were separated by gradient centrifugation employing polysaccharide

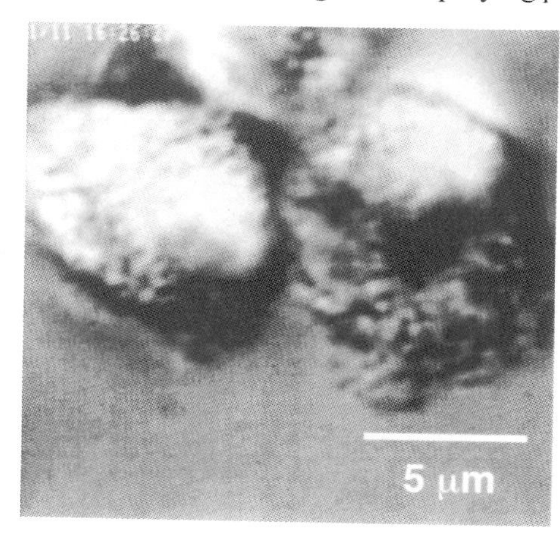

A. B.

Figure 1. Photographs of aggregated polymorphonuclear leukocytes by conventional light microscopy (A) and VEC-DIC microscopy (B) (see No. 2 in Table 1).

(Ficoll 400) and a radiopaque contrast medium (Hypaque; ICN Biomedicals). The resultant cells consisted of 96-98% PMNL. After completion of the gradient centrifugation procedures, the cells were resuspended in culture medium and used for the experiments within 6h.

To examine the interactions between PMNLs and ECs, the coverslip on which ECs were growing was taken out, and placed airtightly (employing grease) on the bottom side of a slide glass with a square window so that a microscopic observation chamber having the cells at the bottom was formed. The slide glass with the observation chamber was mounted on the stage of the inverted microscope and continuously superfused with DMEM at 37°C driven by a rotary pump. Following control observation of the ECs, a diluted PMNL suspension was introduced onto the ECs through the feeding tube. In 5 cases, platelet

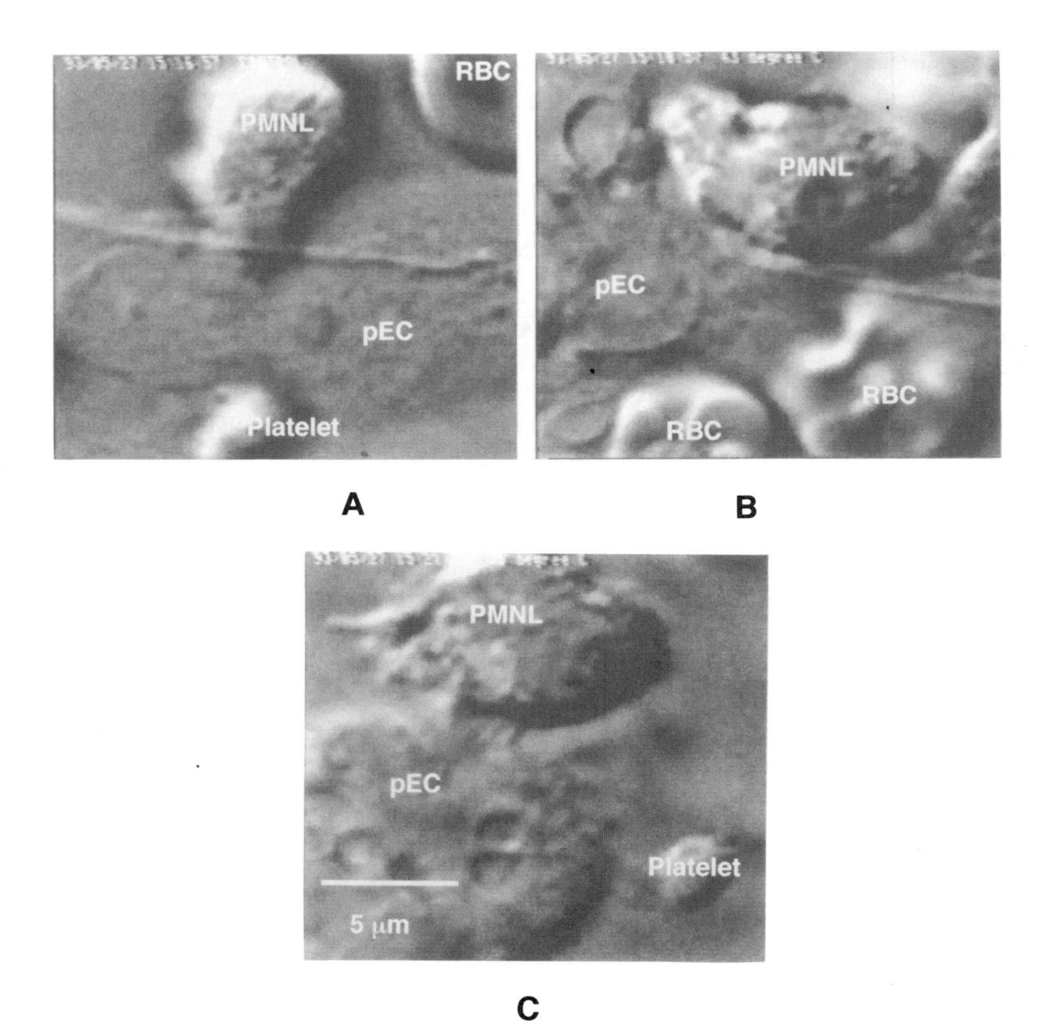

Figure 2. Interactions between a pig brain vascular endothelial cell (pEC) and human PMNL. The lower half of photographs A, B and C is occupied by the pEC, whereas the upper half the free space for superfusing fluid. A PMNL after arrival (A) moved along the edge of the membrane (B) and finally settled down near the nucleus (C). The pEC appeared to beome degenerated judging from its "rocky appearance" with no active movements of the intracellular granules.

activation factor at $<10^{-6}$ g/ml (PAF) was used to facilitate interactions between the ECs and PMNLs[8].

The experimental data were all stored on video tape as sequential images and were subjected to detailed investigation by play-back of the videotape. This enabled us to observe the cell behavior repeatedly at an almost electron microscopic magnification (Fig. 1).

3. RESULTS

Preliminary experiments employing ECs alone confirmed that the EC remained in an unchanged state, exhibiting continuous active granular or intracellular organellar movements within the cytoplasm for more than 3 h despite the light transillumination of the

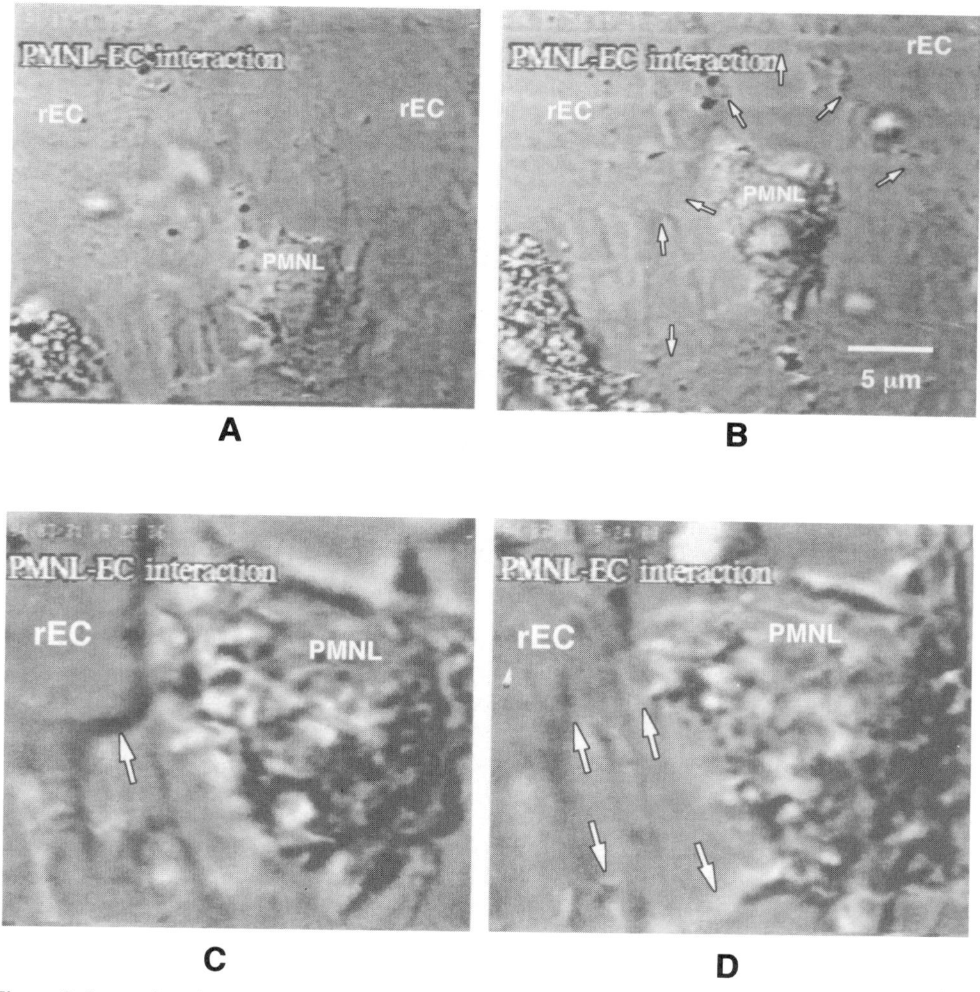

Figure 3. Interactions between rat brain vascular endothelial cells (rECs) and a human PMNL. The PMNL adhered to and flattened over two rECs (A). Note the formation of a wide hole after laceration of the monolayer of membranous spread around the PMNL at 10 min after adhesion (B). Photographs C and D show zoom-up images of the process of membrane laceration indicated by arrows (No. 7).

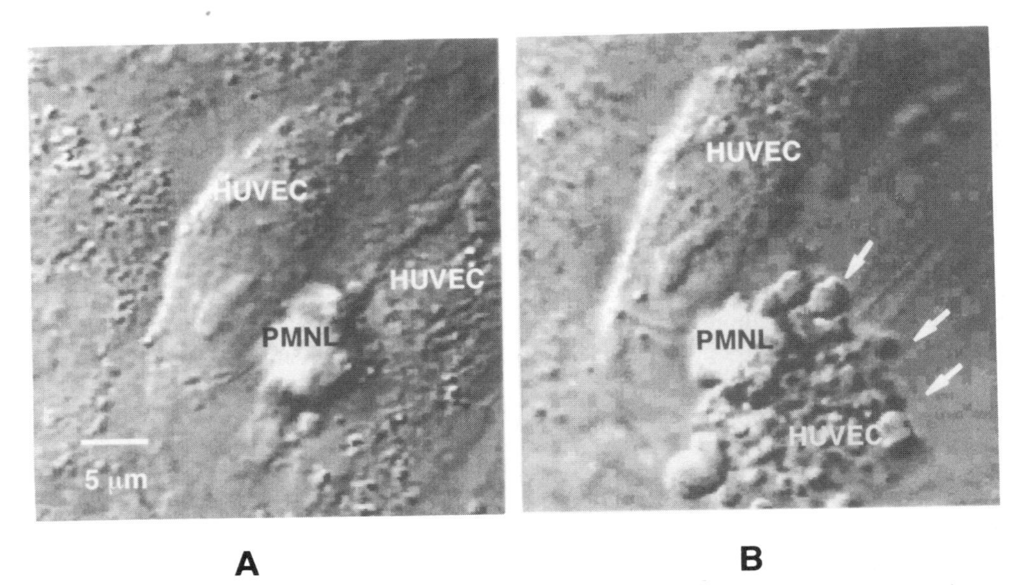

Figure 4. Interaction between a human umbilical cord vein endothelial cell (HUVEC) and human polymorphonuclear leukocyte (PMNL) activated by platelet activating factor. Photograph A: control confluent HUVECs immediately after a PMNL adhered to the right HUVEC. Photograph B: at 30 min after PMNL adhesion. The right HUVEC is retracted, bubbling with blebs (No. 13).

observation chamber. The PMNL alone in the observation chamber appeared to be activated more or less spontaneously and began to move around with pseudopods (Fig. 1; No. 2 in Table 1). Such ameboid movement became more active in the presence of chemotactic agents, e.g., PAF, fMLP, and PMA[9].

Our observations are summarized in Table 1. In the cases of the combined system with ECs and PMNLs, we observed interactions between the PMNLs and ECs almost in all experiments. Even when "little change" occurred within 20 min (No. 8), an interaction between the cells could be predicted to occur if observations were continued for longer. A general outline of our findings was as follows. The PMNL reached the observation chamber upon introduction into the feeding tube, rolled on the EC, and settled on the cell surface. The PMNL began to move with pseudopods near or on the EC surface (Fig. 2; No. 3 in Table 1). The EC occasionally reacted to the PMNL passage by exhibiting local agitation of granules at the site of contact with the PMNL, and increased "ruffling" of the edge of the plasmic membrane[10]. When entrapped by the EC, the PMNL underwent degranulation releasing "smoke-like material", which may consist of lysosomal enzymes and/or activated forms of oxygen[8]. The EC began to react with the PMNL at sites of pseudopodal contact (No. 9), by protruding at the edge of the peripheral plasma membrane within a couple of minutes after PAF introduction, as demonstrated by us in a previous communication[8]. In one case (No. 10), a PMNL attached *en face* to an EC occasionally underwent dynamic rhythmic contraction and swelling like a bellows. After exposure to PMNL attack, some of the EC became lacerated forming a large hole in the endothelial monolayer (Fig. 3), retracted (Fig. 4), shrunken and apparently almost dead, apparently falling into coagulative necrosis (Nos. 5, 6, and 11). However, even at this stage, the PMNL appeared to be alive, struggling to escape from the necrotic cell (No. 11). Finally, the débris stuck together with dying PMNL became detached from the floor and floated in the medium.

Table 1. Summary of observations

No.	Materials	Conditions	Time and state	Observed changes
1	HUVEC	DMEM, VEC microscopy	control observations for 3 h	no change
2	PMNL	without plasma protein	spontaneously activated within 5 min	granular agitation, ameboid
3	pEC + PMNL	DMEM superfused	control observations for 30 min	no change ameboid
4	HUVEC + PMNL	DMEM superfused	immediately	ruffling, local granular agitation ameboid
5	pEC + PMNL	PAF	10 min, connected by single stalk 1 h	ameboid - degranulation - dislodged gap opened, "coagulative necrosis"
6	rEC + PMNL	PAF	10 min, *en face 20 min*	ameboid - degranulation - attacking gap opened, "coagulative necrosis"
7	rEC + PMNL	spontaneous	a few min, *en face 10 min*	flattened - degranulation - attacking contraction, laceration
8	rEC + PMNL	spontaneous	*en face 20 min*	flattened - degranulation - attacking little change
9	HUVEC + PMNL	PAF	10 min, multi-channels spotty protrusions	ameboid - degranulation - swelling (3 h) retraction, detachment, débris mass (2 h)
10	HUVEC + PMNL	PAF	20 min, *en face*	swollen up - rhythmic contractions, necrosis (3 h) agitated granules, retraction (2 h)
11	HUVEC + PMNL	heat (43°C)	30 min, single stalk	"solo dance", long survival retraction, "coagulative necrosis" (1.5 h)
12	HUVEC + PMNL	PAF	5 min, *en face*	polymorphological changes retraction, bubbling (30 min)

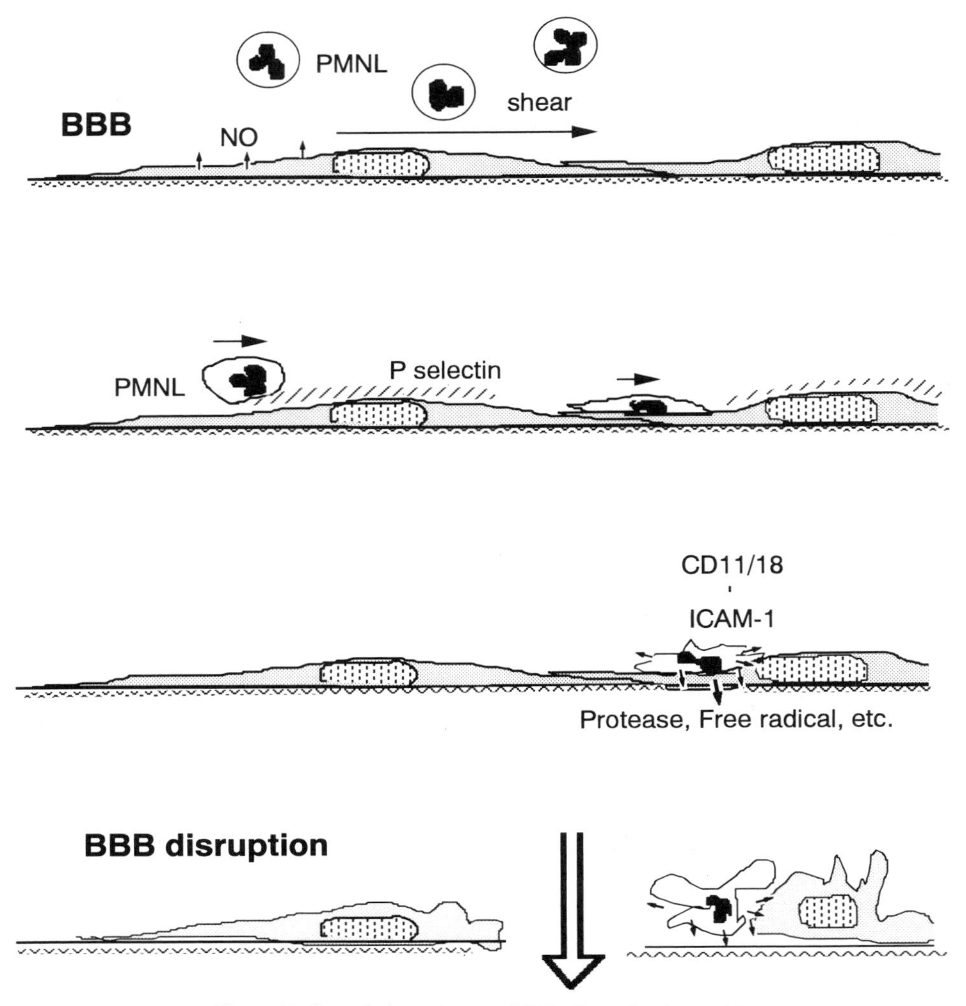

Figure 5. Speculative schema of BBB disruption by PMNL.

REFERENCES

1. Kochanek, P.M., Hallenbeck, J.M. -Polymorphonuclear leukocytes and monocytes/macrophages in the pathogenesis of cerebral ischemia and stroke. *Stroke* 23: 1367-1379, 1992.
2. del Zoppo, G.J., Garcia, J.H. -Polymorphonuclear leukocyte adhesion in cerebrovascular ischemia: Pathophysiologic implications of leukocyte adhesion. In: Granger, D.N., Schmid-Schnbein, G.W. (eds.), *Physiology and Pathophysiology of Leukocyte Adhesion*, Oxford University Press, New York-Oxford, 1995, pp. 408-425.
3. Pozzilli, C., Lenzi, G.L., Argentino, C., Carolei, A., Rasura, M., Signore, A., Bozzao, L., Pozzilli, P. -Imaging of leukocytic infiltration in human cerebral infarcts. *Stroke* 16: 251-255, 1985.
4. Wang, P.Y., Kao, C.H., Mui, M.Y., Wang, S.J. -Leukocytic infiltration in acute hemispheric ischemic stroke. *Stroke* 24: 236-240, 1993.
5. Garcia, J.H., Liu, K.F., Yoshida, Y., Lian, J., Chen, S., del-Zoppo, G.J . -Influx of leukocytes and platelets in an evolving brain infarct (Wistar rat). *Am. J. Pathol.* 144: 188-199, 1994.

6. Matsuo, Y., Onodera, H., Shiga, Y., Nakamura, M., Ninomiya, M., Kihara, T., Kogure, K. -Correlation between myeloperoxidase-quantified neutrophil accumulation and ischemic brain injury in the rat: Effects of neutrophil depletion. *Stroke* 25: 1469-1475, 1994.
7. Terakawa, S., Fan, J., Kumakura, K., Ohara-Imaizumi, M. -Quantitative analysis of exocytosis directly visualized in living chromaffin cells. *Neurosci. Lett.* 123: 82-86, 1991.
8. Haapaniemi, H., Tomita, M., Fukuuchi, Y., Tanahashi, N., Takeda, H., Terakawa, S. -PAF-induced white cell-endothelial cell interactions observed under a VEC microscope. In: M. Tomita, G. Mchedlishvili, W.I. Rosenblum, W.-D. Heiss, Y. Fukuuchi (eds.), Microcirculatory Stasis in the Brain, ICS 1031, Excerpta Medica, Amsterdam, 1993, pp. 185-191.
9. Tomita, M., Fukuuchi, Y., Tanahashi, N., Kobari, M., Terayama, Y., Shinohara, T., Konno, S., Takeda, H., Itoh, D., Yokoyama, M., Terakawa, S., Haapaniemi, H. -Activated leukocytes, endothelial cells, and effects of pentoxifylline: Observations by VEC-DIC microscopy. *J. Cardiovasc. Pharmacol.* 25 (Suppl. 2): S34-S39, 1995
10. Takeda, H., Tomita, M., Fukuuchi, Y., Tanahashi, N., Kobari, M., Yokoyama, M., Ito, D., Terakawa, S.. "Ruffling" of the marginal membranous portion of cultured vascular endothelial cells as observed by VEC-DIC microscopy. *J. Cereb. Blood Flow Metabol.* 15 (Suppl. 1) : S558, 1995.

43

RESPONSE OF BRAIN ENDOTHELIAL CELLS AND BONE-MARROW MACROPHAGES TO LEUKOCYTE-DERIVED EFFECTORS

Lucienne Juillerat-Jeanneret,[1,2] Jun-ichi Murata,[1]
Emanuela Felley-Bosco,[3] and Sally Betz Corradin[4]

[1] Institute of Pathology
Bugnon 27, CH1011 Lausanne, Switzerland
[2] Division of Pneumology
CHUV, CH1011 Lausanne, Switzerland
[3] Institute of Pharmacology and Toxicology
Bugnon 27, CH1011 Lausanne, Switzerland
[4] Institute of Biochemistry
Chemin des boveresses
CH1066 Epalinges, Switzerland

1. SUMMARY

Activation of endothelial cells is involved in the recruitment of blood-borne cells into organs. In the cerebral vascular system, perivascular macrophages of bone marrow origin are found directly underneath the endothelial layer and are postulated to participate in endothelial activation. To evaluate the respective role of endothelial cells and bone marrow-derived macrophages, we compared their responses to inflammatory stimulation. Our results suggest that both endothelial cells and macrophages contribute in mediating inflammatory activation of the blood-brain barrier, but also indicate specific modes of regulation for some functions.

1. RÉSUMÉ

Une activation des cellules endothéliales est nécessaire au recrutement dans les organes de cellules circulantes. Au niveau du système vasculaire cérébral, des macrophages périvasculaires originaires de la moelle osseuse sont localisés dircetement sous la couche endothéliale et participent vraissemblablement à l'activation endothéliale. Dans le but d'évaluer le rôle des cellules endothéliales et des macrophages dans le processu d'activation, nous avons comparé leurs réponses à une stimulation inflammatoire. Nos résultats suggèrent que les deux types de cellules participent au processus inflammatoire au niveau de la barrière hémato-encéphalique, mais que les phénomènes régulatoires sont spécifiques.

Biology and Physiology of the Blood–Brain Barrier, edited by Couraud and Scherman
Plenum Press, New York, 1996

2. INTRODUCTION

During the inflammatory process at the blood-brain barrier, several cell types may interact. These include blood-derived leukocytes, endothelial cells, macrophages (perivascular macrophages, microglial cells, the specific macrophages of the brain, or even in pathological situations, macrophages recruited from the blood) and cells of the central nervous system. Many mediators may participate in the cross-talk between these cells. In the cerebral microvascular system, perivascular macrophages are closely apposed directly underneath the endothelial cells, between the endothelium and the basement membrane. These macrophages are considered to be resident macrophages. Several groups (1,2,3) have demonstrated that these cells are of bone-marrow origin and that they are slowly replaced in normal situations. Perivascular macrophages can express MHC class II and thus present antigens and are thought to be involved in the transport of pathogens into the central nervous system. Together with the endothelium, perivascular macrophages are likely to participate in the inflammatory response of the blood-brain barrier.

In order to understand the respective roles of endothelial cells and perivascular macrophages in mediating inflammation at the blood-brain barrier, we compared their responses to stimulation by inflammatory mediators. Mechanisms regulating the production and release of two cytokines (interleukin-6 and tumor necrosis factor-α) and of nitric oxide, postulated to be involved in the regulation of vascular permeability and function were evaluated. The regulatory role that transforming growth factor-β (TGF-β) plays in the communication network between brain-derived endothelial cells and perivascular macrophages was assessed.

3. METHODS

3.1. Cell Preparation and Cultures

The preparation and characterization of rat brain-derived endothelial cell line EC219 has been described (4). Cells were grown in DMEM medium containing 10% fetal calf serum and 3 mg/ml bovine serum albumin, in collagen R-coated (Serva) plates. Confluent cultures were used for the experiments.

Rat bone marrow-derived macrophages were obtained from precursor cells by in vitro differentiation essentially as described for murine bone marrow-derived macrophages. Prior to experiments, macrophages were maintained for 18 hours in DMEM medium containing 10% fetal calf serum, in the absence of growth factors. These cells expressed the specific human macrophage marker CD68 (monoclonal anti-CD68, Dako). Prior to experiments, culture medium was changed, followed by a 30 min stabilization period, before effectors were added. Culture supernatants and cells were frozen for further determinations.

3.2. Evaluation of Interleukin-6 and Tumor Necrosis Factor-α Bioactivities

Interleukin-6 (IL-6) concentration in EC219 cell supernatants was evaluated as a growth factor for 7TD1 cells as previously described (5). Tumor necrosis factor-α (TNF-α) was measured as a cytotoxic factor for WEHI164-clone 13 cells.

3.3. NO and NO Synthase (NOS)

NO was determined as the stable nitrite derivative with the Griess reagent and NO synthase (NOS) activity was measured in cell extracts by the conversion of arginine to citrulline as previously described (6). Expression of iNOS mRNA was determined by Northern analysis as previously described (7).

3.4. Arginine Transport

The assay of ^{3}H-Arg uptake was performed at room temperature for 3 min in Hepes pH 7.4 buffer containing glucose, Mg^{2+} and Ca^{2+}, in the presence of 125 μM cold Arg for endothelial cells and 6.3 μM cold Arg for macrophages.

3.5. Protein Concentration

Protein concentration was determined in cell extracts with a BCA kit (Pierce) using bovine serum albumin as standard.

4. RESULTS

The production of TNF-α, IL-6 and NO in rat brain-derived endothelial cells and bone marrow-derived macrophages following stimulation with various inflammatory mediators was compared (Table I). From these experiments, the absence of release of TNF-α by activated endothelial cells, their low response to LPS and high reponse to IFN-γ, and their constitutive secretion of IL-6 in comparison with macrophages should be noted. NO is released in response to specific mediators in both endothelial cells and macrophages.

In response to activation, macrophages may produce inhibitory factors, including TGF-β. The role of TGF-β in regulating NO synthesis in brain-derived endothelial cells was not known. The release of NO, iNOS activity and iNOS mRNA expression was thus evaluated 6 and 24 hours post stimulation with TNF-α and IFN-γ, or with TNF-α and IFN-γ in the presence of TGF-β. TGF-β inhibited NO release, iNOS activity and mRNA expression in endothelial cells (Figure 1). TGF-β did not modulate cytokine release (not shown).

Table 1. Secretion of NO, IL-6 and TNF-α in rat brain-derived endothelial cells and bone marrow-derived macrophages. Endothelial cells and macrophages were stimulated with bacterial lipopolysaccharide, LPS (Sigma, 1 μg/ml), murine TNF-α (Boehringer, 100 U/ml) and murine IFN-γ (Genentech, 100 U/ml). TNF-α secretion was measured 4 hours post stimulation; IL-6 and nitric oxide secretion 24 hours post stimulation

	endothelial cells			macrophages		
	NO (μM)	IL-6 (U/ml)	TNF-α(U/ml)	NO (μM)	IL-6 (U/ml)	TNF-α (U/ml)
control	2.6	23 000	0	0	0	0
LPS	3.6	13 000	0	0	0	0
TNF-α	6.6	38 000	0	42.5	3500	8000
IFN-γ	9.9	13 000	0	0	-	20
TNF-α+IFN-γ	24.5	24 000	-	2.5	0	-
LPS+IFN-γ	6.5	4 000	0	130.0	-	-

0: not measurable -: not done

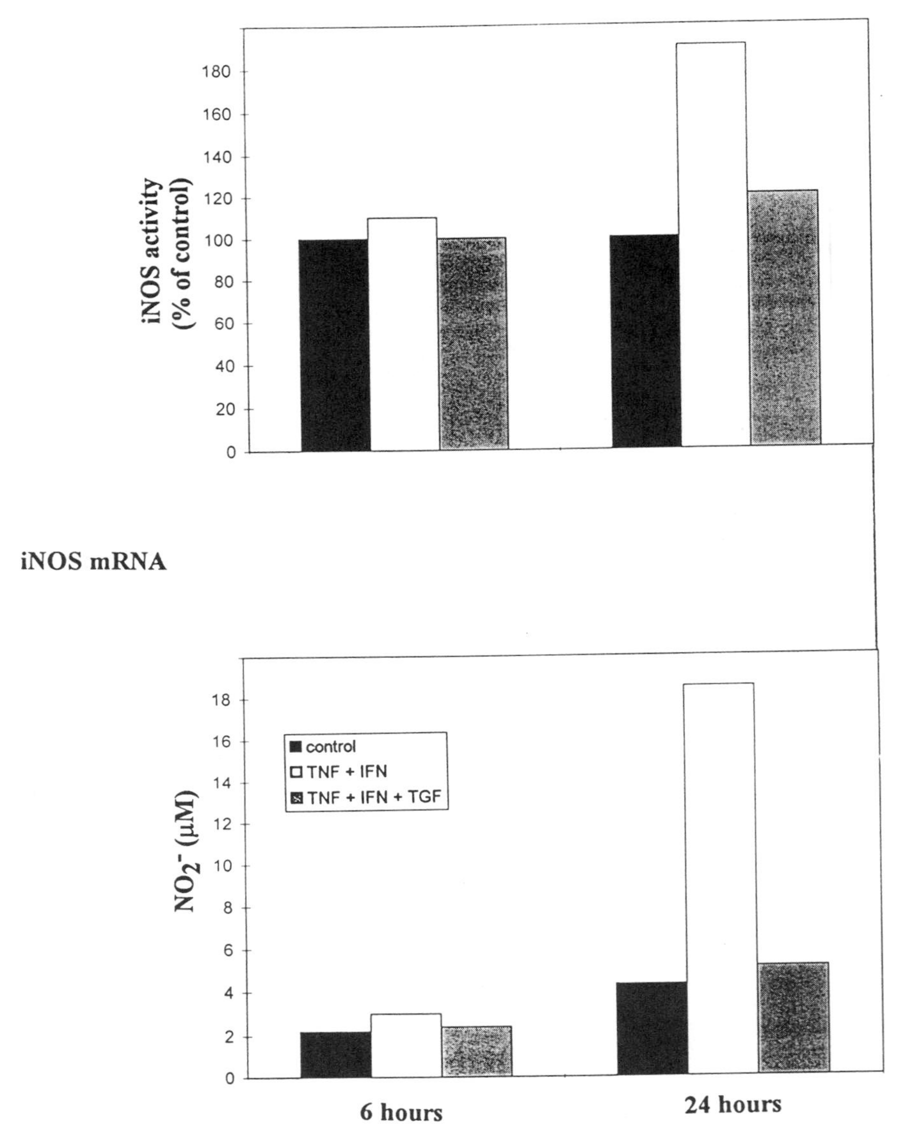

Figure 1. Regulation by TGF-β of NO synthesis in endothelial cells. Confluent EC219 cells were incubated with 100 U/ml each of TNF-α (TNF) and IFN-γ (IFN), with or without 0.5 ng/ml TGF- β_1 (TGF) for 6 or 24 hours. Nitrite was measured in culture supernatants, and NOS activity and mRNA expression in cells. Upper panel: NOS activity, middle panel: iNOS mRNA and lower panel: nitrite concentration.

Since NO is synthesized exclusively from arginine, and thus NO production may also be regulated at the level of substrate availability, we have compared NO release and arginine uptake in endothelial cells and macrophages stimulated with IFN-γ and LPS, in the presence or absence of dexamethasone, a known regulator of inflammation (Figure 2). Arginine uptake was modulated in macrophages, but not in endothelial cells. Moreover, the kinetics of arginine uptake differed in brain-derived endothelial cells and bone marrow-derived macrophages (results not shown).

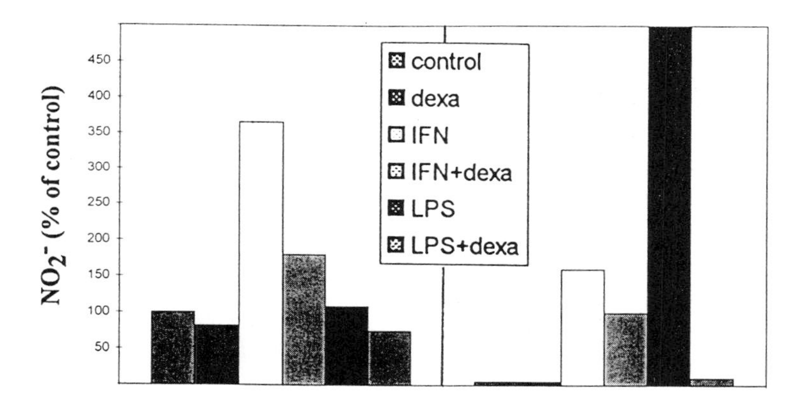

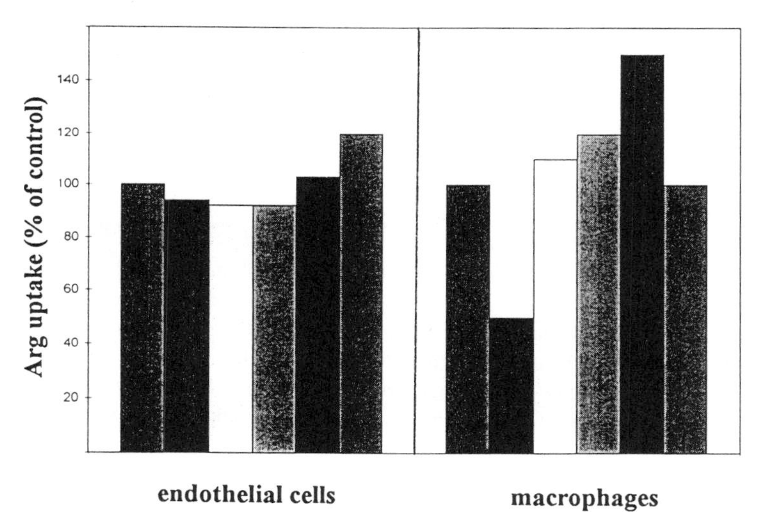

Figure 2. Regulation of Arg-uptake and NO production in endothelial cells and macrophages. Dexamethasone (25 μM) was added simultaneously with LPS (1 μg/ml) or IFN-γ (100 U/ml) (IFN) to confluent EC219 cells or macrophages and arginine uptake or nitrite release was measured 24 hours later. Upper panel: NO release, lower panel: arginine uptake.

5. DISCUSSION

In the cerebral microvascular system, perivascular macrophages of bone-marrow origin are found between the endothelium and the basement membrane. The interactions between endothelial cells and macrophages during regulation of inflammation at the blood brain-barrier are complex and involve a modification of vascular permeability. The inflammatory mediators IL-6, TNF-α and NO are postulated to be involved in regulation of vascular permeability and functions. However, at the blood-brain barrier, cells responsible for the production of these mediators and mechanisms regulating their release are not known. In this report, we compared the response of rat brain-derived endothelial cells and bone marrow-derived macrophages to inflammatory cytokines.

Il-6 and NO were secreted by both endothelial cells and macrophages, while the secretion of TNF-α was restricted to macrophages. The time-course of secretion of the various mediators indicated a specific sequence of events. In response to bacterial products, macrophages will first secrete TNF-α, which in synergism with IFN-γ, potentiates macrophage activation, IL-6 and NO release, and induces endothelial stimulation. Both cells in response to these stimuli synthesize NO, which appears to be a common mediator.

Regulation of the inducible pathway for NO production in brain-derived endothelial cells, in relation to macrophage activation, had not been previously investigated. In the vessel wall, NO may be produced by several cell types, either by NO synthases which are constitutively expressed in endothelial cells and are regulated by intracellular calcium, or upon stimulation, by inducible NO synthase (iNOS) in cells such as macrophages and endothelial cells. iNOS catalyses the production of large amounts of NO which is involved in defense mechanisms and in pathologic processes, including severe hypotension, loss of control of vascular permeability and modified adherence of the vascular wall for blood cells. Both constitutive and inducible NOS utilize arginine as substrate; thus arginine uptake may be limiting for NO release. We show here that this may be involved in regulating NO production in macrophages, but not in endothelial cells.

In response to activation, macrophages may secrete TGF-β (8). TGF-β is a multifunctional cytokine, which can down-regulate inflammation. TGF-β inhibited NO release, but not IL-6 release, in cytokine-stimulated endothelial cells.

It is also known that glucocorticoids such as dexamethasone can inhibit the release of inflammatory cytokines and NO in macrophages. We have previously observed the same effect with the spontaneous release of IL-6 by endothelial cells (6) and in cytokine-activated endothelial cells, dexamethasone completely abolished the release of IL-6 and of NO.

In summary, these results demonstrate a specific mode of regulation for the secretion of and response to inflammatory mediators in brain-derived endothelial cells when compared with bone marrow-derived macrophages. Bacterial products, such as LPS, are involved in macrophages responses, while leukocyte-derived cytokines, such as IFN-γ and TGF-β, are involved in endothelial and macrophage responses. These results suggest a specific function for both cells in mediating inflammatory activation at the blood-brain barrier.

6. ACKNOWLEDGMENTS

We thank Ms P. Fioroni, P. Darekar and P. Dessous L'Eglise Mange for excellent technical assistance. This work was supported by grants from the Swiss National Fund for Scientific Research, the Roche Research Fund and the Swiss Society for Multiple Sclerosis.

7. REFERENCES

1. Perry VH and Gordon S. TINS, 11:273-277, 1988. Macrophages and microglia in the nervous system.
2. Stevens A and Bähr M. J.Neurol.Sci., 118:117-122, 1993. Origin of macrophages in central nervous system.
3. Hickey WF and Kimura H. Science, 239:290-293, 1988. Perivascular microglial cells of the CNS are bone-marrow-derived and present antigen in vivo.
4. Juillerat-Jeanneret L, Aguzzi A, Westler OD, Darekar P and Janzer RC. In Vitro Cell.Dev.Biol. 28A:537-543, 1992. Dexamethasone selectively regulates the activity of enzymatic markers of cerebral endothelial cell lines.
5. Juillerat-Jeanneret L, Fioroni P and Leuenberger Ph. Endothelium, 3:31-37, 1995. Modulation of secretion of interleukin-6 in brain-derived microvascular endothelial cells.

6. Murata JI, Betz Corradin S, Janzer RC and Juillerat-Jeanneret L. Int.J.Cancer, 59:699-705, 1994. Tumor cells suppress cytokine-induced nitric-oxide (NO) production in cerebral endothelial cells.
7. Murata JI, Betz Corradin S, Felley-Bosco E and Juillerat-Jeanneret L. Int.J.Cancer, in press, 1995. Involvement of a transforming growth factor β-like molecule in tumor-cell derived inhibition of nitric oxide synthesis in cerebral endothelial cells.
8. Betz Corradin S, Buchmüller-Rouiller Y, Smith J, Suardet L and Mauël J. J.Leuk.Biol. 54:423-428, 1993. Transforming growth factor β1 regulation of macrophage activation depends on the triggering stimulus.

EXPRESSION OF MACROPHAGE CHEMOTACTIC PROTEIN-1 IN RAT GLIAL CELLS

Charles-Félix Calvo and Michel Mallat

Chaire de Neuropharmacologie
INSERM U114, Collège de France
75231 Paris CEDEX 05, France

SUMMARY

By an *in vitro* chemotaxis assay, we have shown that rat brain macrophages in culture release a soluble factor which stimulates the migration of bone marrow-derived macrophages. This activity was significantly decreased by an immune serum directed against the rat monocyte chemoattractant protein-1 (chemokine MCP-1). By Northern blot analysis, we have shown that MCP-1 is indeed synthetized by brain macrophages in culture, or by astrocytes after *in vitro* activation. We then evidenced an *in vivo* production of MCP-1 in the adult rat brain following injury induced by a local injection of kainic acid. This synthesis is localized in both astrocytes and brain macrophages by immunocytochemistry. Altogether, these results suggest that microglial and astroglial secretion of chemokines could contribute to the mechanism(s) leading to the infiltration of the central nervous system by blood-derived monocytes as observed in several pathologies.

RÉSUMÉ

Dans une étude réalisée chez le rat, nous avons montré que le milieu conditionné de macrophages cérébraux en culture contient une activité chimiotactique s'exerçant sur des macrophages provenant d'une culture de moëlle osseuse. Cette activité est fortement inhibée par un anticorps dirigé contre la chemokine MCP-1 de rat. Par ailleurs, l'expression du messager de cette protéine a été detectée par Northern blot, dans des macrophages cérébraux en culture et dans des astrocytes stimulés par le LPS. Par la suite, nous avons montré que l'expression de MCP-1 était induite *in vivo* lors d'une lésion cérébrale produite par injection d'acide kainic dans le striatum. La synthèse de la protéine a été détectée dans les astrocytes et dans les macrophages cérébraux, suggèrant l'intervention de ces deux types cellulaires dans le (s) mécanisme (s) conduisant à l'infiltration du système nerveux central par des monocytes sanguins observée au cours de certaines pathologies.

Biology and Physiology of the Blood–Brain Barrier, edited by Couraud and Scherman
Plenum Press, New York, 1996

I. INTRODUCTION

I.I. Macrophage Infiltration of the CNS and Possible Mechanisms Involved

The migration of monocytes into the CNS is a phenomenon which is observed in physiological situations during development, but also in response to pathological or experimentally-induced brain insults in adult. During ontogenesis, the CNS is infiltrated by exogenous mononuclear phagocytes which are precursors of the microglial cell population. In the developing parenchyma, microglial cells display an activated phenotype termed ameboid microglia or brain macrophages (BM). The maturation of the brain is associated with the progressive transformation of BM into ramified microglial cells which spread throughout the adult CNS and are considered to be in a "resting" state under physiological conditions [1]. The infiltration of the adult brain by blood-derived monocytes occurs in response to diverse CNS insults including viral infection [2] or experimental injuries [3] and is involved in parenchymal inflammation. The presence of infiltrating monocytes is associated with reaction of resident CNS cells such as astrogliosis and activation of microglial cells, resulting in their transformation from the ramified into the ameboid phenotype [4]. The mechanism (s) by which the monocytes are attracted in the injured adult brain from the perivascular compartment to the parenchyma, remains to be determined, but is likely to be initiated by the establishment of an intracerebral gradient of chemotactic agents. This hypothesis was supported by the detection in the injured brain of cytokines known to exert chemotactic effects on monocytes *in vitro*. [5]. The CNS production of IL1, TNFα and TGFβ was shown in pathologies such as multiple sclerosis, AIDS dementia complex or experimental allergic encephalomyelitis (EAE) [5,2]. Some attempts were made, through intracerebral injections in mice, to demonstrate a direct role of inflammatory cytokines in the recruitment of cells into the CNS [6]. But these mediators, in addition to the morphologic activation of microglia, only induced leukocyte margination, and not cell infiltration of the parenchyma, suggesting that they are not the only mediators of the cell infiltration response. Recent studies performed usually on peripheral tissues *in vitro* or *in vivo* have revealed the existence of the chemokines as a novel class of cytokines and suggested their participation in the mechanism of cell migration into neoplastic tissues [7,8,9]

I.I.I. Chemokines Expression in the CNS. The chemokines are a superfamily of structurally related proteins that regulate the trafficking and activation of mammalian leukocytes. They form two subfamilies whose characteristics aredepicted on the Table: the chemokines α (C-X-C), and β (C-C), based upon first the position of the first two of four conserved cysteine residues found in their primary amino acid sequence and second, the localization of their genes, with each family clustered on a single chromosome. For the most part, members of the chemokines α subfamily preferentially attract and activate neutrophils, whereas chemokines β tend to attract and activate monocytes / macrophages. However, for each subfamily member, the biological activity depends upon specific receptor expression on the target cells [10]. The phenomenon of monocyte recruitment was recently approached by the determination of chemokines α and β expression in the CNS either *in vivo* [11,12] or *in vitro* [13,14]. The expression of MCP-1 has been reported in two studies utilizing the EAE model in either the rat [11] or mouse [12]. In both reports, the clinical development of EAE was paralleled by macrophage infiltration and corresponded to a burst in MCP-1 and IP10 expression [12], or to an induction of MCP-1 with increased levels of CSF-1 in the spinal cord[11]. In one study [12] astrocytes were shown by *in situ* hybridization as the major source of MCP-1 and IP10 in EAE tissues, while neither chemokine was expressed in material from control animal.

Table 1. Cellular origin of human and murine chemokines and their target for chemoattraction

		Origin	Chimioattraction
Chemokines β (C-C) Human / murine χ 17 / χ 11	/MIP-1α	T cells*, B cells*, neutrophils, mast cells, fibroblasts, monocytes	Monocytes, eosinophils, basophils, CD8$^+$
	/MIP-1β	T cells*, B cells*, monocytes	Monocytes, CD4$^+$
	RANTES	T cells*, platelets, monocytes	Monocytes, eosinophils, mast cells, basophils, T cells
	MCP-1/JE	T cells*, B cells*, monocytes, endothelial cells, fibroblasts, keratinocytes, glioma, osteosarcoma	Monocytes, basophils, mast cells, CD4$^+$and CD8$^+$ clones, T cells*
	MCP-2	Osteosarcoma, monocytes	Monocytes, eosinophils, CD4$^+$ and CD8$^+$ clones
	MCP-3	Osteosarcoma, monocytes, mast cells	Monocytes, basophils, eosinophils, CD4$^+$ and CD8$^+$ clones
	/Eotaxine	BAL	Eosinophils
	/C10	P388D1	?
	I309	T cells*	Monocytes, neutrophils
Chemokines α (C-X-C)Human / murine χ 4 / ?	GRO	Monocytes	Neutrophils
	IP10	T cells*, monocytes, endothelial cells, keratinocytes, fibroblasts	Neutrophils, T cells*, monocytes
	PF4	Platelets	Neutrophils, monocytes, fibroblasts
	IL8	Monocytes	Neutrophils, CD4$^+$

*activated cell; BAL=bronchoalveolar lavage fluid; χ=chromosome.
For references, see 7, 8, 9.

These results enlighted the potential role for MCP-1 in the CNS. Indeed this β chemokine was described as a potent monocyte chemoattractant acting *in vitro* [15]. MCP-1 may be produced either constitutively or following activation by a variety of cells including peripheral mononuclear phagocytes [16, 17], endothelial cells [18], fibroblasts [19], smooth muscle cells [20], epithelial cells [21] and some tumor cell lines [15, 22, 23].

In addition to its chemotactic activity, MCP-1 can induce IL1 and IL6 cytokines production in human monocytes *in vitro* [24]. Moreover, MCP-1 has been shown to increase the expression of adhesion molecules on human monocytes [24, 25]. This is important because the expression of adhesion molecules and that of counter-receptors on endothelial cells are crucial for the mechanism of cell transmigration through blood vessel walls, when the blood brain barrier is disrupted in inflammatory states.

II. Results and Discussion

Collectively, the data described above indicate that chemokines can be expressed in the injured CNS, but did not completely characterize the cellular origin of this production. Indeed, the activation of microglia into the BM phenotype is an early response to tissue injuries often associated with astrogliosis. We hypothesize that the microglial activation could be associated with a production of chemotactic factors (cytokines, chemokines) thereby contributing together with astrocytes to the formation and maintenance of chemotactic gradient(s) targeting blood monocytes.

To test this hypothesis, we have used BM from cell culture as a model of activated microglia to assess in a biological assay their ability to release chemotactic factor (s). Brain macrophages were obtained from culture of cerebral cortices of 17-day-old rat embryos [26]. The collected cell population resulted in more than 95 % of cells bearing the macrophage marker ED1 [27], and was used to prepare conditioned medium (CM). In the migratory assay and as a model of macrophagic cells from outside the CNS we have used an homogeneous bone marrow-derived macrophage (BMM) population routinely isolated from rat bone marrow cultures [28]. The potential secretion of a chemotactic activity by BM was thus assessed by counting the number of BMM migrating toward the BM conditioned media in a chemotaxis assay (Boyden chambers). A chemotactic activity was found, associated with a mild chemokinesis, and was assigned to the chemokine MCP-1, constitutively secreted by the BM in culture. This biological activity was significantly inhibited by preincubation of the CM with an antibody to rat MCP-1 prior to being used in the migratory assay (Figure 1). In parallel, *in vitro* expression of MCP-1 was detected by a Northern blot analysis. In BM, a constitutive expression of MCP-1 was detected, at a basal level which could be upregulated by LPS (Figure 2) and to a lesser extend by inflammatory cytokines (not shown). Figure 2 also shows that MCP-1 mRNA is absent from astrocytes in culture, but can be induced within 4 hours following an incubation with LPS.

The demonstration of MCP-1 expression in BM from culture led us to search for an expression of this chemokine in activated microglia *in vivo*, when this phenotype is elicited by kainic acid injection in the adult rat striatum. Indeed, it has been shown that kainic acid can induce neuronal death as well as a rapid microglial response followed by a blood-derived

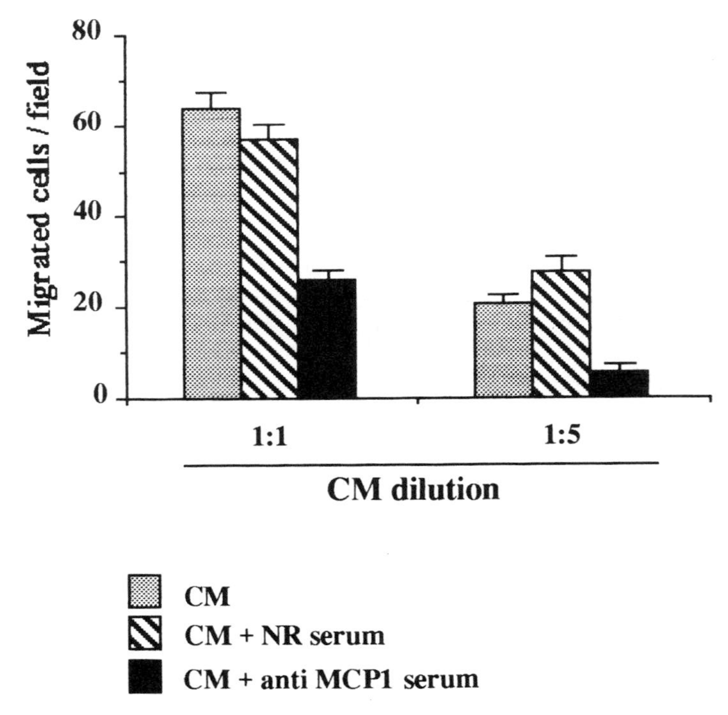

Figure 1. Inhibition with an antibody to rat MCP-1 of the chemotactic activity toward BMM found in the CM of brain macrophages. BM were seeded (1 x 10^6 cells / ml) for 20 hours in minimum medium to obtain the CM which was assayed for BMM attraction in chemotaxis chambers. When indicated, CM was incubated with either non relevant serum (NR), or with a polyclonal serum to rat MCP-1. The results are expressed as the mean number of cells in 10 grid fields of triplicate assays (± SEM) after subtraction of control value. Controls were 16.7 ± 1.4 cells for the medium alone, and 80.6 ± 2.5 cells when zymosan activated rat serum (1 / 100 diluted) was used as positive control.

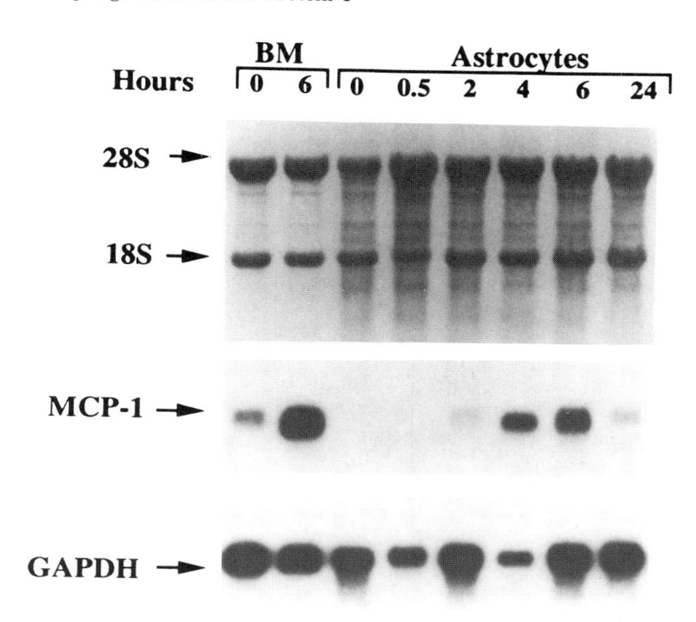

Figure 2. Expression of MCP-1 mRNA in brain macrophages and astrocytes. Total RNA from BM (5.0 x 10^6 cells / dish) or astrocytes (3.0 x 10^6 cells / dish) was extracted following culture with or without LPS (500 ng/ ml) for the time indicated. For each data point, 10 μg of total RNA was subjected to Northern blotting. The membranes were successively probed for MCP-1 [30] and GAPDH.[31] The autoradiograms were exposed for 2 days. (MCP-1) or 1 day (GAPDH).

monocyte infiltration into different CNS regions [29]. In a Northern analysis of total RNA, MCP-1 transcripts appeared markedly induced in the injured striatum during the first 2 days following the injection while they were absent in normal striatum. This transient expression became barely detectable at 7 days and undetectable for up to 30 days post kainic acid injection. Consistent with these results, immunocytochemical staining of MCP-1 allowed us to detect the chemokine in sections of injured tissue as soon as 20 hours following kainic acid injection. This early expression was concomitant with the activation of resident microglial cells, as evidenced by the upregulation of the Mac-1 and ED1 macrophage markers. The distribution of MCP-1 staining was clearly compatible with an expression of the protein by a population of astrocytes and BM in or around the injured area. This was confirmed by double immunostaining of the sections showing colocalization of either MCP-1 and GFAP (glial fibrillary acidic protein, specific marker of astrocytes), or of MCP-1 and ED1 in astrocytes and BM, respectively.

Altogether our study reveal BM as a novel cellular source for MCP-1 and indicate that this protein accounts for the chemoattraction exerted by BM on macrophages freshly derived from an hematopoïetic organ. We suggest that the release of MCP-1 by both astrocytes and BM could favor the initiation and perpetuation of the CNS inflammatory response by forming macrophage chemotactic gradient within the CNS.

ACKNOWLEDGMENTS

We wish to thank Prof. J. Glowinski for his constant help, Dr. D. Hervé for his participation to this work and Dr. T. Yoshimura for providing us with antibodies and probe to the rat material. The work was supported by INSERM and by the Agence Nationale de Recherches sur le SIDA.

REFERENCES

1. Ling, E A., Wong, W C. The origin and nature of ramified and amoeboid microglia : a historical review and current concepts. *Glia* 7 : 9-18, 1993.
2. Gendelman, H E., Lipton, S A., Tardieu, M., Burkrinsky, M I., Nottet, H S L M. The neuropathogenesis of HIV-1 infection. *J. Leukoc. Biol* 56 : 389-398, 1994.
3. Perry, H V., Andersson, P B., Gordon, S. Macrophages and inflammation in the central nervous system. *Trends. Neurosc* 16 : 268-273, 1993.
4. Streit, W J., Graeber, M B., Kreutzberg, G W. Functional plasticity of microglia : a review. *Glia* 1 : 301-307, 1988.
5. Benveniste, E N. Inflammatory cytokines within the central nervous system : sources, function, and mechanism of action. *Am. J. Physiol* 263 : C1-C15, 1992.
6. Andersson, P B., Perry, V H., Gordon, S. Intracerebral injection of proinflammatory cytokines or leukocyte chemotaxins induces minimal myelomonocytic cell recruitment to the parenchyma of the central nervous system. *J. Exp. Med* 176 : 255-259, 1992.
7. Oppenheim, J J., Zachariae, C O C., Mukaida, N., Matsushima, K. Properties of the novel proinflammatory supergene "intercrine" cytokine family. *Annu. Rev. Immunol* 9 : 617-648, 1991.
8. Miller, M D., Krangel, M S. Biology and biochemistry of the chemokines : a family of chemotactic and inflammatory cytokines. *Crit. Rev. Immunol* 12 : 17-46, 1992.
9. Baggiolini, M., Dewalt, B., Mosser, B. Interleukin-8 and related chemotactic cytokines CXC and C C chemokines. *Adv. Immunol* 55 : 97-179, 1994.
10. Murphy, P M. The molecular biology of leukocyte chemoattractant receptors. *Annu. Rev. Immunol* 12 : 593-633, 1994.
11. Hulkower, K., Brosnan, C F., Aquino, D A., Cammer, W., Kulshrestha, S., Guida, M P., Rapoport, D A., Berman, J W. Expression of CSF-1, c-fms, and MCP-1 in the central nervous system of rats with experimental allergic encephalomyelitis. *J. Immunol* 150 : 2525-2533, 1993.
12. Ransohoff, R M., Hamilton, T A., Tanie, M., Stoler, M H., Shick, H E., Major, J A., Estes, M L., Thomas, D M., Tuohy, V K. Astrocyte expression of mRNA encoding cytokines IP-10 and JE/MCP-1 in experimental autoimmune encephalomyelitis. *FASEB. J* 7 : 592-600, 1993.
13. Vanguri, P., Farber, J M. IFN and virus-inducible expression of an immediate early gene, crg-2/IP-10, and a delayed gene, I-Aα, in astrocytes and microglia. *J. Immunol* 152 : 1411-1418, 1994.
14. Hayashi, M., Luo, Y., Laning, J., Strieter, R M., Dorf, M E. Production and function of monocyte chemoattractant protein-1 and other β chemokines in murine glial cells. *J. Neuroimmunol* 60 : 143-150, 1995.
15. Yoshimura, T., Robinson, E A., Tanaka, S., Appella, E., Kuratsu, J I., Leonard, E J. Purification and amino acid analysis of two human glioma-derived monocyte chemoattractants. *J. Exp. Med* 169 : 1449-1459, 1989.
16. Introna, M., Bast, R C., Tannenbaum, C S., Hamilton, T A., Adams, D O. The effect of LPS on expression of the early "competence" genes JE and KC in murine peritoneal macrophages. *J. Immunol* 138 : 3891-3896, 1987.
17. Yoshimura, T., Robinson, E A., Tanaka, S., Appella, E., Leonard, E J. Purification and amino acid analysis of two human monocyte chemoattractants produced by phytohemagglutinin-stimulated human blood mononuclear leukocytes.*J. Immunol* 142 : 1956-1962, 1989.
18. Takehara, K., Leroy, E C., Grotendorst, G R. TGFβ inhibition of endothelial cell proliferation : Alteration of EGF binding and EGF-induced growth-regulatory (competence) gene expression. *Cell.* 49 : 415-422, 1987.
19. Cochran, B H., Reffel, A C., Stiles, C D. Molecular cloning of gene sequences regulated by platelet-derived growth factor. *Cell* 33 : 939-947, 1983.
20. Valente, A J., Graves, D T., Vialle-Valentin, C E., Delgado, R., Schwartz, C J. Purification of a monocyte chemotactic factor secreted by non human primate vascular cells in culture. *Biochemistry* 27 : 4162-4168, 1988.
21. Standiford, T J., Kunkel, S L., Phan, S H., Rollins, B J., Strieter, R M. Alveolar macrophage-derived cytokines induce monocyte chemoattractant protein expression from human pulmonary type II -like epithelial cells. *J. Biol. Chem.* 266 : 9912-9918, 1991.

22. Graves, D T., Jiang, Y L., Williamson, M J., Valente, A J. Identification of monocyte chemotactic activity produced by malignant cells. *Science* 245 : 1490-1493, 1989.
23. Zachariae, C O C., Anderson, A O., Thompson, H L., Appella, E., Mantovani, A., Oppenheim, J J., Matsushima, K. Properties of monocyte chemotactic and activating factor (MCAF) purified from a human fibrosarcoma cell line. *J. Exp. Med* 171 : 2177-2182, 1990.
24. Jiang, Y., Beller, D I., Frendl, G., Graves, D T. Monocyte chemoattractant protein-1 regulates adhesion molecule expression and cytokine production in human monocytes. *J. Immunol* 148 : 2423-2428, 1992.
25. Shyy, Y J., Wickham, L L., Hagan, J P., Hsieh, H J., Hu, Y L., Telian, S H., Valente, A J., Paul Sung, K L., Chien, S. Human monocyte colony-stimulating factor stimulates the gene expression of monocyte chemotactic protein-1 and increases the adhesion of monocytes to endothelial monolayers. *J Clin Invest* 92 : 1745-1751, 1993.
26. Théry, C., Chamak, B., Mallat, M. Cytotoxic effect of brain macrophages on developing neurons. *Eur. J. Neurosci* 3 : 1155-1164, 1991.
27. Dijkstra, C D., Döpp, E A., Joling, P., Kraal, G. The heterogeneity of mononuclear phagocytes in lymphoïd organs : distinct macrophage subpopulations in the rat recognized by monoclonal antibodies ED1, ED2 and ED3. *Immunology* 54 : 589-599, 1985.
28. Tushinski, R J., Oliver, I T., Guibert, L J., Tynan, P W., Warner, J R., Stanley, E R. Survival of mononuclear phagocytes depends on a lineage-specific growth factor that the differentiated cells selectively destroy. *Cell* 28 : 71-81, 1982.
29. Marty, S., Dusart, I., Peschanski, M. Glial changes following an excitotoxic lesion in the CNS. I. Microglia / Macrophages. *Neuroscience* 45 : 529-539, 1991.
30. Yoshimura, T., Takeya, M., Takahashi, K. Molecular cloning of rat monocyte chemoattractant protein-1 (MCP-1) and its expression in rat spleen cells and tumor cell lines. *Biochem. Biophys. Res. Commun* 174: 504-509, 1991.
31. Fort, P., Marty, L., Piechaczyk, M., El Sabrouty, S., Dani, C., Jeanteur, P., Blanchard, J M. Various rat adult tissues express only one major mRNA species from the glyceraldehyde-3-phosphate-deshydrogenase multigenic family. *Nucleic. Acids. Res* 13 : 1431-1442, 1985.

T-CELL LINE INTERACTIONS WITH BRAIN ENDOTHELIAL MONOLAYERS

P. M. G. Munro,[1] L. A. McLaughlin-Borlace,[1] G. Pryce,[1] N. L. Occleston,[2] and J. Greenwood[1]

[1] Department of Clinical Science
[2] Department of Pathology
Institute of Ophthalmology
University College London
EC1V 9EL London, United Kingdom

SUMMARY

In this communication we outline and illustrate some of the many structural and functional similarities that *in vitro* lymphocyte-endothelial interactions share with inflammation *in vivo*. Data presented here suggests, but does not prove, lymphocyte diapedesis *in vitro* to be a parajunctional process also involving enzymatic destruction of the endothelial basement membrane.

RÉSUMÉ

Dans cette communication, nous soulignons et illustrons les similarités structurales et fonctionnelles entre les interactions endothelium-lymphocytes *in vitro* et l'inflammation *in vivo*. Les données présentées ici suggèrent mais ne prouvent pas que la diapédèse des lymphocytes *in vitro* est un processus impliquant la destruction enzymatique de la membrane basale endothéliale.

1. INTRODUCTION

Lymphocyte recruitment at sites of inflammation is a regulated process involving cell-cell and cell matrix interactions. The earliest stages of the recruitment process, namely rolling and stable arrest prior to diapedesis are known from *in vitro* studies to be mediated by endothelial selectins and integrin interactions with their lymphocyte ligands (Luscinskas, Ding and Lichtman, 1995, Berlin, Bargatze, Campbell *et al*, 1995 and Ratner, 1992). The

Biology and Physiology of the Blood–Brain Barrier, edited by Couraud and Scherman
Plenum Press, New York, 1996

subsequent process of diapedesis, involving movement through both the endothelial monolayer and the underlying basement membrane is less well documented.

In this study we describe preliminary findings of a scanning and transmission electron microscopic investigation aimed at determining the route and effects of S-antigen specific T-lymphocyte migration through cytokine activated monolayers of rat brain endothelial cells.

2. MATERIALS AND METHODS

Confluent rat brain endothelial monolayers (BEM's) were prepared using the method of Abbot (1992) and cocultured with S-antigen specific T-cell lines for four hours as described by Greenwood and Calder (1994). Cells were fixed with 3% glutaraldehyde and 1% paraformaldehyde in 0.1M sodium cacodylate buffer pH 7.4, rinsed in PBS and post-fixed for 1 h at 4°C with 1% aqueous osmium tetroxide. Following dehydration through ascending alcohols (50 - 100%, 10 mins), cells were either embedded in araldite for transmission electron microscopy (TEM) or critical point dried, and sputter coated for scanning electron microscopy (SEM). Thin sections were cut using a diamond knife and Reichert Ultracut S microtome, contrasted with uranyl acetate and lead citrate and examined and photographed in a JEOL 1010 TEM at 80kV. Scanning preparations were examined in a JEOL 6100 or Hitachi 520 SEM.

3. RESULTS

In the absence of inflammatory cells BEM's comprise a squamous monolayer of spindle shaped cells producing and covering a single or sometimes double layered basement membrane, Figure 1a. Cells within this monolayer vary in height from 1.5 μm at the nucleus

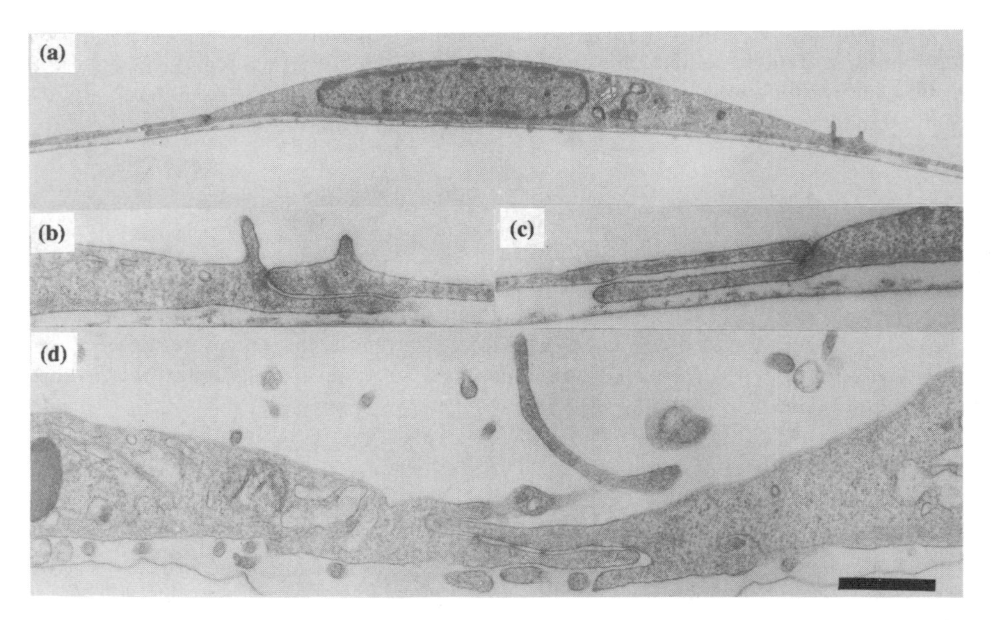

Figure 1. *Endothelial morphology.* Endothelial cells form a distinctive monolayer of squamous cells (a) whose individual members are joined to one another by marginal tight junctions (b-d). *Scale bars: a = 2μm and b-d= 0.5μm.*

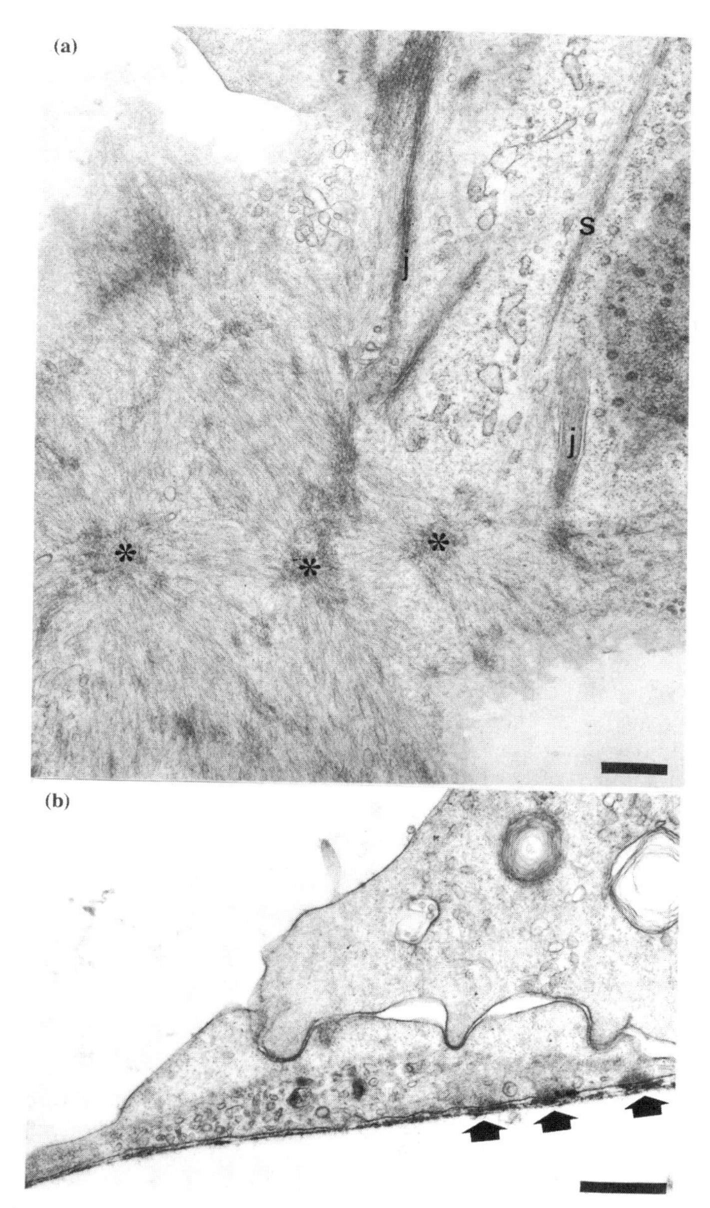

Figure 2. *Cytoskeletal actin organisation.* (a) Grazing *en face* section demonstrating junctional (J), sub-membranous and stress fibre (S) actin arrangements in the parajunctional zone of two adjacent endothelial cells. The asterisks denote focal densities interconnected by actin microfilaments. (b) Vertical section showing forceful probing of the apical endothelial membrane by a lymphocyte. Note the focal concentrations of basal membrane associated actin (arrows) correponds to points of matrix attachment. *Scale bars: a & b = 0.5*μm.

to 0.2 μm at the periphery. Close apposition of neighbouring cell membranes together with increased electron density and a concentration of cytoplasmic actin combine to provide ultrastructural evidence of tight-junctional intercellular contacts being present, Figure 1b-d. In all of these respects primary endothelial cultures closely resemble their *in vivo* counterparts.

Actin, the most prominent cytoskeletal component in primary endothelial cultures is present in sub-membranous locations and is particularly abundant in parajunctional cytoplasm. Some of the actin seen by TEM is organised as stress fibres whose long axes tend to be orientated slightly oblique to that of the cell.

At the present moment we are unable to determine whether sub-membranous actin associated with the apical membrane differs in organisation from that associated with attachment to the basement membrane. As demonstrated by the *en face* section shown in Figure 2a, basal membrane associated cortical actin is organised in a criss-cross fashion. Interestingly, the electron dense intersections can be correlated with focal attachments to the basement membrane, Figure 2b. The same cannot be said for actin associated with the apical membrane.

Lymphocyte-endothelial interactions thought most likely to be associated with diapedesis, as opposed to surface rolling, tended to occur in parajunctional locations and involve lymphocytes with rounded-up rather than flattened morphologies. Such cells were also seen to compress the endothelial cell via a series of protuberances, some involving coated pits, Figure 2b.

It was evident from TEM that transmonolayer diapedesis involved migration through pores whose diameter was less than that of the migrating lymphocyte, Figure 3a. By constricting the lymphocyte, these small diameter pores reduce blood-brain barrier disruption to a minimum by maintaining an effective seal around the invading cell. The apparent tendency for diapedesis to occur in close proximity to normal appearing tight-junctions was more confidently demonstrated by SEM, Figure 3b.

One of the surprising features of these investigations was the relative vigour with which lymphocytes physically probe both endothelial cells and the basement membrane. Figure 2a has already shown how lymphocyte pressure can compress the endothelial cells so that apical and basal membranes are brought together - possibly to the point where they fuse to form a migration pore. As Figure 3c and zymographic analyses of culture medim (not shown) demonstrate, lymphocytes also supplement enzymatic degradation of the basement membrane with 92 and 72 kD gelatinases with equally vigourous probing.

DISCUSSION

It is evident from our observations that primary cultures of rat BEM's retain many of the morphological features of their *in vivo* counterparts and therefore serve as a useful model with which to investigate lymphocyte diapedesis and its consequences. So far we have no specific observations enabling us to determine the precise route of transmonolayer migration. In keeping with *in vivo* observations (Lossinsky, Badmajew, Robson *et al* 1989, Raine,Cannella, Duijvestijn *et al*, 1990, Greenwood, Howes and Lightman, 1994) our data shows lymphocytes probing endothelial cells at parajunctional sites where the underlying endothelial cytoplasm contains cortical cytoskeleton and few organelles. Moreover, the small distance (0.2μm) separating basal and apical membranes in this region might facilitate pore formation via an applied lymphocytic pressure of the type illustrated in Figure 2b.

It is also noteworthy that in Figures 3a,b lymphocytes actually engaged in diapedesis assume a polarised morphology and appear to be constricted as they pass through the endothelial cell monolayer. Constriction by the endothelial cells would clearly minimise

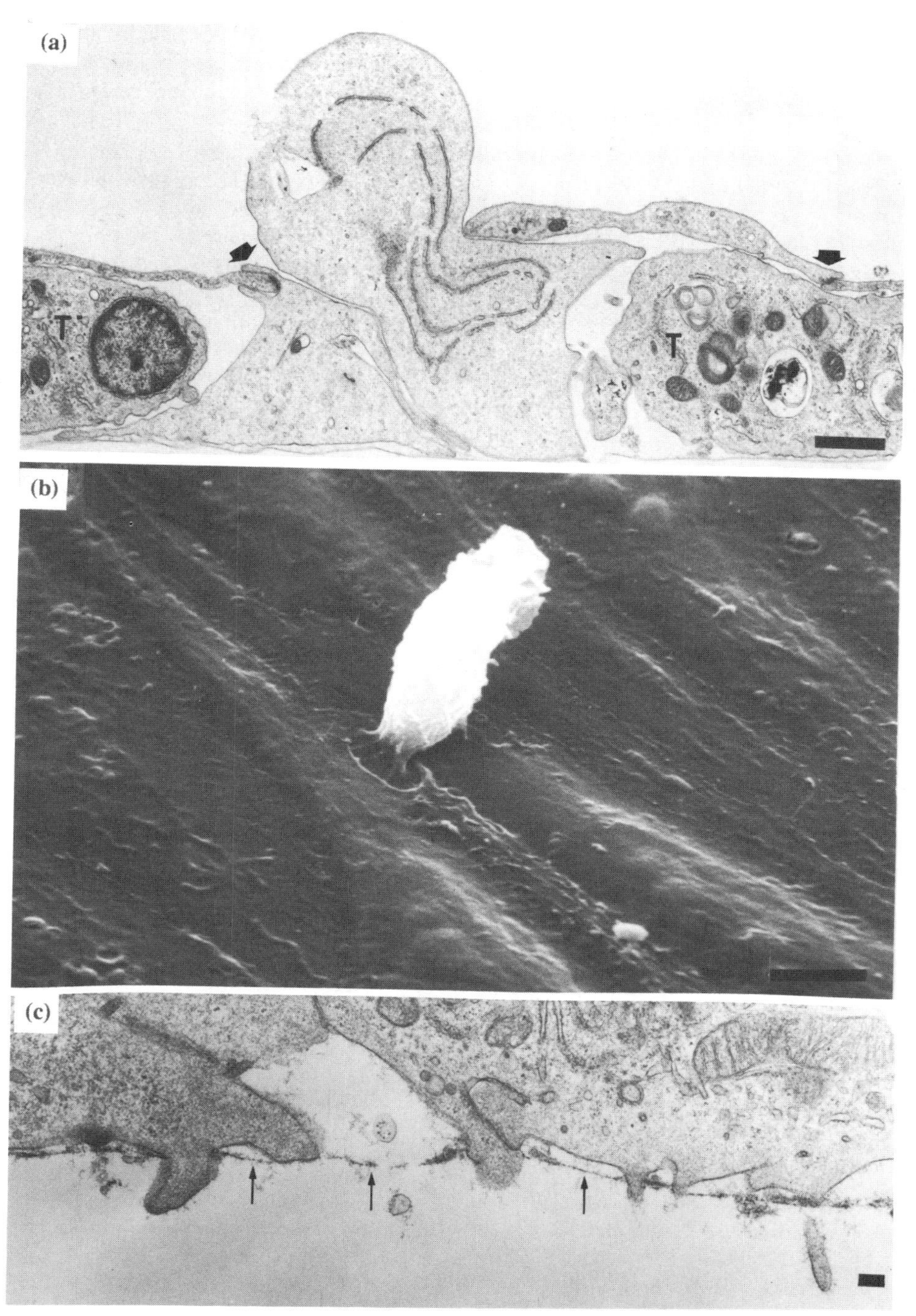

Figure 3. *Lymphocyte diapedesis.* (a) TEM illustrating T-cell penetration of the monolayer at a site adjacent to normal appearing tight junctions (arrows) . Note the absence of basement membrane and the presence of two previously transmigrated T-cells (T). (b) SEM showing parajunctional penetration of an endothelial cells by a a polarised lymphocyte. (c)Transmigrated lymphocyte probing a fragmenting basement membrane (small arrows). *Scale bars: a = 1*μm, *b =* 2μm *and c = 0.2* μm.

blood-brain barrier disruption. Unfortunately, it is not possible to determine by ultrastructural criteria whether the pores through which lymphocytes were seen to squeeze represent limited opening of an interecellular junction or a transcellular pore. Further work, involving serial sectioning is required to resolve this important question.

The absence of basement membrane at sites where more than one lymphocyte have traversed the monolayer is evidence of enzymatic degradation, see Figures 3a & c. Our preliminary zymographic analyses of culture media reveals evidence of both 72kDa (EC 3.4.24.24) and 92kDa (EC 3.4.24.35) gelatinases in both active and proenzymic forms. These two enzymes are of particular importance since they have the capacity to specifically degrade major basement membrane components such as collagen types IV and V and laminin, (Leppert,Waubant and Galardy *et al,* 1995). The 72kDa gelatinase of lymphocytes is VCAM-1 induced (Romanic and Madri, 1994) while its 92kDa counterpart is constitutively expressed but upregulated on activation (Leppert et al, 1995). Both are required for migration through the basement membrane. It is clear from our TEM data, as well as *in vivo* observations (Greenwood, *et al*,1994, Figure 10e) that enzymatic degradation is supplemented by relatively vigorous probing by invading lymphocytes. Immunolabelling will be required to determine whether these microvillus projections bear a surface bound form of gelatinase A as do invasive tumour cells (Sato, Takino, Okada *et al*, 1994).

ACKNOWLEDGMENTS

Presentation of this poster at the 1995 CVCB was made possible by a Wellcome Trust Travel Grant awarded to P. Munro.

REFERENCES

Abbot, NJ, Hughes, CCW, Revest, PA and Greenwood, J. (1992) Development and characterisation of a rat brain capillary endothelial culture: towards an *in vitro* blood-brain barrier. *J. Cell Sci.,* 103, p23.

Berlin, C, Bargatze, RF, Campbell, JJ, von Andrian, UH, Szabo, MC, Hasslen, SR, Nelson, RD, Berg, EL, Erklandsen, SL and Butcher, EC.(1995) α4 integrins mediate lymphocyte attachment under physiologic flow. *Cell,* 80, p411.

Greenwood, J. and Calder, V.L.(1994) Lymphocyte migration through cultured endothelial cell monolayers derived from the blood-retinal barrier. *Immunology,* 80, p401.

Greenwood, J, Howes, R and Lightman, S (1994) The blood-retinal barrier in experimental autoimmune uveoretinitis: leukocyte interactions and functional damage. *Lab. Invest,* 70, p39.

Leppert, D, Waubant, E, Galardy, R, Bunnett NW and Hauser, SL. (1995) T cell gelatinases mediate basement membrane transmigration in vitro *J.Immunol,* 154, p4379.

Lossinsky, A.S., Badmajwew, V., Robson, J.A., Moretz, R.C and Wisniewski, H.M. (1989) Sites of egress of inflammatory cells and horseradish peroxidase transport across blood-brain barrier in a murine model of chronic relaspsing expermental allergic encephalomyelitis. *Acta Neuropathologica,* 78, p351.

Luscinskas, FW, Ding, H and Lichtman, AH. P-selectin and vascular cell adhesion molecule-1 mediate rolling and arrest, respectively, of CD4+ T-lymphocytes on tumour necrosis α-activated vascular endothelium. (1995) *J. Exp.Med,* 181, p1179.

Raine, CS, Canella, B, Duijvestijn, AM, Cross,AH (1990) Homing to central nervous system vasculature by antigen-specific lymphocytes. II. Lymphocyte/endothelial cell adhesion during the initial stages of autoimmune demyelination. *Lab. Invest,* 63, p476

Ratner, S. (1992) Lymphocyte migration through extracellular matrix. *Invasion Metastasis,* 12, p82.

Romanic, A.M. and Madri, J.A. (1994) The induction of 72kD gelatinase in T-cells upon adhesion to endothelial cells is VCAM-1 dependent. *J.Cell Biology,* 125, p1165.

Sato, H, Takino, T, Okada, Cao, J, Shinagawa, A, Yamamoto, E and Seiki, M. (1994) A matrix metalloproteinase expressed on the surface of invasive tumour cells. *Nature,* 370, p61

DEVELOPMENT AND CHARACTERIZATION OF SV40 LARGE T IMMORTALIZED ENDOTHELIAL CELLS OF THE RAT BLOOD-BRAIN AND BLOOD-RETINAL BARRIERS

P. Adamson, G. Pryce, V. Calder, and J. Greenwood

Department of Clinical Ophthalmology
Institute of Ophthalmology
University College London
London EC1V 9EL
United Kingdom

SUMMARY

In a series of recent reports an immortalised rat brain endothelial cell line (RBE4) has been described that has proved to be a valuable resource for studying the properties of cerebral endothelia. Using a different approach, we have developed immortalised endothelial cell lines derived from primary cultures of both rat brain and rat retina. Cells were immortalised using virus containing supernatants generated from producer cells containing a replication deficient SV40 retrovirus encoding the tsa58 large T-antigen and a neomycin selectable marker. Parent lines of brain (IO/GP8) and retinal (IO/JG2) which were resistant to exposure to G418 were expanded and cloned by limiting dilution. Using morphological criteria one brain (IO/GP8/3) and one retinal (IO/JG2/1) clone was expanded for further investigation. The clones stained positive for the presence of large-T-antigen. Fixed cell ELISA and flow cytometry demonstrated that the immortalised cells expressed antigens specific to CNS-derived endothelial cells and could be induced to produce other immunologically important molecules such as MHC class II, ICAM-1 and VCAM-1.

RÉSUMÉ

Une Série de récents rapports mentionnent l'utilisation d'une lignée immortalisée de cellulles endothéliales, isolée à partir du cerveau de rat, comme modèle cellulaire

Biology and Physiology of the Blood–Brain Barrier, edited by Couraud and Scherman
Plenum Press, New York, 1996

approprié pour l'étude in vitro, des propriétés de l'endothélium cérébral. En utilisant une approche différente, nous avons développé de notre côté, deux lignées immortalisées de cellules endothéliales de rat dérivées de cultures primaires à partir du cerveau, d'une part et de la rétine, d'autre part. Ces cellules ont été immortalisées par l'utilisation d'une souche déficiente du retrovius SV40, contenu dans des surnageants de culture de cellules infectées et codant pour l'antigène T sous contrôle du gène néomycine. Deux lignées parentales furent isolées, (IO/GP8) et (IO/JG2) à partir du cerveau et de la rétine, respectivement, après culture en présence de G418- Elles furent utilisées pour l'expansion et le clonages par dilution limite. En utilisant alors des critères morphologiques, nous avons dérivé deux lignées IO/GP8/3 à partir du cerveau et IO/JG2/1 à partir de la rétine, pour de futures investigations. Ces deux clônes se révèlent positifs pour l'expression de l'antigène T. De plus, des ELISA réalisés sur cellules fixées et des études complémentaires de cytométrie de flux confirment l'expression des antigènes spécifiques de cellules endothéliales du système nerveux central. Aussi, est-il possible sur ces cellules d'induire l'expression d'autres molécules d'intérêt immunologique comme les molécules de classe II du CMH, ICAM-1 et VCAM-1.

1. INTRODUCTION

The specialised endothelial cells of the CNS vasculature form a selective cellular interface between the blood and CNS parachema. Unlike many other vascular beds the endothelial cells which line CNS blood vessels are characterised by their ability to form tight intercellular junctions and coupled with minimal pinocytosis and pores yield impermeable vessels which restrict the passage of small molecules into the CNS. This restriction has led to the terms blood-brain and blood retinal barriers. The specialisation of CNS endothelia is not restricted to the maintenance of a physical barrier since it has been observed that the endothelial cell surface of cells derived from the CNS is considerably less adhesive for circulating leucocytes than cells derived from other vascular beds (Hughes et. al., 1988; Wang et. al., 1993). In addition CNS endothelia show expression of surface molecules which are absent from non-CNS endothelia such as the high affinity glucose transporter GLUT-1, transferrin receptor and the product of the multidrug resistance gene P-glycoprotein (Pgp). Until recently, investigations were restricted to in vivo studies but the advent of techniques to isolate and culture cells of both the blood brain and blood-retinal barriers has allowed increased experimental flexibility. However these cells are phenotypically unstable in long term culture and relatively few cells are isolated. The instability of cells in culture has led most investigators to restrict in vitro studies to primary culture cells. Now that techniques are readily available to immortalise cells, the development of immortalised cell lines from both blood-brain and blood-retinal barriers will be of considerable use in investigating the properties of both the blood-brain and blood-retinal barriers. However for this approach to be of value immortalised cells must retain the original characteristics of primary culture cells over many passages. Recently an immortalised cell line derived from rat brain endothelium (RBE4) using the adenovirus E1A gene has been described which retains many of the characteristics of primary culture endothelium and is proving to be a useful resource in a number of investigations (Roux et. al., 1994). However since the process of immortalisation is based on the random chromosomal integration of the immortalising gene it is important that a number of cell lines are constructed. Here we describe the development and characterisation of immortalised rat endothelium derived from both brain and retina.

2. METHODS

2.1. Preparation of Rat Brain, Retinal, and Aortic Endothelium

Endothelial cells were isolated from the brain and retina of specific pathogen free (SPF) 4-6 week old female Lewis rats (Charles River Ltd., Kent, U.K) and cultured according to previously described methods (Greenwood, 1992; Abbott et al., 1992). These techniques routinely produce primary cultures of >95% purity. Briefly, rat retina or chopped cerebral cortex was dispersed by enzymatic digestion and microvessel fragments separated from other material and single cells by density dependent centrifugation on Percoll. The microvessel fragments were washed and plated on to collagen coated plastic tissue culture flasks. Growth media consisted of Ham's F-10 medium (Sigma) supplemented with 17.5% plasma derived serum (Advanced Protein Products Ltd., West Midlands, UK), 7.5μg/ml endothelial cell growth supplement (Advanced Protein Products Ltd.), 80μg/ml heparin, 2mM glutamine, 0.5μg/ml vitamin C, 100U/ml penicillin and 100μg/ml streptomycin (all Sigma). The cultures were maintained at 37°C in 5% CO2 and media replaced every 3 days until the formation of monolayers.

Aortic endothelium was isolated by the method described by McGuire and Orkin (1987). Rat aorta was removed by dissection, cut into small pieces (2-5 mm) and placed luminal side down onto collagen-coated 24 well plates in endothelial cell growth medium. The media consisted of RPMI media supplemented with 20% foetal calf serum, 7.5μg/ml endothelial cell growth supplement (Advanced Protein Products Ltd.), 80μg/ml heparin, 2mM glutamine, 0.5μg/ml vitamin C, 100U/ml penicillin and 100μg/ml streptomycin. After 3 days the explants were removed and outgrowing cells were expanded and passaged by trypsinisation. At confluence the cells had the "cobblestone" morphology characteristic of large vessel endothelium, expressed von Willebrand factor and grew in medium containing D-valine (a capacity lacking in fibroblasts and smooth muscle cells). Cells were used after passage 3 which is the earliest stage at which sufficient cells were available for experimentation.

2.2. SV40 Large T Immortalisation of Brain and Retinal Endothelium

A replication deficient SV40 retrovirus was produced from the producer cell line SVU19.5 (a generous gift from Dr. P.S. Jat, Ludwig Institute, London UK) in which a packaging defective mouse moloney leukemia provirus was present. The retroviral vector encoded a full length large T antigen containing both tsa58 and U19 mutants as well as the neo^r selectable marker (aminoglycoside phosphotransferase I). The supernatant derived from the producer cell line was passed through a 0.45μm filter to remove unwanted producer cells and added to primary cultures of rat brain or retinal endothelium plated 2-3 days prior to the transfection. 200μl of virus in 2ml of media containing 8μg/ml polybrene (Aldrich, Dorset, UK) was added to the cells. This was left for 4 h at 37°C in a CO_2 incubator, gently shaking the flask at 15 minute intervals. Following this incubation media was removed and 5ml of fresh medium added and cultured overnight. Cells were then maintained in the incubator as normal for 48 hrs. Cells were then plated into selective media containing 200 μg/ml G418 (geneticin, Gibco) and immortalized parent lines were obtained by selection of resistant colonies. Clones were acquired from parent lines by trypsinization and plating into 96 well plates at a concentration of 0.33 cell per well. From these plates a number of clones were obtained and on morphological criteria a single clone from each parent line was expanded.

2.3. Detection of Endothelial Surface Antigens by ELISA, Immunohistochemistry, and Flow Cytometry

Primary cultures and immortalised lines of rat brain and rat retinal endothelial cells were seeded at confluent density onto 96 well plates which had previously been coated 0.05% type IV collagen. Before plating the collagen was fixed in ammonia vapour and plates washed twice with HBSS. Cells were cultured for 3 days before use in experiments. After experimental treatments cells were washed four times in ice-cold Hanks buffered salt solution (HBSS) and fixed with 0.1% glutaraldehyde in phosphate buffered saline A (PBSA) for 10 min at room temperature. Aldehydes were subsequently quenched with 50mM Tris-HCl, pH7.5 for 20 min at room temperature. Primary antibodies were diluted in 100μl HBSS containing 100μg/ml normal rabbit IgG and 4mg/ml bovine serum albumin and then incubated with cells for 45 min at 37°C. Cells were washed 4 times with PBSA containing 0.2% Tween-20 and incubated with biotinylated anti-mouse-IgG (1:700; Amersham Int., UK) for 45 min at 37°C. Cells were again washed 4 times with PBSA containing 0.2% Tween-20 and incubated with streptavidin-horse radish peroxidase (1:700; Amersham Int., UK) for 45 min at 37°C. Cells were washed 4 times in PBSA containing 0.2% Tween-20 and incubated with 100μl tetramethylbenzidine in citrate-acetate buffer (pH5) for 10 min. Reactions were stopped by the addition of 50μl 1M sulphuric acid and product quantitated by optical density at 450nm. For immunohistochemical studies immortalized brain and retinal EC were seeded onto LabTek chamber slides (Gibco/BRL) and grown to confluence. Cells were fixed and permeabilised as for ELISA and antigens visualised with anti-mouse-FITC. Flow cytometry of confluent retinal or brain EC cultures was performed on a FACScan (Becton-Dickinson FACScan, Oxford, UK). After washing cell monolayers were dissociated in 0.2% EDTA and cells resuspended in phosphate buffered saline. 5×10^4 cells per vial were incubated for 1 h with primary antibody followed by a further 1 hr incubation with FITC conjugated rabbit anti-mouse IgG F(ab)2 antibody (FITC-RAMIgG) in the presence of 20% normal rat serum. After washing twice, cells were resuspended in phosphate buffered saline and analysed. Unstained cells were used to set the parameters, and cells stained with FITC-RAMIG alone were used to set background control.

3. RESULTS

3.1. Morphology

Endothelia: The primary cultures of brain and retinal endothelia exhibited spindle shaped morphology characteristic of these cells (Fig 1). The morphology of the parent lines differed from the primary cultures in that the brain endothelia (IO/GP8) had a slightly more cobblestone morphology. With the retinal parent line, however, the morphology remained predominantly spindle shaped although was different from the primary cultures in that they were more thickened (Fig.1). The clones used in subsequent investigations were designated IO/GP8/3 for the brain derived endothelial and IO/JG2/1 for retina.

3.2. Expression of tsa58 Large T-Antigen in Immortalised EC

All immortalised retinal and brain endothelial cells selected in G418 all showed strong nuclear staining with mouse anti-T-antigen IgG. No staining with mouse anti-T-antigen was observed in primary cultures (Fig.2).

3.3. Expression of Major Histocompatability Antigens

All EC cultures showed basal expression of MHC class I (OX-18) which was induced by treatment with either IFN-γ or TNFα. Primary cultures of retinal or brain EC as well as all T-antigen expressing parent lines and clones showed no basal expression of MHC class II. Cultures treated for rat recombinant IFN-γ but not TNFα showed strong induction of MHC class II I-A (OX-6) and MHC class II I-E (OX-17) (Table 1).

3.4. Expression Of Specific CNS-Endothelial Markers

Primary cultures of retinal and brain endothelium plus all T-antigen expressing endothelial cell lines and derived clones showed basal expression of the CNS-endothelium specific markers GLUT-1, P-glycoprotein and transferrin receptor but these antigens were not observed on aortic endothelium. In addition the phenotype-discriminating OX-43 which is only found on peripheral endothelia was not present on primary cultures or immortalised endothelial cells but was present on aortic endothelium (Table 1).

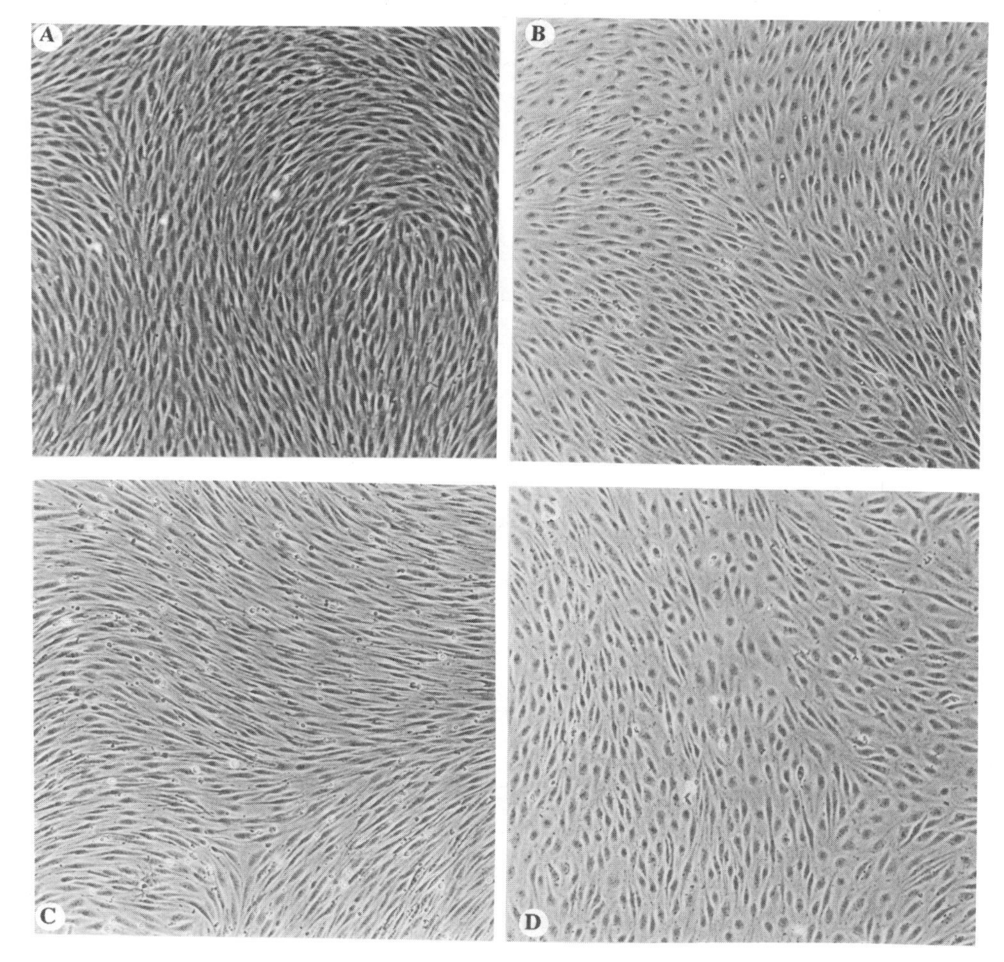

Figure 1. Morphology of primary cultures of rat brain and retinal endothelial cell and G418 resistant clones. (A) Primary culture of rat retinal endothelial cells. (B) G418 resistant rat retinal endothelial cells (IO/JG2/1). (C) Primary culture of rat brain endothelial cells. (D) G418 resistant rat brain endothelial cells (IO/GP8/3).

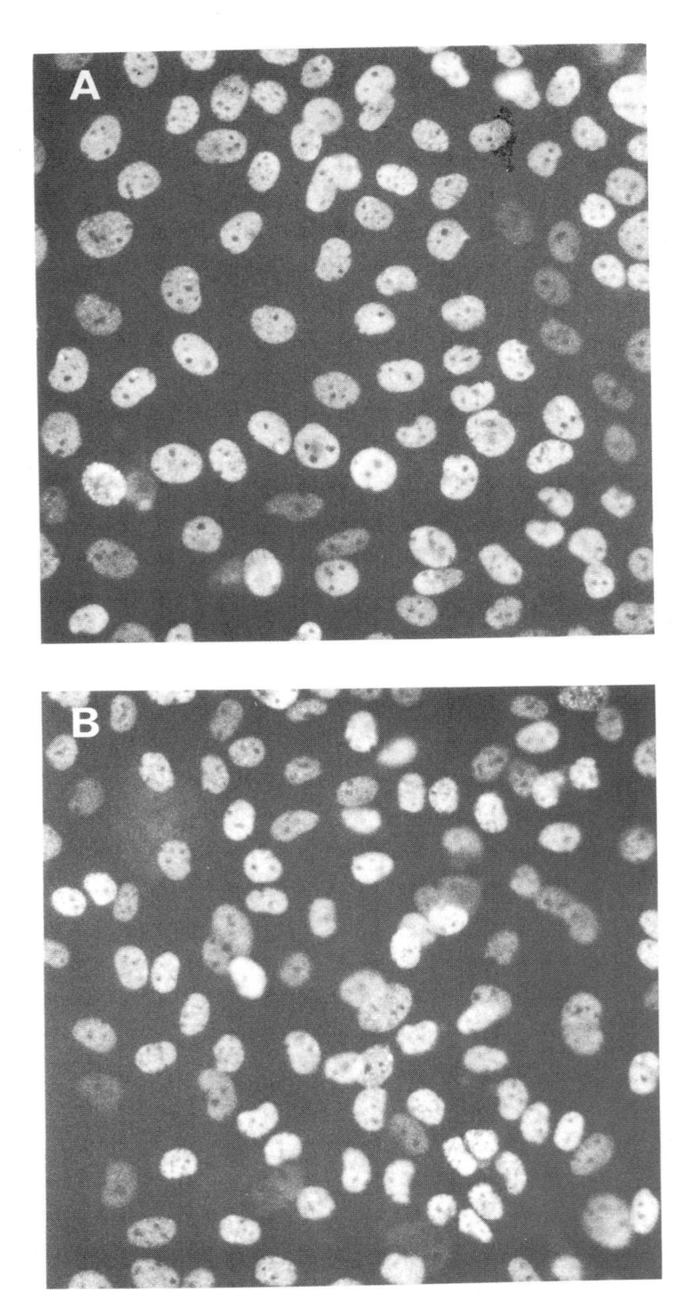

Figure 2. Immunohistochemical localisation of SV40 large T-antigen in G418 resistant Endothelial cell clones. Cells were fixed in 3% paraformaldehyde, permeabilised and stained and stained with mouse anti-T-antigen IgG followed by anti-mouse-FITC. (A) IO/JG2/1 (B) IO/GP8/3.

Table 1. Expression of antigens on primary cultures of rat endothelium and immortalized cell lines. Cells were plated into 96 well plates and were cultured for 3 days after which cells were fixed and expression of antigens determined by ELISA. Experiments in which cells were stimulated with either 200U/ml IFN-γ or 200U/ml TNFα were cultured in the presence of the cytokines for between 24 and 48 hrs. + = High level expression, (+) = low level expression, - = no expression. ND = not determined

Antigen	1° REC			IO/JG2 Parent			IO/JG2/1 Clone			1° BEC			IO/GP8 Parent			IO/GP8/3 Clone			Aortic EC		
	Basal	IFN	TNF	Basal	IFN	TNF	Basal	IFN	TNF	Basal	IFN	TNF	Basal	IFN	TNF	Basal	IFN	TNF	Basal	IFN	TNF
P-glycoprotein (JSB-1)	+	nd	nd	+	nd	nd	+	nd	nd	+	nd	nd	+	nd	nd	+	nd	nd	-	nd	nd
GLUT-1	+	nd	nd	+	nd	nd	+	nd	nd	+	nd	nd	+	nd	nd	+	nd	nd	-	nd	nd
Transferrin Rcp. (OX-26)		nd	nd	+	nd	nd	(+)	nd	nd	+	nd	nd	(+)	nd	nd	(+)	nd	nd	-	nd	nd
RECA-1	+	nd	nd	nd	nd	nd	+	nd	nd	+	nd	nd	nd	nd	nd	+	nd	nd	nd	nd	nd
ICAM-1 (3H8/1A29)	+	+	+	+	+	+	+	+	+	+	+	+	+	+	+	+	+	+	+	+	+
VCAM-1 (5F10)	-	+	+	-	+	+	-	+	+	-	+	+	-	+	+	-	+	+	+	+	+
Non-CNS EC (OX43)	-	nd	nd	-	nd	nd	-	nd	nd	-	nd	nd	-	nd	nd	-	nd	nd	+	nd	nd
CD44 (OX-50)	+	nd	nd	+	nd	nd	+	nd	nd	+	nd	nd	+	nd	nd	+	nd	nd	+	nd	nd
MHC class I (OX-18)	+	+	+	+	+	+	+	+	+	+	+	+	(+)	+	+	+	+	+	+	+	+
MHC class II I-A (OX-6)	-	(+)	-	-	+	-	-	+	-	-	+	-	-	+	-	-	+	-	-	+	-
3H12B	+	+	+	(+)	+	+	(+)	+	+	+	+	+	(+)	+	+	+	+	+	+	+	+
4A2	+	+	+	+	+	+	+	+	+	+	+	-	+	+	+	+	+	+	+	+	+

3.5. Expression of Adhesion Molecules

All endothelial cell cultures constitutively expressed ICAM-1 (1A29 and 3H8 antisera) which could be further induced by treatment with 200U/ml IFN-γ or 200U/ml TNFα for 24 hrs (Table 1). Primary cultures of endothelial cells, and all T-antigen expressing cells showed no basal expression of VCAM-1 (5F10). VCAM-1 was induced in both primary cultures of retinal and brain endothelial cells and T-antigen expressing derivatives after treatment for 24 or 48 hrs with 200U/ml IFN-γ or 200U/ml TNFα but was noticeably stronger at 48 hrs (Table1).

4. DISCUSSION

Cells from both rat brain and retina which have been immortalized with SV40 large T-antigen expressed endothelial adhesion molecules and MHC molecules in response to proinflammatory cytokines and showed a similar morphology to primary cultures. In contrast to human cells where transfection of the SV40 large T antigen only results in an extended life span of endothelial cell cultures, rodent cells appear to be truly immortalised and appear stable in culture (>30 passages) Cells derived from primary cultures of both brain and retinal endothelial cells expressing large T-antigen remained anchorage dependent and unable to grow on coated glass. These cells can be grown in large scale culture, and in combination with other immortalised cell lines such as the RBE4 cells will greatly aid biochemical studies of the blood-brain barrier.

5. REFERENCES

1. Abbott, N.J., Hughes, C.C.W., Revest, P.A and Greenwood, J. (1992) J. Cell. Sci. 103: 23.
2. Greenwood, J. (1992) J. Neuroimmunol. 39: 123.
3. Wang, Y., Greenwood, J., Calder, V. and Lightman, S. (1993) J. Neuroimmunol. 48: 161.
4. McGuire P.G. and Orkin R.W. (1987) Lab. Invest. 57: 94.
5. Hughes C.C.W., Male D.K. and Lantos P.L (1988) Immunology 64: 677.

GLYCERALDEHYDE-3-PHOSPHATE DEHYDROGENASE BINDS TO THE CYTOPLASMIC DOMAIN OF INTERCELLULAR ADHESION MOLECULE-1

Christian Fédérici, Luc Camoin, Maurice Hattab, A. Donny Strosberg, and Pierre-Olivier Couraud

CNRS UPR-0415, ICGM
22 rue Méchain, 75014 France

SUMMARY

To elucidate the mechanisms of the transendothelial migration of leukocytes, we attempted to identify the cellular proteins capable of interaction with the cytoplasmic domain of the adhesion molecule intercellular adhesion molecule-1 (ICAM-1), in a rat brain microvessel endothelial cell line (RBE4 cells). A 27-residues peptide, corresponding to the cytoplasmic domain of rat ICAM-1, was synthesized and coupled to CNBr-activated Sepharose 4B. By affinity chromatography and elution with the soluble synthesized peptide, several ICAM-1 interacting proteins were purified from the RBE4 cell cytosol. By microsequencing, we could identify two of these proteins: the cytoskeletal protein β-tubulin (55 kDa) and the glycolytic enzyme glyceraldehyde-3-phosphate dehydrogenase, GAPDH (39 kDa), which participates in microtubule network assembly. Experiments carried out with purified GAPDH showed that the enzyme directly interacts with the ICAM-1 matrix. A series of synthetic ICAM-1 C-terminal peptides were tested for their ability to bind GAPDH and affect the glycolytic enzyme activity. The 15 amino-acid C-terminal peptide was found to completely block enzyme activity. In addition, GAPDH binding to the affinity gel could be competed by NAD^+. Our results suggest that GAPDH may interact with the C-terminal moiety of the ICAM-1 intracellular domain and that this interaction may affect the metabolic state of endothelial cells. The relationship between leukocyte adhesion and endothelial metabolism remains to be investigated.

RÉSUMÉ

Dans le but de préciser les mécanismes impliqués dans la migration transendothéliale des leucocytes, nous avons cherché à identifier, dans la lignée RBE4 de cellules endothéliales

Biology and Physiology of the Blood–Brain Barrier, edited by Couraud and Scherman
Plenum Press, New York, 1996

cérébrales de rat, les protéines cellulaires capables d'interagir avec le domaine cytoplasmique de la molécule d'adhérence ICAM-1. Un peptide de 27 résidus, correspondant au domaine cytoplasmique de la molécule ICAM-1 de rat, a été synthétisé et couplé à un support de Sépharose 4B. Cette matrice d'affinité a permis, à partir du cytosol de cellules RBE4, la purification de protéines capables de s'associer spécifiquement avec ICAM-1, l'élution se faisant grâce au peptide synthétisé soluble. Par microséquençage, nous avons pu identifier deux des protéines majoritaires: la β-tubuline (55 kDa) et l'enzyme glycolytique glyceraldehyde-3-phosphate deshydrogenase, GAPDH (39 kDa) qui participe à l'assemblage du réseau des microtubules. Des expériences effectuées avec la GAPDH purifiée montrent qu'elle interagit directement avec le support ICAM-1. Des peptides correspondant au domaine C-terminal d'ICAM-1 ont été testés pour leur capacité de se lier à la GAPDH, ainsi que pour leur effet inhibiteur de l'activité glycolytique de l'enzyme. Le peptide constitué des 15 acides aminés C-terminaux d'ICAM-1 bloque de manière totale cette activité. De plus, le NAD^+ permet l'élution de la GAPDH retenue sur le support ICAM-1. Nos résultats suggèrent que la GAPDH interagit avec la moitié C-terminale du domaine intracellulaire d'ICAM-1 et que cette interaction pourrait affecter le métabolisme des cellules endothéliales. La relation entre adhérence leucocytaire et métabolisme endothélial reste à être explorée.

INTRODUCTION

Endothelial cell adhesion molecules, acting as receptors for leukocyte integrins, mediate adhesion of leukocytes to vascular endothelium (1, 2). During inflammation responses, the up-regulated expression of many of these molecules initiates the transendothelial migration of leukocytes (3). Intercellular adhesion molecule-1 (ICAM-1, CD54), one of the specific ligands for the β2 integrins Mac-1 ($\alpha_M\beta2$) and LFA-1 ($\alpha_L\beta2$), is a transmembrane glycoprotein of the immunoglobulin family. As suggested for other adhesion molecules, ICAM-1 might participate in signal transduction through association of its cytoplasmic domain to cytoskeleton proteins. Such an interaction was recently reported by Carpén (4) with the actin-binding protein α-actinin. In a previous study, we have demonstrated that the tyrosine phosphorylation of the actin-binding protein cortactin was generated through activation of $p60^{src}$ by the cross-linking of ICAM-1 at the surface of brain microvessel endothelial cells (5). Here, RBE4 cytosolic proteins were isolated by affinity chromatography on the synthetic cytoplasmic domain of ICAM-1.

I. GAPDH AND β–TUBULIN: TWO CELLULAR PROTEINS INTERACTING WITH ICAM-1

The synthetic peptide C27, corresponding to the cytoplasmic domain of rat ICAM-1 (Table1), was coupled to CNBr-activated Sepharose-4B. The cytosolic fraction obtained from ~10^8 RBE4 cells was loaded on this affinity matrix. Bound proteins were eluted using the C27 soluble peptide at concentration of 2 mg/ml. The fraction was dialyzed for 5 hours at 4°C prior to freeze-drying, and analyzed by tricine-SDS-PAGE. Five major bands of molecular mass of 39, 45, 55, 71 and 89 kDa were detected. After blotting on polyvinylidene (PVDF) membrane, Coomassie blue-stained bands were excised and sequenced on a protein sequencer (model 473, Applied Biosystems). The N-terminal sequences of proteins corresponding to the 39 kDa and 55 kDa bands were determined as VXVGVNGFGRIGRL and MREIVHIQAGQXGNQI (in both sequences, only one residue, X, could not be unambiguously identified). Partial amino-acid sequences comparisons with protein data bank led to

Table 1. Amino acid sequence of synthetic ICAM-1 peptides, C27 corresponding to the cytoplasmic tail of rat ICAM-1 and the truncated peptides C10, C15, C18, C23. Controls unrelated to synthetic peptides are V19, corresponding to the cytoplasmic domain of VCAM-1, the lysine- and arginine-rich peptide (KR) and a 26-aa peptide (26)

C10:	LKLKVQAPPP
C15:	AQEEALKLKVQAPPP
C18:	LQKAQEEALKLKVQAPPP
C23:	IRIYKLQKAQEEALKLKVQAPPP
C27:	RQRKIRIYKLQKAQEEALKLKVQAPPP
KR:	LEDARRLKAIYEKKK
26:	LRNEIQEVKLEEGNAGKFRRARFLRY
V19:	RKANMKGSYSLVEAQKSKG

the identification of these two ICAM-1 interacting proteins: the monomer of the tetrameric glycolytic enzyme GAPDH and the cytoskeleton protein β-tubulin, respectively.

II. SPECIFICITY OF GAPDH RETENTION ON ICAM-1 SEPHAROSE

C27 and two synthetic ICAM-1 C-terminal truncated peptides (C15, C10) (Table 1), were tested for their ability to elute specifically purified GAPDH (0.5 mg) from ICAM-1 Sepharose. Complete GAPDH elution was observed with C27 and C15, whereas elution efficiency of C10 was limited (Fig.1). Irrelevant peptides (the cytoplasmic domain of the adhesion molecule VCAM-1 (V19) and the peptide KR with similar overall charge to C27) failed to elute the enzyme. Similar experiment carried out with purified β-tubulin showed a low level of specific retention, suggesting that GAPDH mediates to a great extent the tubulin binding observed in experiment with cytosolic fraction. In addition, micromolar (5μM) concentrations of NAD^+, co-substrate and modulator of the allosteric enzyme GAPDH, completely eluted GAPDH. These results suggest that (i) the interaction between GAPDH and the cytoplasmic tail of ICAM-1 requires the 15 C-terminal amino-acids of ICAM-1, (ii) the NAD^+- and ICAM-1 binding sites may overlap or interfere following conformational changes induced by NAD^+ binding to the enzyme.

III. INHIBITORY EFFECT OF TRUNCATED ICAM-1 PEPTIDES ON GAPDH ACTIVITY

GAPDH, one of the key enzymes of glycolysis, catalyzes, after NAD^+ binding, the oxidative phosphorylation of D-glyceraldehyde-3-phosphate in the presence of inorganic phosphate. C27 and truncated peptides (C23, C18, C15 and C10) were tested for their effect on GAPDH activity. C27, C23 and C10 (60μM) failed to modify the maximal activity of the enzyme, like irrelevant peptides used as controls (KR and a 26 residues peptide). On the

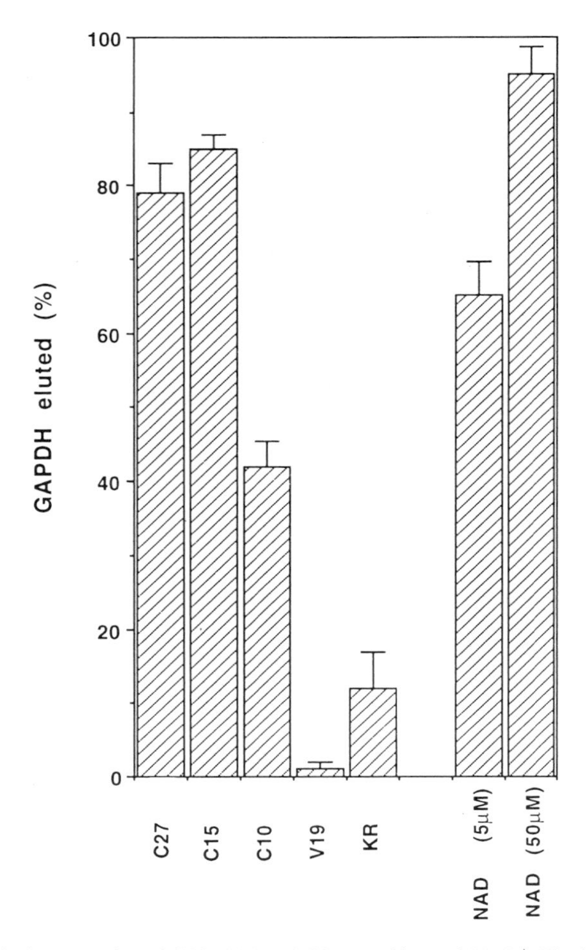

Figure 1. GAPDH elution capacity of C10, C15 and C27 peptides and NAD$^+$. Elution of purified GAPDH from ICAM-1 Sepharose was carried out with C27, the truncated peptides C10 and C15, the cytoplasmic domain of VCAM-1 (V19) and the lysine- and arginine-rich peptide (KR) (all at 2 mg/ml) or NAD$^+$. Quantification was performed by densitometry scanning of the band corresponding to GAPDH monomer; results are expressed as means ± S.E.M. (n=3), with100% corresponding to GAPDH eluted by NaCl 0.5M.

opposite, C18 and C15 (60µM) efficiently inhibited GAPDH activity (Fig.2A), in a dose-dependent manner (Fig.2B).

CONCLUSION

In conclusion, we have identified GAPDH, in the cytosolic fraction of immortalized rat brain microvessel endothelial cells, as an ICAM-1 binding protein. The capacity of NAD to elute GAPDH from ICAM-1 Sepharose and the C15 inhibitory effect on enzyme activity suggest that NAD$^+$- and ICAM-1 binding sites may overlap. Since GAPDH is known to bind to tubulin, we conclude that ICAM-1 may associate, via its C-terminal domain, with the microtubule network. Further studies performed in RBE4 cells are necessary to define more

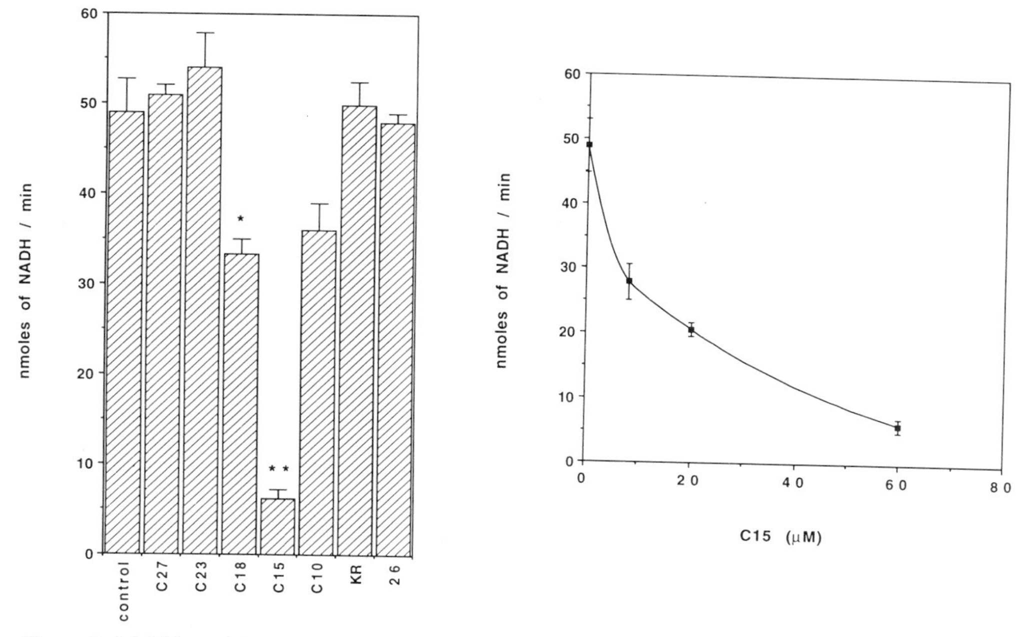

Figure 2. Inhibition of GAPDH glycolytic activity by ICAM-1 peptides. The glycolytic activity of GAPDH was calculated according to the increase of NADH absorbance at 340nm. A. Comparison of C10, C15, C18, C23 and C27 peptides. All peptides were used at the same concentration of 60 μM. Irrelevant peptides are the lysine- and arginine-rich KR peptide and a 26-aa peptide (26). Results are expressed in nmoles of NADH/min, as means ± S.E.M. (n=3). Inhibition by C15 and C18 peptides were significantly different from control (*: $p < 0.02$; **: $p < 0.001$). B. Dose-dependent inhibition of GAPDH glycolytic activity by C15. Results are expressed in nmoles of NADH/min, as means ± S.E.M. (n=3).

precisely the interplay between ICAM-1 and GAPDH, its possible regulation by leukocyte adhesion to endothelial cells and the role it might play on metabolic state of endothelial cells.

REFERENCES

1. Springer, T.A., Adhesion receptors of the immune system, *Nature* 346, 425-434, 1990.
2. McEver, R.P., Leukocyte-endothelial cell interactions, *Curr. Op. Cell Biol.* 4, 840-849, 1992.
3. Gearing, A.J.H., and Newman, W., Circulating adhesion molecules in disease, *Immunol. Today* 14, 506-512, 1993.
4. Carpén, O., Pallai, P., Staunton, D.E., and Springer, T.A., Association of intercellular adhesion molecule-1 (ICAM-1) with actin-containing cytoskeleton and α-actinin, *J.. Cell Biol.* 118, 1223-1234, 1992.
5. Durieu-Trautmann, O., Chaverot, N., Cazaubon, S., Strosberg, A.D., and Couraud, P.O., Intercellular adhesion molecule-1 activation induces tyrosine phosphorylation of the cytoskeleton associated protein cortactin in brain microvessel endothelial cells, *J. Biol. Chem.* 269, 12536-12540, 1994.

48

HYPERTENSIVE VERSUS NORMOTENSIVE MONOCYTE ADHESION TO CULTURED CEREBRAL MICROVASCULAR ENDOTHELIAL CELLS

R. M. McCarron, Y. Yoshihide, M. Spatz, and J. Hallenbeck

Stroke Branch, NINDS, NIH
Bethesda, Maryland 20892

I. SUMMARY

Hypertension may predispose to the development of thrombotic vascular disease by promoting interactions between circulating blood monocytes (mo) and vascular endothelial cells (EC). The adherence of mo from spontaneously hypertensive (SHR), normotensive Wistar-Kyoto (WKY) and Sprague-Dawley (SD) rats to cultures of activated and untreated homologous cerebral microvascular EC was examined. There were no significant differences in the percent of monocytes from all three strains adhering to untreated EC. Mo binding was inhibited (approx. 50%) by antibodies against β2-integrin molecules (CD11a, CD11b, CD18) and intercellular adhesion molecule-1 (ICAM-1). Treatment of EC with LPS, IL-1β, TNFα or IFNγ dose-dependently upregulated mo adhesion. The level of syngeneic and allogeneic mo adhesion to stimulated SHR EC was significantly more upregulated than was observed with identically treated WKY or SD EC ($p<0.001$). Mo adhesion to stimulated EC was inhibited (10-20%) by antibodies to the β1-integrin, VLA-4 and antibodies to the β2-integrins and ICAM-1 (20-25%). These results demonstrated that treated SHR EC exhibited significantly higher adhesivity and suggest that hypertension may enhance responsiveness of endothelium to factors which promote mo adhesion. Subsequent mo interactions, via cytokines, may be important in the genesis of stroke.

I. RÉSUMÉ

L'hypertension artérielle peutr favoriser l'interaction des monocytes circulants (Mo) avec les cellules endothéliales des vaisseaux sanguins (CE) et ainsi prédisposer au développement de la thrombose vasculaire. Afin d'étudier ce phénomène, nous avons examiné l'adhérence des Mo issus de rats présentant une hypertension spontanée (souche SHR) et de rat normotendus (Wistar-Kyoto WKY et Sprague-Dawley SD) sur des cultures de cellules

Biology and Physiology of the Blood–Brain Barrier, edited by Couraud and Scherman
Plenum Press, New York, 1996

endothéliales microvasculaires homologues activées ou non. Le nombre de Mo adhérant sur les cultures de cellules endothéliales est identique, quelle que soit la souche de rats étudiée. L'interaction des Mo est inhibée à 50% par les anticorps anti-b2-intégrine (CD11a, CD11b, CD18) et par la molécule d'adhésion intercellulaire ICAM-1 .Le traitement des CE par le LPS, l'IL-1b, le TNFa ou l'IFNg augmente l'adhésion des Mo de façon dose-dépendante. Cette induction de l'interaction des Mo syngéniques et allogéniques sur des cultures d'EC stimulées de rats SHR est plus importante que celle observée avec des cultures d'EC stimulées de rats WKY ou SD (P). L'interaction entre les Mo et les EC stimulées est inhibée par des anticorps anti-intégrine-b1, VLA-4 (10-20%) et par des anticorps anti-intégrine-b2 ou l'ICAM-1 (20-25%). Ces résultats montrent que des cultures de cellules endothéliales microvasculaires de rats SHR présentent une plus grande adhésivité, et suggèrent que l'hypertension pourrait induire la réponse de l'endothélium vasculaire à des facteurs promouvant l'adhésion des monocytes. De telles interactions, par l'intermédiaire de cytokines, peuvent jouer un rôle important dans la genèse des thromboses cérébrales.

II. INTRODUCTION

Adhesion of peripheral blood mo to vascular endothelium are an important initial event in response to inflammatory responses in a variety of disorders including MS, EAE and cerebral ischemic stroke[1-3]. These interactions are mediated by molecules such as LFA-1, Mac-1, VLA-4 and sialyl-Lewis-x expressed by mo and ICAM-1, E-selectin, VCAM-1 and GMP-140 expressed by EC. The expression of these molecules by vascular EC can be up-regulated by factors such as IL-1β, TNFα, IFNγ and LPS.

ICAM-1 expression by cultured cerebromicrovascular EC from SHR rats was more sensitive to cytokine-induced upregulation than cultures from WKY and SD rats[4]. This finding suggests a mechanism by which hypertension may function as a risk factor for stroke (*i.e.*, by its affect on responsiveness of EC to factors regulating interactions with mo resulting in increased *in vivo* perivascular mo accumulation).

III. METHODS

III.I. Animals

SHR, WKY and SD rats were used to isolate EC and mo. All procedures were performed in accordance with guidelines in the NIH Guide for the Care and Use of Laboratory Animals.

III.II. EC Cultures

The isolation and cultivation of cerebromicrovascular EC from rats were performed with some modification as previously described[5]. All cultures were >95% positive for Factor VIII. EC were incubated in media alone or media containing recombinant IL-1β, TNFα, IFNγ or LPS; all cytokines contained <10 pg/ml endotoxin. ICAM-1 expression by EC cultures was quantitated by ELISA[6].

III.III. Monocytes

Peripheral blood from rats was centrifuged (700 x g) on Ficoll-Hypaque density gradients followed by hyperosmotic Nycodenz density gradients. Mo were washed in PBS,

0.5% BSA, 2% EDTA to remove platelets and were >90% pure (assessed by their ability to phagocytyze latex beads and by FACS analysis using ED-1 antibody.

III.IV. Monocyte Adhesion

^{51}Cr-labeled mo were added to EC monolayers (5x10^4/50 μl/well) and incubated (37°C) for 30 min. Non-adherent cells were removed, adherent cells were lysed (2% Triton-X) and radioactivity counted. The % adherent cells were calculated by: Triton-X cpm/total cpm added. In blocking experiments, cells were preincubated for 20 min with 10-30 μg/ml of antibody (Ab); inhibition was calculated by the formula: [%adhesion (no Ab) - %adhesion (+Ab)] /% adhesion (no Ab) x 100%.

III.V Antibodies: Anti-rat ICAM-1, anti-rat VLA-4, anti-rat LFA-1α chain (CD11a), anti-rat LFA-1β chain (CD18) and anti-rat Mac-1α chain (CD11b) were obtained from Seikagaku America, Inc (Rockville, MD). Monoclonal anti-rat ED-1 and ED-2 were obtained from Biosource Int'l. (Camarillo, CA).

IV. RESULTS

Adhesion of resting mo from SHR, WKY and SD rats to untreated cultures of syngeneic cerebral microvascular EC (21.7%, 22.5% and 20.9%, respectively) were not significantly different. Activation of EC by culture with IL-1β, TNFα, IFNγ or LPS up-regulated ICAM-1 expression on EC from all three strains. In addition, similar treatment also upregulated adhesion of syngeneic and allogeneic mo (Table 1). As shown in Table 1, treatment of SHR EC resulted in a greater upregulation of both syngeneic and allogeneic mo adhesion than was observed with WKY or SD EC.

Table 1. Monocyte adhesion to cultured cerebrovascular endothelial cells

Mo	EC treatment††	Adhesion (%)† SHR EC	WKY EC	SD EC
SHR	None	18.3	19.6	17.9
	LPS	38.7	34.0*	29.6*
	TNF + IL-1	48.5	41.4*	36.5*
	IFN	30.7	26.8*	25.1*
WKY	None	16.9	19.1	20.0
	LPS	32.2	29.4*	26.7*
	TNF + IL-1	35.2	31.9*	28.4*
	IFN	24.3	23.5*	22.1*
SD	None	17.6	21.2	19.4
	LPS	32.4	28.2*	26.7*
	TNF + IL-1	35.0	29.7*	29.2*
	IFN	23.8	23.5*	21.2*

†Data from four separate experiments were pooled and presented as percent adhesion.
††Cerebrovascular EC cultures were incubated for 20 h with indicated factors at the following concentrations: 100 ng/ml LPS; 100 U/ml TNF + 40 U/ml IL-1; or 100 U/ml IFN.
*Indicates significant (2-way ANOVA and Bonferroni correction; p <0.001) differences between SHR EC and WKY or SD EC in the levels of "up-regulated monocyte adhesion" (calculated as percent increased adhesion according to the formula: 100% x [% adhesion (test) - % adhesion (control)]/% adhesion (control) for each respective mo/EC combination.

Treatment of SHR, WKY or SD mo with antibodies against the β2 integrin CD18 inhibited binding to untreated EC monolayers (54.7%, 46.5% and 49.8%, respectively). Antibodies against ICAM-1 also inhibited adhesion of SHR, WKY or SD mo (50.7%, 51.6%, and 52.0%, respectively) to monolayers of untreated EC.

The binding of mo to stimulated EC was inhibited to a lesser extent by antibodies to CD18 and ICAM-1 (19.4-24.9%). Although antibodies to VLA-4 had no affect on mo adhesion to untreated EC, they partially inhibited (7.9-16.3%) binding of mo to treated EC. No differences in the percent inhibition by antibody treatment were observed between SHR, WKY or SD EC cultures.

V. DISCUSSION

The findings demonstrate that the degree of upregulated adhesion of mo to EC monolayers was greater for SHR EC than was observed for WKY or SD EC. Characterization of these adhesive interactions revealed LFA-1/ICAM-1, Mac-1/ICAM-1 and VLA-4/VCAM-1 adhesion pathways are only partially responsible for mediating adhesive interactions. The significantly higher mo adhesivity expressed by treated SHR EC can not be accounted for by any of the integrins examined in these experiments. The lack of complete inhibition of adhesion by antibodies to these adhesion molecules indicate a role for additional molecules on mo and/or EC in these interactions. Interestingly, the findings also demonstrate that SHR mo adhesion to both syngeneic and allogeneic activated EC monolayers was significantly more up-regulated than WKY or SD mo. Thus, mo from hypertensive animals are more sensitive to factors responsible for the increased adhesion observed with all EC.

In summary, the findings suggest that: 1) hypertension enhances responsiveness of endothelium to factors which promote mo adhesion; and 2) peripheral blood mo from hypertensive animals exhibit enhanced adhesion to endothelium. Enhanced mo interaction with endothelium may be important in the genesis of stroke and indicate mechanism(s) by which hypertension may function as a stroke risk factor.

VI. REFERENCES

1. Clark, W. M., K.P. Madden, R. Rothlein and J.A. Zivin. Reduction of central nervous system ischemic injury in rabbits using leukocyte adhesion antibody treatment. *Stroke* 22:877-883, 1991.
2. Sobel, R.A., M.E. Mitchell and G. Fondren. Intercellular adhesion molecule-1 (ICAM-1) in cellular immune reactions in the human central nervous system. *Am J. Pathol.* 136:1309-1316, 1990.
3. McCarron, R.M., L. Wang, M.K. Racke, D.E. McFarlin and M. Spatz. Cytokine-regulated adhesion between encephalitogenic T lymphocytes and cerebrovascular endothelial cells. *J. Neuroimmunol.* 43:23-30, 1993.
4. McCarron, R.M., L. Wang, A.-L. Siren, M. Spatz and J.M. Hallenbeck. Adhesion molecules on normotensive and hypertensive rat brain endothelial cells. *Proc. Soc. Exp. Biol. Med.* 205:257-262, 1994.
5. Spatz, M., J. Bembry, R.F. Dodson, H. Hervonen and M.R. Murray. Endothelial cell culture derived from isolated cerebral microvessels. *Brain Res.* 191:577-582, 1980.
6. McCarron, R.M., L. Wang, A.-L. Siren, M. Spatz and J.M. Hallenbeck. Monocyte adhesion to cerebromicrovascular endothelial cells derived from normotensive and hypertensive rats. *Am. J. Physiol.* 267:H2491-H2497, 1994.

THE EFFECT OF INFLAMMATORY AGENTS UPON THE BLOOD-RETINAL BARRIER

S. D. Bamforth, H. M. A. Towler, S. L. Lightman, and J. Greenwood

Department of Clinical Ophthalmology
Institute of Ophthalmology, University College London
Bath Street, London, EC1V 9EL, United Kingdom

1. SUMMARY

The blood-retinal barrier (BRB), like that of the blood-brain barrier (BBB), forms a selective interface between the blood and the neural parenchyma. During inflammatory diseases of the retina there is a large scale increase in leucocyte infiltration and breakdown of the BRB. It is not entirely clear, however, what the causative factors are in BRB disruption but cytokines, and other inflammatory agents, have been implicated. Here we will discuss the effects of the cytokines interleukin-1ß (IL-1ß), tumour necrosis factor-α (TNF-α), interleukin-6 (IL-6) and interleukin-2 (IL-2), arachidonic acid (AA) and the eicosanoids, leukotriene B_4 (LTB_4) and prostaglandin E_2 (PGE_2), and histamine, upon the rat BRB when administered intravitreally or intra-arterially.

1. RÉSUMÉ

La barrière rétine-sang, comme celle du cerveau-sang, forme un interface sélectif entre le sang et le parenchyme neural. Durant les maladies inflammatoires de la rétine, il ya une forte augmentation de l'infiltration de leucocytes et une rupture de la barrière rétine-sang. Néanmoins, les facteurs qui provoquent la lésion de la barrière rétine-sang ne sont pas encore très clairs, mais des cytokines et des composants inflammatoires ont été impliqués. Ici, nous discuterons chez le rat des effets de cytokines comme l'interleukine-1ß (IL-1ß), le "tumour necrosis factor-α" (TNF-α), l'interleukine-6 (IL-6) et l'interleukine-2 (IL-2), l'acide arachidonique, les éicosanoïdes, le leukotriène B_4 (LTB_4) et la prostaglandine E_2 (PGE_2), et l'histamine, après injection dans l'humeur vitrée où injection intra-arterielle.

2. INTRODUCTION

The BRB is comprised of two components: the vascular endothelium situated in the anterior of the retina, and the retinal pigment epithelium located at the posterior aspect of

Biology and Physiology of the Blood–Brain Barrier, edited by Couraud and Scherman
Plenum Press, New York, 1996

the retina which overlies the highly fenestrated vessels of the choriocapillaris. The BRB plays an active part in the pathogenesis of inflammatory eye diseases. As with the BBB, the BRB is instrumental in recruiting inflammatory cells from the circulation and during inflammatory disease a frequent consequence of this is a breakdown in barrier integrity. One of the main problems associated with retinal inflammatory disorders, such as uveitis, is the formation of vasogenic oedema that arises as a result of increased BRB permeability, leading to tissue damage and a loss of vision (Greenwood et al., 1995). However it is not clear which inflammatory agents are responsible for causing increased BRB permeability, and whether they play a direct or indirect role.

We have investigated the effect of a variety of inflammatory agents on the integrity of the BRB.

3. METHODS

Male Lewis rats each received an intravitreal injection, or intra-arterial infusion, of the inflammatory agent to be studied. At various time points post injection (PI) the morphological integrity of the retina was evaluated by one or more of the following methods. Light and electron microscopy was used to monitor the influx of any inflammatory leucocytes and to examine their interaction with retinal endothelial cells. To visualise any increase in BRB permeability, the large molecular weight tracer horseradish peroxidase (HRP) was infused via the femoral vein and allowed to circulate for 10 min before the animals were perfusion fixed and the eyes processed for microscopy, as previously described (Greenwood et al., 1994). The permeability of the BRB, or BBB, was measured quantitatively by introducing the small molecular weight tracer [^{14}C]-mannitol into the circulation at a rate sufficient to maintain steady plasma levels (Luthert et al., 1986). During the 20 min infusion sequential blood samples were taken and measured for blood-gas levels. At the end of the experiment, the vasculature was flushed free with saline and the animal decapitated. The eyes were rapidly dissected, the retina was carefully removed as previously described (Lightman et al., 1987), weighed and solubilised, along with samples of blood plasma. Scintillant was added and the ß emissions counted using a liquid scintillation analyser. Any increase in barrier permeability could then be expressed by the ratio of radioactivity (g tissue)$^{-1}$ to radioactivity (ml blood plasma)$^{-1}$ (Rt/Rp).

4. RESULTS

4.1. Cytokines

Following the intravitreal injection of IL-1ß (1-2 x10^3 U) (Bamforth et al., 1996) a few mononuclear (MN) leucocytes were detected in the lumen of blood vessels at 2 h PI. More leucocytes were observed at 4 h PI, both MN and polymorphonuclear (PMN) in appearance, and the cellular infiltrate increased to give a peak between 24 and 48 h, with leucocytes present in large numbers within the retinal vessels, retinal parenchyma and vitreous (Figure 1), and adhering to the endothelium (Figures 2 & 3). This inflammation persisted until 72 h, and had completely resolved by 7 days.

Increased BRB permeability was demonstrated histologically by the extravasation of HRP. The HRP reaction product was observed flooding the basement membrane of retinal blood vessels at 48 PI (Figure 4).

A distinct biphasic pattern of BRB breakdown to [^{14}C]-mannitol was detected following the intravitreal injection of IL-1ß, as expressed by an increased Rt/Rp ratio. The initial disruption occurred at 4 h PI ($p<0.001$), which had returned to control values by 12

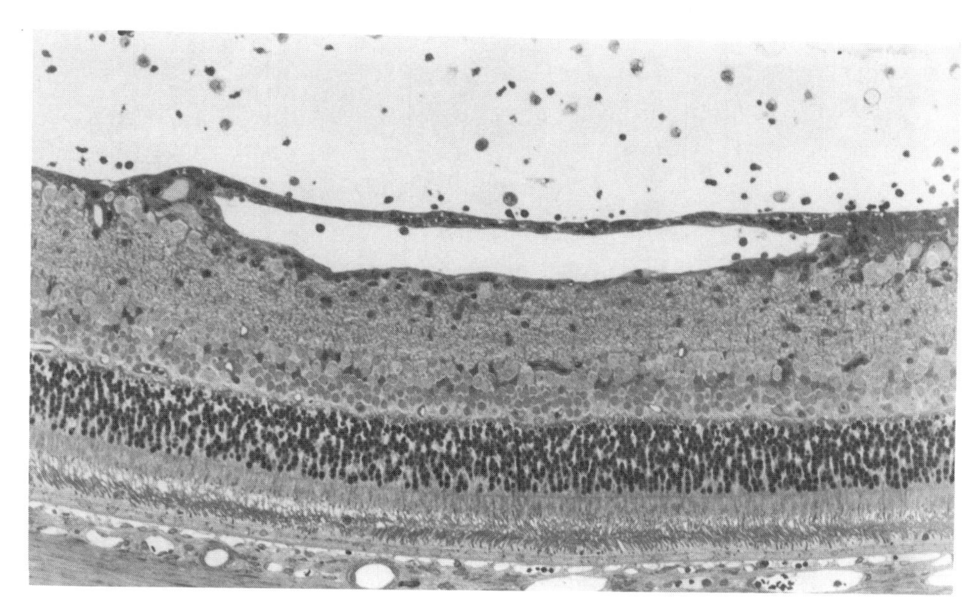

Figure 1. Toluidine-blue stained resin section of an inflamed retina following intravitreal injection of IL-1ß. Magnification x90.

h PI. This was then followed by a second increase in BRB permeability at 24 h PI ($p<0.01$) (Figure 5), that persisted through to 48 h PI. By 72 h the BRB disruption had resolved as no further increase in permeability was detected.

Following injection of TNF-α (2 x10^4 U) into the vitreous (Bamforth et al., 1994), only occasional leucocytes were observed within the retinal vessels, retinal parenchyma and vitreous. A monophasic increase in BRB permeability to [^{14}C]-mannitol was detected, however, with a significant rise in the Rt/Rp values from 1 to 5 days PI ($p<0.01$) (Figure 5). No extravasated HRP was observed within the retina at 5 days PI.

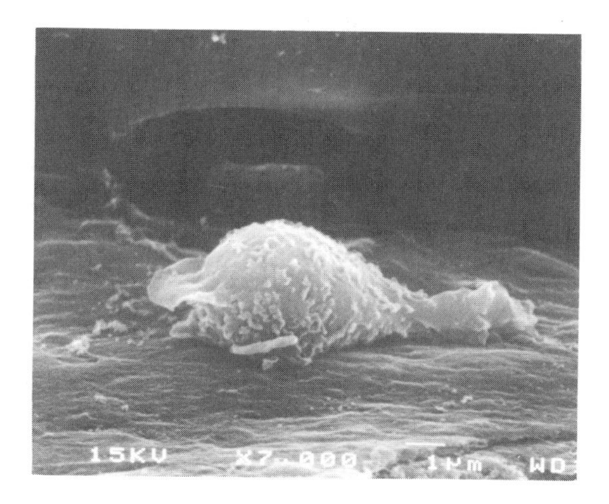

Figure 2. Scanning electron micrograph at 24 h PI IL-1ß of a leucocyte adhering to the endothelium of a large blood vessel, probably a vein. Magnification x7000.

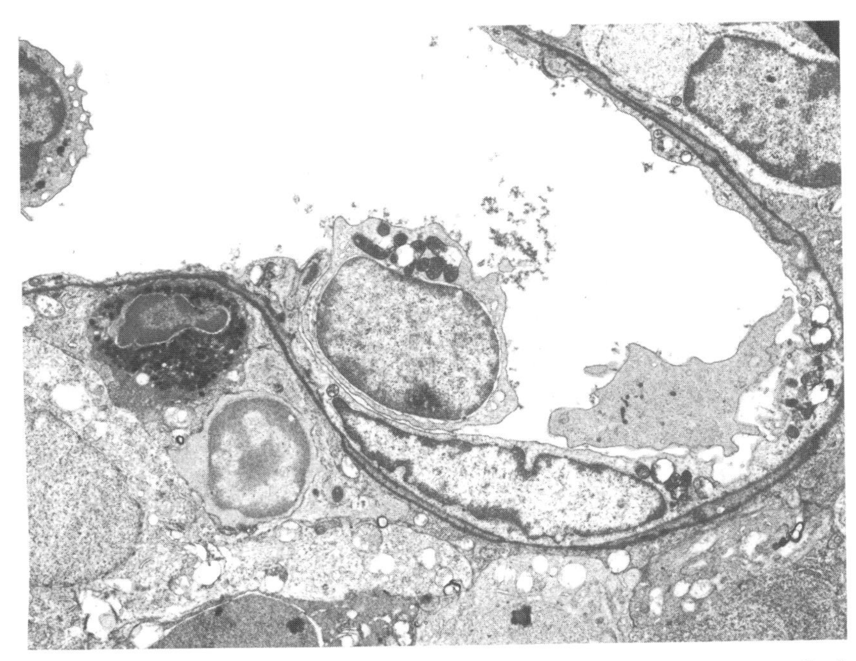

Figure 3. Transmission electron micrograph (TEM) at 12 h PI IL-1ß of a MN leucocyte adhering to the endothelium of a retinal vessel. A previously migrated PMN leucocyte can be seen beneath the basement membrane. Magnification x6000.

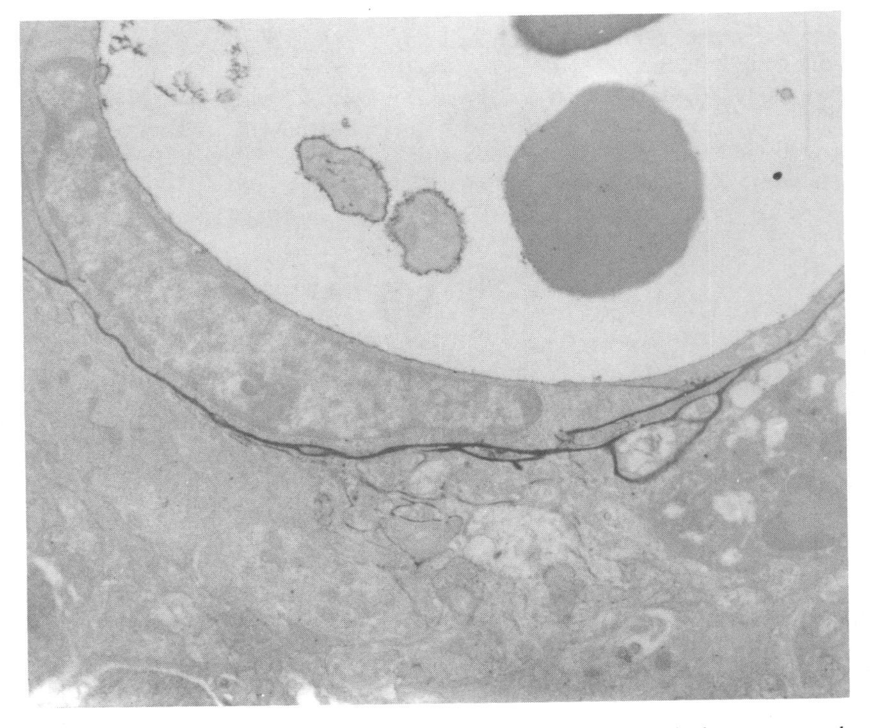

Figure 4. TEM of an unstained section illustrating extravasated HRP flooding the basement membrane of a blood vessel at 48 h PI IL-1ß. Magnification x12,000.

Intravitreal administration of IL-6 ($1x10^3$ U) induced only an occasional MN leucocyte to adhere to the endothelium of retinal vessels. No increase in BRB permeability was detected over a 4 day time period. Similarly, following the administration of various doses of IL-2, intravitreally, or intra-arterially via the internal carotid artery, no significant breakdown of the BRB or BBB was detected after a one hour time course.

4.2. Arachidonic Acid and Eicosanoids

AA and LTB_4 appeared to have the effect of increasing BRB permeability up to 2 h following intravitreal administration, but not following PGE_2 (Towler et al., 1993) (Figure 5). HRP extravasation, however, was not seen within the retina following LTB_4 injection.

4.3. Histamine

The intravitreal administration of histamine (10^{-3} or 10^{-4} M) had a significant effect on the BRB when the permeability was measured immediately after the injection was given ($p<0.05$) (Figure 5). No further increase in BRB permeability was detected up to 72 h PI. The intracarotid infusion of the same doses of histamine also had a significant effect on increasing the permeability of the BRB, and also the BBB, when measured immediately after histamine administration. The permeability of the BRB, but not the BBB, remained significantly higher than the controls when measured 15 min after the intracarotid histamine infusion.

4.4. Hyperosmolar Arabinose

As a positive control, a hyperosmolar solution (1.8M) of arabinose was infused into the right internal carotid which caused a significant disruption of the BRB (Figure 5).

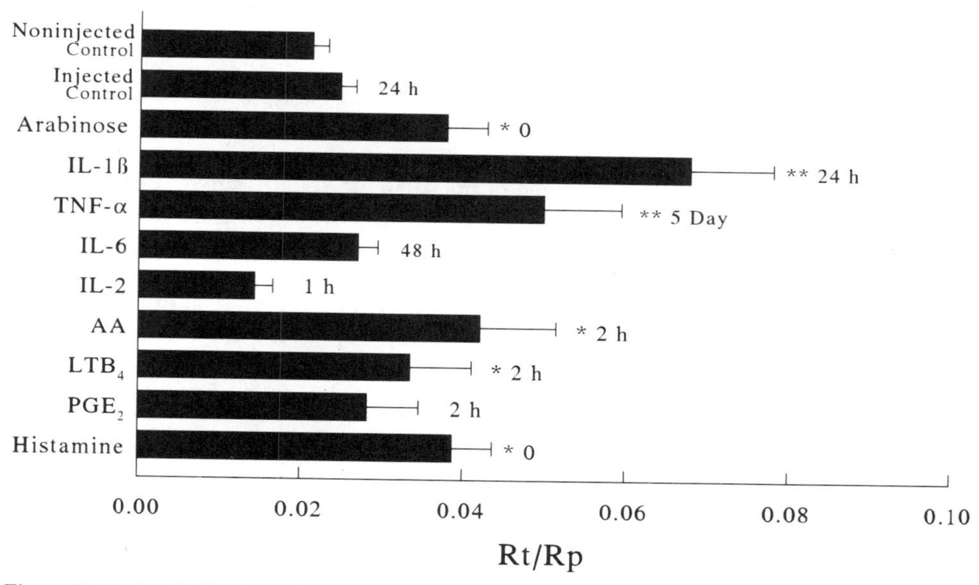

Figure 5. Maximal BRB disruption following intravitreal injection of an inflammatory agent. Time of maximal opening PI is shown for each agent. The opening of the BRB to intravascular administration of hyperosmolar arabinose is given for comparison. Data shown are means ± SEM. Significant differences from controls $*p<0.05$; $**p<0.01$.

5. DISCUSSION

It appears that certain inflammatory agents have the ability to affect the blood-retinal and blood-brain barriers of the rat, but in differing degrees and through apparently different mechanisms.

IL-1ß, when administered intravitreally, appears to act indirectly on the BRB by inducing a cellular infiltrate which subsequently causes barrier breakdown. The cellular inflammatory response to IL-1ß is similar to that seen following the intravitreal injection of IL-1ß into the rabbit vitreous (Claudio et al., 1994). However, the biphasic opening of the BRB in response to IL-1ß has not been described before. The initial increase in BRB permeability corresponded with the appearance of leucocytes within the retina. The second phase of opening at 24 to 48 h PI occurred when maximal extravasated leucocytes were present.

In contrast to IL-1ß, the intravitreal injection of TNF-α is able to induce BRB breakdown, but without a large scale concomitant cellular infiltration. This implies that TNF-α has a more direct effect on the BRB or tissue resident cells which produces a more prolonged opening of the barrier.

The administration of IL-6 or IL-2 produced no effect on the BRB at the doses examined, although a previous study has demonstrated an inflammatory response accompanied by the breakdown of blood-ocular barriers following an intravitreal injection of IL-6 (Hoekzema et al., 1992). The difference in response could be attributed to the different species of cytokine used in each study.

The intravitreal injection of IL-2 did not increase significantly the permeability of the BRB after one hour. However, IL-2 injected into the vitreous of the rabbit has been shown to cause an inflammatory response and breakdown of the blood-ocular barriers (Samples et al., 1993). These effects were first increased at 24 h, and were maximal at 5 days PI. This indicates that IL-2 does not directly affect the barrier but induces a more chronic effect within the eye, similar to that seen with TNF-α. The intravenous administration of IL-2 also appears to be unable to acutely disrupt the BRB or BBB of the rat, a finding which has been previously reported in the mouse (Banks et al., 1992).

The eicosanoids are considered to be potent mediators within ocular inflammatory disorders (Bazan et al., 1990). Intravitreal AA and LTB_4, at the doses studied, only had a mild effect on increasing BRB permeability after 2 h which, as for TNF-α, could not be detected by HRP. This could be due to the formation of pores at the intercellular junctions being of a size that will allow extravasation of the small molecular weight tracer mannitol but not the large tracer HRP (Lightman and Greenwood, 1992).

In the literature there is conflicting data over the action of histamine on the BRB (Greenwood, 1992). We have shown that histamine is able to produce a transient increase in the permeability of the BRB when it is administered to either the abluminal or luminal side. The rapidity with which histamine causes an increase in the permeability of the blood-CNS barriers studied indicates that its effects are rapid and likely to be due to a direct affect on the vascular endothelium. Both AA and LTB_4 are also likely to act directly upon the BRB although the response is not as rapid as with histamine.

In the rat certain inflammatory agents, such as IL-1ß, appear to exert their effects indirectly on the BRB through the induction of a cellular infiltrate, whereas others, such as histamine, are able to rapidly and directly induce transient barrier opening. Other mediators such as TNF-α may also be acting directly but do not exert their effects until later.

6. REFERENCES

Bamforth, S.D., Greenwood, J. and Lightman, S. (1994) The effect of TNF-α on blood-retinal barrier permeability. Immunology 83(Suppl):54 (Abstract)

Bamforth, S.D., Lightman, S. and Greenwood, J. (1996) The effect of interleukin-1ß on blood-retinal barrier permeability, leukocyte recruitment and retinal morphology. Am. J. Pathol. (Submitted).

Banks, W.A. and Kastin, A.J. (1992) The interleukins-1α, -1β and -2 do not acutely disrupt the murine blood-brain barrier. Int. J. Immunopharmacol. 14:629-636.

Bazan, N.G., Toledo de Abreu, M., Bazan, H.E.P. and Belfort, Jr. (1990) Arachidonic acid cascade and platelet-activating factor in the network of eye inflammatory mediators: therapeutic implications in uveitis. Int. Ophthalmol. 14:335-344.

Claudio, L., Martiney, J.A. and Brosnan, C.F. (1994) Ultrastructural studies of the blood-retina barrier after exposure to interleukin-1β or tumor necrosis factor-α. Lab. Invest. 70:850-861.

Greenwood, J. (1992). Experimental manipulation of the blood-brain and blood-retinal barriers. In: M.W.B. Bradbury (Ed.), Physiology and pharmacology of the blood-brain barrier. Springer-Verlag, New York. pp. 460-486.

Greenwood, J., Howes, R. and Lightman, S. (1994) The blood-retinal barrier in experimental autoimmune uveoretinitis: leukocyte interactions and functional damage. Lab. Invest. 70:39-52.

Greenwood, J., Bamforth, S.D., Wang, Y. and Devine, L. (1995). The blood-retinal barrier in immune-mediated diseases of the retina. In: J. Greenwood, D. Begley, & M. Segal (Eds.), New concepts of a blood-brain barrier. Plenum Press, London. (In press)

Hoekzema, R., Verhagen, C., van Haren, M. and Kijlstra, A. (1992) Endotoxin-induced uveitis in the rat. The significance of intraocular interleukin-6. Invest. Ophthalmol. Vis. Sci. 33:532-539.

Lightman, S.L., Palestine, A.G., Rapoport, S.I. and Rechthand, E. (1987) Quantitative assessment of the permeability of the rat blood-retinal barrier to small water-soluble non-electrolytes. J. Physiol. 389:483-490.

Lightman, S. and Greenwood, J. (1992) Effect of lymphocytic infiltration on the blood-retinal barrier in experimental autoimmune uveoretinitis. Clin. exp. Immunol. 88:473-477.

Luthert, P.J., Greenwood, J., Lantos, P.L. and Pratt, O.E. (1986) The effect of dexamethasone on vascular permeability of experimental brain tumours. Acta Neuropathol. (Berl) 69:288-294.

Samples, J.R., Boney, R.S. and Rosenbaum, J.T. (1993) Ocular inflammatory effects of intravitreally injected interleukin-2. Curr. Eye. Res. 12:649-654.

Towler, H.M.A., Bamforth, S.D., Greenwood, J. and Lightman, S. (1993) The effect of intravitreal leukotrienes and prostaglandins on the blood-retinal barrier of the Lewis rat. Invest. Ophthalmol. Vis. Sci. 34:1426 (Abstract)

50

IDENTIFICATION OF THE PENUMBRA IN AN ISCHEMIC RAT MODEL BY MRI AFTER INJECTION OF A SUPERPARAMAGNETIC CONTRAST AGENT

D. Ibarrola, H. Seegers, A. Väth, A. François-Joubert, M. Hommel, M. Décorps, and R. Massarelli

INSERM U438, RMN Bioclinique
CHU A. Michallon, BP 217
38043 Grenoble Cedex 9, France

SUMMARY

One of the most important issues for the therapy of cerebral stroke is the early detection of the ischaemic site and the identification of the penumbra surrounding the core of the lesion. In this context, we have studied the potential of a superparamagnetic contrast agent on a rat model of Middle Cerebral Artery occlusion (MCA-o). T_2-weighted NMR images acquired after injection showed the ischaemic lesion as soon as one hour following the MCA-o. The use of the contrast agent revealed, on the diffusion-weighted images, a cortical zone where the signal intensity was reduced in comparison with the images before injection. This cortical zone may correspond to the penumbra, as also suggested by perfusion autoradiography.

RÉSUMÉ

La détection précoce de la région ischémique et de la pénombre qui l'entoure est particulièrement importante pour le diagnostic et l'initiation d'une thérapie. Nous avons étudié l'effet de l'injection d'un produit de contraste superparamagnétique sur un modèle d'occlusion de l'Artère Cérébrale Moyenne (o-ACM) chez le rat. Les images par RMN pondérées en T_2, après injection, permettent de détecter la lésion dès la première heure suivant l'o-ACM. Sur les images pondérées en diffusion, la présence de l'agent de contraste met en évidence une zone essentiellement corticale où l'intensité du signal est réduite par rapport aux images avant injection et qui pourrait correspondre à la pénombre, comme cela semble être confirmé par une étude autoradiographique de perfusion.

Biology and Physiology of the Blood–Brain Barrier, edited by Couraud and Scherman
Plenum Press, New York, 1996

1. INTRODUCTION

The early detection of an ischaemic cerebral insult and of the penumbra zone which surrounds it is of considerable interest for diagnosis and further therapeutic treatment. We have studied the effect of the injection of a superparamagnetic agent, AMI-227, with a long blood half-life, on T_2- and Diffusion-weighted imaging of a rat brain model of ischaemia.

Focal cerebral ischaemia is due to a significant, localized, reduction of cerebral blood flow in one hemisphere. Two zones can be distinguished in the ischaemic region: the core and the penumbra. In the core the damage is irreversible while, in the penumbra, defined as the parenchyma surrounding the core, nerve cells (i.e., neurons and astrocytes) are not irreversibly damaged. It corresponds to an Œdematous zone where blood perfusion, although reduced, is still present. In this region, cells are electrically silent but they remain viable for some time. They may be saved by reperfusion and drugs may prevent the extension of the infarction in this zone.

The aim of the present study was to assess the potentials of a superparamagnetic contrast agent for the early detection by MRI *in vivo* of ischaemic brain infarction and, in particular, for the identification of the penumbra zone.

2. MATERIAL AND METHODS

2.1. Animal Model

Acute focal cerebral ischaemia was induced in Sprague Dawley rats (males, 280-300g) by means of a Middle Cerebral Artery occlusion (MCA-o) following a modification (1, 2) of the technique described by Koizumi et al. (3). MCA-o is obtained by introducing an intraluminal nylon filament into the carotid artery. This filament is advanced intracranially up to the right MCA bifurcation. All sources of blood from the internal carotid artery, the anterior and the posterior cerebral artery are thus occluded dramatically reducing blood flow into the MCA. The technique is potentially reversible.

2.2. NMR Experiments

Two different Proton MRI techniques have been used. The first is the T_2-weighted spin echo sequence, particularly sensitive to diffusing water. The technique allows the visualization of late vasogenic Œdema which corresponds to an accumulation of water in the extracellular space (4). T_2 Magnetic Resonance Imaging (MRI) was performed with a 2.35 T horizontal magnet (MSL, Bruker) and a T_2-weighted sequence (TE=80 ms, TR=2000 ms, slice thickness=1 mm, matrix=128*128, FOV=3 cm).Two groups of rats (n=5) were considered. The animals in the first group were non-injected controls. The animals of the second group were given intraveinously the contrast agent about 30 minutes following MCA-o. For each rat, sets of 10 coronal slices were acquired covering the whole of the ischaemic territory. These sets were measured every 30 minutes, over a period ranging from 40 minutes to 5 hours following the occlusion and then 24 hours following MCA-o.

The second technique was a diffusion-weighted sequence. This sequence is more sensitive to the diffusion of the protons of water and appears to be specific for the detection of cytotoxic Œdema, where the diffusion of water molecules is more restrained. One group of rats (n=6) was studied. Five coronal slices were acquired before and after the injection of the contrast agent. The MRI parameters were kept identical to the T_2-weighted sequence (with b=1200 $s.mm^{-2}$).

For all experiments, a circular surface coil was used for excitation as well as for detection. The rats were positioned in a stereotaxic head holder. Anaesthesia was induced by means of halothane inhalation. Body temperature was kept at 37°C with a heating blanket.

2.3. Contrast Agent

A superparamagnetic Ultra Small Particles Iron Oxide agent (AMI-227, Guerbet, Aulnay-sous-Bois; rat blood half life $T_{1/2}$=300 min in plasma at 0,47T and 37°C) was used as intravascular contrast tracer. AMI-227 induces field gradients around the vessels leading to a reduction of the signal intensity with both the MRI sequences which we have applied. Animals received a dose of 200 μmol Fe/kg (0,6 ml) injected into the saphena vein.

2.4. Autoradiography

At the end of the MRI experiment, 50 mCi (555 Bq) of ^{99m}Tc-HMPAO (HexaMethyl-PropyleneAmineOxime) was injected into the saphena vein (0,6 ml). The radioactive tracer, commonly used as a marker for tissue perfusion (5), crosses easily the blood-brain-barrier and remains trapped within the cells. Five minutes later, the rats were sacrified and the brains were excised and immediatly frozen. Twenty μm thick section were taken with a cryotome and exposed for 2 hours on a photographic film (Kodak X-OMAT).

3. RESULTS

3.1. T_2–Weighted Images

Figure 2 shows typical results (a total of 10 animals were studied in this experiment) obtained for a control rat and an injected rat. The images represent samples taken at different times following MCA-o. In the control rat, the signal intensity did not vary significantly during the early stages of the ischaemia. In the vascularized tissue of the injected rat, the contrast agent induced rapidly a signal decrease which was proportional, to some extent, to the tissue perfusion. Thus, during early stages of blood deprivation, the ischaemic tissue appeared relatively hyperintense with respect to the normally perfused tissue which was made hypointense by the presence of the contrast agent in the blood vasculature (Figure 1). Five hours following the MCA-o, inside the ischaemic zone, some hypointensities were observed which became strongly hypointense after 24 h. Such hypointensities were presumably due to the accumulation of the contrast agent in the extracellular space, as a result of the rupture of the blood-brain barrier.

The variation of signal intensity in the core of the ischaemic lesion following the MCA-o is shown in Figure 2. The results, expressed in terms of a relative contrast between ischemic and contralateral hemispheres, show that the variations in signal intensity in controls are not statistically significant during the first 4.5 hours following MCA-o. In contrast, after injection of AMI-227, a statistically significant signal hyperintensity was observed as soon as one hour following occlusion.

3.2. Diffusion–Weighted Images

Typical results obtained on 3 coronal slices measured on one of the rats (n=5) are shown in Figure 3. The images on the left show the measurements performed before

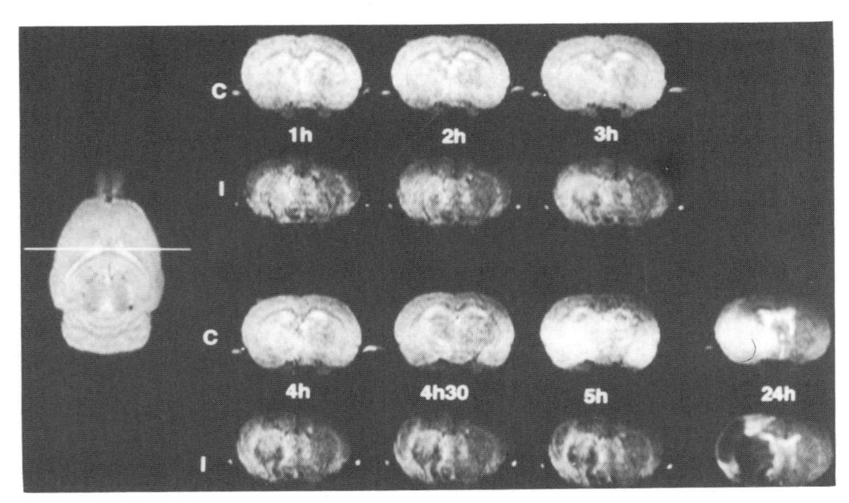

Figure 1. T_2-weighted images of rat brain. C: Control rats (n=5); I: Injected rats with the superparamagnetic contrast agent (n=5). Evolution of the lesion as a function of time.

injection, while the images on the right have been obtained 90 minutes following MCA-o, immediately after contrast agent injection.

In the the lesioned hemisphere, before injection, we observed a relative signal hyperintensity. This area presumably covers the cytotoxic oedematous core as well as the penumbra. After injection of AMI-227, a decrease of signal intensity in the perfused tissue was observed, as it was previously shown in the T_2-weighted SE images. However, the injection of the superparamagnetic contrast agent reduced as well the extent of the hyperintense area detected in diffusion-weighted SE images. This is particularly true within the cortical area (Figure 4, arrows). The reduced extension reflects the presence of the contrast agent in those regions which are still perfused, thus delimiting presumably the penumbra.

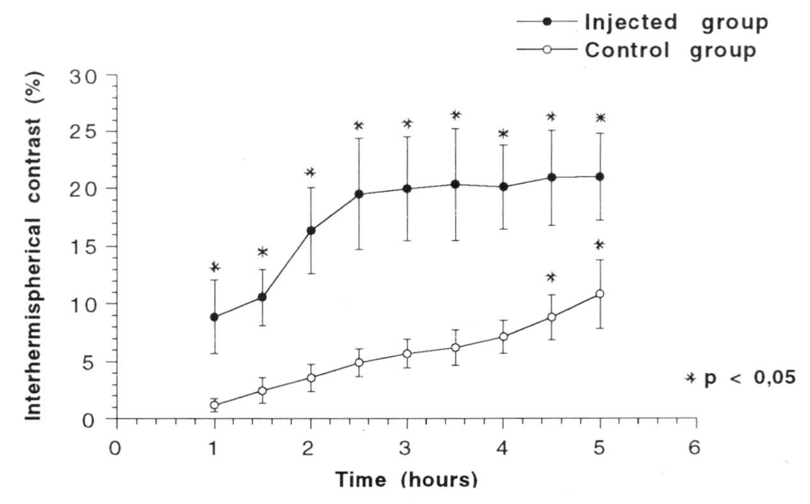

Figure 2. Relative contrast between ischemic and controlateral hemispheres. Injected and control group (n=5). Non parametric Mann Whitney test (p<0,05)

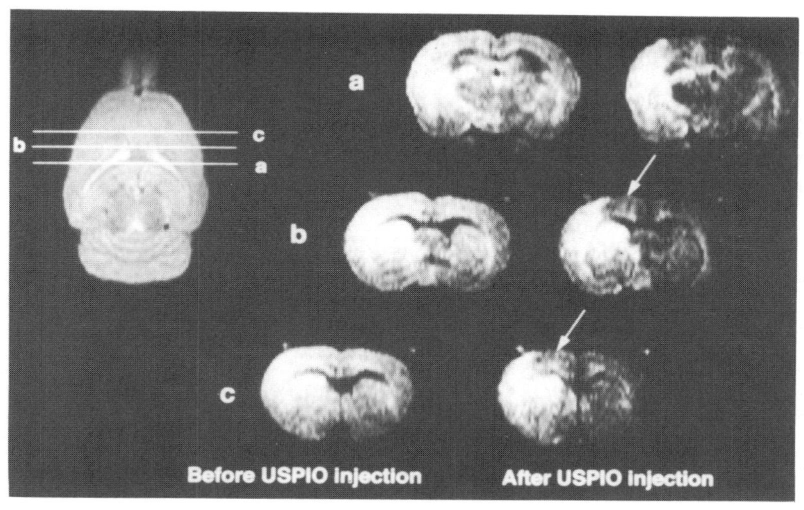

Figure 3. Diffusion-weighted images of rat brain before (60 min following the MCA-o) and after contrast agent injection (90 min following the MCA-o). The arrows show the hypointense area appeared after injection.

3.3. Autoradiography

To confirm the hypothesis, rats were injected with ^{99m}Tc-HMPAO, a well known radioactive agent to measure blood flow (6). Figure 4 shows a typical autoradiographical result which we have obtained on one of the injected rats. The healthy hemisphere, on the right, is intensely dark on the autoradiographic film, while in contrast part of the ischaemic hemisphere appears clear, indicating the absence of perfusion. A cortical area in the ischaemic hemisphere exhibits intermediate darkening, suggesting reduced perfusion. This cortical area correlates well with the cortical one detected, in the same rat, with diffusion-weighted images (Figure 4, arrows).

4. DISCUSSION

Several authors have suggested that MRI diffusion-weighted imaging may be, at present, the most appropriate technique available to detect variations in cellular water movements *in vivo* (6). It has also been suggested that the early detection of hyperintensities in cerebral ischaemic tissue, by this MRI sequence, may correspond to the appearance of a cytotoxic Œdema (7).

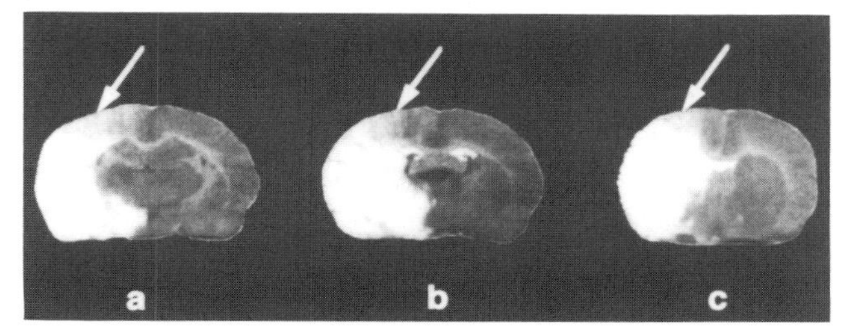

Figure 4. ^{99m}Tc-HMPAO Autoradiography of coronal rat brain section. The arrows show an intermediate gray area indicating a reduced perfusion.

The present results show that Diffusion-weighted imaging may be used to measure the extent of the Œdema at a relative early time after MCA-o. The injection of a superparamagnetic agent lowers the signal intensity by circulating in the cerebral blood vasculature. Consequently the non vascularized region of the cerebrum will appear relatively hyperintense. The data show that the signal detected was noticeable with both T_2 and diffusion-weighted sequences. In particular, with the latter, the signal was detected earlier with a sharper contrast.

The rationale behind the experiments herewith presented was to show that the combination of a diffusion-weighted sequence and the injection of the superparamagnetic contrast agent should, by difference, give the extension of an Œdematous zone where blood circulation, although reduced, is still possible, i.e., the "penumbra" zone. The results indicated that this was the case and the injection of ^{99m}Tc-HMPAO confirmed the possible use of this MRI approach to detect the penumbra *in vivo*.

Clinical trials are previewed to apply this technique to human cerebrovascular pathologies.

5. CONCLUSION

A rat model of MCA occlusion has been applied to study the ischaemic lesion by means of MR imaging techniques. The use of a contrast agent allows the early detection of an ischaemic lesion on conventional T_2-weighted spin echo images. In diffusion-weighted images, the ischaemic lesion is normally hyperintense, but the core of the lesion can not be differentiated from the penumbra. In conjunction with the use of a contrast agent, diffusion-weighted images reveal a cortical zone where the signal intensity is reduced. This cortical zone corresponds presumably to the penumbra, as also suggested by perfusion autoradiography.

ACKNOWLEDGMENTS

Part of the study was supported with funds given by he Région Rhône-Alpes. Our warmest thanks to the Laboratoire Guerbet for the gift of AMI-227.

6. REFERENCES

1. Roussel SA, van Bruggen N, King MD, Houseman J, Williams SR and Gadian DG. *NMR in Biomed* 7, 21, 1994
2. Zea Longa E, Weinstein PR, Carlson S and Cummins R. *Stroke* 20, 84, 1989
3. Koizumi J, Yoshida Y, Nakazawa T and Ooneda T. *Jpn. Stroke*, 8, 1, 1986
4. Mintorovitch J, Moseley ME, Chileuitt L, Shimizu H, Cohen Y and Weinstein PR. *MRM* 18, 39, 1991
5. Hoffman TJ, Corlija M, Chaplin SB, Volkert WA and Holmes RA. *J Cereb Blood Flow, Metab.* 8, S38, 1988
6. van Bruggen N, Cullen BM, King MD, Doran M, Williams SR, Gadian DG and Cremer JE. *Stroke* 23, 576, 1992
7. Kohno K, Hoehn-Berlage M, Mies G, Back T and Hossman KA. *MRI*, 13, 73, 1995

HYPOXIA AND REOXYGENATION OF A CELLULAR BARRIER CONSISTING OF BRAIN CAPILLARY ENDOTHELIAL CELLS AND ASTROCYTES

Pharmacological Interventions

H. Giese,[1] K. Mertsch,[1] R. F. Haselof,[1] F. H. Härtel,[2] and I. E. Blasig[1]

[1] Forschungsinstitut für Molekulare Pharmakologie
10315 Berlin, Germany
[2] Institut für Biologie
Humboldt-Universität zu Berlin
10099 Berlin, Germany

SUMMARY

Blood-brain barrier (BBB) has been neglected in pharmacological interventions of ischemic brain although it can be reached easily after systemic administration of a drug. Brain capillary endothelial cells (BCEC) may contain NMDA receptors so that the antagonist MK-801 was studied to protect BBB function. Oxygen deficiency is a main limitation during ischemia known to generate free radicals. During hypoxia and reoxygenation, an increase of radical-induced lipid peroxidation in both BCEC and astrocytes (AC) was found, accompanied by disturbances of BBB function. Therefore, the radical scavenging lazaroid U83836E was also studied. Upon hypoxia, the permeability of the barrier (BCEC and AC, cultured separately on the two sides of a filter) increased. This effect was intensified during the following reoxygenation. MK-801 and U83836E reduced the hypoxia-induced increase of permeability.

RÉSUMÉ

La barrière hémato-encéphalique (BBB) a été négligée dans des interventions pharmacologiques du cerveau ischémique bien qu'il ne soit pas difficile de le démontrer après application d'un médicament. Des cellules endothéliales de cerveau (BCEC) possèdent des récepteurs NMDA de sorte que l'antagoniste MK-801 a été appliqué pour protéger la fonction de la barrière hémato-encéphalique. Un manque d'oxygène est la limitation essentielle

Biology and Physiology of the Blood–Brain Barrier, edited by Couraud and Scherman
Plenum Press, New York, 1996

pendant l'ischémie en produisant des radicaux libres. Pendant l'hypoxie et la reoxygenation une augmentation des taux de la péroxidation des lipides induite par des radicaux dans des BCEC et des astrocytes (AC) a été trouvée,ainsi que des modifications de la fonction de la BBB. De plus, l'aminostéroide U83836E, qui est un agent de spin-trap a été étudié. Sous hypoxie la perméabilité de la barrière formée de BCEC et de AC cultivés séparément sur les deux cotés d'un filtre a été augmentée, et le phénomène s'est intensifié après la réoxygenation.suivante. MK-801 et U83836E ont réduit le taux d'accroissement de la perméabilité provoqué par l'hypoxie.

INTRODUCTION

During cerebral ischemia, an enhanced extracellular glutamate concentration[1] and overstimulation of excitatory amino acid (EAA) receptors, particularly of the N-methyl-D-aspartate (NMDA)[2] type, is observed. Ca^{2+} entry through NMDA receptor channels triggers several metabolic events, including release of amino acids[3] and formation of free radicals[4] which may lead to breakdown of BBB. The BBB is comprised of BCEC with tight junctions occluding paracellular transport and restricting the passage of ions and polar solutes between blood and brain.[5] AC are essential for the development of BBB properties in cultured BCEC.[6] Little is known about the presence of NMDA receptors in BCEC[7,8] and glia. NMDA activated currents[9] and expression of mRNA encoding the NMDAR2B subunit[10] have been described for Bergmann glial cells of mouse and rat brain. The presence of NMDA receptors has been suggested for Muller glial cells of chicken retina,[11] which may share with Bergmann glial cells a similar developmental origin.[12] But up to now, there is no evidence for NMDA receptors on type-1[13] and type-2 AC.[14] The processes of AC form a cellular environment around brain capillaries, which might exert a protective action on glutamate induced injury of BCEC during hypoxia. Glial cells provide the compartment in which glutamate is converted [15] and detoxified,[16] as found in bovine and rat oligodendrocytes and, especially, in AC. There is little activity in neurons.[17] Moreover, AC show an uptake of neuronal glutamate, which is a critical mechanism maintaining the physiological EAA transmission. Pharmacological inhibition of the uptake results in rapid potentiation of cell responses to EAA[19] and may lead to delayed excitoxicity.[20] The glutamate uptake function is inhibited in ischemia/reperfusion injury,[18] an effect exerted by both EAA and free radicals.[2] Finally, the generation of free radicals may play an important role in hypoxia-induced functional disturbances of BCEC.[23,24] Thus, NMDA receptor antagonists such as MK-801,[21] as well as antioxidatively acting lazaroids such as U83836E,[22] could be of therapeutic relevance to protect the integrity of the BBB. The aim of the present study was to utilize mono-layers of BCEC grown in coculture with AC on a permeable filter to assess hypoxia/reoxygenation-induced changes in permeability through the cell barrier and the effect of MK-801 and U83836E.

MATERIAL AND METHODS

Cell cultures, permeability studies and hypoxia/reoxygenation have been described earlier.[23,24] Briefly, primary porcine BCEC and secondary AC from neonatal rats were separately cultured on the two sides of a filter (0.4 μm, rat tail collagen coated Millicell-CM; inserts in 6-well plates) in DMEM, 10% FBS, 2 mM glutamine, 1 ng/ml bFGF, 100 ng/ml heparin, 250 μg/ml CPT-cAMP, 10 μg/ml IBMX, and 50% AC conditioned medium. Permeation and hypoxia/reoxygenation studies were performed at confluence of the mono-layers (after about 10 days). 10 min after addition of Na-fluorescein into the donor chamber

(BCEC side) the fluorescence in the receptor chamber was measured (excitation 488 nm, emission 512 nm) to calculate the permeability coefficient.[25] An analogous procedure was chosen when radioactively labelled substances were applied (scintillation measurement). For hypoxia/reoxygenation, cells were washed with PBS (Ca^{2+}, Mg^{2+}, pH 7.2); 2 ml of the PBS were added per insert and well, respectively (6-well plates, rotating water bath, 37 °C). The plates were gassed with 95% N_2/5% CO_2 (hypoxia) followed by 30 min with 95% O_2/5% CO_2 (reoxygenation) or with 95% O_2/5% CO_2 (normoxic control). Thiobarbituric acid reactive substances (TBARS, measured by HPLC using malondialdehyde standard),[24] glutamate,[26] glutathione (GSH), glutathione disulfide (GSSG),[27] protein,[24] and radical scavenger activity (ESR spin trapping)[28] were determined.

RESULTS AND DISCUSSION

BCEC and AC, grown in coculture on the filter, exhibited the typical morphology resembling that when cultivated in wells; BCEC expressed γ-glutamyl transpeptidase and alkaline phosphatase, AC the specific marker glial fibrillary acidic protein. The transendothelial electrical resistance, characterizing tightness of the BCEC monolayer, was 358±39 ohm·cm². The permeability of nonpermeable markers (fluorescein, sucrose) and of dichlorkynurenic acid was low and comparable with literature.[25,29-32] In contrast, the lipid-soluble MK-801 (Boehringer Ingelheim) showed a significantly higher permeability comparable with literature data.[21] Enhancement of permeability of the low-molecular marker fluorescein often used to characrize the transcellular permeability indicates deterioration in the monolayer integrity, alterations of nonspecific transcellular pathways or possible toxicity to cells.[32]

After hypoxia and reoxygenation, the permeation of fluorescein increased significantly compared to the normoxic control (Fig. 1A). Under hypoxic conditions, this effect was significantly reduced by 10 μM MK-801 (Fig. 1B). MK-801 is an selective antagonist of the NMDA receptor,[21] a subclass of the glutamate receptor, and has been shown to prevent

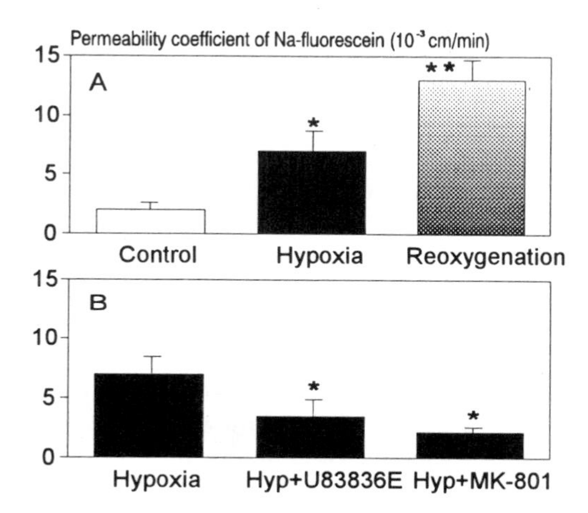

Figure 1. The effect of hypoxia (120 min), posthypoxic reoxygenation (30 min) and of U83836E and MK-801 (10 μM each) on the permeability of low molecular weight compounds through a barrier consisting of brain capillary endothelial cells and astrocytes, separately cultured on the two sides of a filter. Mean±SEM; n=3; * $p<0.05$, ** $p<0.01$, compared to normoxic control conditions (control, A) and hypoxic conditions (hypoxia, B), respectively.

Table 1. Glutathione status and radical-induced lipid peroxidation (TBARS, thiobarbituric acid reactive substances) of brain capillary endothelial cells (BCEC) and astrocytes (AC) and the effect of hypoxia (120 min 95% N_2/5% CO_2) and reoxygenation (30 min)

Parameter measured (nmol/mg protein)	BCEC	AC
Glutathione, reduced	42.0±3.0	61.2±1.6***
Glutathione, oxidized	not detectable	1.25±0.2
TBARS after normoxia	0.79±0.07	1.24±0.04***
TBARS after hypoxia	1.04±0.10$^{\#\#}$	1.22±0.11*
TBARS after reoxygenation	1.34±0.10$^{\#\#\#}$	1.70±0.08$^{\#}$,***

Mean±SEM; n=5; *$p<0.05$ and ***$p<0.001$, compared to BCEC; $^{\#}p<0.05$, $^{\#\#}p<0.01$ and $^{\#\#\#}p<0.001$, compared to TBARS (thiobarbituric acid reactive substances) after 150 min normoxia.

BBB breakdown following focal cryogenic brain injury, a well-studied model of vasogenic brain edema.[8] Glutamate, released in high amounts in the central nervous system during hypoxia, can act via the NMDA receptor.[34] Excessive activation of this receptor has been implicated in cerebral ischemic injury by promotion of calcium entry into neurons.[1] Data regarding the existence of NMDA receptors on BCEC are controversial. No receptor binding sites were detectable in membrane preparations of cerebral ovine microvessels.[7] It has been assumed that the protective effects of NMDA receptor antagonists against ischemia and anoxia are unlikely to be mediated via direct actions on the microvasculature. On the other hand, studies on isolated rat brain capillaries suggested that the activation of NMDA receptors is responsible for triggering the stimulation of the ornithine decarboxylase/polyamine cascade and for BBB breakdown following focal cryogenic injury.[8] It has been concluded that capillary NMDA receptors may be important in BBB breakdown and vasogenic brain edema observed in neuropathological disorders associated with abnormal elevations of extracellular glutamate such as ischemia, anoxia and hypoglycemia. As demonstrated in separate experiments (data not shown), BCEC may release glutamate during hypoxia and reoxygenation, resulting in micromolar concentrations in the incubation solution. These concentrations are several orders of magnitude higher than the binding constant of the NMDA receptor.[8]

In ischemic/hypoxic disturbances free radicals may play an important role. Table 1 shows that both BCEC and AC possess a powerful antioxidative defence system as indicated by high contents of GSH and low GSSG. During hypoxia/reoxygenation radical induced lipid peroxidation (TBARS, detected as malondialdehyde using HPLC) was enhanced indicating damage of membrane lipids and, hence, injury of cellular integrity as described in separate experiments performed under comparable conditions.[24] Comparing both cell types, a stronger radical defence and a lower membrane lipid peroxidation upon hypoxia was observed in AC. The data of Fig. 1A confirm *in vivo* observations suggesting that hypoxia and reoxygenation may result in radical-induced functional disturbances of the enthothelial cell barrier. This assumption is supported by experiments with the antioxidant U83836E (10 µM, Upjohn). U83836E prevented hypoxia-induced loss in tightness of the cell barrier (Fig. 1B) and was found to be a powerful ·OH-scavenger measured ESR spectroscopically (half maximum effect at 5µM). Acknowledgements. We thank Dr. J. Oehlke for performing experiments with radioactive compounds, for glutamate measurements and Michael Hausding for his help in preparing the manuscript. Supported by DFG SFB507 and BMFT/BMBF BEO 03110015A/B.

REFERENCES

1. S.M. Rothman, and J.W. Olney, Glutamate and the pathophysiology of hypoxic-ischemic brain damage, Ann. Neurol., 19:105-111 (1986).
2. A. Volterra, D. Trotti, and G. Racagni, Glutamate uptake is inhibited by arachidonic acid and oxygen radicals via two distinct and additive mechnisms, Molec. Pharmacol., 46:986-992 (1994).
3. S. Rehncrona, E. Westerberg, B. Akesson, and B.K. Siesjo, Brain cortical fatty acids during and following complete and severe incomplete ischemia, J. Neurochem., 38:84-93 (1982).
4. W. Cao, J.M. Carney, A. Duchon, R.A. Floyd, and M. Chevion, Oxygen free radical involvment in ischemia and reperfusion injury to brain, Neurosc. Lett., 88:233-238 (1988).
5. L.L. Rubin, D.E. Hall, S. Porter, K. Barbu, C. Cannon, H.C. Horner, M. Janatpour, C.W. Liaw, K. Manning, J. Morales, L.I.. Tanner, J. Tomaselli., and F. Bard, A cell culture model of the blood-brain barrier, J. Cell Biol., 115:1725-1735 (1991).
6. R.C. Janzer, and M.C. Raff, Astrocytes induce blood-brain barrier properties in endothelial cells, Nature (London), 325:235-257 (1987).
7. P.M. Beart, K.-A.M. Sheehan, and D.T. Manallack, Absence of N-methyl-D-aspartate receptors on ovine cerebral microvessels, J. Cereb. Blood Flow. Metab., 8:879-882 (1988).
8. H. Koenig, J.J. Trout, A.D. Goldstone, and Ch.Y. Lu, Capillary NMDA receptors regulate blood-brain barrier function and breakdown, Brain Res., 588:297-303 (1992).
9. T. Müller, J. Grosche, C. Ohlemeyer, and H. Kettenmann, NMDA-activated currents in Bergmann glial cells, Neuroreport, 4:671-674 (1993).
10. J.M. Luque, and J.G. Richards, Expression of NMDA 2B receptor subunit mRNA in Bergmann glia, Glia, 13:228-232 (1995).
11. A.-M. Lopez-Colome, A. Ortega, and M. Romo-de-Vivar, Excitatory amino acid-induced phosphoinositide hydrolysis in Müller glia, Glia, 9:127-135 (1993).
12. I. Sommer, C. Langenaur, and M. Schachner, Recognition of Bergmann glial and ependymal cells in the mouse nervous system by monoclonal antibody, J. Cell Biol., 90:448-458 (1981).
13. L.A. McNaughton, and S.P. Hunt, Regulation of gene expression in astrocytes by excitatory amino acids, Brain Res. Mol. Brain Res., 16:261-266 (1992).
14. D.J. Wyllie, and S.G. Cull-Candy, A comparison of NMDA receptor channels in type-2 astrocytes and granule cells from rat cerebellum, J. Physiol. (London), 475:95-114 (1994).
15. A.J. Patel, A. Hunt, R.D. Gordon, and R. Balazs, The activities of different neural cell types of certain enzymes associated with the metabolic compartmentation of glutamate, Dev. Brain Res., 4:3-11 (1982).
16. A.M. Benjamin, Ammonia in metabolic interactions between neurons and glial, in glutamine, glutamate, and GABA in the central nervous system, (Hertz L., Kvamme E., McGeer, E.G., and Schousbe A., eds), pp. 399-414. Alan R. Liss, New York, (1983).
17. F.A. Tansey, M. Farooq, and W. Cammer, Glutamine synthetase in oligodendrocytes and astrocytes: New biochemical and immunocytochemical evidence, J. Neurochem., 56:266-272 (1991).
18. F.S. Silverstein, K. Buchanan, and M.V. Johnston, Perinatal hypoxia-ischemia disrupts striatal high affinity [^{3}H] glutamate uptake into synaptosomes, J. Neurochem., 47:1514-1619 (1986).
19. S. Saweda, M. Higashima, and C. Yamamoto, Inhibition of high affinity uptake augment depolarizations of hippocampal neurons induced by glutamate, kainate, and related compounds, Exp. Brain Res., 60:323-329 (1985).
20. G.J. McBean, and P.J. Roberts, Neurotoxicity of L-glutamate and D,L-threo-3-hydroxyaspartate in the rat striatum, J. Neurochem., 44:247-254 (1985).
21. E.H.F. Wong, J.A. Kemp, T. Priestley, A.R. Knight, G.N. Woodruff, and L.L. Iversen, The anticonvulsant MK-801 is a potent N-methyl-D-aspartate antagonist, Proc. Natl. Acad. Sci. USA, 83:7104-7108 (1986).
22. G. Sutherland, N. Haas, and J. Peeling, Ischemic neocortical protection with U74006F dose-response curve, Neurosci. Lett., 149:123-125 (1986).
23. H. Giese, K. Mertsch, and I.E. Blasig, Effect of MK-801 and U83836E on a porcine brain capillary endothelial cell barier during hypoxia, Neurosci. Lett., 191:169-172 (1995).
24. K. Mertsch, T. Grune, A. Ladhoff, and I.E. Blasig, Hypoxia and reoxygenation of brain endothelial cells *in vitro*: comparison of biochemical and morphological response, Cell. Mol. Biol., 41:243-253 (1995).
25. J.B.M.M. Van Bree, A.G. De Boer, M. Danhof, L.A. Ginsel, and D.D. Breimer, Characterization of an "*in vitro*" blood-brain barrier: effects of molecular size and lipophilicity on cerebrovascular endothelial transport rates of drugs, J. Pharmacol. Exp. Ther., 247:1233-1239 (1988).
26. J. Oehlke, S. Savoly, and I.E. Blasig, Utilization of endothelial cell monolayers of low tightness for estimation of transcellular transport characteristics of hydrophilic compounds, Eur. J. Pharmac. Sci., 2:365-372 (1994).

27. O.W. Griffith, Determination of glutathione and glutathione disulfide using glutathione reductase and 2-vinylpyridine, Anal. Biochem., 106:207-212 (1980).
28. R.F. Haseloff, I.E. Blasig, H. Meffert, and B. Ebert, Hydroxyl radical scavenging and antipsoriatric activity of benzoic acid derivatives, Free Rad. Biol. Med., 9:111-115 (1990).
29. M.-P. Dehouck, St. Meresse, P. Delorme, J.C. Fruchart, and R. Cecchelli, An easier, reproducible, and mass-production method to study the blood-brain barrier *in vitro*, J. Neurochem., 54:1798-1801 (1990).
30. U. Jaehde, R. Masereeuw, A.G. De Boer, G. Fricker, J.F. Nagelkerke, J. Vonderscher, and D.D. Breimer, Quantification and visualization of the transport of octreotide, a somatostatin analogue, across monolayers of cerebrovascular endothelial cells, Pharmaceut. Res., 11:442-448 (1994).
31. K. Ohno, K.D. Pettigrew, and S.I. Rapoport, Lower limits of cerebrovascular permeability to nonelectrolytes in the conscious rat, Am. J. Physiol., 235:H299-H307 (1978).
32. W.M. Pardridge, D. Triguero, J. Yang, and P.A. Cancilla, Comparison of *in vitro* and *in vivo* models of drug transcytosis through the blood-brain barrier, J. Pharmacol. Exp. Ther., 253:884-891 (1990).
33. K.L. Audus, F.L. Guillot, and J.M. Braughler, Evidence for 21 aminosteroid association with the hydrophobic domains of brain microvessel endothelial cells, Free Rad. Biol. Med., 11:361-371 (1991).
34. J.T. Greenamyre, J.M.M. Olson, J.B. (Jr) Penney, and A.B. Young, Autoradiographic characterization of N-methyl-D-aspartate-, quisqualate-, and kainate-sensitive glutamate binding sites, J. Pharmacol. Exp. Ther., 233:254-263 (1985).

52

EFFECTS OF HYPOXIA ON AN *IN VITRO* BLOOD-BRAIN BARRIER

M. Plateel, M. P. Dehouck, G. Torpier, J. P. Fruchart, R. Cecchelli, and E. Teissier

INSERM, Unité 325
Institut Pasteur
59019 Lille
France

I. ABSTRACT

Using a cell culture model of the BBB, we investigated the brain capillary ECs response to hypoxia. ECs contain the same free-radical protecting enzymes and glutathione as isolated capillaries. These anti-oxidant enzymes and GSH are not induced by astrocytes. These results clearly show that ECs could be a relevant model to study the effects of hypoxia at the BBB level. In the time course of hypoxia, free radical detoxifying enzymes and GSH level decrease. Concomittant rearrangements of F-actin of the ECs and a decrease in the ATP level might explain the paracellular permeability increase of the monolayer. An apoptotic process is detected by *in situ* end labeling of DNA. This suggests that hypoxia alters some functions of the ECs monolayer and increases the susceptibility of the BBB to further oxidative damages.

I. RÉSUMÉ

Utilisant un modèle in vitro de BHE, nous avons étudié la réponse des CEs à l'hypoxie. Elles présentent les mêmes taux d'enzymes de protection des radicaux libres et de glutathion que les capillaires. Ces enzymes anti-oxydantes et le GSH ne sont pas induits par les astrocytes. Ces résultats montrent que les CEs de la coculture constitue un modèle d'étude pertinent des effets de l'hypoxie au niveau de la BHE. Au cours de l'hypoxie, les enzymes détoxifiantes des radicaux libres et le taux de GSH diminuent. La réorganisation de la F-actine des CEs concomitante d'une diminution des taux d'ATP peuvent expliquer une augmentation de la perméabilité paracellulaire de la monocouche. De plus, un processus apoptotique est détecté par un marquage *in situ* de l'ADN. Ceci suggère que l'hypoxie fragilise la BHE et augmente ainsi sa sensibilité aux attaques oxydantes survenant lors de la réoxygénation.

Biology and Physiology of the Blood–Brain Barrier, edited by Couraud and Scherman
Plenum Press, New York, 1996

Table 1. Specific activities of anti-oxidant enzymes and GSH content in homogenates of isolated fresh capillaries, brain capillary ECs alone or in coculture with astrocytes, and pericytes

Samples	GSH-Perox	GSH-Reduct	Catalase	SOD	GSH
Capillaries	20.48±2.19	20.16±2.00*	46.45±4.89**	32.98±3.9	17.67±0.19
ECs	22.28±1.85	33.19±2.95	33.61±3.25	27.49±2.36	15.65±0.59
ECs coc	25.50±1.20	30.95±2.26	37.42±3.15	26.57±2.05	16.47±1.19
Pericytes	13.64±1.14°°°	15.73±1.62°°°	16.49±1.92°°°	7.97±0.92°°°	4.77±0.61°°°

(GSH-peroxidase) In nmol of hydrogen peroxide reduced/mg protein/min. (GSH-reductase) In nmol of NADPH oxidized/mg protein/min. (Catalase) In µmol of hydrogen peroxide consumed/mg protein/min. (SOD) In units of SOD/mg protein, (1 unit inhibits the rate of cytochrome C reduction by 50%, in the presence of xanthine-xanthine oxidase, pH 7.8). (GSH) in nmol/mg protein, measured in HPLC with a Au/Hg working electrode, set at + 0.15V.

II. RESULTS

II.I. A Relevant Model

Our aim was first to check if co-cultured brain capillary ECs [1] could be a relevant model for studying the effects of oxygen variations on the BBB.

That is why the antioxidant profiles and the GSH levels of brain capillary ECs alone or in coculture with astrocytes, as well as whose of pericytes were compared with those obtained with freshly isolated microvessels (Table 1).

Our results demonstrate that the activities of the four anti-oxidant enzymes of the *in vitro* model were close to those found in freshly isolated capillaries. The brain capillaries ECs could be a relevant model to study the effects of oxygen variations on the BBB.

II.II. Effects of Hypoxia

The activity of all free radical detoxifying enzymes slowly decreased with time during hypoxia and this decrease became highly significant after 24 and 48 hrs (Table 2).

A decrease in the concentration of GSH (Table 2) accompanies the decrease in the ATP level (Table 3). When the permeability of the monolayer to small molecules was assessed, a dramatic increase in the permeability to sucrose was noted after 48 hrs of hypoxia (Table 3).

Concomitant with alterations of actin filaments, interendothelial cell gaps in the monolayer could be seen and became more accentuated in 48 hrs. Detection of DNA breakages by using in situ end labeling od DNA confirmed the programmed cell death of endothelial cells.

Table 2. Effect of hypoxia on specific activities of anti-oxidant enzymes and on GSH level of brain capillary ECs

ECs	GSH-Perox	GSH-Reduct	Catalase	SOD	GSH
Control	100.39±10.53	100.08±3.82	100.18±4.68	100.71±5.45	99.59±0.40
Hypoxia 12h	88.47±6.81	109.62±0.93	98.32±8.26	82.76±20.69	71.48±0.00**
Hypoxia 24h	85.58±3.51*	82.53±3.46**	72.09±1.31***	65.00±5.00*	68.67±0.00**
Hypoxia 48h	72.14±2.41***	70.32±2.44***	63.91±5.63***	48.73±8.79***	58.23±8.24**

Enzyme activities and GSH level were assessed as described in Table 1. Results are given as percentages of values found at time 0.

Table 3. Time course of the effect of hypoxia on the permeability of brain capillary EC monolayers and effect of hypoxia on ATP level of brain capillary ECs

ECs	Pe Sucrose Normoxia	Pe Sucrose Hypoxia	ATP
0h	0.42±0.12	0.42±0.12	100,71±6
12h	0.71±0.09	0.27±0.04	109±6
24h	0.64±0.03	0.37±0.04	76±6**
48h	0.85±0.44	2.55±0.59**	49±6***

Brain capillary ECs cocultured for 12 days with astrocytes were incubated in the presence of coculture conditioned medium in normoxia or in hypoxia and permeability assays were carried out at the indicated times (12, 24 or 48 hrs) by adding [^{14}C]sucrose tracer to the luminal compartment. For ATP, the results are given as percentages of values found at time 0.

III. CONCLUSION

During hypoxia, we have demonstrated [2] first, a decrease in the anti-oxidant profile which could increase the susceptibility of the brain capillary ECs to further oxidant injury and second, a reorganization of the cytoskeleton of the ECs probably due to the limitation of energy supply, leading to an increase in the permeability of the monolayer.

REFERENCES

Dehouck M.P., Méresse S., Delorme P., Fruchart J.C., and Cecchelli R. (1990) An easier, reproductible, and mass-production method to study the blood-brain barrier in vitro. J. Neurochem. 54 , 1798-1801.

Plateel M., Dehouck M.P., Torpier G., Cecchelli R. and Teissier E. (1995) Hypoxia increases the susceptibility of the blood-brain barrier endothelial cell monolayer. J. Neurochem. in press.

53

EXTRACELLULAR MATRIX PROTEINS IN CEREBRAL VESSELS IN CHRONIC HYPERTENSION

Sukriti Nag, Dan Kilty, and Shruti Dev

Division of Neuropathology
University of Toronto
The Playfair Neuroscience Research Unit
Toronto, Canada M5T 2S8

SUMMARY

This study demonstrates progressive mural thickening of cerebral arterioles in chronic hypertension due to medial hyperplasia and due to overexpression of extracellular matrix proteins such as laminin, collagen IV and fibronectin. These changes are most marked in the stage II and III permeability lesions in which mural deposition of serum proteins in arterioles also occur. Since overexpression of extracellular matrix proteins occurs in cerebral vessels in conditions unassociated with hypertension, it may represent a nonspecific adaptive response of cellular components of the vessel wall to a variety of stimuli.

RÉSUMÉ

Notre étude montre que l'hypertension vasculaire chronique s'accompagne d'un épaississement de la paroi des artérioles cérébrales dû à une hyperplasie cellulaire de la media des vaisseaux ainsi qu'à une surexpression des proteines de la matrice extracellulaire tels que la laminine, le collagene 4 et la fibronectine. Ces modifications de l'histologie des artérioles cérébrales sont plus marquées au cours des stades 2 et 3 du développement des lésions vasculaires accompagnant l'hypertension, au cours desquelles a aussi lieu le depôt des protéines sériques sur la paroi des artérioles. Néanmoins, la surexpression de certaines protéines de la matrice extracellulaire dans les vaisseaux cérébraux peut être observée dans des conditions indépendantes de l'hypertension, suggérant que ce phénomène pourrait découler d'une réaction cellulaire vasculaire non spécifique face à differents facteurs.

Biology and Physiology of the Blood–Brain Barrier, edited by Couraud and Scherman
Plenum Press, New York, 1996

INTRODUCTION

Vascular hypertrophy and vascular remodelling are well documented findings in cerebral vessels in humans [2] and experimental studies of chronic hypertension [1,3,7,9]. In both hypertrophy and remodelled arteries there is an increase in the extracellular matrix in vessel walls. Another consistent finding in chronic hypertension is the occurence of multifocal areas of blood-brain barrier (BBB) breakdown in grey matter areas such as the cortex and basal ganglia [7]. Based on their histology, the areas of BBB breakdown were described as belonging to 3 stages [7]. Severe mural thickening is a feature of intracerebral arterioles in stage II and stage III lesions.

The present immunohistochemical study was undertaken to determine whether the glycoproteins-fibronectin, laminin and collagen IV contribute to increased mural thickness of cerebral vessels in experimental chronic hypertension. The areas of BBB breakdown were localised using antisera to rat serum proteins.

METHODS

Chronic hypertension was induced in female Wistar rats as described previously [8]. Normotensive and hypertensive rats were sacrificed at weekly intervals over a 9 week period following the surgery to induce hypertension. Coronal sections taken through the frontal lobes and through the occipito-temporal lobes were evaluated by light microscopy, following staining with hematoxylin-eosin. Adjacent sections were used for immunohistochemistry using the indirect streptavidin-biotin-peroxidase method. Sections were treated with 0.5% pepsin for 20 min at 37^0C, prior to an overnight incubation at 4^0C in antisera to fibronectin (Gibco BRL), laminin (Collaborative Biomedical Products) and collagen IV (Collaborative Biomedical Products).

RESULTS

Normotensive Rats

Brains did not show vascular or parenchymal changes. Collagen IV and fibronectin immunoreactivity was observed in the adventitia and basement membrane of meningeal vessels and in the basement membrane of all intracerebral vessels. Laminin immunoreactivity was observed in the endothelial basement of all vessels and in the extracellular matrix between the medial smooth muscle cells of meningeal arteries and intracerebral arterioles.

Hypertensive Rats

Medial hyperplasia and expansion of the extracellular matrix was observed in pial arteries and intracerebral arterioles diffusely in the brain. Immunohistochemistry showed increased immunoreactivity in these vessels with antisera to laminin, collagen IV and fibronectin (Figure 1). Rats having a mean maximum systolic blood pressure in the range of 150-200 mm Hg showed a mild degree of changes while changes of a moderate degree were observed in rats having mean maximum systolic blood pressures above 200 mm Hg.

Four rats having mean maximum systolic blood pressures above 220 mm Hg showed focal areas of BBB breakdown to endogenous serum proteins and fibronectin in the cerebral cortex and basal ganglia. The areas of BBB breakdown in the cerebral cortex tended to occur in the frontal cortex in the boundary zone of supply of the anterior and middle cerebral arteries and in the occipital cortex in the boundary zone of supply of the anterior, middle and posterior cerebral arteries.

The histological findings in the areas of BBB breakdown were similar to our previous observations [7]. Early or stage I lesions showed serum proteins in arteriolar walls with extravasation into the surrounding neuropil and no tissue changes. Stage II lesions were most frequently encountered in this study and showed focal areas of BBB breakdown in the cerebral cortex associated with spread of endogenous serum proteins into the underlying white matter and via the corpus callosum into the white matter of the opposite hemisphere. Cystic necrosis was observed in the central portion of the area of BBB breakdown accompanied by a macrophage response (Figure 2A). Astrogliosis was observed in the lesions but was particularly prominent at the margin of the lesion and the adjacent cortex and in the edematous white matter. Stage III permeability lesions consisted of glial scars or cystic cavities traversed by blood vessels. Marked mural thickening with accompanying overexpression of laminin, collagen IV and fibronectin was observed in arterioles in and adjacent to stage II and stage III permeability lesions (Figure 2B).

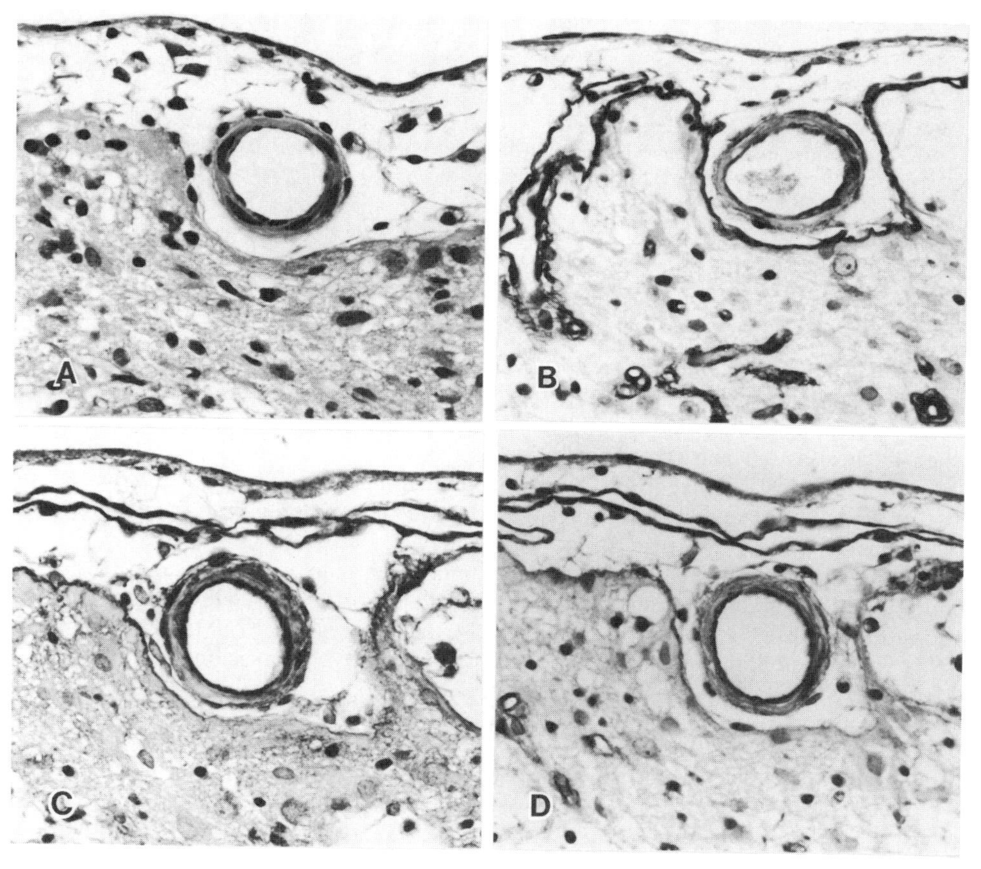

Figure 1. A-D: Serial sections of a pial artery overlying a Stage III permeability lesion from a hypertensive rat having a mean maximum systolic blood pressure of 280 mm Hg at the time of sacrifice. Note the moderate degree of medial hyperplasia and mural thickening (A) due to overexpression of laminin (B), fibronectin (C) and collagen IV (D). (A), HE Stain. (A-D) x 440.

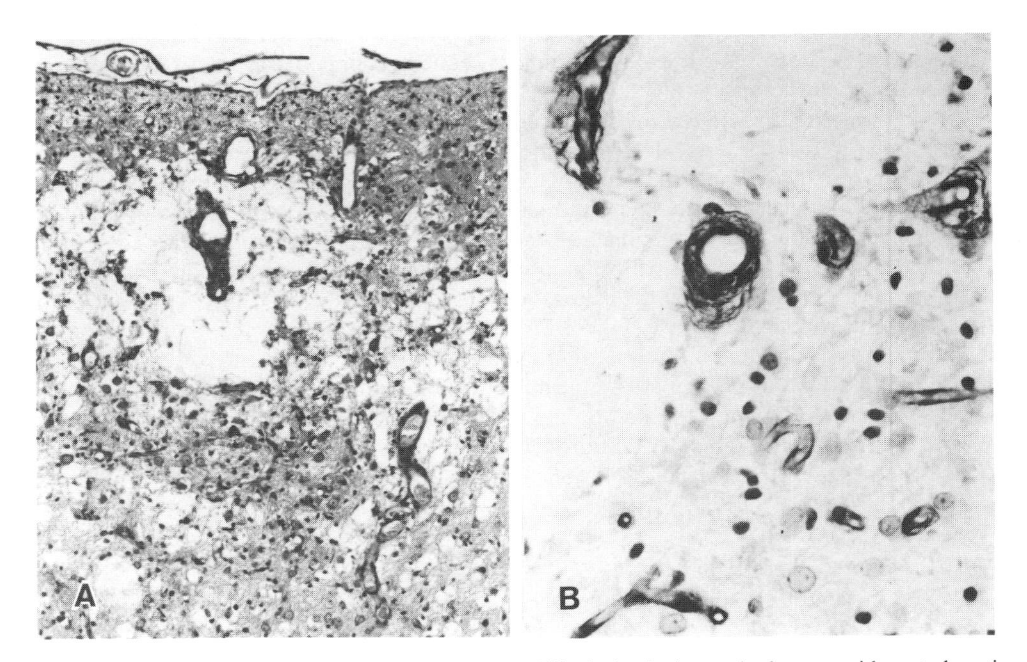

Figure 2. A: Hypertensive rat showing a stage II permeability lesion in the cerebral cortex with central cystic necrosis. The arterioles in this lesion show moderate to severe mural thickening. **B**: Severe mural thickening and overexpression of laminin in an intracerebral arteriole from a stage III permeability lesion in the cerebral cortex. A x 115; B x 460.

DISCUSSION

This study demonstrates progressive mural thickening of cerebral arterioles in chronic hypertension due to medial hyperplasia and due to overexpression of extracellular matrix proteins such as laminin, collagen IV and fibronectin. These changes were most marked in the stage II and III permeability lesions in which mural deposition of serum proteins in arterioles also occurred. Mural thickening and overexpression of extracellular matrix proteins in intracerebral arterioles was observed concomitant with the development of hypertension.

In hypertension, overexpression of extracellular matrix proteins maybe due to vascular distension and stretching or due to the direct effect of increased levels of circulating peptides. Both these factors are known to cause overexpression of extracellular matrix proteins in noncerebral vessels in hypertension[5,6,10,11]. Overexpression of extracellular matrix proteins occurs in cerebral vessels in other conditions such as brain tumors[4] and the acquired immunodeficiency syndrome[12], in the absence of hypertension. Thus overexpression of extracellular matrix proteins may represent a nonspecific adaptive response of cellular components of the vessel wall to a variety of stimuli.

ACKNOWLEDGMENTS

This work was supported by the Heart and Stroke Foundation of Ontario Grant No. B2401.

REFERENCES

1. Baumbach, G.L.,and Heistad, D.D., Remodeling of cerebral arterioles in chronic hypertension. *Hypertension* 13:968, 1989.
2. Chester, E.M,, Agamanolis, D.P., Banker, B.Q., Victor, M., Hypertensive encephalopathy: A clinicopathologic study of 20 cases. *Neurology* 28:928, 1978.
3. Fredriksson, K., Nordborg, C., Kalimo, H., Olsson, Y., Johansson, B.B., Cerebral microangiopathy in stroke-prone spontaneoulsy hypertensive rats. An immunohistochemical and ultrastructural study. *Acta Neuropathol. (Berl.)* 75:241, 1988.
4. Giordana, M.T., Germano, I., Giaccone, G., Mauro, A., Migheli, A., Schiffer, D., The distribution of laminin in human brain tumors:an immunohistochemical study. *Acta Neuropathol.(Berl.)* 67:51, 1985.
5. Himeno, H., Crawford, D.C., Hosoi, M., Chobanian, A.V., Brecher, P., Angiotensin II alters aortic fibronectin independently of hypertension. *Hypertension* 23;823, 1994.
6. Kato, H., Suzuki, H., Tajima, S., Ogata, Y., Tominaga, T., Sato, A., Saruta, T., Angiotensin II stimulates collagen synthesis in cultured vascular smooth muscle cells. J *Hypertension* 9:17, 1991.
7. Nag, S., Cerebral changes in chronic hypertension: combined permeability and immunohistochemical studies. *Acta Neuropathol.(Berl.)* 62:178, 1984.
8. Nag, S., Cerebral endothelial mechanisms in increased permeability in chronic hypertension. *Adv in Exp. Med. and Biol.* 331:263, 1993.
9. Nag, S., Robertson, D.M., Dinsdale, H.B., Morphological changes in spontaneously hypertensive rats. *Acta Neuropathol.(Berl.)* 52:27, 1980.
10. Riser, B.L., Cortes, P., Zhao, X., Bernstein, J., Dumler, F., Narins, R.G., Intraglomerular pressure and mesangial stretching stimulate extracellular matrix formation in the rat. *J. Clin. Investig.* 90:1932, 1992.
11. Takasaki, I., Chobanian, A.V., Sarzani, R., Brecher, P., Effect of hypertension on fibronectin expression in rat aorta. *J. Biol. Chem.* 265:21935, 1990.
12. Taruscio, D., Malchiodi Albedi, F., Bagnato, R., Pauluzzi, S., Francisci, D., Cavaliere, A., Donelli, G., Increased reactivity of laminin in the basement membranes of capillary walls in AIDS brain cortex. *Acta Neuropathol. (Berl.)* 81:552, 1991.

THE EFFECT OF NICOTINE PRETREATMENT ON THE BLOOD-BRAIN BARRIER PERMEABILITY IN NICOTINE-INDUCED SEIZURES

A. S. Diler,[1,2] G. Üzüm,[1] J. M. Lefauconnier,[2] and Y. Z. Ziylan[1]

[1] Department of Physiology
Faculty of Medicine
Istanbul University Istanbul
Capa, Istanbul, 34390 Turkey
[2] INSERM U26 Unité de Neuro-Pharmaco-Nutrition
Hôpital F. Widal
200 Rue du Fg. St Denis, 75010 Paris, France

SUMMARY

Nicotine is a toxic substance which because of its lipid solubility can cross the blood brain barrier. It has several different actions in the CNS; one of which is neuroexcitation, where it can result in seizure activity. Based on the observations that nicotine pretreatment ameliorated blood flow and glucose utilisation in caudate putamen on rats whose mesostriatal dopamine system had been cut and that nicotine pretreatment rendered animals less susceptible to nicotine induced seizures than saline administered controls, we conducted this set of experiments where we investigated the protective effect of nicotine pretreatment on the BBB permeability in nicotine induced seizures. Administration of saline or subseizure producing dose of nicotine (1mg/kg i.p.) was followed by seizure producing doses of nicotine (2, 5 or 8 mg/kg, i. p.). Intravenous technique was used to calculate the unidirectional blood to brain transfer constant (Kin) for six different brain regions, with [^{3}H] α-AIB as a tracer. Mean Kin in brains of all acute nicotine groups (2, 5 or 8 mg/kg) increased by 83.94%, 182.6% and 265% respectively. Twenty one days chronic nicotine pretreatment prevented the rise in Kin AIB to 2 mg/kg acute nicotine and partially ameliorated the disturbed BBB to 5 and 8 mg/kg.

RÉSUMÉ

La nicotine est une substance toxique qui, du fait de son caractère lipophile, peut traverser la barrière hémato-encéphalique. Elle a différents effets au niveau du système

Biology and Physiology of the Blood–Brain Barrier, edited by Couraud and Scherman
Plenum Press, New York, 1996

nerveux central, dont la neuroexcitation pouvant conduire à un état de choc. Des études montrent que chez des rats, dont le sytème dopaminergique mésostriatal est lésé, un prétraitement avec de la nicotine augmente le flux sanguin et le métabolisme du glucose dans le caude putamen. De même, un prétraitement à la nicotine rend les animaux moins sensibles à l'effet de choc dû à cette substance que des animaux témoins prétraités avec du sérum physiologique. Nous avons étudié l'influence de cet effet protecteur de la nicotine sur la perméabilité de la barrière hémato-encéphalique au cours des états de choc provoqués par la nicotine. L'administration d'une solution saline de doses de nicotine inférieures à celles produisant un effet de choc (1 mg/kg, i.p.), ou de doses conduisant à l'apparition d'états de choc (2, 5 et 8 mg/kg) a été réalisée. La technique d'injection intraveineuse a été utilisée pour calculer la constante de transfert du sang vers le cerveau (Kin) du traceur 3H AIB dans 6 régions différentes du cerveau. Le Kin moyen pour les groupes d'animaux traités avec 2,5 et 8 mg/kg de nicotine augmentait respectivement de 83.94, 182.6 et 265%. Un prétraitement de 21 jours par la nicotine prévient totalement l'augmentation du Kin apres injection de 2mg/kg de nicotine et partiellement pour les doses de 5 et 8 mg/kg.

INTRODUCTION

Nicotine is a tertiary amine composed of a pyridine and pyrolidine ring. Because of its lipid solubility it can readily pass into the brain paranchyma crossing the blood brain barrier (BBB). This explains nicotine's direct effects in the CNS. It has several different actions in the CNS; one of which is neuroexcitation, where it can result in seizure activity. However nicotine pretreatment decreases the sensitivity to nicotine induced seizures (1).

The pretreatment with nicotine may have beneficial effect on BBB lesions caused by convulsions. Our previous studies with Evans Blue (EB) dye have confirmed that pretreatment with nicotine might play a protective role on the integrity of BBB in convulsions induced by nicotine (2).

Thus in this study, we used a BBB radiotracer technique, which allows precise quantitative expression of changes in barrier permeability. We conducted the present study (i) to explore the BBB permeability alterations during convulsions, which can not be detected by large molecular weight tracers such as EB dye; (ii) to compare the qualitative and quantitative data for the integrity of BBB in convulsions induced by nicotine and thus verify the protective role of chronic nicotine administration on BBB integrity. An abstract of this work in part has been published (2).

MATERIALS AND METHODS

Male Sprague-Dawley rats (Iffa Credo, France) weighing 225-300g were used. Animals were studied in two groups; sham pretreated (controls) and nicotine pretreated. Each group consisted 4-7 animals. Saline or subseizure producing dose of nicotine (1 mg/kg/day i. p.) were injected twice daily for 21 days before the permeability measurements. The regional cerebrovascular permeability was determined following the administration of seizure producing doses of nicotine; 2, 5 or 8 mg/kg (i. p.) to 3 weeks nicotine pretreated rats or to controls.

I.V. technique was utilised to calculate the unidirectional Blood to Brain transfer constant (Kin) in cortex, striatum, hippocampus, thalamus-hypothalamus and cerebellum. [3H] α-AIB was the choice of BBB tracer. Whole blood, arterial plasma and weighed regions ofthe brain were prepared for radioactivity counting as previously described by Ziylan et al (3). Regional unidirectional Blood to Brain transfer constant (Kin) were calculated for [3H] α-AIB from the tissue and plasma radioactivity data using the following equation based on the works of Ohno et al and Blasberg et al (4, 5);

$$Kin = CbrT - VCwbT / {}_0\int^T Cpl\ dt,$$

where Cbr is the amount of tracer in the brain per unit mass of tissue (dpm g^{-1}) at the time T, T is the duration of the experiment (min), Cpl is the arterial plasma concentration (dpm ml^{-1}), V is the regional blood volume and Cwb is the tracer concentration in the final blood (dpm ml^{-1}).

Regional blood volume was determined separetely for all groups as [14C] Dextran (M W 79 000) space for 1 min as the value dpmg^{-1} / dpm ml^{-1}. All the substances were obtained from Sigma (St Louis, MO).

Data are presented as means ± S. D. Comparisons were assessed by ANOVA followed by Mann-Whitney U test. A p level less than 0.05 was considered to be statistically significant.

RESULTS

All control rats given single bolus dose injections of nicotine rendered seizure activity within 15 seconds. No marked difference in seizure pattern was observed in rats receiving 2, 5 or 8 mg/kg. All the nicotine doses used acutely (2, 5 or 8 mg/kg), made significant increases in Kin for AIB, indicating a disruption of BBB integrity with varying degrees; 83.94%, 182.6% and 265% respectively.

Twenty one days chronic pretreatment prevented the rise in Kin to 2 mg/kg acute nicotine injection. The slight increase in Kin AIB was found to be insignificant (Fig. 1).

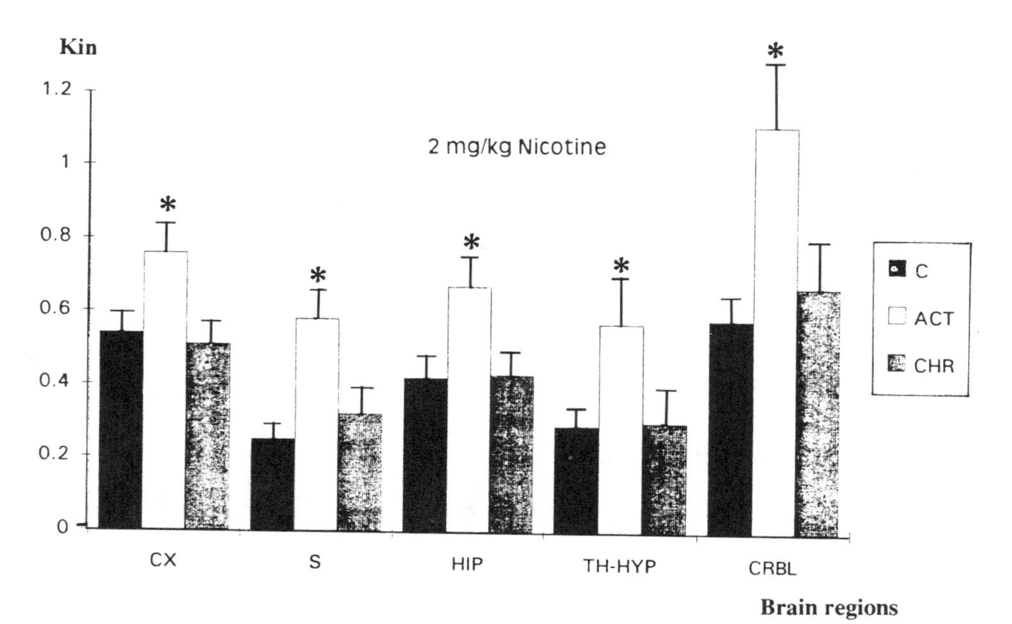

Figure 1. Regional kin values (ml g^{-1} min^{-1} x 10^3) are mean +/- S.D. *Significant difference (p, u.0) between controls and corresponding brain regions. (C=control, ACT=acute-2, 5, 8 mg/kg nicotine i.p., CHR=chronic-nicotine pretreatment 1mg/kg followed by acute doses of nicotine.

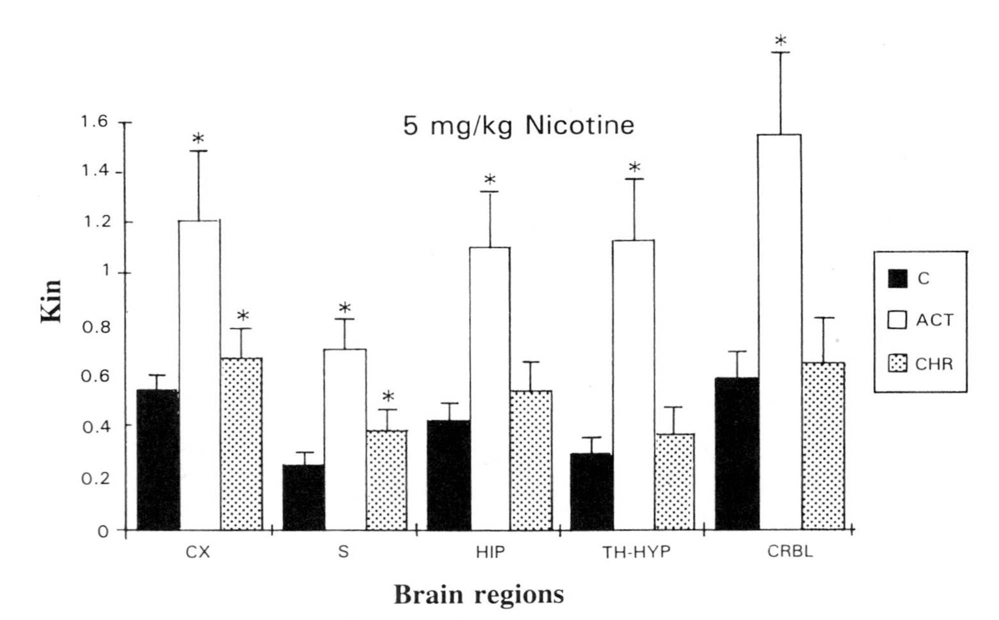

Figure 2. Regional kin values (ml g^{-1} min^{-1} x 10^3) are mean +/- S.D. *Significant difference (p, u.0) between controls and corresponding brain regions. (C=control, ACT=acute-2, 5, 8 mg/kg nicotine i.p., CHR=chronic-nicotine pretreatment 1mg/kg followed by acute doses of nicotine.

We have also demonstrated a marked amelioration on the disturbed barrier permeability with chronic pretreatment to 5 and 8 mg/kg nicotine. This recruitment was not total but partial, rendering a relative decrease. Although the chronic nicotine pretreatment suppressed the Kin enhancement occuring with acute nicotine up to a certain level, the results were still higher than their corresponding control brain regions. Permeability changes as well as the degree of amelioration showed regional variability. With 5mg/kg nicotine, cortex and striatum showed enhanced Kin AIB ($p < 0.05$, U. 0). Other regions showed somewhat insignificant increases (Fig. 2).

With 8 mg/kg nicotine, only hippocampus showed an insignificant increas in Kin AIB, whereas increament in the remaining brain regions were statistically significant ($p < 0.05$, U. 0) (Fig. 3).

DISCUSSION

Nicotine is one of the most widely used substance by humans. It can cross the BBB and affect different functions in the CNS (6, 7). Nicotine's effect in the CNS can be ranged from neurotoxicitation to neuroregulation (7, 8).

A relatively new aspect of nicotine's action has been put forward by Owman et al (9). In their studies they have employed nicotine pretreatment on rats whose meso-striatal dopamine system had been transsected. They have found amelioration in blood flow and glucose utilisation in caudate putamen on the side where meso-striatal dopamine system had been transsected. Following this study De Fiebre et al (1) showed that nicotine pretreatment rendered animals less susceptible to nicotine induced seizures than saline pretreated controls.

Moreover, in a previous study we have also found evidence that pretreatment with nicotine might be beneficial on the BBB- the prerequisite for maintaining microenvironment

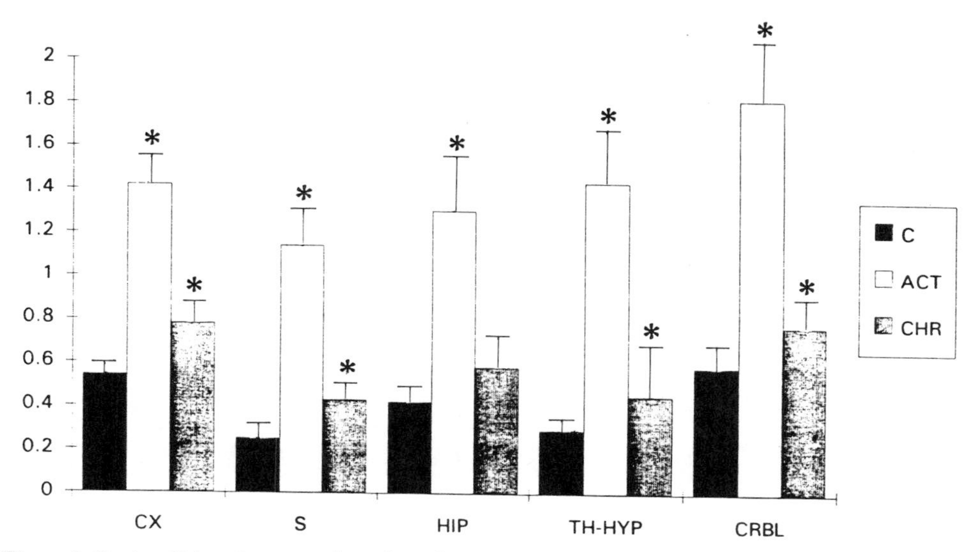

Figure 3. Regional kin values (ml g^{-1} min^{-1} x 10^3) are mean +/- S.D. *Significant difference (p, u.0) between controls and corresponding brain regions. (C=control, ACT=acute-2, 5, 8 mg/kg nicotine i.p., CHR=chronic-nicotine pretreatment 1mg/kg followed by acute doses of nicotine.

homeostasis within CNS- lesioned by convulsions. Thus other circulating neurotoxic and neuroactive subtances, which are normally excluded from the brain, may get access. This may indicate the importance of BBB in the occurrence of seizures.

Since our data depict that repeated administration of subconvulsive nicotine doses ameliorates the perturbation of BBB permeability in nicotine induced seizures, one would anticipate that the neurotoxic effect of nicotine might be defrayed by pretreating subjects with nicotine for several days.

Our results verify our previous findings where we had shown protection of nicotine pretreatment on BBB permeability (see introduction) as well as on seizure intensity and latencies to seizures.

ACKNOWLEDGMENTS

This project has been supported by European Science Foundation fellowship- to Dr. *A. Sarper Diler* - (ESF) PFT, SVF/93/48/T.

REFERENCES

1. De Fiebre C, Collins AC (1988) Decreased sensitivity to nicotine induced seizures as a consequence of nicotine pretreatment with Long-Sleep and Short-Sleep mice. Alcohol, 5: 55-61.
2. Üzüm G, Curgunlu S, Hacialioglu M, Ercan S, Diler AS, Ziylan YZ (1993) Protective effect of chronic administration of nicotine on neuronal and vascular functions in the brain. Eur J Neurosci. Suppl. 6, 93: S358.
3. Ziylan YZ, Lefauconnier JM, Bernard G, Bourre JM (1989) Regional alterations in Blood to Brain Transfer of α–aminoisobutyric acid and sucrose after chronic administration and withdrawal of dexamethasone. J Neurochem. 52: 684-690.

4. Ohno K, Pettigrew KO, Rapoport SI (1978) Lower limits of cerebrovascular permeability to nonlectrolytes in the conscious rat. Am J Physiol. 253, H299-307.
5. Blasberg RG, Fenstermacher JD and Patlak CS (1983) Transport of α–aminoisobutyric acid across brain capillary and cellular membranes. J Cereb Blood Flow Metab. 3: 8-32.
6. Barrett JE (1983) Interrelationships between behaviour and pharmacology as factors determining the effects of nicotine. Pharmacol. Biochem. Behav.19: 1027-1029.
7. Clarke PBS (1987) Nicotine and smoking; a perspective from animal studies. Psychopharmacology, 92: 135-143.
8. Pomerleau OF, Rosecrans J (1989) Neuroregulatory effects of nicotine. Psychoneuroendocrinology, 92: 135-143.
9. Owman C, Fuxe K, Janson AM, Kåhström J (1989) Studies of protective actions of nicotine on neuronal and vascular functions in the brain of rats; comparison between sympathetic noradrenergic and mesostriatal dopaminergic fiber systems and the effect of a dopamine agonist. In: Progress in Brain Research, Vol. 79, Chapter 26, (Nordberg A, Fuxe K, Holmsted B, Sudwall A eds.) Elsevier Science Publishers B. V., 267-276.

ASYMMETRICAL CHANGES IN BLOOD-BRAIN BARRIER PERMEABILITY DURING PENTYLENETETRAZOL-INDUCED SEIZURES IN RATS

Baria Öztaş

Department of Physiology
Istanbul Faculty of Medicine
University of Istanbul
Çapa, 34390, Istanbul, Turkey

SUMMARY

The asymmetrical breakdown of the blood-brain barrier to Evans-blue was studied in female rats subjected to pentylenetetrazol induced seizures. The animals were divided into two groups: control group (Group I): eight female rats were injected with 0.2 ml of saline into the femoral vein. Pentylenetetrazol (PTZ) group (Group II): twelve female rats were injected 80 mg/kg PTZ (i.v.). The amount of release dye in homogenized cerebral hemispheres was determined by spectrophotometric analysis. The mean value for Evans-blue dye was found to be 1.81±0.5 mg % in the left hemisphere and 1.40±0.4 mg % in the right hemisphere after PTZ induced seizures. This difference betwen left and right hemispheres was found to be significant ($p<0.05$). The findings suggest that disruption of the blood-brain barrier during PTZ induced seizures is asymmetric between right and left hemispheres.

RÉSUMÉ

La rupture assymétrique de la barrière hémato-encéphalique a été étudiée à l'aide du bleu Evans chez des rats femelles chez lesquelles ont été déclenchées des crises épileptiques induites par le pentylenetetrazol (PTZ). Les animaux ont été divisés en deux groupes : un groupe témoin (groupe 1) composé de 8 rats femelles recevant une injection intraveineuse (veine fémorale) de 0.2 ml de solution physiologique; un groupe (groupe 2) composé de 13 rats femelles recevant une injection intraveineuse de PTZ (80 mg/kg). La quantité de colorant présent dans les hémisphères est déterminée après homogénéisation du tissu cérébral par analyse spectrophotometrique.Chez les rats traités, la valeur moyenne de bleu Evans présent dans l'hémisphère gauche est de 1.81 +/- 0.5 mg % et de 1.40 +/- 0.4 mg% dans l'hémisphère

Biology and Physiology of the Blood–Brain Barrier, edited by Couraud and Scherman
Plenum Press, New York, 1996

droit. La différence entre les deux hémisphères est significative ($p<0.05$). Ces résultats suggèrent que la rupture de la BHE durant les crises épileptiques induites par le PTZ est différente dans les deux hémisphères cérébraux.

INTRODUCTION

Anatomic brain asymmetries have been observed post-mortem and have been measured by a variety of technique(9). Functional asymmetry has also described in the rodent barrel cortex(4). CT reports of reversed asymmetries in patient with schizophrenia could imply a structural basis for these disorders(8), but other studies have failed to find abnormal asymmetries in these patient groups(1). The goal of the present study was to compare the permeability of the blood-brain barrier (BBB) in the left hemisphere and in the right hemisphere under control conditions and during epileptiform seizures.

MATERIAL METHOD

The experiments were carried out on adult female Wistar rats. Evans-blue was used as an indicator of BBB permeability. Evans-blue (4 ml of a 2% solution in saline/kg) was injected into the femoral vein, and five minutes later, PTZ (80 mg/kg i.v.) was administered. Mean arterial blood pressure was recorded continously during the experiments. The animals were divided into two groups: control group (Group I): eight female rats were injected with 0.2 ml of saline into the femoral vein. PTZ group (Group II), twelve female rats were injected 80 mg/kg PTZ. Approximately twenty min after the injection of saline or PTZ, all rats were killed by perfusion of saline via the left heart ventricle. A quantiative estimation of homogenized left and right hemispheres that released the dye were measured by a spectrophotometer as described previously(6).

RESULT

Evans-blue dye was 0.25 ± 0.01 mg % in left hemisphere and 0.27 ± 0.01 mg % in right hemisphere in group I. This difference between left hemisphere and right hemisphere was not found to be significant. BBB lesions were present in 80% of PTZ injected rats. The mean value for Evans-blue dye was fonud to be 1.81 ± 0.5 mg % in the left hemisphere and 1.40 ± 0.4 mg % in the right hemisphere after PTZ induced seizure. This difference between left and right hemispheres was found to be significant ($p<0.05$) (Figure 1).

DISCUSSION

The finding of the present study has shown that asymmetric breakdown of the BBB had been seen between in the left and right hemispheres in the animals treated with PTZ. Nitsch and Klatzo have shown that BBB breakdown was bilateral and confined to anatomically limited brain areas during PTZ induced seizures in the rabbits(5), but the asymmetric breakdown have been seen in the kianic acid induced seizure in some of the rabbits(5). Arato et al. have shown that the asymmetry of the serotonergic mechanisms were in the orbital frontal cortex in women and men. This asymmetry for the right side was higher in women than men(2). Biochemical studies have revealed sex difference in serotonergic system of the rat brain(3). On the other hand, the influence of serotonin on blood brain barrier permeability

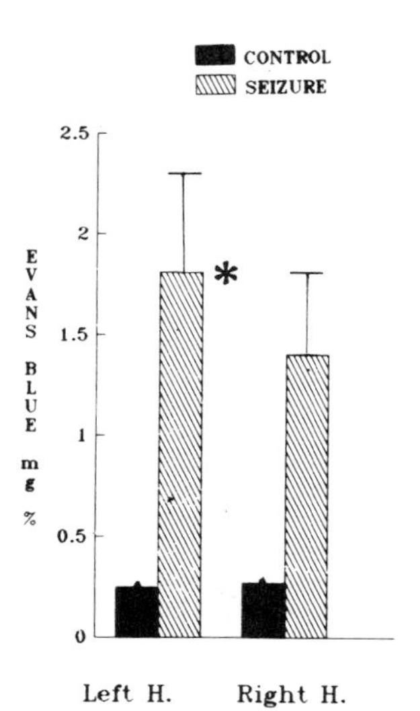

Figure 1. Evans-blue content (mg%) in the left and right hemispheres in female rats after pentylenetetrazol induced seizures. *$p<0.05$: vs left hemisphere using Student's t-test.

was shown in many studies(7). But we did not find any study that it was related to hemispheric asymmetry of the serotonergic mechanisms in the rat brain.

REFERENCES

1. Andreasen N., Dennert J., Olsen S. and Damasio A.A. Am. J. Psychiatry 139: 427-430 (1982).
2. Arato M., Frescka E., Tekes K., MacCrim mon D.J.: Acta Psychiatry Scand 84: 110-111 (1991).
3. Fischette C.T., Biegon A. and McEwen B.S.: Science 222: 333-335 (1983).
4. McCasland J.S., Carvel G.E., Simons D.J. and Woolsey T.A.: Somatosensory and Moor. Research 8: 111-116 (1991).
5. Nitch C. and Klatoz I.: J. Neurol Sciences 59: 305-322 (1983).
6. Öztaß B., and Kaya M.: Pharm. Res. 23: 41-46 (1991).
7. Sharma H.S. and Dey P.K.: Neuropharmacology 25: 181-187 (1986).
8. Tsai L., Nasrallah H. and Jacoby C.: Arch. Gen. Psychiatry 90: 1286-1289 (1983).
9. Zipursky R.B., Lim K.O. and Pfefferbaum A.: Psychiatry Res. 35: 71-89 (1990).

INFLUENCE OF INTRACAROTID HYPERTHERMIC SALINE INFUSION ON THE BLOOD-BRAIN BARRIER PERMEABILITY IN ANESTHETIZED RATS

Baria Öztaş and Mutlu Küçük

Department of Physiology
Istanbul Faculty of Medicine
University of Istanbul
Çapa/Istanbul, 34390 Turkey

SUMMARY

Retrograde infusion of hyperthermic saline solution at 43°C into the left external carotid artery increases cerebrovascular permeability to Evans-blue albumin in the left cerebral hemisphere in normothermic Wistar rats. Isotonic saline solutions at 37°C for Group I and at 43°C for Group II were infused for 30-s at a constant rate of 0.12 ml/s into the left external carotid artery. All animals receiving hyperthermic saline perfusion, had disturbed blood-brain barrier permeability. Based on visual inspection, disruption grade in the left hemispheres of 6 out of 10 animals were 3+. Mean values for Evans-blue dye were found to be 0.28±0.06 mg% in left hemisphere after normothermic saline infusion (Group I), and 2.4±0.5 mg% in the same hemisphere after hyperthermic saline infusion (Group II). The difference was found to be significant between Group I and Group II ($P<0.01$).

RÉSUMÉ

La perfusion par voie rétrograde d'une solution saline à 43°C, dans la carotide externe gauche de rats normaux Wistar conduit à une augmentation de la perméabilité vasculaire cérébrale pour l'albumine/bleu Evans. Deux groupes de rats ont été perfusés pendant 30 s à un débit constant de 0.12 ml/s avec une solution saline isotonique à 37° C (rats temoins) ou à 43° C. Tous les animaux perfusés avec la solution à 43° C présentent des perturbations de la perméabilité de la BHE. Une observation visuelle a permis de déterminer que le degré de rupture de la BHE dans les hémisphères cérébraux gauches de 6 animaux sur 10 atteignait le niveau 3+.Le passage du bleu Evans est de 0.28 +/- 0.06 mg% dans l'hémisphère gauche

Biology and Physiology of the Blood–Brain Barrier, edited by Couraud and Scherman
Plenum Press, New York, 1996

des rats témoins et de 2.4 +/- 0.5 mg% dans l'hémisphère gauche des rats perfusés avec la solution hyperthermique. La différence entre les deux groupes est significative ($p<0.01$).

INTRODUCTION

The walls of cerebral microvessels represent a cellular membrane between blood and brain with special morphological and physiological properties. The blood-brain barrier (BBB) at the cerebrovascular endothelial cells prevents passage of water-soluble drugs, proteins and electrolytes from blood to brain(1). Osmotic opening of the BBB therefore has been used clinically to enhance entry of water soluble drugs from blood into the brain tumors(3). A few experimental methods of reversible disruption of the BBB have also been described. Using a rat model, Spigelman et al. have shown that the ability of an intracarotid infusion of sodium dehydrocholate reversibly disrupted the BBB(10). Recently, we have reported that retrograde infusion of hypothermic saline solution (8±1°C) into the external carotid artery of normothermic Wistar rats, reversibly increased cerebrovascular permeability to Evans-blue albumin in the ipsilateral hemisphere(6). On the other hand, in patient who have died as a result of heat stroke, cerebral edema is a common pathological findings. Biochemical tracer investigations have provided contradictory results on whether hyperthermia increases or decreases the permeability of the BBB(5,7). Profound induced hyperthermia (42-43°C) has been used for several years to treat cancer patients(2). The present experiments were therefore designed to answer the following question: If hyperthermic isotonic saline solution is infused for 30 seconds at a constant rate of 0.12 ml.sn-1 into the left external carotid artery, will it modify BBB permeability?

METHODS

The experiments were carried out on adult male Wistar rats. They were anesthetized with diethyl ether. The left external and internal carotid arteries were exposed. A cannulae was placed in the external carotid for injection of the carotid bifurcation while the distal portion of the external carotid artery was ligated. A 4 ml/kg body weight dose of Evans-blue solution(4) was injected intravenously 5 minutes prior to carotid artery infusion. In group I (n=6) a filtered isotonic saline solution at a temperature of about 37°C was infused for 30-s at a constant rate of 0.12 ml.s-1 into the left external carotid artery with an infusion pump (90 A Harvard Apparatus)(8). In group II (n=10) the isotonic saline solution at 43°C was infused. At the end of experiments, approximately 20 min after Evans-blue injection, all rats were killed by perfusion of saline solution through the heart under diethyl-ether anesthesia. A quantative estimation of homogenized brain that released the dye was measured by a spectrophotometer as described previously(5,6). Data expressed as means±SD., and statistical analysis was performed by Student's t-test.

RESULTS

The BBB permeability was disturbed in all animals receiving hyperthermic saline perfusion at 43°C (Group II). Disruption grade in the left hemispheres of 6 out of 10 animals was most evident in cerebral regions which are supplied by the left internal carotid artery. The mean values for Evans-blue dye in infused and noninfused hemispheres are shown in figure I. The mean value for Evans-blue dye was found to be 0.28±0.06, 0.32±0.08 and 0.34±0.08 mg% in the left hemisphere, right hemisphere and cerebellum respectively after

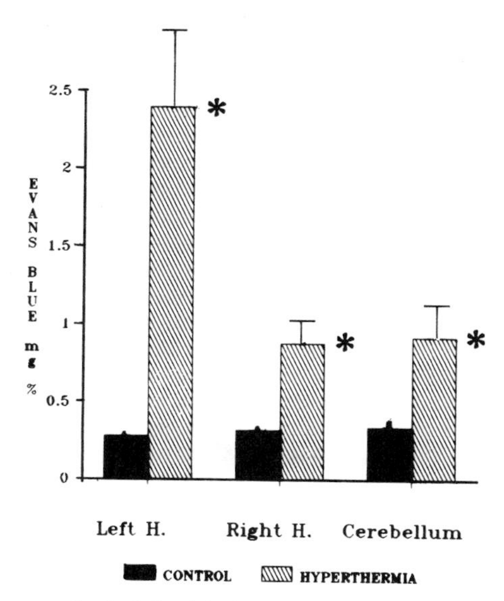

Figure 1. Evans-blue mg% content in the left, right hemispheres and cerebellum in the control and 20 min after the left internal carotid infusion of hyperthermic saline solution (43°C) *$p<0.01$ (vs. control using Student's t-test).

a 30-s infusion of isotonic saline solution into the left external carotid artery at 37°C (Group I). The BBB permeability difference between normothermic (37°C) and hyperthermic (43°C) saline infusion is significant ($p<0.01$). In the hyperthermic group (43°C) we obtained excellent BBB disruption as shown by 3+ Evans-blue staining of the infused cerebral hemisphere of 6 out of 10 animals. Quantitation of the extravasated Evans-blue dye by spectrophotometric analysis was found to be 2.4±0.5; 0.88±0.3 and 0.92±0.4 mg% in the left hemisphere, right hemisphere and cerebellum respectively after 30-s infusion of isotonic saline solution into the left external carotid artery at 43°C (Group II) (Figure 1). The difference was found to be significant between Group I and Group II for all brain regions i.e., left hemisphere, right hemisphere and cerebellum ($p<0.01$).

DISCUSSION

The major finding of this study is that permeability of the BBB to albumin after injection of intracarotid hyperthermic (43°C) saline solution is increased significantly in cerebral regions supplied by the left internal carotid artery. Although there have been many studies about osmotic opening of the BBB (8,9,10), this is the first study to examine the effect of hyperthermic saline infusion into the external carotid artery on the BBB permeability. The mechanism by which osmotic opening occurs is largely by shrinkage of the endothelial cells and the separation of tight junctions although alterative mechanisms, including vesicular enhancement have also been suggested(1). But, the detailed mechanism of the effect of the intracarotid hyperthermic saline infusion on the BBB dysfunction is not known.

REFERENCES

1. Greenwood J (1991), Neuroradyology 3:95-100.
2. Hynynen K, Lulu BA (1990), Invest. Radiology 25:824-834.
3. Neuwelt EA, Frenkel EP, Diehl J, et al (1980), Neurosurgery 7:44-52.
4. Öztaş B, Kaya M (1994), Epilepsy Res 19:221-227.
5. Öztaş B (1995), Pharmacol, Bicohem Behav (in press).
6. Öztaş B, Küçük M (1995), Neurosci. Letters 190:203-206.
7. Preston E (1982), Acta Neuropathol 57:255-262.
8. Robinson PJ, Rapoport SI (1990), J Cereb Blood Flow and Metab 10:153-161.
9. Salahuddin TS, Johansson BB, Kalimo H, Olsson Y (1988), J Neuropathol 77:5-13.
10. Spigelman MK, Zappula RA, Malis LI, et al. (1993), Neurosurgery 12:606-612.

57

RECOVERY FROM ACUTE EXPERIMENTAL AUTOIMMUNE ENCEPHALOMYELITIS (EAE) CHARACTERIZED BY ENDOTHELIAL CELL UNRESPONSIVENESS TO CYTOKINES AND PERICYTE ACTIVATION

Paula Dore-Duffy, Roumen Balabanov, and Ruth Washington

Wayne State University School of Medicine
Department of Neurology
Division of Neuroimmunology
421 E. Canfield, Rm. 3124 Elliman
Detroit, Michigan 48201

I. INTRODUCTION

Experimental allergic encephalomyelitis (EAE) is an antigen-driven T-cell-mediated autoimmune disease in animals that closely resembles the human disease multiple sclerosis (MS). In rats, EAE is characterized by an acute clinical phase followed by recovery [1]. The mechanism by which animals recover from the paralytic disease is unknown. It is thought that cytokine-mediated regulation by T-suppressor cells plays an important role [2-4]. Transforming growth factor beta-1 (TGFβ-1), is a naturally occurring immunoregulatory protein [5] which has emerged as a leading candidate in immunoregulation of EAE [6-14]. The mechanisms by which TGFβ1 ameliorates the incidence and severity of EAE is not well understood, but a primary effect on macrophage and/or T-cell-mediated immune function or on the blood brain barrier (BBB) has been suggested [6,9,13-18]. In a previous study, we have reported that TGFβ1 inhibits interferon gamma (IFNγ)-mediated activation of rat central nervous system (CNS) microvessel endothelial cells (EC) and that TGFβ1 pretreated EC are unresponsive to subsequent stimulation [19]. In the present study, we examine in culture, CNS microvessels from Lewis rats through induction, clinically apparent disease, and in the recovery phase of EAE for the expression of EC activation antigens and for sensitivity to further activation with IFNγ. Expression of activation antigens is sequential and precedes appearance of clinical symptoms. Results further indicate that recovery from EAE is, in part, characterized by EC unresponsiveness to IFNγ, and vascular pericyte reactivity.

Biology and Physiology of the Blood–Brain Barrier, edited by Couraud and Scherman
Plenum Press, New York, 1996

II. MATERIALS AND METHODS

Induction of EAE in Lewis Rats

Female Lewis rats were injected subcutaneously with 0.05 ml of an emulsion of Myelin Basic Protein-Complete Freunds Adjuvant (MBP-CFA) containing 25 μg guinea pig MBP and 100 μg *Mycobacterium butyricum* (DIFCO Laboratories, Detroit, MI) into one hind foot pad. Control rats were similarly injected with CFA alone. Rats were observed daily for clinical signs of EAE and scored as follows: 0= no symptoms; 1= loss of tail flacidity; 2= paresis; 3= hind limb paralysis with or without incontinence.

General Reagents and Cytokines

Human TGF-β1, purified from platelets, was purchased from Sigma Chemicals, St. Louis, MO. Rat rIFN–γ was purchased from Amgen, Thousand Oaks, CA. Aliquots were stored at 4°C then diluted to desired concentration in PBS pH 7.2 just before use. TNFα was purchased from Genzyme, Cambridge, M.A.

Preparation of CNS Microvessels

Microvessels were prepared according to [20-23] analysis of relative fluorescence intensity by laser cytometry [19]. Stained coverslips were inverted and mounted on standard glass slides then analyzed on the Meridian ACAS 470 laser cytometer. An attenuated laser beam focused at about 1 μm PMT (set to read negative for isotype control antibody) excites the fluorochrome. The emission is recorded by a 16 bit microcomputer to produce false color images of fluorescent intensity (FI) and distribution. A minimum of 20-50 randomly selected microvessels per slide from two-three slides per determination were analyzed. Results are expressed as average FI. Transmission-light and fluorescence-light images were recorded using the 100X oil objective. Positive fluorescence is considered to be that above antibody control levels.

III. RESULTS

Endothelial Activation in EAE

CNS microvessels were isolated from MBP + CFA immunized Lewis rats and from CFA injected control rats at 0,3,7,12,20 and 30 days post inoculation (PI). Microvessels were stained for immunologically reactive EC activation antigens and analyzed by laser cytometry (Table 1). CNS microvessels were found to express enhanced levels of two of the three activation markers tested (class II and ICAM-1) by 7 days PI. Only 29% of the microvessel fragments expressed class II Ag and (31%) of vessel fragments expressed increased ICAM-1. MHC class II Ag appeared to be expressed in discrete zones along an individual fragment. Expression of antigen in *in vitro* cytokine treated microvessels is uniformly distributed down the length of the fragment.

MHC class II, Class I and increased ICAM-1 are expressed on the 12th day PI. During the recovery phase, microvessel expression of all EC activation markers disappeared. Detectable MHC class II and ICAM-1 was quantitated by laser cytometry and found to be associated with cells along the vessel length. Double immunostaining of fragments for MHC class II antigen and Factor VIII, a marker for EC showed that

Table 1. Kinetic expression of EC activation markers

Days PI	Antigen	MFI (M±SD)
0 (MBP)	MHC Class II	201±30
	MHC Class I	466±42
	ICAM-1	395±43
3 (MBP)	Class II	403±39
	Class I	400±62
	ICAM-1	419±51
3 (CFA)	Class II	396±46
	Class I	422±49
	ICAM-1	424±29
7 (MBP)	Class II	1399±212*
	Class I	436±48
	ICAM-1	888±60*
7 (CFA)	Class II	432±60
	Class I	458±111
	ICAM-1	479±136
12 (MBP)	Class II	1569±219*
	Class I	921±71
	ICAM-1	1302±120*
12 (CFA)	Class II	443±39
	Class I	492±81
20 (MBP)	Class II	995±55*
	Class I	464±39
	ICAM-1	927±74
30 (MBP)	Class II	701±29
	Class I	492±64
	ICAM-1	690±33

* Statistically significantly different. $p<0.01$.

Factor VIII positive vessels exhibit a different pattern of staining. This suggests that the MHC class II positive cells may not be EC. Staining for alpha muscle specific actin (αMSA), considered to be a microvessel marker for pericytes [23], confirm the probable identification of the MHC class II positive cell as a pericyte. Further, dual staining of pericytes and the basement membrane using antibody which recognizes fibronectin or laminin followed by confocal analysis shows that the MHC class II positive cell lies beneath the basement membrane.

Endothelial Cell Unresponsiveness during Recovery

We have questioned whether microvessel sensitivity to cytokine-mediated activation and down regulation (TGFβ). [19]

Changes during EAE

CNS microvessels were isolated from animals 0, 3, 7, 12, 20 and 30 days PI. Microvessels were placed in culture overnight in endotoxin-free medium plus 10% FCS with or without IFNγ (500 U/ml); IFNγ + TGFβ (10 ng/ml); or TGFβ alone. Following incubation, microvessels were washed then stained using immunocytochemical techniques for ICAM-1, MHC class II and MHC class I antigens. Results indicate show that EC are unresponsive to IFNγ during the recovery phase (20 and 30 days PI).

IV. DISCUSSION

In the present report we show that brain microvessel endothelium becomes activated sequentially during the induction and clinical phases of EAE. EC activation subsides during development of the recovery phase. Further, during recovery EC become unresponsive to restimulation with IFNγ. The recovery phase is also associated with the appearance of MHC class II positive cells which lie beneath microvessel basement membrane and are probably pericytes

V. REFERENCES

1. Paterson PY, Swanborg RH. (1988): Demyelinating diseases of the central and peripheral nervous systems. In: M. Samster (ed), Immunological Diseases, 4th ed., Little Brown, pp. 1877.
2. Bernard CCA. (1977): Suppresser cells prevent experimental autoimmune encephalomyelitis in mice. Clin. Exp. Immunol. 29:100-111.
3. Karpus WJ, Swanborg RH. (1989): CD4+ suppressor cells differentially affect the production of IFNγ by effector cells of experimental autoimmune encephalomyelitis. J. Immunol. 143:3492-3501.
4. Zamvill SS, Steinman L. (1990): The T lymphocyte in experimental allergic encephalomyelitis. Ann. Rev. Immunol. 8:579-632.
5. Massaque J. (1990): The transforming growth factor-β family. Ann. Rev. Cell Biol. 6:597-620.
6. Karpus WJ, Swanborg RH. (1991): CD4+ suppressor cells inhibit the function of effector cells of experimental autoimmune encephalomyelitis through a mechanism involving transforming growth factor beta. J. Immunol. 146:1163-1168.
7. Johns LD, Flanders KC, Ranges GE, Sriram S. (1991): Successful treatment of experimental allergic encephalomyelitis with transforming growth factor beta-1. J. Immunol. 147:1792-1796.
8. Racke MK, Cannella B, Albert P, Spron M, Raine CS, McFarlin DE. (1992): Evidence of endogenous regulatory function of transforming growth factor-β1 in experimental allergic encephalomyelitis. Int. Immunol. 5:615-620.
9. Stevens DB, Gould KE, Swanborg RH. (1994): Transforming growth factor-β1 inhibits tumor necrosis factor-α/lymphotoxin production and adoptive transfer of disease by effector cells of autoimmune encephalomyelitis. 51:77-83.
10. Mustafa M, Vingsbo C, Olsson T, Issazadeh J, Ljungdahl A, Holmdahl R. (1994): Protective influences on experimental autoimmune encephalomyelitis by MHC Class I and Class II alleles. J. Immunol. 153:3337-3344.
11. Santambrogio L, Hochwald GM, Leu CH, Thorbecke GJ. (1993): Antagonistic effects of endogenous and exogenous TGF-beta and TNF on auto-immune diseases in mice. Immunopharma. & Immunotoxicol. 15:461-78.
12. Santambrogio L, Hochwald GM, Saxena B, Leu CH, Martz JE, Carlino JA, Ruddle NH, Palladino MA, Gold LI, Thorbecke GJ. (1993): Studies of the mechanisms by which transforming growth factor-beta (TGFβ) protects against allergic encephalomyelitis. Antagonism between TGFβ and tumor necrosis factor. J. Immunol. 151:7307-7315.
13. Miller A, Lider O, Roberts AB, Sporn MB, Weiner HL. (1992): Suppressor T cells generated by oral toleration to myelin basic protein suppress both *in vitro* and *in vivo* immune responses by the release of transforming growth factor β after antigen-specific triggering. Proc. Natl. Acad. Sci. 89:421-425.
14. Fabry H, Topham DJ, Fee D, Herlein J, Carlino JA, Hart MN, Sriram S. (1995): TGF-β2 decreases migration of lymphocytes in vitro and homing of cells into the central nervous system in vivo. J. Immunol. 155:325-332.
15. Racke MK, Dhib-Jalbut S, Canella B, Albert PS, McFarlin DE. (1991): Evidence of endogenous regulatory function of transforming growth factor-β1 in experimental allergic encephalomyelitis. Int. Immunol. 5:615-620.
16. Miller A, Al-Sabbagh A, Santos LM, Das MP, Weiner HL. (1993): Epitopes of myelin basic protein that trigger TGF-beta release after oral tolerization are distinct from encephalitogenic epitopes and mediate epitope-driven by bystander suppression. J. Immunol. 125:7307-7315.
17. Johns LD, Sriram S. (1993): Experimental allergic encephalomyelitis: Neutralizing antibody to TGFβ1 enhances the clinical severity of the disease. J. Neuroimmun. 47:1-7.

18. Santos LM, Al-Sabbagh A, Londono A, Wiener HL. (1994): Oral tolerance to myelin basic protein induces regulatory TGF-beta-secreting T-cells in Peyer's patches of SJL Mice. Cell. Immunol. 157:439-447.
19. Dore-Duffy, P, Balabanov R, Washington R, Swanborg R. (1994a): Transforming growth factor beta-1 inhibits cytokine-induced CNS endothelial cell activation. Molec. Chem. Neuropath. 22:161-175.
20. Bowman PD, Betz AJ, Ard D, Wolinsky JS, Penney JB, Shiver RR, Goldstein GW. (1981): Primary culture of capillary endothelium from rat brain. In Vitro. 17:353-362.
21. Joo F Karnushina I. (1973): A procedure for the isolation of capillaries from rat brain. Cytobios. 8:41-48.
22. Dore-Duffy P, Washington R, Balabanov R. (1994b): Cytokine-mediated activation of cultured CNS microvessels: A system for examining antigenic modulation of CNS endothelial cell activation. Molec. Chem. Neuropath. 22:161-175.
23. Herman I, D'Amore PA. (1985): Microvascular pericytes contain muscle and nonmuscle actins. J. Cell. Biol. 101:43-52.

58

TIME-COURSE OF DEMYELINATION AND BLOOD-BRAIN BARRIER DISRUPTION IN THE SEMLIKI FOREST VIRUS MODEL OF MULTIPLE SCLEROSIS IN THE MOUSE

A. M. Butt,[1] S. Kirvell,[1] R. D. Egleton,[2] S. Amor,[3] and M. B. Segal[1]

[1] Division of Physiology, U.M.D.S.
St. Thomas' Hospital
Lambeth Palace Road
London SE1 7EH
[2] Biomedical Sciences Division
King's College London
[3] Department of Immunolgy
Rayne Institute, U.M.D.S.
St. Thomas' Hospital, London, United Kingdom

1. SUMMARY

Increased permeability of the blood-brain barrier (BBB) is a prominent feature of multiple sclerosis (MS). However, the cause of the BBB leak and its role in the pathogenesis of MS are unknown. The present study addressed these questions in the Semliki Forest virus (SFV) model of MS in optic nerves of Balb/C mice, using the unidirectional transfer coefficient for [^{14}C]mannitol as a measurement of BBB permeability and immunolabelling with anti-myelin basic protein (MBP) to determine the time-course and extent of demyelination. In SFV, there was a significant increase in BBB permeability in the optic nerve, prior to the onset of demyelination. The BBB leak was blocked by treatment with cimetidine, an antihistamine which acts on H_2 receptors at the BBB. Preliminary results suggested that cimetidine may also have delayed the onset or reduced the level of demyelination in the optic nerve. The results support a role for histamine in BBB leak in the SFV mouse model of MS and indicate that BBB leak may be integral to the pathogenesis of demyelination.

1. RÉSUMÉ

Une plus grande perméabilité de la BHE est une importante caractéristique de la sclérose en plaques. Toutefois, on ne connait pas la cause de cette fuite de la BHE ni son rôle

Biology and Physiology of the Blood–Brain Barrier, edited by Couraud and Scherman
Plenum Press, New York, 1996

dans la pathogénicité de cette maladie. Cette étude avait pour but d'élucider ces questions au moyen du modèle virus semliki forest (SFV) de la sclérose en plaques dans les nerfs optiques des souris Balb/C, en utilisant le coefficient de transfert unidirectionnel du

[^{14}C]mannitol pour mesurer la perméabilité de la BHE, et l'immunomarquage par la protéine basique anti-myeline (MBP) pour définir le temps de passage et le degré de démyelinisation. Avec le SFV, on notait une perméabilité accrue de la BHE dans le nerf optique, avant le début de la démyélinisation. La fuite de la BHE était bloquée par traitement par la cimétidine, un anti-histaminique qui agit sur les récepteurs H2 de la BHE. Les résultats préliminaires montrent que la cimétidine a abaissé également le taux de démyélinisation ou en a retardé le début dans le nerf optique. Ces résultats suggèrent que l'histamine joue un rôle dans la fuite de la BHE dans le modèle SFV-souris de sclérose en plaques et indique que celle-ci pourrait être le facteur essentiel de la pathogénèse de démyélinisation.

2. INTRODUCTION

It is now clear that leakage of the BBB is an important component in MS and in animal models of MS[1]. Magnetic resonance imaging (MRI) show that the BBB leaks once every 2-3 months in relapse-remitting MS[2,-4]. Clinical relapse is associated with MRI detected BBB "leak" in MS[5] and experimental allergic encephalomyelitis (EAE) in rats and guinea-pigs[6-8]. Cytochemical analysis of plaques from MS patients supports the proposal that BBB leak is associated with demyelinating lesions[9-11]. In animal studies, radioisotope and histological studies confirm that BBB disruption is associated with the inflammatory process[12-18]. The cause of the BBB leak is unknown, but a number of studies implicate histamine in the disease process[19]. In humans, mast cell activation has been shown to be increased in MS[20], and elevated levels of histamine have been found in the brain and CSF of MS patients[21] and in EAE[22]. Elevated brain levels of histamine are associated with blood-brain barrier opening and oedema formation in animal experiments[23]. This parallels MRI studies of BBB opening in MS which also measures the leak of ions across the BBB and the subsequent formation of brain oedema. Using the measurement of electrical resistance, we have shown that histamine acts on H_2 receptors to mediate BBB opening and that its action is blocked by cimetidine[24]. Significantly, BBB disturbances have been reported in the SFV model of MS in the mouse[25,26] and we have shown recently that histamine mediated BBB opening in SFV[27,28]. In the present paper, we have investigated the relationship between BBB leak and demyelination in the optic nerve in the SFV model of MS in Balb/c mice. SFV causes T-cell mediated demyelination in the CNS[29-32], and was the prefered model for this study because it causes pathophysiological changes in the optic nerve which resemble those in patients with multiple sclerosis[33-34].

3. METHODS

3.1. Animals

Three- to four-week old male mice of the Balb-C strain were inoculated i.p. with 5000 plaque forming units of SFV strain strain A7(74) in 0.1 ml of 0.75% BSA. In controls, mice were injected i.p. with BSA. At 3-35 days post-inoculation (PID), optic nerves were examined to determine either the extent of demyelination by immunocytochemistry or the permeability of the BBB to mannitol.

3.2. Cimetidine Treatment

A further two experimental groups, of SFV-infected and non-infected mice, were treated with i.p. injections of cimetidine at 2 $mg.kg^{-1}$ daily for 5 days (PID 0-5), or with saline to serve as controls. Permeability of the BBB to mannitol was measured at a single time-point at PID 5. The extent of demyelination was determined at PID5, 14 and 21.

3.3. Imunocytochemistry

Mice were killed by cervical dislocation and their optic nerves removed for immunocytochemistry. Intact optic nerves were fixed by immersion in 4% paraformaldehyde in phosphate buffered saline (PBS) for 30 min at room temperature. Optic nerves were washed in PBS, dehydrated in a graded series of alcohols and infiltrated with polyester wax at 37°C. Sections were cut at 7μm, and after dewaxing through a descending series of alcohols, sections were allowed to stand for 15 min in PBS prior to immunostaining. Consecutive serial sections were labelled with the polyclonal rabbit antineurofilament antibody NF200 (Sigma) at 1:100 to label axons, and polyclonal rabbit (Dakopatts) anti-myelin basic protein (MBP) at 1:100 to label myelin sheaths. Labelling was visualized using anti-rabbit IgG conjugated with fluorescein. Following 2 washes with PBS, sections were mounted in citifluor. Sections were examined using an epifluorescence Leitz microscope.

3.4. Measurement of BBB Permeability

Mice were anaesthetized with midazolam (2$mg.kg^{-1}$) and fetanyl/fluanisone (0.3 $ml.kg^{-1}$) i.p. and then heparinized i.p. (10,000 $U.kg^{-1}$). Mice were then perfused via the heart with a Ringer/albumin plasma substitute containing 10 $\mu Ci.ml^{-1}$ of [^{14}C]mannitol. After 20 min the mice were the mice were killed and the optic nerves were removed and prepared for β-counting on a LKB Rackbeta 1219 scintillation counter. The unidirectional transfer coefficient (K_{in}) for mannitol from blood to nerve was calculated using the single time point method.

4. RESULTS

4.1. Time–Course of Demyelination

Changes in MBP labelling in optic nerve sections following SFV inoculation are illustrated in figure 1 and summarised in Table 1. In control optic nerves, MBP labelling was compact and homogeneous throught the tissue. At post-inoculation day (PID) 5-7 there was macrophage infiltration and perivascular cuffing, with early signs of myelin disruption appearing as vacuolation of MBP labelling (Fig. 1a). At PID 14 large areas of demyelinated lesions were apparent (Fig. 1b); labelling of consecutive sections with NF200 confirmed that axons were preserved within the lesions. By PID 21 there was remyelination and only small lesions were apparent (Fig. 1c). At PID 28-35 recovery was almost complete and MBP labelling was compacted, on the whole, but there appeared to be an increase in the number of small focal lesions (Fig. 1d), compared to PID 21.

4.2. Blood-Brain Barrier Permeability

Changes in the K_{in} for mannitol following SFV inoculation are summarised in Table 1. There was a significant increase in the K_{in} for the optic nerve by 76% ($p<0.05$) at PID5

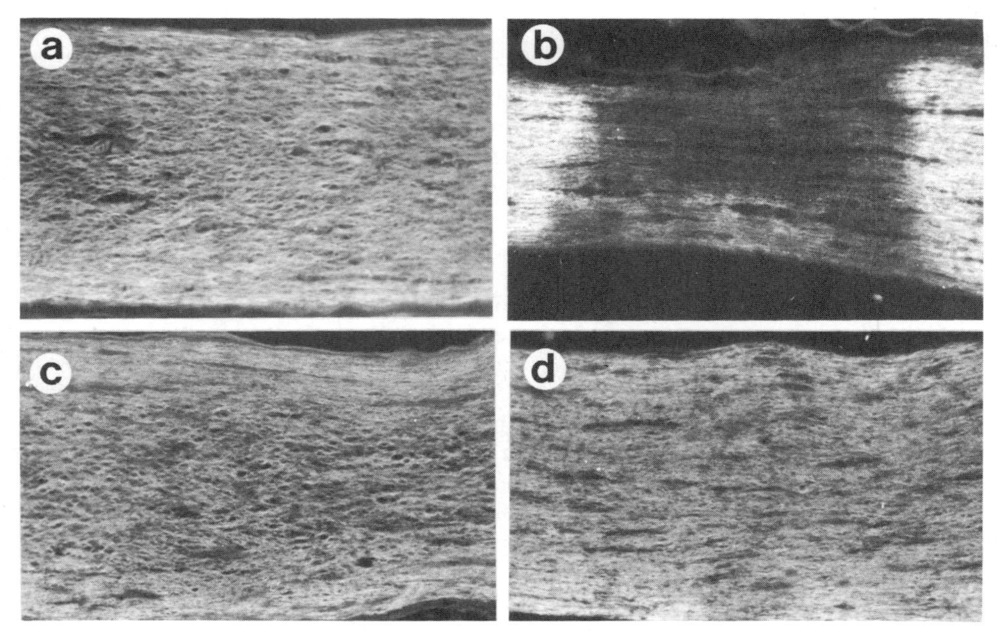

Figure 1. Time-course of demyelination in the optic nerve of SFV infected mice, using MBP immunolabelling. (a) At PID 7 MBP labelling appeared vacoulated. (b) At PID 14 lareg lesions were evident. (c) At PID 21 lesions were remyelinated, on the whole. (d) At PID 28-35 small focal lesions were evident.

which recovered to a level slightly higher than in controls by PID 7-10. There was evidence for a second increase in the K_{in} at PID 21, by 25-30% (not significantly higher than controls, p=0.09), which was followed by complete recovery by PID28. The results are related to the time-course of demyelination in Table 1.

4.3. Cimetidine Treatment

The effects of cimetidine treatment are summarised in Table 2. Treatment with cimetidine completely blocked the increase in K_{in} in the optic nerve at PID 5. Immunolabelling of optic nerve sections with MBP revealed little evidence of demyelination at PID

Table 1. Relation between BBB leak and the time-course of demyelination in SFV infected mice. There was a biphasic increase in BBB permeability which preceded demyelination.

PID	% increase in BBB permeability	Extent of demyelination
0	0%	-
5-7	76%	+
14	10%	+++
21	30%	+
28	10%	++
35	0%	+

Key for demyelination: - compact MBP labelling with no sign of demyelination; + vaculation of myelin sheaths; ++ small focal lesions; +++ large areas of demyelination

Table 2. Effect of cimetidine treatment on BBB leak and demyelination in SFV infected mice. Cimetidine blocked the BBB leak in SFV and reduced the level of demyelination (cf. table 1).

PID	% increase in BBB permeability	Extent of demyelination
0	0%	-
5-7	0%	-
14		+
21		++

5. At PID14 there was some vacuolation in the MBP labelling, and small focal lesions were observed at PID 21. Large lesions comparable to those in mice not treated with cimetidine were not observed.

5. DISCUSSION

The results demonstrate that BBB leak during SFV infection was an early feature of the pathogenesis of demyelination. The leak was completely blocked by cimetidine, an H2 receptor antagonist, and blocking the BBB leak may have delayed or reduced the level demyelination. There was evidence that there were two peaks in BBB leak and demyelination in SFV, and in both cases BBB leak preceded demyelination. The first BBB leak was at PID 5, prior to the earliest signs of myelin disruption, but coincident with macrophage infiltration and perivascular cuffing. Macrophage infiltration across the BBB is an essential component of the pathogenesis of SFV[29,30,35], and the results of the present study suggest that histamine may have decreased macrophage infiltration and subsequently demyelination. Histamine can cause BBB leak, possibly by constricting endothelial cells[24], and this may facilitate infiltration of T-cells. However, it is perhaps more reasonable that mast cells, which are the likely source of histamine in situ and are located on the brain side of the BBB, were activated as part of a more general inflammatory response. For example, infection of the endothelial cells by the virus may have led to BBB leak[36], or mast cells may have been activated by MBP[37]. Alternatively, the release of other factors such as TNFα by infiltrated t-cells may have caused mast cells to release their contents. This may be the the case for the second BBB leak which was secondary to demyelination. Nevertheless, the evidence is that first BBB leak was mediated by histamine, and that blocking the leak with cimetidine resulted in an apparent decrease in myelination. It is noteworthy that in Balb/C mice the susceptibility to SFV and EAE has been linked to histamine sensitivity[37,38], and has been correlated with the number of mast cells in the brain[40]. In summary, the results support a role for histamine in BBB leak in SFV, but further experiments are required to confirm that BBB leak is directly related to demyelination. These results may have some importance for the development of successful treatment strategies of MS in humans.

ACKNOWLEDGMENTS

Supported by the Multiple Sclerosis Society of Great Britain and Northern Ireland

6. REFERENCES

1. Poser, C.M. (1993). The pathogenesis of multiple sclerosis. Additional considerations. J. Neurol. Sci. 115 (suppl.): S3-15.
2. Harris, J.O., Frank, J.A., Patronas, N., McFarlin, D.E. & McFarland, H.F. (1991). Serial gadolinium-enhanced magnetic resonance imaging scans in patients with early, relapsing-remitting multiple sclerosis: implications for clinical trials and natural history. Ann. Neurol. 29: 548-55.
3. Stone, L.A., Smith, M.E., Albert, P.S., Bash, C.N., Maloni, H., Frank, J.A. & McFarland, H.F. (1995). Blood-brain barrier disruption on contrast-enhanced MRI patients with mild relapsing-remitting multiple sclerosis: relationship to course, gender, and age. Neurol. 45: 1122-6.
4. Willoughby EW *et al., (1989). Ann. Neurol. 25: 43.*
5. Kermode, A.G., Thompson, A.J., Tofts, P., MacManus, D.G., Kendall, B.E., Kingsley, D.P., Moseley, I.F., Rudge, P. & McDonald, W.I. (1990). Breakdown of the blood-brain barrier precedes symptoms and other MRI signs of new lesions in multiple sclerosis. Pathogenic and clinical implications. Brain 113: 1477-89.
6. Hawkins, C.P., Munro, P.M., MacKenzie, F., Kesselring, J., Tofts, P.S., du Boulay, E.P., Landon, D.N. & McDonald, W.I. (1990). Duration and selectivity of blood-brain barrier breakdown in chronic relapsing experimental allergic encephalomyelitis studied by gadolinium-DPTA and protein markers. Brain 113: 365-78.
7. Namer, I.J., Steibel, J., Poulet, P., Armspach, J.P., Mohr, M., Mauss, Y. & Chambron, J. (1993). Blood-brain barrier breakdown in MBP-specific T cell induced allergic encephalomyelitis. A quantitative in vivo MRI study. Brain 116: 147-59.
8. Seeldrayers, P.A., Syha, J., Morrissey, S.P., Stodal, H., Vass, K., Jung, S., Gneiting, T., Lassmann, H., Haase, A. & Hartung, H.P. (1993). Magnetic resonance imaging investigation of blood-brain barrier damage in adoptive transfer experimental autoimmune encephalomyelitis. J. Neuroimmunol. 46: 199-206.
9. Gay, D. & Esiri, M. (1991). Blood-brain barrier damage in acute multiple sclerosis patients. An immunocytochemical study. Brain 114: 557-72.
10. Kwon, E.E. & Prineas, J.W. (1994). Blood-brain barrier abnormalties in longstanding multiple sclerosis lesions. An immunohistochemical study. J. Neuropath. Exp. Neurol. 53: 625-36.
11. Lassmann, H., Suchanek, G. & Ozawa, K. (1994). Histopathology and the blood-cerebrospinal fluid barrier in multiple sclerosis. Ann. Neurol. 36 (Suppl.): S42-6.
12. Butter, C., Baker, D., O'Neill, J.K. & Turk, J.L. (1991). Mononuclear cell trafficking and plasma protein extravasation into the CNS during chronic relapsing experimental allergic encephalomyelitis in Biozzi AB/H mice. J. Neurol. Sci. 104: 9-12.
13. Daniel, P.M., Lam, D.K. & Pratt, O.E. (1981). Changes in the effectiveness of the blood-brain barrier and blood-spinal cord barriers in experimental allergic encephalomyelitis. J. Neurol. Sci. 52: 211-9.
14. Juhler, M., Barry, D.I., Offner, H., Konat, G., Klinken, L. & Paulson, O.B. (1984). Blood-brain and blood-spinal cord barrier permeability during the course of experimental allergic encephalomyelitis in the rat. Brain Res. 302: 347-55.
15. Koh, C.S., Gausas, J. & Paterson, P.Y. (1993). Neurovascular permeability and fibrin deposition in the central nervous neuraxis of Lewis rats with cell-transferred experimental allergic encephalomyelitis in relationship to clinical and histopathological features of the disease. J. Neuroimmunol. 47: 141-5.
16. Kristensson, K. & Wisniewski, H.M. (1977). Chronic relapsing experimental allergic encephalomyelitis. Studies in vascular permeability changes. Acta Neuropath. 39: 189-94.
17. Sternberger, N.H., Sternberger, L.A., Kies, M.W. & Shear, C.R. (1989). Cell surface endothelial proteins altered in experimental allergic encephalomyelitis. J. Neuroimmunol. 21: 241-8.
18. Zlokovic, B.V., Skundric, D.S., Segal, M.B., Colover, J., Jankov, R.M., Pejnovic, N., Lackovic, V., Mackic, J., Lipovac, M.N. & Davson, H. (1989). Blood-brain barrier permeability changes during acute allergic encephalomyeletis induced in the guinea pig. Met. Brain Dis. 4: 33-40.
19. Prineas, J.W., Kwon, E.E., Cho, E.-S. & Sharer, L.R. (1984). Continual breakdown and regeneration of myelin in progressive multiple sclerosis plaques. Ann. N.Y. Acad. Sci. 436: 11-32.
20. Rozniecki, J.J., Hauser, S.L., Stein, M., Lincoln, R. & Theoharides, T.C. (1995). Elevated mast cell tryptase in cerebrospinal fluid of multiple sclerosis patients. Ann. Neurol. 37: 63-6.
21. Tuomisto, L., Kilpelainen, H. and Riekkinen, P. (1983). Histamine and histamine-N-methyltransferase in the CSF of patients with multiple sclerosis. Agents and Actions 13: 255-7.
22. Orr, E.L. & Stanley, N.C. (1989). Brain and spinal cord levels of histamine in Lewis rats with acute experimental autoimmune encephalomyelitis. J. Neurochem. 53: 111-8.
23. Joo, F. (1993). The role of histamine in brain oedema formation. Functional Neurology 8: 243-50.

24. Butt, A.M. & Jones, H.C. (1992). Effect of histamine and antagonists on electrical resistance across the blood-brain barrier in rat brain-surface microvessels. Brain Res. 569: 100-5.
25. Colover, J. (1988). Immunological and cytological studies of autoimmune demyelination and multiple sclerosis. Brain Behav. Immun. 2: 341-5.
26. Parsons, L.M. & Webb, H.E. (1982). Blood brain barrier and immunoglobulin G levels in the cerebrospinal fluid of the mouse following peripheral infection with the demyelinating strain of Semliki Forest virus. J. Neurol. Sci. 57: 307-18
27. Egleton, R.D., Amor, S., Butt, A.M. & Segal, M.B. (1994). Changes in blood-brain barrier permeability to mannitol during Semlik Forest virus infections in the mouse. J. Physiol. 479: 101P.
28. Egleton, R.D., Dawson, J., Butt, A.M., Amor, S. & Segal, M.B. (1994). Cimetidine reduces blood-brain barrier changes in animal models of multiple sclerosis. J. Physiol. 480: 11P
29. Fazakerley, J.K., Amor, S. & Webb, H.E. (1983). Reconstitution of Semliki Forest virus infected mice induces immune mediated pathological changes in the CNS. Clin. Exp. Immunol. 52: 115-120.
30. Jagelman, S., Suckling, A.J., Webb, H.E. & Bowen, E.T.W. (1978). The pathogenesis of avirulent Semliki Forest virus infections in athymic nude mice. J. Gen. Virol. 41: 599-607.
31. Kelly, W.R., Blakemore, W.F., Jagelman, S. & Webb, H.E. (1982). Demyelination induced in mice by avirulent Semliki Forest virus. II. An ultrastructural study of focal demyelination in the brain. Neuropathol. App. Neurobiol. 8: 43-53.
32. Tansey, E.M., Pessoa, V.F., Fleming, S., Landon, D.N. & Ikeda, H. (1985). Pattern and extent of demyelination in the optic nerves of mice infected with Semliki Forest virus and the possibility of axonal sprouting. Brain 108: 29-41.
33. Tansey, E.M., Allen, T.G.J. & Ikeda, H. (1986). Enhanced retinal and optical nerve excitability associated with demyelination in mice infected with Semliki Forest virus. Brain 109: 15-30.
34. Tremain, K.E. & Ikeda, H. (1983). Physiological deficits in the visual system of mice infected with Semliki Forest virus and their correlation with those seen in patients with demyelinating disease. Brain 106: 879-895.
35. Subak-Sharpe, I, Dyson, H. & Fazakerley, J. (1993). In vivo depletion of CD8+ T cells prevents lesions of demyelination in Semliki Forest virus infection. J. Neurosci. Res. 35: 445-51.
36. Soilu-Hanninen, M., Eralinna, J.P., Hukkanen, V., Roytta, M., Salmi, A.A. & Salonen, R. (1994). Semliki Forest virus infects mouse brain endothelial cells and causes blood-brain barrier damage. J. Virol. 68: 6291-8.
37. Brenner, T., Soffer, D., Shalit, M. & Levi-Schaffer, F. (1994). Mast cells in experimental allergic encephalomyelitis: characterization, distribution in the CNS and in vitro activation by myelin basic protein and neuropeptides. J. Neurol. Sci. 122: 210-3.
38. Linthicum, D.S. & Frelinger, J.A. (1982). Acute autoimmune encephalomyelitis in mice. II. Susceptibility is controlled by the combination of H-2 and histamine sensitization genes. J. Exp. Med. 156: 31-40.
39. Teuscher, C., Blankenhorn, E.P. & Hickey, W.F. (1987). Differential susceptibility to actively induced experimental allergic encephalomyelitis and experimental allergic orhitis among BALB/c substrains. Cell. Immunol. 110: 294-304.
40. Yong, T., Bebo, B.F., Sapatino, B.V., Welsh, C.J., Orr, E.L. & Linthicum, D.S. (1994). Histamine-induced microvascular leakage in pial venules: differences between the SJL/J and BALB/c inbred strains of mice. J. Neurotrauma 11: 161-71.

BLOOD-BRAIN BARRIER PERMEABILITY CHANGES DURING SEMLIKI FOREST VIRUS-INDUCED ENCEPHALOMYELITIS IN THE BALB/C MOUSE

A Role for Histamine?

R. D. Egleton,[1] A. M. Butt,[1] S. Amor,[2] and M. B. Segal[1]

[1] Physiology Department, U.M.D.S
[2] Immunology Department, U.M.D.S
St Thomas' Hospital
London SE1 7EH, United Kingdom

1. SUMMARY

We have measured blood-brain barrier permeability changes during Semliki Forest Virus infections in the Balb/c mouse, and shown that the permeability changes have a biphasic response peaking at post inoculation days 5 and 14. The histamine H_2 receptor antagonist, cimetidine reduced the permeability changes at post inoculation day 5.

1. RÉSUMÉ

Nous avons mesuré les modifications de la perméabilité de la barrière hémato-encéphalique au cours de l'infection de souris Balb/c par le virus semliki Forest. Nous avons montré que la perméabilité de la BHE est modifiée en deux phases avec des maximums 5 et 14 jours après l'inoculation du virus. L'admistration de cimétidine, antagoniste du récepteur histaminergique H2, réduit l'augmentation de la perméabilité apparaîssant 5 jours après l'inoculation du virus.

2. INTRODUCTION

Multiple sclerosis (MS) is a central nervous system (CNS) disease characterised by focal demyelination. The aetiology of MS is unknown, though, it is possible that virus infection may play a role; antibodies to viruses including canine distemper and measles [1],

Biology and Physiology of the Blood–Brain Barrier, edited by Couraud and Scherman
Plenum Press, New York, 1996

have been found in the characteristic oligoclonal immunoglobulin bands of MS patients cerebrospinal fluid[2]. Semliki Forest Virus (SFV) is a *Togaviridae* of the *Alphavirus* group[3], which induces encephalomyelitis in a number of mice species. The A7(74) strain is non-virulent, and is characterised by demyelinated lesions centred on cerebral microvessels[4].

Reversible damage to the blood-brain barrier (BBB) has been shown in MS patients using gadolinium-DTPA enhanced magnetic resonance imaging (MRI) [5], and is also a common feature in animal models such as chronic relapsing experimental allergic encephalomyelitis (CR-EAE) in the guinea-pig [6].

The actual mechanism of BBB opening is unclear, however MS is an immune mediated disease, which involves T-cell and Mast cells. Mast cells are commonly found associated with MS lesions [7], so it is likely that histamine may play a role in the disease. Histamine has previously been shown to modulate both BBB permeability [8], and electrical resistance [9] in the rat via an H_2 mediated mechanism. Histamine H_2 antagonists also reduce the clinical signs and demyelination in the Lewis rat model of EAE [10].

In this study, we have measured changes in BBB permeability to ^{14}C mannitol during the time course of SFV infections in the Balb/c mouse. We have also studied the effect of the histamine H_2 antagonist, cimetidine, on the observed permeability changes.

3. METHODS

Pathogen free, three to four week old Balb/c mice were obtained from Bantin and Kingman. The mice were injected interperitoneally (i.p) with 5000 plaque forming units of Semliki Forest Virus (SFV) strain A7 74. At post inoculation days (PID) 3, 5, 7, 10, 14, 21, 28, and 35, mice were anaesthetised with 2 $mg.kg^{-1}$ midazolam hydrochloride (Hypnovel, Roche), 0.3 $ml.kg^{-1}$ of fentanyl/fluanisone (Hypnorm, Janssen) and heparinised at 100,000 $U.kg^{-1}$. The animals were then perfused via an inter ventricular cannulae in the left ventricle and the right atrium was sectioned. The perfusate consisted of an oxygenated Krebs-Henseleit Ringer with 4 % BSA and 14 mM glucose, containing $10\mu Ci.100ml^{-1}$ ^{14}C mannitol (Amersham). The perfusion was carried out for 20 minutes, after which the animal was killed by cervical dislocation and the brain and spinal cord rapidly removed. During the perfusion samples from both the inflow (arterial) and outflow (venous) ringer were taken and measured for glucose consumption, pH, pCO_2 and pO_2 . Brain samples were taken for adenosine triphosphate measurement using a kit obtained from Sigma, and compared to non-perfused control decapitated animals. Brain and spinal cord samples were weighed and prepared for β-counting by homogenisation and solublisation with soluene (Packard). Three ml of Liguiscint liquid scintillant (Amersham) was added to each sample and triplicate Ringer samples and the samples were counted on a LKB Rackbeta 1219 β-counter.

The uptake coefficient for ^{14}C mannitol was calculated using a simple two compartment model. The results were compared using the Student's t-test.

4. RESULTS

4.1. Monitoring of Perfusion

In perfused and non-perfused control mice, there was no significant difference in either the water content or the ATP levels of the brain (Table 1). In perfused mice the mean arterial (inflow)pO_2 was 585 ± 26 mmHg, and the glucose levels were 14.30 ± 0.03 mM(Table 1).

Table 1. The perfusion parameters, for the perfusion of Balb/c mice, each was measured in 6 mice. The partial pressures of both O_2 and CO_2 are measured in mmHg, and the glucose concentration is in mM. In the brain there is no significant difference between the non-perfused and perfused animal, for ATP or water content. In the perfusion fluids, there is a reduction in both the O_2 and glucose content in the venous compared to the arterial, and a corresponding increase in CO_2, indicating metabolism is occuring in the animal

Brain		Water content (%)			ATP content ($\mu M.g^{-1}$)	
Non-perfused		80.67 (0.72)			2.32 (0.29)	
Perfused		80.30 (0.39)			2.51 (0.34)	
Fluids	pO_2	range	pCO_2	range	Glucose	range
Arterial	585±26	520-710	38.6±1.7	35.2-40.5	14.3±0.03	13.3-15.7
Venous	162±12	149-183	58.3±2.5	49.6-75.3	13.3±0.04	12.7-14.9

4.2. BBB Changes

The permeability of the BBB to ^{14}C mannitol was measured during the time course of SFV infections at PID 3, 5, 7, 10, 14, 21, 28, and 35, the results were expressed as a % change of control permeability.

At PID 3, there is an increase of 37.4 % in the permeability of the brain ($p<0.01$), though there is no significant change in the spinal cord. At PID 5, the permeability in both the brain and spinal cord peaked with values of 74.2 and 40.8 % respectively both of which were significantly higher than control permeability ($p<0.001$, in both cases). The permeability to mannitol was at control levels at PID 10, and was again significantly higher in both brain and spinal cord ($p<0.05$) by PID 14 with values 31.4 % and 21.5 % higher than controls. In the brain this increase in permeability was maintained until PID 35, whilst in the spinal cord the permeability was at control levels by PID 21 (Table 2).

SFV infected animals were also treated with the histamine H_2 receptor antagonist cimetidine from PID 1-4, and the permeability to mannitol measured at PID 5. There was a significant reduction in the permeability to mannitol in both the brain and spinal cord when compared to the PID 5 mice ($p<0.05$ in the brain and $p<0.001$ in the spinal cord) (Table 2).

5. DISCUSSION

We have measured penetration of the low molecular weight marker mannitol into the brain and spinal cord of control non-infected Balb/c mice, and studied the affect of SFV infections (a viral induced encephalomyelitis) on mannitol permeability. Further we have studied the effect of a histamine H_2 receptor antagonist on the observed changes.

Table 2. Percentage change in BBB permeability to ^{14}C mannitol from control value, during SFV infections in the BALB/c mouse, each represents 6-8 mice. There are two peaks in permeability during SFV, one at PID 5 and the other at PID 14. The histamine H_2 receptor antagonist cimetidine reduces thethe permeability changes seen at PID 5

	PID 3	PID 5	PID 10	PID 14	PID 35	Cimetidine
Brain	37.4	74.2	-2.4	31.4	28.6	45.9
Spinal cord	12.5	40.8	7.2	21.7	-13.1	-1.9

The method presented here is an adaptation of the technique used by Preston *et al.*[11]. In this study, we have shown that ATP levels and water content, were not different in non-perfused and perfused brain (Table 1), indicating that the perfusion technique does not compromise the BBB. The levels of pO_2 and pCO_2, are in the same range as seen with other saline perfusion systems, ensuring adequate O_2 delivery to cerebral endothelial cells.

In this study, it can be seen that SFV infections lead to an increase in mannitol permeability. Over the time course of the infection, the permeability changes in both the brain and spinal cord show a biphasic response. The initial response peaks in both brain and spinal cord at PID 5, and is the maximal permeability increase shown in this model. This corresponds to the peak virus titre in the brain [12] and to the initial entry of IgG positive cells into the brain [13]. The second permeability change commences at PID 14, and corresponds with the onset of demyelination and other histological changes (Butt *et al.*, this meeting.)

Cimetidine when given from PID 1-4, leads to a reduction in permeability in the brains of PID 5 animals, and blocks the permeability change in the spinal cord completely, indicating that the first permeability phase is at least partially modulated by histamine via a H_2 receptor mediated mechanism.

6. REFERENCES

1. Allen, I.V., Brankin, B., 1993. J. Neuropathol. Exp. Neurobiol. 52(2), 95-105.
2. Thompson, E.J., Kaufmann, P., Shortman, R.C., *et al.*, 1979. B.M.J. 1, 16- 17.
3. Topley and Wilson's Principles of Bacteriology, Virology and Immunology Eigth edition volume 4, 574-580. Arnold, London, 1990.
4. Fazakerly, J.F., Pathak, S., Scallan, M., *et al.*, 1993. Virology. 195, 627-637
5. Miller, D.H., Rudge, P., Johnson, G., *et al.*, 1988. Brain. 111, 927-939.
6. Egleton, R.D., Segal, M.B., 1994. J. Physiol. 475, 71P.
7. Kruger, P.G., Nyland, H.I., 1995. Med. Hypotheses. 44 (1), 66-69.
8. Boertje, S.B., Ward, S., Robinson, A., 1992. Res. Comm. Chem. Path. Pharm. 76 (2), 143-154.
9. Butt, A.M., Jones, H.C., 1992. Brain Res. 569, 100-105.
10. Dietsch, G.N., Hinrichs, D.J., 1989. J. Immunol. 142(5), 1476-1481.
11. Preston, J.E., Al-Sarraf, H., Segal, M.B., 1995. Dev. Brain Res. 87,69-76.
12. Balluz, I.M., Glasgow, G.M., Killen, H.M., *et al.*, 1993. Neuropathol. Applied Neurobiol. 19, 233-239.
13. Parsons, L.M., Webb, H.E., 1992. Neuropathol. Applied Neurobiol. 18, 351-359.

60

HIV-1 ENCEPHALITIS IS A CONSEQUENCE OF VIRAL INFECTION AND NEUROIMMUNE ACTIVATION

Yuri Persidsky and Howard E. Gendelman

Department of Pathology and Microbiology
University of Nebraska Medical Center
Omaha, Nebraska 68198-5215

I. SUMMARY

The neuropathogenic mechanism(s) of human immunodeficiency virus type 1 (HIV-1) encephalitis was investigated utilizing a severe combined immunodeficiency disease (SCID) mouse system where virus-infected monocytes are stereotactically placed into the cerebral cortex and deep nuclei. Aspects of HIV-1 induced central nervous system (CNS) disease including viral (p24 antigen) expressing macrophages and multinucleated cells, marked astrogliosis, and microglia activation were present in this animal model system as in HIV-infected human brain tissue. A marked neuroimmune activation correlated with numbers of HIV-1 infected monocytes found in affected SCID brains. These changes were accompanied by abnormalities within the blood-brain barrier (BBB) (endothelial cell expression of adhesion molecules, hypertrophy of astrocyte foot processes and perivascular edema). Similar results have already been demonstrated in human autopsy specimens of lentivirus affected brain tissue. These data suggest that HIV-1 encephalitis is consequence of macrophage activation with accompanying BBB impairment(s).

I. RÉSUMÉ

Les méchanismes neuropathogéniques de l'encéphalite provoquée par le virus de l'immunodéficience humaine de type 1 (HIV-1) ont été étudiés en ulilisant des souris immunodéficientes SCID (severe combined immunodeficiency disease) dans lesquelles des monocytes infectés par le HIV-1 ont été introduits dans le cortex cérébral et les noyaux profonds par stéréotaxie. Différents aspects de la maladie provoquée par le HIV-1 dans le système nerveux central, à savoir des macrophages exprimant de l'antigène viral (p24), des cellules multinucléées, une astrogliose marquée et une activation de la microglie ont été observés dans ce modèle animal comme dans les cerveaux humains infectés par le HIV-1. Une activation neuroimmune marquée est corrélée avec le nombre de monocytes infectés

Biology and Physiology of the Blood–Brain Barrier, edited by Couraud and Scherman
Plenum Press, New York, 1996

par le HIV-1 trouvés dans les cerveaux des souris SCID infectées. Ces changements sont accompagnés par des anomalies de la barrière hémato-encéphalique (expression de molécules d'adhésion par les cellules endothéliales, hypertrophie des pseudopodes des astrocytes et oedème périvasculaire). Des résultats similaires ont déjà été démontrés dans des prélèvements d'autopsie de cerveaux humains dans des cas d'infections par des lentivirus. Ces résultats suggèrent que l'encéphalite provoquée par le HIV-1 est la conséquence de l'activation des macrophages accompagnée de l'altération de la barrière hémato-encéphalique.

II. METHODS

Monocytes were purified by centrifugal elutriation and infected with HIV-1_{ADA} (a macrophage tropic strain) at a multiplicity of infection (MOI) of 0.1. After 7 days of viral infection SCID mice were inoculated intracerebrally with infected or control uninfected cells. Animals were sacrificed 3-28 days after cell treatments. Mouse brain was processed for morphologic, immunocytochemical and/or molecular assays. Immunocytochemistry was performed on 5 μm cryostat or paraffin sections to assay numbers and distribution of human monocytes (CD68), viral antigens (HIV-1 p24 expressing cells), cellular activation (MHC class II), human and mouse cytokines [tumor necrosis factor alpha (TNFα), interleukin 1 beta (IL-1β) and interleukin 6 (IL-6)], mouse microglia/macrophages [F4/80 and very late antigen 4 (VLA-4)], astrocyte [glial fibrillary acidic protein (GFAP)] and adhesion molecules [vascular cell adhesion molecule 1 (VCAM-1), intercellular adhesion molecule 1 (ICAM-1), endothelial cell adhesion molecule 1 (ELAM-1)]. Seven human brains (5 cases of HIV-1 encephalitis and associated dementia and two virus-negative controls) were obtained at necropsy. HIV-1 encephalitis was defined pathologically by the presence multinucleated giant cells, microglial nodules, astrocytosis and myelin pallor. HIV-associated dementia was a premortem clinical diagnosis made by an attending neurologist or AIDS specialty physician.

III. RESULTS/DISCUSSION

Brains of SCID mice inoculated with HIV-1-infected monocytes showed an increase in number and distribution of astrocytes as compared to the animals inoculated saline or with control uninfected cells. Perivascular edema and hypertrophy of astrocyte foot processes forming close contacts with basement membrane of brain microvessels were detected by light and electron microscopy in virus-infected tissue. Altered foot processes were strongly immunoreactive for GFAP including areas around small capillaries. A marked cytoplasmic accumulation of glycogen was within the reactive astrocytes most pronounced surrounding areas of virus-infected human monocytes. These changes closely paralleled the wide-spread cerebral reactive astrocytosis. Both mouse perivascular macrophages and microglia showed morphological signs of activation (increased branching and cytoplasmic distentions). Many of these cells also expressed VLA-4 and IL-6 antigens. Select brain microvascular endothelial cells (EC) expressed VCAM-1. Astrocytosis, resident microglia and EC activation developed in response to HIV-1-infected monocytes. These virus containing cells also expressed MHC class II. A clear relationship between the numbers of HIV-1 p24 antigen positive monocytes and the level(s) of reactive astrogliosis was demonstrated. Indeed, only minimal changes in numbers of astrocytes and microglia were found in and around areas of uninfected cells or the needle tract devoid of human cells.

Perivascular edema, thickening of microvessel walls and increased GFAP immunoreactivity of astrocyte foot processes were in brains of 5 patients with HIV-1 encephalitis without opportunistic infections and/or malignancies. Such changes were accompanied by an intensive astrocytosis and perivascular accumulation of CD68-positive macrophages many of which contained large levels of viral p24 antigens. These macrophages/microglia expressed membrane MHC class II antigens. Cytoplasmic staining for IL-1β and IL-6 were abundant. In addition, increased number of microglial cells (detected by combined CD68 or RCA-1 immunostaining) also contained HLA-DR antigens. Expression of adhesion molecules (E-selectin and VCAM-1) was in brain microvascular EC. Functional alterations in the BBB (microvascular EC, vessel wall, astrocyte processes) developed with the activation and accumulation of macrophages and microglia. Signs of BBB impairment, macrophage/microglia activation, and adhesion molecule expression on EC were absent in 2 HIV-negative human brains. These data, taken together, strongly suggest that BBB abnormalities occur both in human brains with HIV-1 encephalitis and in SCID mice inoculated intracerebrally with human monocytes. This animal model system for this HIV-1-associated neurological disease may soon be utilized for drug testing in order to improve mental function in affected AIDS patients. Increased permeability of BBB could be an important factor accelerating the development of HIV-1-associated dementia.

61

EVALUATION OF BLOOD-BRAIN BARRIER RUPTURE IN A RAT GLIOMA MODEL BY USING A CONTRAST AGENT FOR MAGNETIC RESONANCE IMAGING

G. Le Duc,[1] C. Rémy,[1] M. Péoc'h,[2] M. Décorps,[1] and J. F. Le Bas[3]

[1] INSERM U438
[2] Laboratoire d'Anatomie Pathologique
[3] Unité IRM
CHU A. Michallon, BP 217
38043 Grenoble Cedex 9, France

SUMMARY

Brain tumor visualization has been improved significantly by Magnetic Resonance Imaging (MRI).This work shows that Sinerem®, a superparamagnetic contrast agent can bring new information about rat brain tumor physiology, more particularly BBB permeability, neovascularization and edema. Sinerem® induces a strong hyposignal mainly in some central patches and in the periphery of the tumor, both corresponding to vessels. Furthermore, Sinerem® slowly diffuses from vessels to extravascular space due to BBB breakdown.

RÉSUMÉ

L'utilisation d'agents de contraste en IRM a considérablement amélioré le diagnostic des tumeurs cérébrales. Ce travail met en évidence les résultats obtenus à partir d'un agent de contraste superparamagnétique sur un modèle de gliome intracérébral chez le rat. Après injection de Sinerem®, les tumeurs présentent des tâches centrales et une couronne périphérique en hyposignal. Cet hyposignal prononcé semble correspondre d'abord aux vaisseaux puis aux zones de rupture de la B.H.E: l'agent de contraste diffuse alors depuis les vaisseaux vers l'espace extracellulaire.

1. INTRODUCTION

Brain tumor visualization has been improved significantly by Magnetic Resonance. Imaging (MRI). Contrast agent administration is helpful in separating tumors from surround-

Biology and Physiology of the Blood–Brain Barrier, edited by Couraud and Scherman
Plenum Press, New York, 1996

ing edema: Dotarem® diffuses out of the intravascular space at the site of BBB breakdown and tumor clearly appears from surrounding tissue[1]. However, Dotarem® rapidly diffuses because of its small size (2 nm) and results do not allow the characterization of BBB rupture. The purpose of this work was to determine if Sinerem®, a new promising superparamagnetic contrast agent[2-3] could bring new information about brain tumors physiology, more particularly BBB disruption, neovascularization and edema.

2. MATERIAL AND METHODS

2.1. Animal Model

Intracerebral glioma [4-5] was induced by stereotactic injection of 10^4 C6 cells in the right caudate nucleus of female Sprague-Dawley rats (200-220 g body weight). The rats died within 3 or 4 weeks latter.

2.2. N.M.R. Experiments

T_2 weighted (T_R=2000 ms, T_E=80 ms) or T_1 weighted (T_R=500 ms, T_E=34 ms) images (slice thickness=1 mm) were acquired using a spin-echo sequence on a 2.35 Tesla magnet (MSL, BRUKER). For each rat, 3 transverse and 10 coronal slices were acquired before and after contrast agent injection (15 min, 1-2-3-4-5-6 h).

2.3. Contrast Agent

The vascular contrast agent used for this study was Sinerem® (Laboratoire Guerbet-Aulnay sous Bois). The mean particle size was 20 nm. Sinerem® was intravenously injected at a dose of 200 μmoles Fe/kg (blood half-life = 4.5 hours). Dotarem® (Laboratoire Guerbet-Aulnay sous bois) was injected at a dose of 0.1 mmol/kg and was used as a reference contrast agent.

2.4. Histology

At the end of MRI experiments, rats were sacrificed. The brains were fixed by in situ transcardiac perfusion, removed and then embedded in paraffin. Five micrometers sections were cut and stained with Masson trichrom.

3. RESULTS

3.1. Imaging Experiments

The figure 1 shows typical results obtained for 2 tumoral rats after Sinerem® injection (a) and Dotarem® injection (b).

The figure 2 shows quantitative results obtained from controlateral and tumoral hemispheres of a rat brain 15 min after Sinerem® injection on T_2 weighted images. The variation (ΔI) of signal intensity (I) in a defined area before and after contrast agent injection was expressed as follows:

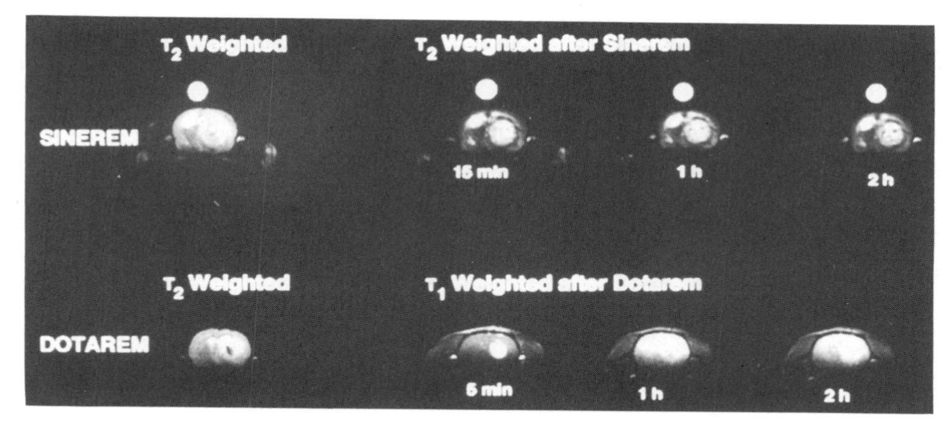

Figure 1. Tumoral rat brains coronal images (a) after Sinerem® injection (T_2 weighted images) and (b) after Dotarem® injection (T_1 weighted images).

$$\Delta I(\%) = \frac{[I]before - [I]after}{[I]before} \times 100$$

Fifteen minutes after injection, Sinerem® induced a highly reproducible decrease of MRI signal (51.6% in the cortex, 42.6% in the caudate nucleus) in contralateral brain. Then, the MRI signal slowly returns to its initial level (2,4%/h). In the tumor, the signal decrease is lower (36.4%) than in normal tissue. In addition, a peripheral signal was observed: its

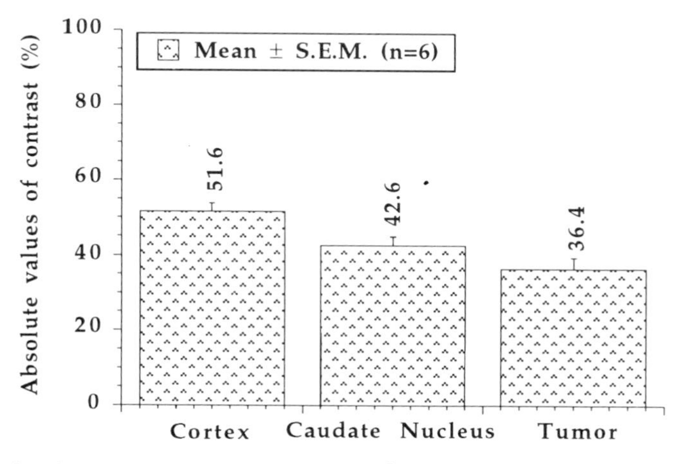

Figure 2. Relative signal variation 15 min after Sinerem® injection for caudate nucleus, cortex, and tumor.

thickness slowly increased with an inward progression after Sinerem® injection. Furthermore, a progressive apparition of dark patches was also observed in the central part of the tumor, their size increasing with time. On the other hand, these tumors showed a strong hypersignal after Dotarem® injection on T_1 weighted images.

3.2. Histology

The figure 3 shows a typical section of a tumoral brain, three weeks after cells implantation. Different structures can be observed from the center to the periphery of the tumor. The center (N) is necrotic. A thin layer of pseudo-palissadic tumorous cells can be found around this area and present a high density with a tight interstitium. They are surrounded by a large area containing tumorous cells exhibiting an edematous and loose interstitium (T). The transition between tumoral and normal brain is characterized by a thin edematous layer of orientated cells with elongated nuclei (E).

4. DISCUSSION

Due to water diffusion within the susceptibility gradients, Sinerem® induced a strong decrease of signal in normal tissue (51.6% in cortex, 42.6% in caudate nucleus) as well in tumoral tissue (36.4%). The difference between normal and tumoral tissue might indicate a lower blood volume in tumor. The heterogeneity of Sinerem® effect on tumor, 15 min after injection, would indicate that vessels distribution is more important in the tumor periphery. Histological findings led us to raise the following explanations concerning NMR results. The hyposignal concerning peripheral area and dark patches might be due to a BBB rupture and a slow passage of particles in the extracellular space. The inward progression of peripheral hyposignal might be due to a passive diffusion in the edematous area which present a loose interstitium. Particles do not enter the central part of the tumor (where they are stopped by pseudo-palissadic cells) and the normal peritumoral tissue (where the cells are more packed).

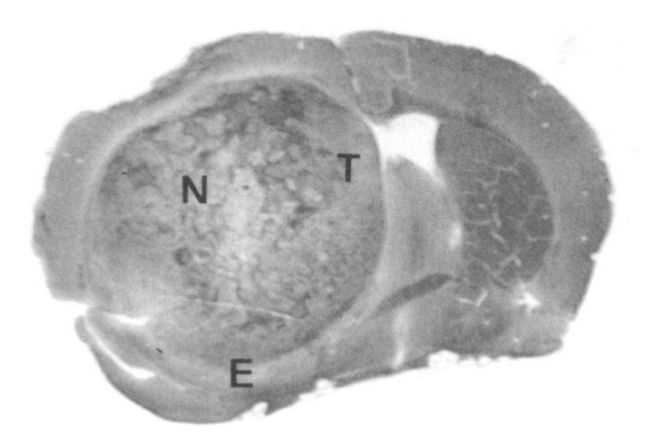

Figure 3. Typical aspect of tumoral rat brain on Masson trichrom section (X10).

5. CONCLUSION

Due to the particule size, Sinerem® seems to remain mainly intravascular immediately after injection and then slowly diffuse from vessels to extravascular space due to BBB breakdown. Thus, the use of these particles in MRI could bring two kinds of informations:

- Just after injection, Sinerem® could allow the quantification of tumor blood volume, which is an important information for radiotherapy (the better vasculari-zation of the tumor, the better efficiency of radiotherapy).
- Later on, Sinerem® could help us to characterize the degree of BBB permeability, which could be useful for the choice of chemotherapy.

ACKNOWLEDGMENTS

This work was supported by Laboratoire GUERBET, MGEN, Ligue contre le Cancer, Espoir, Région Rhône-Alpes, FRMF.

6. REFERENCES

1. Félix R., Schorner W., Laniado M. et al. Radiology (1985);156: 681-688.
2. Chambon C., Clément O., Le Blanche A. et al. Mag. Res. Imaging (1993); 11: 509-519.
3. Weissleder R., Elizondo G., Wittenberg J. et al. Radiology (1990);175: 489-493.
4. Kobayashi N., Allen N., Clendenon N., et al. J. Neurosurg (1980); 53: 808-815.
5. Benabid A., Rémy C., Chauvin C. in Biology of Brain Tumors (1986). Printed in the Netherlands. Walkers M.D., Thomas D.G.T. (eds).

62

DYNAMICS OF PERMEABILITY OF BLOOD-TUMOR BARRIER IN PATIENTS WITH BRAIN GLIOMAS

A. I. Svadovsky, V. V. Peresedov, T. N. Sharipova, V. S. Shubin, and S. M. Lognikova

Institute of Neurology RAMS
Moscow, Russia

SUMMARY

The problem of brain gliomas therapy is one of the most complex parts of neurobiology and medicine. The purpose of this study was to investigate the permeability of BTB in patients with brain hemispheric gliomas. We investigated the Blood Tumor Barrier (BTB) permeability in brain gliomas of patients in dynamic conditions by combined paraclinical methods: CT, MRI, gammaencephalography (GE), Felgenhauer's coefficient (FC). Our data suggested that selective permeability of BTB in brain gliomas was invariable even with rIL-2 therapy. We considered that peritumoral oedema (PTE) consists of a serous liquid with a minimal albumin content. The efficiency of immunotherapy is inversely proportional to presence, size and spreading of PTE on CT-MRI.

RÉSUMÉ

Nous avons étudié la perméabilité de la barrière sang-tumeur (BTB) dans des gliomes cérébraux de patients par une méthode dynamique en combinant plusieurs méthodes paracliniques: CT, MRI, gamma-encéphalographie (GE), et coefficient de Felgenhauer (FC).

Nos résultats suggèrent que la perméabilité sélective de la BTB des gliomes cérébraux est invariable, même avec une thérapie par l'IL-2. Nous pensons que l'oedème péritumoral (PTE) est constitué de liquide séreux contenant un taux minimal en albumine. L'efficacité de l'immunothérapie est inversement proportionnelle à la présence, à la taille, et à l'étalement du PTE vu en CT-MRI. Le problème de la thérapie des gliomes cérébraux est l'un des plus complexes en neurobiologie et en médecine. Le but de ce travail est d 'étudier la perméabilité de la BTB chez des patients ayant des gliomes cérébraux hémisphériques.

Biology and Physiology of the Blood–Brain Barrier, edited by Couraud and Scherman
Plenum Press, New York, 1996

MATERIALS AND METHODS

We examined the permeability of BTB in 22 patients, 16 males and 6 females. Ages of patients were between 27 and 62 years. In 17 patients, the level of BTB disorders was estimated by FC, according to Albumin or CSF IgG in mg/l/Albumin or plasma IgG in g/l (normal values 5-8). The patterns and the time of accumulation of the radionuclide (RN) ^{99m}Tc(Technetium) were assessed by GE. All studies were performed before surgery, after one , seven or ten days and at the end of the treatment course, with yeast rIL-2 therapy (NPO "Biotech", Russia). We performed total and subtotal tumor debulking in most patients. An histological verification showed low grade gliomas in 9 patients and high grade gliomas in 13 patients. rIL-2 therapy was made with a total dose of 10-12 millions U by intravenous perfusion. In addition, the data of light microscopy and electron microscopy (EM) [GEM-100B, Japan] from biopsied tissues after a first operation or after reoperation (5 cases) were analyzed. GE data were processed on PC by "Goldrada" program.

RESULTS

Different levels of RN accumulation , depending on anaplasia grades, were observed in our cases. Low grade gliomas (astrocytomas) deposited very few RN, and oppositely, high grade gliomas accumulated more RN. We saw a dissociation between sizes of areas accumulation and CT and MRI pictures. After presurgery treatment by rIL-2, GE pictures showed as a rule significant changes in accumulation of RN in whole tumor, including a better visualisation of margins of gliomas. However, after therapy with rIL-2, Ge in a dynamic mode (after open surgery) showed a picture of increased BBB in the perifocal zone of brain gliomas (in normal tissue) as a ring (or semiring) as well as a deposition of ^{99m}Tc within substrates, if removal was subtotal. Such a picture can be interpreted as the production around a brain glioma of a zone of acute inflammation, resulting in a demarcation between normal and pathological tissue, rather than a tumor limphocytic infiltration and lysis of abnormal cells. BTB permeability on FC revealed no disturbances in the Albumin content in base line probes. After rIL-2 therapy, FC values also revealed no changes of permeability, but GE revealed an increased permeability (as mentioned above). Moreover, in 15 patients, absolute values of FC were less than border low coefficients. In the meantime, in some cases, we observed intrathecal synthesis of IgG after treatment by rIL-2. FC values were unchanged whatever the presence, size and spreading of peritumoral oedema on CT and MRI. In most cases, EM showed normal tight junctions. However, we revealed some signs demonstrating an increased microvessel permeability as pinocytosis, forming big vacuoles in the endothelial cells and in other cells. Besides, enlargement of perivascular space, depth, delamination and focal sponginess of the basal membrane were observed. In repeated biopsies, we saw necrosis of the tumor tissue, widely spread lymphocytic infiltration and numerous macrophages. In this case, we noted an active penetration of lymphocytes between tumor cells on the EM. In separate cases, we observed the consolidation with macrophages and tumor cells on the local part of its cell membrane.

CONCLUSION

Albumin content in PTE is minimal, even less than normal. Selective accumulation of RN in tumor remains still unclear. It is probably connected with functional neoplasmic cell activity. Some reports described the increase of PTE after immunotherapy before open surgery. Taking this and our data in account, we suggest that a main part of the PTE is serous liquid, where the albumin content is minimal. We think that the low albumin content in PTE is a specific reaction of the brain gliomas, and that a significant albumin content could lead to intensive PTE development. FC values seem to depend on the presence, size and spreading of PTE on the CT-MRI. We observed better clinical results in the cases presenting an inverse proportion of presence, size and spreading of PTE on CT-MRI.

63

EFFECTS OF INDUCTION OF INDUCIBLE NITRIC OXIDE SYNTHASE ON BRAIN CORTICAL BLOOD FLOW AND BLOOD-BRAIN BARRIER

N. Suzuki, Y. Fukuuchi, A. Koto, Y. Morita, T. Shimizu, M. Takao, and M. Aoyama

Department of Neurology
Keio University
Tokyo, Japan

SUMMARY

Recently, nitric oxide (NO) has been found to act a crucial role in the development of the endotoxin shock. This study aims to observe vascular changes and to elucidate mechanisms of development of the brain damage in the endotoxin shock.

Five male Sprague-Dawley rats, weighing 250-300g, were anesthetized with α-chloralose and urethane, and ventilated with room air. The brain cortical blood flow (CoBF) in the parietal cortex supplied by the right middle cerebral artery was recorded by a laser-Doppler flowmeter. CoBF and systemic blood pressure were continuously recorded before and after an intraperitoneal injection of E.coli lipopolysaccharide (4mg/kg b.wt.). Three hours after the injection of LPS, the horseradish peroxidase was injected intravenously. One hour later, the animals were deeply anesthetized and perfused with 5% glutaraldehyde. The whole brain was dissected and frozen sections of the deep frontal cortex were incubated for horseradish peroxidase (HRP) activity, using a method of Graham and Karnovsky. They were post fixed in osmic acid, then processed and observed with an electron microscope. An electron microscopic NADPH-diaphorase histochemistry was also performed in these sections. The blood pressure was significantly decreased ca.30 min after LPS administration and a sustained increase (ca.30%) in CoBF was observed.

The deposits of HRP were observed as dark clusters at the outer adjacency of the endothelial cells. NADPH-d positive deposits were observed in the nuclear membranes, endoplasmic reticula and mitochondria in the smooth muscle as well as in the endothelia.

The NOS induced in the endotoxin shock may affect the function of the endothelium and the smooth muscle, which constitute BBB, and act as a crucial mediator in developing the septic encephalopathy.

Biology and Physiology of the Blood–Brain Barrier, edited by Couraud and Scherman
Plenum Press, New York, 1996

RÉSUMÉ

Il a été établi récemment que l'oxyde nitrique NO jouait un rôle crucial dans le développement des chocs septiques dus aux endotoxines. Cette étude a pour but d'étudier les modifications vasculaires et d'élucider les mécanismes qui provoquent des lésions du cerveau par suite de ce phénomène.

Cinq rats males Sprague-Dawley, pesant de 250 à 300 g, ont été anesthésiés par l'α-chloralose et l'uréthane, sous ventilation à l'air ambient. La circulation corticale du cerveau (CoBF) dans le cortex pariétal, assurée par l'artère cérébrale centrale droite, a été enregistrée au moyen d'un fluxmètre laser-Doppler. La CoBF et la pression systémique artérielle ont été enregistrées en continu avant et après une injection intrapéritonéale de liposaccharide de E. coli, à raison de 4 mg/kg de poids corporel. Trois heures après l'injection de LPS, la peroxydase de raifort était injectée par voie intraveineuse. Une heure plus tard, les animaux ont été anesthésiés plus profondément et perfusés avec du glutaraldéhyde à 5%. Le cerveau entier a été disséqué, et des coupes congelées du cortex frontal profond ont été incubées pour révéler l'activité de la peroxydase de raifort (HRP), selon la méthode de Graham et Karnovsky. Les coupes ont été ensuite fixées à l'acide osmique et étudiées par microscopie électronique. Une étude histochimique par microscopie électronique avec la NADPH-diaphorase a été également réalisée sur ces coupes.

La pression artérielle diminue significativement environ 30 mn après l'administration de LPS, et on observe une CoBF accrue (environ 30%) et persistante.

Les dépôts de HRP apparaissent comme des amas noirs sur le pourtour des cellules endothéliales. Les dépôts positifs de NADPH-d apparaissent sur les membranes nucléaires, le réticulum endoplasmique, et les mitochondries, dans le muscle lisse comme dans l'endothélium.

On en déduit que le NO induit dans les chocs septiques dus aux endotoxines doit affecter la fonction de l'endothélium et du muscle lisse, qui constituent la BHE, et agir comme un médiateur crucial du développement de l'encéphalopathie septique.

1. INTRODUCTION

The endotoxin shock caused by generalized gram-negative bacilli infection is one of the fatal condition in sepsis. The septic encephalopathy has been known as an altered brain function related to the presence of microorganisms or their toxins in the blood. The pathophysiology of septic encephalopathy has not been fully elucidated. Recently, nitric oxide (NO) has been found to play a crucial role in the development of the endotoxin shock (1,2). NO, synthesized from L-arginine by a constitutive calcium/calmodulin dependent NO synthase (cNOS) in the vascular endothelium, plays a role in the physiological regulation of blood flow and blood pressure(3). A different NOS, calcium/calmodulin independent and inducible by immunological stimuli such as lipopolysaccharide (LPS) and cytokines in the phagocytic cells including macrophages, neutrophils, and Kupffer cells in the liver (4,5). Moreover, it has been proposed that NO contributes to the hypotension induced by LPS in the rat.

Thus, the role of inducible, calcium/calmodulin-independent NOS (iNOS)is indubitably crucial in the development of the endotoxin shock. This study aimed to explore the pathophysiology of the brain microcirculatory damages in endotoxin shock with two aspects. First, the localization of iNOS is studied in the cerebral blood vessels by means of histochemistry. Secondly, the effects of iNOS on brain cortical blood flow and blood brain barrier are investigated by means of blood flow measurement and histochemistry, respectively.

2. MATERIALS AND METHODS

2.1. Study for Localization of NADPH-Diaphorase

A total of 10 male Sprague Dawley rats weighing 250-300g were used. Under the pentobarbital anesthesia, E.coli LPS (4mg/kg body weight, Sigma) was injected intraperitoneally. Three hours after the injection of LPS, 5 animals were perfused with a mixture of 4% paraformaldehyde and 0.1% glutaraldehyde. After the fixation, the whole brain was dissected and the major cerebral arteries were carefully prepared for NADPH-diaphorase histochemistry with nitro blue tetrazolium for formazan production. The specimens were observed and photographed in a light microscope with an immersion lens. The remaining 5 rats were served as control in which no LPS was injected.

2.2. Study for Ultrastructural Localization of NADPH-Diaphorase

The other 5 male rats were also treated with E.coli LPS. Three hours after the administration of LPS, the animals were perfused with 4% paraformaldehyde and 1% glutaraldehyde. The major cerebral arteries were dissected and processed for NADPH-diaphorase histochemistry with 2-(2′-benzothiazolyl)-5-styryl-3-(4′-phthalhydrazidyl) tetrazolium salt (BSPT) for electron microscope (6). After osmification with 2% OsO_4, the specimens were observed under an electron microscope (JEOL-1200 EXII, acc. volt. 75kV). Other 5 rats were served as control which had no LPS administration.

2.3. Study of Effects of iNOS Induction on Blood Brain Barrier and Cerebral Blood Flow

The other 5 male Sprague-Dawley rats were anesthetized with α-chloralose and urethane, and ventilated with room air. The brain cortical blood flow (CoBF) in the parietal cortex supplied by middle cerebral artery was recorded by laser-Doppler flowmeter (ALF 21, Advance). The blood flow and the systemic blood pressure were continuously recorded before and after an intraperitoneal injection of E.coli LPS (4mg/kg), the same dose as applied in the histochemical study. The blood flow was continuously observed for 3 hours after LPS administration. Three hours after the injection of LPS, the horseradish peroxidase (Sigma type II, 25mg/ml) was injected intravenously. One hour later, the animals were deeply anesthetized and perfused with 5% glutaraldehyde. The whole brain was dissected and sections of the cortex were processed to see horse radish peroxidase activity, using a method of Graham and Karnovsky, post fixed in osmic acid, and prepared for electron microscopic observation.

3. RESULTS

3.1. Localization of iNOS in the Cerebral Vessel Wall

In control animals, NADPH-diaphorase activity was demonstrated in the endothelium and perivascular nerve terminals and fibers of the adventitia. But, only a few activity was observed in the smooth muscle cells. On the other hand, in the LPS injected animals, NADPH-diaphorase activity was observed even in the smooth muscle cells.

In electron microscopic study of non LPS treated control animals, NADPH-diaphorase positive material was observed on the nuclear envelope and the endoplasmic

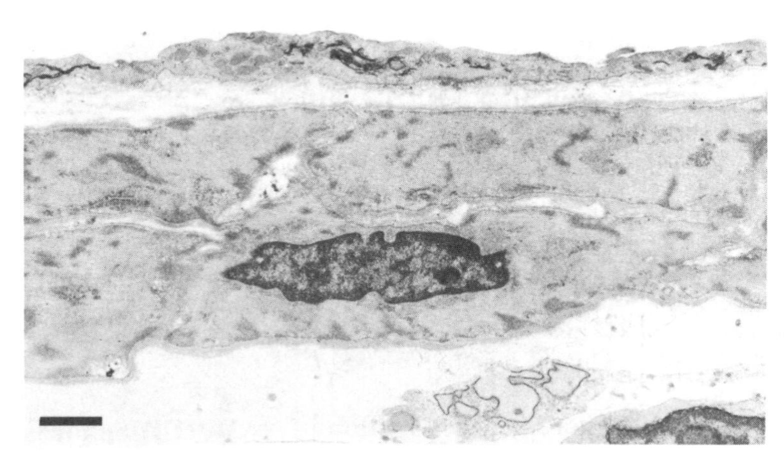

Figure 1. NADPH-diaphorase activity in the cerebral arterial wall of the control rat. NADPH-diaphorase positive materials appear on the nuclear envelopes and the endoplasmic reticulum in the endothelium. Negative NADPH-diaphorase activity is found in the smooth muscle cells. Scale bar: 1 μm.

reticulum in the endothelium. However, no positive material was observed in the nuclear membrane of the smooth muscle cells (Fig.1). NADPH-diaphorase positive linear figures were observed in the axon of the perivascular nerve fibers. In some regions, a few NADPH-diaphorase positive deposits were observed in the outer membrane of the mitochondria. In the LPS treated animals, strongly positive materials were observed in the nuclear membrane, endoplasmic reticulum, and mitochondria of the smooth muscle cells (Fig.2). In the perivascular nerve fibers,positive materials were also markedly augmented after LPS treatment.

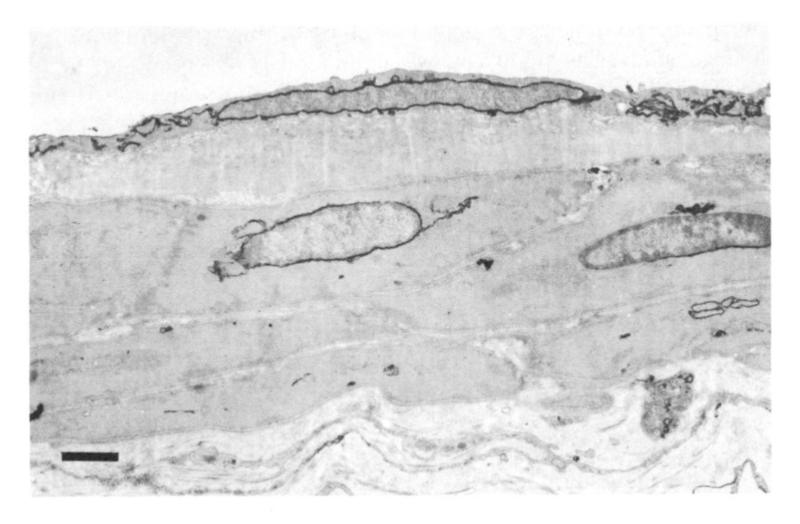

Figure 2. NADPH-diaphorase activity in the cerebral arterial wall of the LPS-treated rat. Positive materials are remarkable in the nuclear membranes, endoplasmic reticulum and mitochondria of the smooth muscle cells. Notably, positive materials are markedly augmented in the endothelial cells and perivascular nerve fibers. Scale bar: 1 μm.

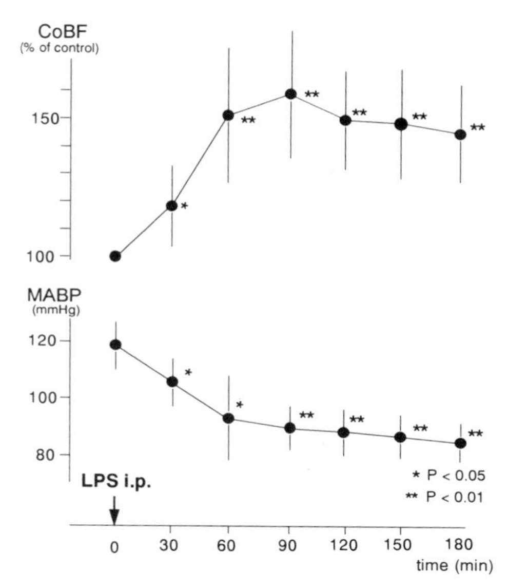

Figure 3. Effects of LPS on systemic blood pressure and brain cortical blood flow. Significant decrease in blood pressure appeared after LPS administration then lowered blood pressures were sustained the end of the observation. While, brain cortical blood flow increased significantly after LPS administration. Statistics was performed with ANOVA followed by modified t-test of Bonferoni.

3.2. Effects of Induction of iNOS on Cortical Blood Flow and Blood Brain Barrier

The blood pressure began to decrease at about 17min after the peritoneal injection of LPS and showed a maximal decrease of 37mmHg at 84min after the injection. Significantly low blood pressure sustained until the end of the observation. The CoBF, on the other hand, began to increase at about 16min after the injection and took a maximal increase of 72% at 86min. The increased CoBF sustained (Fig 3). No significant changes were observed in the physiological parameters before and after the administration of LPS (Table 1).

After LPS administration, HRP deposits were observed as dark clusters at the outer adjacency of the endothelial cells (Fig 4), which suggests destruction of the barrier mechanism of the endothelial cells.

Table 1. Values of physiological parameters before and after administration of lipopolysaccharide

	Control condition	180 min after LPS i.p.
PaO_2 (torr)	101.0 ± 11.8	97.5 ± 11.4
$PaCO_2$ (torr)	30.1 ± 4.2	26.4 æ 3.3
pH	7.525 ± 0.029	7.491 æ 0.019

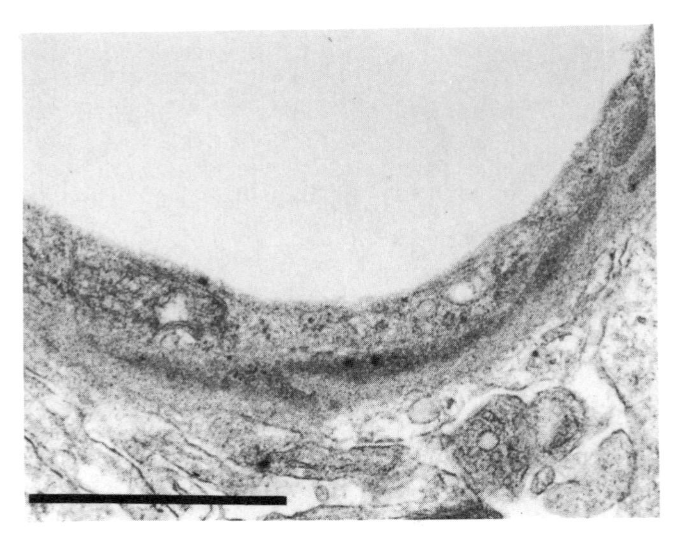

Figure 4. Effect of LPS administration on BBB. HRP deposit accumulated at the outer adjacency of the endothelial cells in the brain parenchyma. Scale bar: 1µm.

IV. DISCUSSION

Several recent studies have suggested an important role of NO in hypotension induced during endotoxin shock. The induction of a calcium-independent NOS activity has been demonstrated in vivo in various rat tissues 3 to 6 hr after exposure to LPS/endotoxin (7,8). In humans, higher plasma nitrite/nitrate levels have been demonstrated in septic patients, and this correlates with decreased systemic vascular resistance (9). In experiments, competitive NO synthase inhibitors have shown reversal of the vascular effects of LPS (10,11). Pretreatment of LPS-injected rats with dexamethasone reduces hypotension, vascular hypoactivity and induction of the calcium independent NOS activity in vivo (12,13).

Thus, the role of iNOS in the endotoxin shock has been demonstrated both in experimental and clinical studies. However, the pathophysiology of the central nervous system involvement in septic shock has not been fully elucidated. It has been apt to be referred as a component of the "multiple organ failure". The clinical diagnosis of septic encephalopathy is one of exclusion. Altered brain function in the febrile patients is caused by a number of conditions other than sepsis per se, e.g., meningoencephalitis, stroke, metabolic failure, etc. However, even after excluding these pathological conditions, dysfunctions of the central nervous system have been known to accompany with the septic shock.

This study demonstrated destruction of the barrier function of the endothelial cells in the cerebral blood vessels by LPS administration. It is plausible that the brain dysfunction in the septic shock is attributable to the disturbance of the blood brain barrier. Indeed, the cerebral blood flow was observed to increase by approximately 72% after administration of LPS with concomitant decrease in systemic blood pressure.

In previous studies, the induction of iNOS in various organs in LPS treated animals was explored by biochemical or enzymatic methods (14,15). The expression of iNOS mRNA in LPS treated animals was also demonstrated (16). However, there has been no previous report regarding iNOS induction of the brain in LPS treated animals.

NADPH-diaphorase activity was utilized as a marker for iNOS in the present study. It is well known that all three types of NOS, i.e., neuronal, inducible and endothelial types, have been shown to exhibit NADPH-diaphorase activity following aldehyde fixation (17,18). Although only 51% homology in these three types of NOS was found, the NADPH binding site was found to be conserved. It is exactly this part of the enzyme that is thought to be responsible for the diaphorase activity. NADPH-diaphorase has been utilized as a marker for the presence of iNOS in glia (17,19). NADPH-diaphorase activity is not observed in the smooth muscle cells of the cerebral blood vessels in the steady state. In the LPS treated animals, however, NADPH-diaphorase activity appeared on the nuclear membrane, the endoplasmic reticulum and the outer membrane of the mitochondria of the smooth muscle cells as a membrane bound state. It is natural that the appearance of NADPH-d activity in the smooth muscle cells after LPS-treatment is attributed to the induction of iNOS. However, the iNOS with 130kD MW is known to be localized in soluble partition in contrast to neuronal NOS in the membrane partition. This discrepancy might be attributed to the followings. First, an another type of NOS which belongs to the membrane partition might be induced concomitantly (20). Second, it is possible that the detection for NADPH-diaphorase activity in the cytosol is limited. Third, the myristoylation has occurred in N-terminal in iNOS in this condition.

The induction of Ca-dependent NOS by LPS has been demonstrated in the rat ileum and caecum (14). Indeed, NADPH-d activity in the endothelial cells and perivascular nerve fibers, which is accounted as endothelial NOS and neuronal NOS, respectively, was augmented after LPS treatment in the present study.

Induction of iNOS by LPS caused a significant decrease in blood pressure and a sustained increase in CoBF in the present study. The mechanism of the blood flow increase by LPS administration is obscure. It is tempting to assume that cerebral vasodilatation caused by induction of iNOS in the smooth muscle cells had overcome the systemic hypotension.

Concomitant induction of cNOS in the endothelial cells might also be involved, had overcome the systemic hypotension. A destruction of the barrier function of the endothelial cells was suggested by leakage of HRP from the cerebral vessels after LPS administration. Taken together, an extreme vasodilatation and a disturbance of BBB function might have caused a hyperemic and/or edematous state in the brain. Such pathological conditions would possibly manifest clinical feature of the encephalopathy.

In conclusion, the iNOS induced in the endotoxin shock may affect the function of the endothelium and the smooth muscle, which constitute BBB, and act as a crucial mediator in developing the encephalopathy accompanying septic shock.

REFERENCES

1. Fleming I, Gray GA, Julou-Schaeffer, Parratt JR, Stoclet J-C: Incubation with endotoxin activities the L-arginine pathway in vascular tissue. Biochem. Biophys. Res. Commun. 171:562-568, 1990
2. Beasley D, Schwartz JH, Brenner BM: Interleukin 1 induces prolonged L-arginine dependent cyclic guanosine monophosphate and nitrite production in rat vascular smooth muscle cells. J. Clin. Invest. 87:602-608, 1991
3. Rees DD, Palmer RMJ, Moncada S: Role of endothelium-derived nitric oxide in the regulation of blood pressure. Proc. Natl. Acad. Sci. USA 86:3375-3378, 1989
4. Marletta MA, Yoon PS, Iyengar RI, Leaf CD, Wishnok JS:Macrophage oxidation of L-arginine to nitrite and nitrate: nitric oxide is an intermediate. Biochemistry 27:8706-8711,1988.
5. Billiar TR, Curran RN, Stuehr DJ, West MA, Bentz BG, Simmons RL: An L-arginine-dependent mechanism mediates Kupffer cell inhibition of hepatocyte protein synthesis in vitro. J. Exp. Med. 169:1467-1472, 1989

6. Wolf G, Wurdig S, Schunzel G:Nitric oxide synthase in rat brain is predominantly located at neuronal endoplasmic reticulum: an electron microscopic demonstration of NADPH-diaphorase activity. Neurosci. Lett. 147:63-66, 1992
7. Rees DD, Cellek S, Palmer RMJ, Moncada S: Dexamethasone prevents the induction by endotoxin of a nitric oxide synthase and the associated effects on vascular tone: an insight into endotoxin-shock. Biochem. Biophys. Res. Commun. 173:541-547,1990
8. Czabo C, Mitchell JA, Gross SS, Thiemermann C, Vane JR:Nifedipine inhibits the induction of nitric oxide synthase by bacterial lipopolysaccharide. J. Pharmacol. Exp. Ther.265:674-680, 1993
9. Ochoa JB, Udekwu AO, Billiar TR, Curran RD, Cerra FB, Simmons RL, Peitzman AB: Nitrogen oxide levels inpatients after trauma and during sepsis: Ann. Surg. 214:621-624, 1991
10. Kilbourn RG, Jubran A, Gross S, Griffith OW, Levi R, Adams J, Lodato R: Reversal of endotoxin-mediated shock by NG-methyl-L-arginine, an inhibitor of nitric oxide synthesis. Biochem. Biophys. Res. Commun. 172:1132-1138, 1990
11. Gray G, Schott C, Julou-Schaeffer G, Fleming I, Parratt JR, Stoclet J-C: The effect of inhibitors of the L-arginine/nitric oxide pathway on endotoxin induced loss of vascular responsiveness in anaesthetized rats. Br. J.Pharmacol 103:1218-1224, 1991
12. Wright C, Rees D, Moncada S: Protective and pathological role of nitric oxide in endotoxin shock. Cardiovasc. Res. 26:48-57, 1992
13. Thiemermann C, Czabo C, Mitchell JA, Vane JR: Vascular hyporeactivity to vasoconstrictor agents and hemodynamic decompensation in hemorrhagic shock is mediated by nitric oxide. Proc. Natl. Acad. Sci. USA 90:267-271, 1993
14. Salter M, Knowles RG, Moncada S: Wide spread distribution, species distribution and changes in activity of Ca2+-dependent and Ca2+-independent nitric oxide synthesis. FEBS Lett. 291:145-149, 1991
15. Mitchell JA, Kohlhaas KL, Sorrentino R, Warner TD, Murad F, Vane JR: Induction of endotoxin of nitric oxide synthase in the rat mesentery: lack of effect on action of vasoconstrictors. Br. J. Pharmacol. 109:265-270, 1993
16. Liu S, Adcock IM, Old RW, Barnes PJ, Evans TW:lipopolysaccharide treatment in vivo induces widespread tissue expression of inducible nitric oxide synthase mRNA. Biochem. Biophys. Res. Commun. 196:1208-1213, 1993.
17. Galea E, Feinstein DL, Reis DJ: Induction of calcium-independent nitric oxide synthase activity in primary rat glial cultures. Proc. Natl. Acad. Sci. USA 89:10945-10949,1992
18. Kichener PD, Bourreau J-P, Diamond J: NADPH-diaphorase histochemistry identifies isolated endothelial cells at sites of traumatic injury in the adult rat brain. Neuroscience 53:613-624, 1993
19. Wallace MN, Bisland SK: NADPH-diaphorase activity in activated astrocytes represents inducible nitric oxide synthase. Neuroscience 59:905-919, 1994
20. Foesterman U, Schmidt HHW, Pollock JS, Sheng H, Mitchell JA, Warner TD, Nakane M, Murad F: Isoforms of nitric oxide synthase:characterization and purification from different cell types. Biochem. Pharmacol. 42:1849-1857, 1991

INDEX